The efforts of countless agriculturalists over the past ten thousand years have led to the improvement of a wide range of agricultural crops. In this major new work, Lloyd Evans provides an integrated view of their domestication, adaptation and improvement, bringing together genetic diversity, plant breeding, physiology and aspects of agronomy. Considerations of yield and yield potential provide continuity throughout the book. Food, feed, fibre, fuel and pharmaceutical crops are all discussed. Cereals, grain legumes and root crops, both temperate and tropical provide many of the examples, but pasture plants, oilseeds, leafy crops, fruit trees and others are also considered.

Lloyd Evans was born in New Zealand, where he received his initial training in science and agriculture before doing his doctorate in soil science at Oxford. In 1956, after a post-doctoral fellowship in the phytotron at the California Institute of Technology, he joined the CSIRO Division of Plant Industry in Canberra, Australia. He was involved in the design and use of CERES, the Australian phytotron. Much of his work there has been on the physiology of crop yield, and on the control of flowering by daylength. His research has been recognized by election to fellowship of the Australian Academy of Science, the Royal Society of London and of the New Zealand and Norwegian national academies.

Particularly through his membership of the Technical Advisory Committee of the CGIAR, and close involvement with several of the international centres of agricultural research, the author has had many opportunities to observe the diverse agricultural problems of developing countries.

Crop evolution,
adaptation and yield

L. T. Evans

CAMBRIDGE
UNIVERSITY PRESS

Published by the Press Syndicate of the University of Cambridge
The Pitt Building, Trumpington Street, Cambridge CB2 1RP
40 West 20th Street, New York, NY 10011–4211, USA
10 Stamford Road, Oakleigh, Melbourne 3166, Australia

First published 1993
First paperback edition 1996

Printed in Great Britain at the University Press, Cambridge

A catalogue record for this book is available from the British Library

Library of Congress cataloguing in publication data
Evans, L. T.
Crop evolution, adaptation, and yield / L. T. Evans.
        p.   cm.
Includes bibliographical references and index.
ISBN 0-521-22571-X
1. Crops – Evolution.   2. Crops – Adaptation.   3. Crop yields.
I. Title.
SB106.O74E93   1993
633 – dc20   92-28314   CIP

ISBN 0 521 22571 X hardback
ISBN 0 521 29558 0 paperback

Preface

When the economist Walter Rostow asked and was told what I was trying to write he called it, not altogether pejoratively, 'a megalomaniac book'. So you, potential reader, might well ask why any author should attempt to cover so much ground, given the hazards of venturing into fields beyond his expertise, and why he should expect you to join him?

My reasons are that the yield of crops is a fascinating subject of great historical significance and still central to our global housekeeping; that too often has this many-sided subject been viewed from only one vantage point; that the nature of greater yields is widely misunderstood yet has implications for plant breeding strategies, development policy and environmental management; and that further increases in crop yields will be vital to the welfare of at least those four out of every five people on Earth in AD 2000 who will be living in developing countries.

Given the scope of the subject, a less hazardous option would have been to edit a book by the requisite range of experts, but a coherent overall perspective often eludes that approach. The perspective presented here has grown out of a preoccupation with the physiology of crop yield combined with opportunities to see something of the agricultural problems in many environments and regions of the world.

Several areas of lively current research which have received ample attention elsewhere, such as genetic engineering and crop modelling, are only lightly touched on here. Nor is this a book about plant breeding, for which excellent texts are already available. Rather, it is a book about the nature of advances in agriculture. Some crops, like the perennial trees so important in the humid tropics, get a raw deal, but I have tried to balance the claims of temperate and tropical crops in both developed and developing countries.

ix

The introductory chapter provides an overview of the whole field, especially for those wanting a closer look at only one corner of it. However, my hope is that the synthetic approach followed here may lead to more fruitful interactions between disciplines, between the two agricultures – the traditional and the modern – and between the two cultures within biology itself, of *Drosophila* and *Arabidopsis* on the one hand and of wheat and soybeans on the other.

L.T. Evans
October, 1991

Acknowledgements

Many colleagues have provided me with insights and information, drawn my attention to out-of-the-way publications, clarified issues or remoulded my interpretations, among them David Andrews, Roger Austin, Paul Biscoe, Derek Byerlee, T.T. Chang, Dana Dalrymple, Tina David, Don Duvick, Tony Fischer, Wayne Hansen, Diane K. Johnson, Gurdev Khush, Gerald Stanhill and Danny Zohary. My unbroken series of arguments over many years with Otto Frankel about the nature and improvement of crop plants has been an enduring stimulus.

He and John Passioura, Richard Groves, Derek Byerlee, Tony Brown, T.T. Chang, Howard Rawson, Rod King, Ian Wardlaw, David Bagnall, Rex Oram, Richard Richards, Jim Davidson, Peter Randall, John Freney, John Kirk and John Zwar have all read and improved individual chapters. I am also grateful to the anonymous referees who read and commented on the whole text, to Alan Crowden and his predecessors at Cambridge University Press for their patience and skill in the arts of keeping authors at it, and to Lynn Davy for her helpful editing.

Here in Canberra, I am indebted to Carol Murray for tracking down difficult references, to Renata Sawa and Pat Riddell for endless word processing, to Anne Warrener for illustrations and to Cheryl Blundell for so patiently checking everything.

This book and I owe more than can be measured to the CSIRO Division of Plant Industry and to my colleagues in it. I also acknowledge the support of the Rockefeller Foundation for the writing of the final chapter during a residency at the Villa Serbelloni in Bellagio, the only part of the overall process which has not required great forbearance by my wife, Margaret.

I am also grateful to the original authors and/or publishers for permission to use quotations and to republish many of the figures in the book, and to Nancy Sawer for allowing me to use her delightful crop vignettes.

CHAPTER 1

Introduction

*Few scientists think of agriculture as
the chief, or the model science. Many,
indeed, do not consider it a science at
all. Yet, it was the first science – the
mother of sciences; it remains the
science which makes human life
possible; and it may well be that,
before the century is over, the success or failure of Science as a whole
will be judged by the success or failure of agriculture*

André & Jean Mayer (1974)

Preamble

The success or failure of agriculture may be judged in many ways but the most significant criteria will continue to be the adequacy, reliability and quality of our food supplies as the human population continues to increase. Most of our food comes from a small number of crop plants, and the improvement of their yield has been the main source of greater food production in recent years, and will continue to be so over the next decades. It remains the key to greater global carrying capacity, to freedom from hunger, to enhanced development in Third World countries and, in the light of the quotation above, perhaps even to the judgement of science.

In many developed countries, however, this latter proposition is likely to be little recognized and even less accepted. The proportion of their population engaged in agriculture is small, as is the proportion of income spent on food, the supply of which is varied, reliable, of high quality and taken for granted by the largely urban populations. Surpluses rather than deficits are the problem, so both the public and their political leaders see little need for research towards still greater crop yields. Currently, in fact, there are strong pressures in the opposite direction associated with the government subsidies that exacerbate these surpluses from high yields, the consumption of scarce resources in their production, and their environmental consequences.

In this context, agricultural research in these countries has been almost too successful. At the end of the eighteenth century the Reverend Thomas Malthus had expounded his view that whereas human populations could increase geometrically, their means of subsistence could increase only arithmetically. So he would have been surprised by the subsequent rise in the yields of wheat in England and rice in Japan, illustrated in Figure 1.1. Indeed,

1

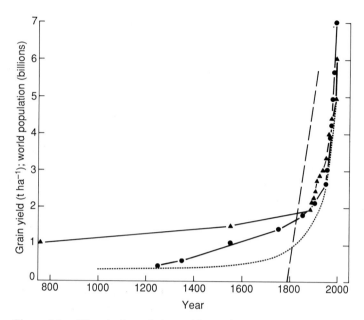

Figure 1.1.   Historical trends in world population (......), and in the grain yield of wheat in England (●) and of brown rice in Japan (▲) compared with the limiting rate of improvement assumed by Malthus (1798) (——). Based on data assembled by Borrie (1970), Gavin (1951), Matsuo (1959), Stanhill (1976) and FAO Production Yearbooks.

for several European countries in recent years, food supplies could have increased geometrically whereas population has increased only arithmetically, if at all. Such a reversal of the Malthusian metric surely highlights the success of agricultural science, as does the downward trend of 1% per annum in the real price of wheat and other staple foods over the past 100 years. Nevertheless, Calder wrote in 1967 that 'agriculture is simply failing us', and went on to suggest that the world would have to rely on alternative food-producing systems such as aquaculture, single-cell protein and synthetic foodstuffs. Since Calder made his claim, the world has not only continued to rely as heavily as ever on agriculture for food, feed and fibre, but has also looked to it increasingly for fuel, feedstocks (Ng *et al.*, 1983) and pharmaceutical precursors.

Why then must further increases in yield be sought? In most developed countries arable land continues to be lost from agriculture for urban development, highways, airfields, water catchment, recreational parks, conservation areas, afforestation, etc., so yields from the remainder must be made able to increase. Indeed, only rising yields will permit the diversification discipline when yields were often only 3–4-fold and fell as low as 1.8-fold, which happened for wheat in England in the thirteenth century (Slicher van Bath, 1963). No wonder Peter Kalm, a student of Linnaeus, was so impressed by maize when told it returned at least 300-fold that he called it 'the lazy man's grain' (Oxholm & Chase, 1974). Millets are supposedly so-called because thousands of grains are harvested for each one sown. Even wheat and barley could return 'an hundred fold' as in the parable of the sower (Matthew 13:8), but such high ratios probably referred to individual plants although modern record crops of wheat can return ratios of more than 160.

As long as the ratio of seed harvested to seed sown was an important criterion of yield, cereals may well have been unconsciously selected for abundant tillering, large inflorescences, small grains and weak seed dormancy. However, as the supply of arable land came under greater pressure, yield per unit area per crop became a more important criterion, and remains so. Much of the discussion in this book is concerned with it and with characteristics enhancing it, which may be quite different from those giving a high seed return ratio.

Rising pressure on the supply of arable land may, where climate and water supply permit, lead to more intensive multiple cropping. In tropical regions particularly, more frequent cropping is an important component of greater crop production, and under these circumstances yield per hectare per crop becomes less important than yield per hectare per day. With this latter criterion of yield come new selection pressures, particularly on the speed of reproductive development and its sensitivity to seasonal daylengths. Among Philippine rice varieties, for example, grain production per hectare per day has increased more strikingly than grain yield per hectare per crop (Evans *et al.*, 1984; Khush, 1987).

As pressures on the area of arable land continue to increase – as they will do for many years to come – irrigation, fertilizers, pesticides and other inputs are used to an ever greater extent as substitutes for land by raising yield per hectare. When these other agricultural resources in turn become limiting, crop yields may be assessed not only in terms of land area but also in terms of the amount of water or phosphorus or energy used in their production. In the longer term, the capacity to yield under high ultraviolet radiation or atmospheric CO_2 or pollution levels may also be important. In countries where wages are high, output per worker may be of as much consequence as yield per hectare. Geertz (1963) points out that although both Japan and Indonesia raised the yield of rice per hectare, only Japan also raised the output per farm worker, thereby avoiding 'agricultural involution'.

Thus, yield criteria for crops in the future, as for those in the past, may differ from those currently being emphasized, and as they change so may the yield-limiting processes and the characteristics for which crops are selected. In self-sustaining life support systems for space exploration, for example, yield per unit volume per day is an important criterion.

Moreover, given the preoccupation with crop yield throughout this book, it should be made clear at the outset that the raising of yield as measured by any of the criteria discussed above is not always the most urgent task of the plant breeder. Simply maintaining yields at previous levels in the face of onslaughts by new forms of pest and disease organisms may be more urgent, as may better survival in adverse conditions or the reduction of labour or other costs. Nevertheless, many surveys, like that for rice by Hargrove (1977), have found greater yield potential to be the plant breeder's most common objective.

Crop yields and the world's food supply

In the next chapter we focus on the crucial part played by the improvement of crop yields in raising world food production in recent years. As Figure 1.1 illustrates for both wheat in England and rice in Japan, it is only in the past century or so that the national yields of these staple cereals have risen rapidly, from less than 2 to more than 6 tonnes per hectare, while world population has risen from less than 2 to approaching 6 billion people over the same period.

Throughout the preceding millennia the food requirements of the growing world population could be met by putting more land under crops. Comprehensive statistics for crop area, yield and production in the world were not collected regularly until the International Institute of Agriculture in Rome, beginning in 1908 and followed by the United Nations Food and Agriculture Organization (FAO) after World War II, began publishing them. Only then could global housekeeping begin in earnest, although some earlier writers had attempted to estimate world production, most notably Crookes for wheat in 1898. Estimates of crop areas and yields are also made by other agencies now, to circumvent some of the problems of official statistics, while dietary surveys extend the production figures to estimates of human consumption, taking account of post-harvest losses and alternative uses such as animal feed.

However, it is the variation in food consumption between social groups, and between years or even seasons, that is particularly significant in assessing malnutrition and the need for more food. Changes to official nutritional standards (e.g. FAO/WHO, 1973) also have a profound effect not only on estimates of the number of hungry and malnourished people in the world, but also on the relative emphasis given to energy as against protein supplies. In

Contents

For Margaret, Nicholas, John and Catherine

of land use and the protection of vulnerable areas and of biological diversity which are likely to become more prominent issues in the future. Most deforestation is due to the spread of agriculture. The cost-price squeeze experienced by farmers continues its apparently inexorable course, and rising land prices can accompany cheaper food only by the further raising of yields. More intensive crop production will also be needed to enhance the stability and quality of food production, and will be encouraged by the rise in atmospheric CO_2 level. Above all, while developed countries may take assured food supplies for granted, this is not the case in developing countries. In these, most of the people still live in rural areas, and food accounts for a high proportion of their income. The yield of their crops is a subject of great interest: feast or famine depend upon it. Moreover, their populations are still growing rapidly, and rising yields are needed to meet local food requirements. Food aid and trade may help out in crises, but are no substitute for local self-sufficiency when transport and communication systems are poorly developed. Only with a substantial rise in crop yields are they likely to achieve their demographic transition.

Thus, the yields of crops, and ways of increasing them further, are likely to be as important in the long-term future as they have been in the past. They have been considered in many books dealing with individual crops, countries or regions or from the perspective of particular disciplines, quite apart from the writings of a long line of prophets of doom, whose prophecies have been overtaken even before they were forgotten (e.g. Paddock & Paddock, 1968).

The aim of this book is to attempt a more comprehensive approach and to develop a more rounded picture of crop yield and its improvement. Although it draws heavily on the staple food crops for examples, many others both temperate and tropical are also referred to. Several disciplines are invoked – while recognizing the hazards of such a broad approach – but the central perspective is that of a crop physiologist rather than of a plant breeder. The core of the book deals with the physiological nature of crop domestication, adaptation and improvement, and the picture that emerges has important implications. It bears, for example, on the relation between plant breeding and agronomy, on the role of inputs and on the likelihood of significant advances in yield under low input or adverse conditions. Enthusiasm for plant breeding programmes for such conditions would be tempered by a better understanding of the nature of what has been achieved so far, as would many criticisms of 'the Green Revolution'. The perspective also bears on the relative advantages of breeding for wide adaptability as against more specific adaptation, on the reliability of yields, and on strategies for agricultural research. Empirical selection for yield will no doubt continue to succeed in the future as it has so effectively in the past, despite the periodic claims that we have reached the yield plateau. But if the promise of biotechnology and molecular plant breeding is to be translated into yield gains, the nature of the yield-limiting processes in crops needs to be better understood.

Because the book ranges rather widely, this introductory chapter traverses its overall structure and arguments. Before that, however, we should consider the various criteria of crop yield.

Criteria of crop yield

When our ancestors harvested the fruits, seeds or tubers of wild plants, they were probably more concerned with the variety, ease of harvest or preparation, and transportability of the foods available to them than with their yield. Since the beginnings of agriculture about ten thousand years ago, however, yield has been an important attribute of crops, although the yield criterion of most immediate interest has changed.

For those who gathered grain from the wild cereal stands in the Fertile Crescent of the Near East, the most appropriate criteria were likely to have been how quickly the grain could be harvested before it was shed on the ground, and with what effort. Harlan (1967) gathered over a kilogram of clean seed per hour from wild wheat stands in Turkey, using only his hands or a flint-bladed sickle and concluded that 'A family group ... working slowly upslope as the season progressed, could easily harvest wild cereals over a three-week span or more and, without even working very hard, could gather more grain than the family could possibly consume in a year ... a very attractive alternative to living by the chase'. Harlan would have gathered about 50 calories for each calorie he spent on harvest, but not all wild plants yield such an abundant return. Those with larger seeds, with more seeds per inflorescence, or with more compact or uniformly ripening panicles, features that increase the speed of harvest, may have been preferred. For example, although oats, wheat and barley are found together in wild stands, wheat and barley can be harvested more quickly than oats with its smaller grains and less compact and evenly maturing panicles, as Ladizinsky (1975) has shown, and they were domesticated sooner.

Once the sowing of crops became established practice, a new criterion of yield became important, namely the ratio of seed harvested to seed sown. This is the measure of yield mentioned in the Bible, by Roman writers such as Columella, and throughout the Middle Ages (cf. Slicher van Bath, 1963; Titow, 1972). In fact, this is the measure of crop yield that has been used throughout most of the period since agriculture began, and was used by Malthus (1798) as a metaphor for the power of geometric increase in populations. Its significance, especially in times of shortage, was that the grain set aside for sowing had to be foregone as food, entailing considerable

Crop evolution, adaptation and yield

turn these influence the relative importance attached to energy-rich crops such as the cereals vis-à-vis the higher-protein legumes. Both the statistics and the nutritional standards have their shortcomings, and these should be borne in mind throughout this book. Figure 1.2 illustrates the geography of dietary energy supplies per head.

Although increase in crop area has been the main source of increased food production since agriculture began, this is no longer the case in the more densely populated developed countries, in some of which the arable area is beginning to decline. Even in these countries, however, cultivation may be extended to new areas as a result of plant breeding and new agronomic practices. For example, the breeding of varieties more tolerant of adverse soil or climatic conditions may extend the arable area, while the introduction of minimum tillage techniques sometimes allows steeper slopes to be safely cultivated. It is no simple matter, therefore, to estimate how much additional land could be brought into cultivation, and on this important issue optimists and pessimists differ almost as much as they do on the estimation of ultimate crop yields.

Our perception of the driving force in agricultural development also influences our expectations of crop yields. In the Malthusian analysis that was so crucial to both Darwin and Wallace, population growth is dependent on increase in the food supply. By contrast, Boserup (1965, 1981) argues that population growth is the independent variable and a major factor in determining agricultural progress. She believes, as does Clark (1967), that this is a more realistic and fruitful assumption, and she presents examples of its operation. Howell (1987) suggests that it was population pressure which led to the spread of agriculture through Neolithic Europe, and many other examples of it being a driving force in agricultural change have been suggested (cf. Flannery, 1972). The drive to increase food production in Asia in the 1960s, often referred to as the 'Green Revolution', and the efforts currently being made in Africa to reverse the decline in per capita food consumption associated with rapid population growth, illustrate Boserup's thesis. However, there are also instances where population pressure is not the driving force (see, for example, Bronson, 1977). Modern European agriculture is a case in point.

Besides greater area under crops and greater yields, two other components contribute to rising food production. The first is of greatest significance in more tropical regions where either rainfall or irrigation can support crop growth throughout the year. Under these circumstances, intensification of cropping, i.e. harvesting more crops each year from the same piece of land, may enhance local food supplies as much as increased yields can, and may be easier to realize. The contribution made by crop intensification is not easy to

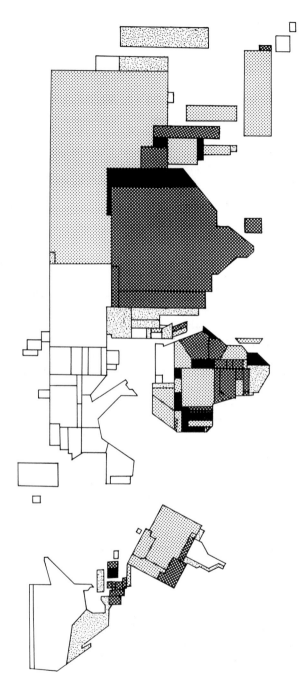

Figure 1.2. The geography of dietary energy supplies per capita, 1979–81 (adapted from FAO, 1987). The relative size of countries is shown in proportion to their population, while their hatching indicates average calorie intake, ranging from over 3000 (no hatching) through progressively heavier hatching to less than 2000 (solid black) kilocalories per head per day.

extract from official statistics. Unlike greater yield per crop, it depends on the shortening of crop life cycles, but shares with it the need for greater use of inputs such as irrigation, fertilizers, pesticides and herbicides.

The other component of rising food production is the shift towards relatively greater dependence on a small number of staple food crops. These tend to displace other traditional crops because of their higher yielding capacity, wider adaptability or greater reliability, and despite the possible disadvantages of a reduced variety of foodstuffs, lower nutritional status or more vulnerable and depletive farming systems. Of the large number of species domesticated by mankind, many are no longer cultivated or have become much less significant, such as Job's tears (*Coix lachryma-jobi*) and several species of *Amaranthus*. The world's food supply is now dominated by three cereals and a few other crops (Figure 1.3). Even the major pulses are being displaced by cereals in countries, such as India, where they are important both nutritionally and for their role in the farming systems. The staple crops are, more and more, being selected for performance in environments beyond their previous range: for example wheat and potatoes for the hot tropics, rice for cooler areas and maize for drier ones. So significant are these few staple crops to mankind that the basis of their productivity and adaptability merits a greater understanding than we currently have.

Although increases in arable area have been more important in the past, and increases in cropping intensity may be important in the future, it is greater yield per crop which has been the key to greater food production in recent decades and is likely to remain so in the coming ones. The extent to which yields of the major cereals have risen in many developed countries has been a remarkable achievement. Malthus (1798) suggested that even 'the most enthusiastic speculator' could not expect British wheat production to increase by more than the then current level of production in a period of 25 years. The yield of wheat in England at that time was about 1.6 t ha^{-1}, and since 1950 it has increased at a faster rate than Malthus allowed his most enthusiastic speculator, as may be seen in Figure 1.1. This rate of improvement may seem modest by comparison with many other improvements, such as the functions per computer chip (Meindl, 1987), the efficiency of illumination (Starr & Rudman, 1973), or the yield of penicillin (Aiba *et al.*, 1973), yet it remains the key to the solution of the world food problem.

The why, when, where, what and how of domestication

Chapter 3 begins with a consideration of hunter-gatherer societies, as a background to the processes of plant domestication which began about 10 000 years ago. Whatever our views of hunter-gatherers – and these have ranged from Rousseau's 'noble savages' all the way to the 'living fossils'

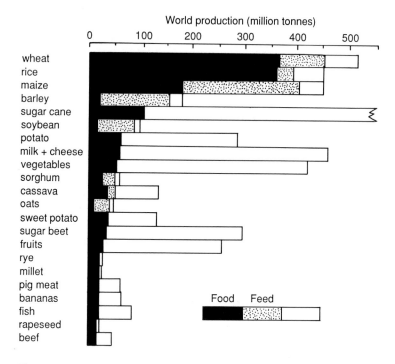

Figure 1.3. World production of the major foods in 1987. The overall length of each bar is proportional to the harvested weight of the commodities, based on data from the FAO Production Yearbook; the dotted bars indicate the amount used for feed, where known (Sarma, 1986), and the solid bars the amount available for food, in terms of estimated edible dry weight. The harvested weight of sugar cane is about one billion tonnes.

portrayed by Victorian biologists and then back to 'the original affluent society', in Sahlins' (1968) phrase – the development of agriculture was a crucial step in human evolution because in domesticating plants and animals mankind was also domesticated. The area of land needed to support each person was greatly reduced, settlements could be permanent, and complex civilizations became possible.

As keen observers of nature, hunter-gatherers were probably well aware of the facts of plant life long before they were put to use in agriculture. Although aboriginal people in Australia did not domesticate any plants or practise cultivation, they recognized the relation between seeds and plants, and when gathering wild millet they sometimes scattered seed about to ensure next

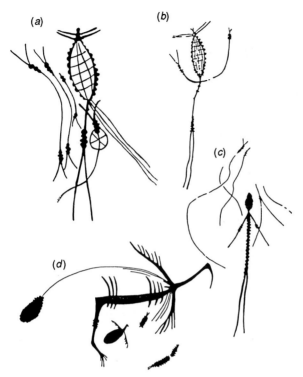

Figure 1.4. A rare example of the early depiction of food plants in rock art: yams (*Dioscorea* spp.) at Kakadu National park, N.T., Australia, albeit as partly human (*a–c*) or animal (*d*) figures (Chaloupka, 1983).

year's harvest. Likewise, when they collected wild yams they usually returned the upper part with the vine to the hole from which it had been dug. This was supposedly to conceal their 'theft' from the yam spirit (Figure 1.4), but it also served to ensure another generation of yams. Like other hunter-gatherers they occasionally 'managed' their wild food plants with fire and even irrigation, and therefore practised husbandry, and sometimes even cultivation, but without the longer-term commitment implied by reliance on agriculture.

Why, then, did they not turn to agriculture sooner? Sahlins' phrase suggests that their life style, so long as there was enough land to sustain it, had many attractions and the shift to agriculture may have been made reluctantly, under pressure of population. It was not seen as an unmitigated blessing; similar thoughts are still voiced by Texan or Australian pastoralists when cropping first becomes established in their vicinity.

Did agriculture evolve independently in different regions and in different forms with whatever suitable plants were at hand? This clearly occurred in the repeated and independent domestications of rice, yams, the cottons, beans, peppers and cucurbits. However, many archaeologists have been reluctant to concede this for the invention of agriculture (Trigger, 1989), preferring the alternative hypothesis that it was invented in one region and then diffused to others with variations along the way. Childe (1934) viewed the establishment of agriculture in the Fertile Crescent of the Near East as a 'Neolithic Revolution'. The north-westerly spread of agriculture through Europe, in the course of which plants that were regarded as weeds in the Fertile Crescent – such as rye – were also domesticated as crops, has now been traced. Near Eastern domesticates, both plant and animal, as well as agricultural implements such as the plough, also spread in an easterly direction to the Indus valley and on to China, as well as southwards to Ethiopia, taking linguistic affiliations with them (Renfrew, 1988, 1989).

It was natural that European archaeologists, immersed in 'the glory that was Greece and the grandeur that was Rome', should focus on the antecedents of their civilization in the Near East, particularly in view of the archaeological richness due in no small part to the favourable conditions for preservation in more arid areas (Harris, 1972). However, Sauer (1952), a geographer and also a diffusionist, chose south-east Asia as the most likely cradle of agriculture, and root and tuber crops as the most likely first domesticates, grown in forested areas by sedentary fisher-folk. He argued that forested areas were more readily cleared for cultivation than were perennial grasslands before the plough was invented, and that bush fallow could be much older than Near Eastern agriculture although much less likely to yield archaeological remains. Thus, although most discussions of the changes wrought by domestication have been in terms of the cereals and pulses so prominent in the Fertile Crescent, we should not overlook the quite different processes that may have been operative in early bush fallow agriculture in the tropics. We therefore begin discussion of the geography of domestication in SE Asia rather than the usual starting point, the Near East. The archaeological data from SE Asia are still slender, as is to be expected in such environments, particularly in view of the extensive post-Pleistocene submergence of coastal areas with the rise in sea level, but they are beginning to appear.

Nevertheless, the Near East remains by far the most compelling example of a cradle of domestication and early agriculture, with a striking spatial coincidence between archaeological sites and present distributions of the wild progenitors, a comprehensive suite of domesticated species – both animal and plant, cereal and legume – and the most convincing evidence of contemporaneous transitions from wild to domesticated forms. While this far surpasses the archaeological evidence from Africa and Latin America, the beginnings of indigenous forms of agriculture in those regions are also becoming clearer.

Figure 1.5. The eight centres of origin of crop plants as proposed by N.I. Vavilov in 1926 (after Harlan, 1975).

Earlier writers such as Darwin and de Candolle recognized that many crop plants had been domesticated in association with early centres of civilization, but it was Vavilov (1926, 1951) who proposed a relatively small number of centres of origin for most of the world's domesticated plants (Figure 1.5). Vavilov's notion, that the areas of greatest genetic diversity for a particular crop coincided with its likely area of origin or domestication, was a logical inversion of the 'age and area' hypothesis of J.C. Willis (Evans, 1979). However, various objections to it have been raised. His centres of diversity may, for example, be zones of more recent adaptive radiation or hybridization. Some of his centres remain well defined, such as the Fertile Crescent and North China, but others seem more diffuse, non-centres in Harlan's (1971) terminology, while the domestication of crops such as rice, sorghum, beans, yams, cotton and bananas appears to have occurred over wide areas, and more than once.

Far more species were gathered than were domesticated. For the world as a whole perhaps one species of higher plant per hundred was gathered and one per thousand domesticated, but these proportions vary from region to region, and with plant form and family. American Indians in the Sonoran desert gathered almost 20% of the plants in their area and the Australian Alyawara up to 30%. Many plant families have no domesticates, whereas the Solanaceae, Cucurbitaceae and Musaceae have a high proportion.

The nature of the changes associated with the domestication of plants has been the subject of prolonged debate, as it still is with the domestication of animals (Hemmer, 1990). Charles Darwin, against the currents of his time, was profoundly interested in the differences between wild organisms and their domesticated relatives. His two volumes on *'The Variation of Animals and Plants under Domestication'* (1868) pioneered discussion of the ways in which crop plants have been selected and their yield improved, and they remain a remarkable source of insight and balanced judgement. Darwin considered that the greatest changes wrought by selection under domestication were on size and form, particularly of the organs valued most by man. Schwanitz (1966, 1967) also emphasizes the gigantism of domesticates, but the increase in size is less striking in agricultural than in many horticultural domesticates.

A still unresolved question is the extent to which adaptedness to agriculture was modified by domestication in addition to the well recognized morphological and physiological changes. Are the homologous series of variations documented by Darwin, Vavilov, Schwanitz and others merely the observable tip of a genetic iceberg, the unseen part consisting of many changes accumulated over the silent millennia since domestication which combine to make modern varieties more amenable to agriculture? Darwin (1868) was not sure. He refers at one point to 'the little which man has effected, by incessant efforts during thousands of years, in rendering the plants more productive or the grains more nutritious than they were in the time of the old Egyptians'. By contrast, Alphonse de Candolle (1882) considered that domesticates were more profoundly different from their wild progenitors than from one another, as indicated in the epigraph to Chapter 3. The question is of importance not only in relation to the nature of domestication but also, for example, in relation to the value of using wild relatives as a source of desired characteristics in plant breeding programmes. Many plant breeders prefer to seek these in elite varieties rather than in wild relatives, as Duvick's (1984a) survey shows, in the belief that the long process of adapting genomes to agricultural conditions may be set back too much by such wide crosses unless they are inescapable. This issue relates, in turn, to the controversy over the 'genetic debts' due for access to the genetic resources of wild relatives on the one hand and to improved varieties on the other (Kloppenburg & Kleinman, 1987).

Recently domesticated crops, such as guayule, jojoba, and several species of lupin, offer some insights on this question. The very fact that two wild desert plants could, with little selection, become productive crops responsive to fertilizers and irrigation suggests that domestication may not involve profound changes in at least some plants. This is what Darwin (1868) concluded from Buckman's apparently quick improvement of the parsnip and Vilmorin's of the carrot.

Such considerations are relevant to the search for new crops, especially for sources of new pharmaceuticals or feedstocks. If domestication involves profound adjustments of the genome beyond the usual list of homologous characters, then it would be more promising to augment a low content of the target compound by selection in an established crop, as was done with sugar beet in the eighteenth century. On the other hand, if domestication is not much more than meets the eye, then it would be more effective to begin domesticating a wild plant with a high content of the desired compound. Modern methods of genome analysis should soon resolve the question of how much domesticates do in fact differ from their wild relatives.

Adaptation and the ecology of yield

Three complementary but quite different sets of processes relate to the physiology and genetics of yield. First there are those influencing adaptedness to agriculture and changed during the taming of wild plants, the subject of Chapter 3. Then there are those involved in the spread of crops beyond the cradles of domestication to different environments, through modification of their adaptive responses, the subject of Chapter 4. Finally, there is the raising of yield potential within each environment, the subject of Chapter 5. The first took place up to 10 000 years ago at the dawn of civilization, mostly in what are now less developed countries. The second probably began soon after, but its pace has accelerated over the past 500 years. The third has been largely confined to this century and, like the world population, has risen most rapidly in recent decades, (Figure 1.1).

A few domesticated species have remained endemic to their centre of origin (such as *Digitaria iburua*) or nearby (such as *Eragrostis tef* and West African rice, *Oryza glaberrima*) but many others are now grown around the world and in quite different environments from those where they originated. Some of these spread slowly, allowing adaptation to occur along the way.

Crop plants brought back from voyages of discovery or colonization, however, were often exposed to substantial changes in environmental conditions and frequently failed unless their introduction was repeated or based on a range of collections. Columbus presented Queen Isabella with sweet potatoes on his return to Europe in 1493 and by 1521 they had spread to the Far East via Africa (Burkill, 1951). By contrast, the first potatoes brought to Europe in 1570 were of the Andean subspecies in which tuberization was adapted to short days, with the result that they matured too late except in southern Europe. It wasn't until a century later that clones adapted to long days were introduced and the crop had any agricultural impact in Europe. A less well known example concerns the wild Australian yam, *Dioscorea transversa*, which was partly domesticated by Melanesians while in Queensland, in an effort to replace their noble yams. When taken with them back to Melanesia on their return, the Australian species failed to form tubers and was rejected because it 'does not know the seasons' (Barrau, 1979).

Not knowing the seasons has been a problem for many introduced plants, for tuber and root crops no less than for cereals and pulses. The crops have often grown well enough but their reproductive development has failed, being closely adapted to seasonal conditions at the place of origin. Response to daylength has been the control mechanism most commonly in need of modification in many crops as they have spread, especially to higher latitudes. Rice, soybean, sorghum and cotton offer striking examples. In many cases the requirement for short days has been relaxed by selection but in others, such as wheat, the requirement for long days has sometimes been enhanced as the crop moved to higher latitudes.

Responses to other environmental factors have also been modified as the crops spread. Comparisons of wild and cultivated cowpeas, for example, indicate that the cultivars are less sensitive to daylength for flower initiation, to temperature for germination, to relative humidity for pod dehiscence and to control of germination by hard-seededness (Lush & Evans, 1981). In general, the responses to temperature and other environmental factors have been less profoundly modified than those to daylength, but even small shifts in cold tolerance can have substantial agricultural consequences.

Adaptation to water stress – whether by escape, avoidance or tolerance – has also played its part in the wider distribution of crops, and continues to receive a great deal of research attention. Escape can be engineered by a shortening of the life cycle, and modification of the time to maturity has played a major role in adapting crops to new environments. Perennial shrubs such as castor bean (Narain, 1974) and cotton have been converted into annuals, and the life cycles of annuals have been shortened. Three hundred years ago Indian cotton was a perennial shrub, frost sensitive and confined to the tropics, where it flowered in winter. But in the eighteenth century, after day-neutral forms were found and plants were selected for shorter life cycles, cotton could be grown at higher latitudes (Hutchinson, 1976; Stephens, 1976).

The optimal length of the life cycle is the outcome of compromise. Shortening it allows single crops to be grown at higher latitudes and in drier areas, and more crops per year to be grown in the tropics. On the other hand, the yield per crop is often positively correlated with its duration, at least up to a point and depending on the environmental conditions. In optimizing crop duration, much depends on the balance between the various stages of the life cycle (vegetative, reproductive, and storage) as controlled by daylength and other environmental signals.

Old rice varieties in the tropics, for example, were short day plants whose response to daylength had been unconsciously selected so that many of them flowered just at the end of the wet season, with the result that grain filling and ripening took place in sunny conditions. With deep water rices, flowering occurred at the end of the floodwater crest. The long vegetative phase resulting from these daylength responses allowed the crops slowly to accumulate enough nutrients from low-fertility soils to ensure a reasonable yield. When varieties less sensitive to daylength were first selected, to permit out-of-season crops under irrigation, the first ones, like IR8, had a long juvenile stage during which they could not initiate flowering, making high yields possible even in the absence of control of the life cycle by daylength. However, as the level of fertilizer use has risen, the length of the juvenile stage has been reduced but not that of the subsequent stages in the life cycle (Evans *et al.*, 1984). As this example shows, there is a continuing interplay between the climatic environment, soil conditions and crop selection, leading to changes in the relative durations and rates of the various stages in the life cycle of the crop, and therefore also in the significance of the various yield components.

For all the major crop plants, production beyond their centre of origin is now vastly greater, and yields higher, than within it. Mainly this is because much of the cropping beyond the centre of origin is under higher input agriculture in developed countries, and often under climatic conditions conducive to higher yields. Some pests and diseases may be less troublesome than at the centre of origin, and the spread of crops to new environments can expose more diversity within the population than was evident at the centre, thereby promoting selection and change.

The ecology of yield is a subject on which we have far less information than we need. The central problem is the old one of the confounding of factors, because temperature, irradiance, daylength, water supply, etc. tend to vary simultaneously in the field and causation is difficult to determine. For various processes or for the growth of single plants, the effects of individual climatic factors and their interactions can be studied in phytotrons, but for the yield of crops this can be done on only a limited scale. It has, however, been helpful in dissociating the effects of irradiance and temperature on yield in a few crops. Both are powerful determinants of yield, with complex interactions between them. High temperatures, for example, may be disastrous under low irradiance but favourable when irradiance, water and nutrient supplies are great enough to overcome the disadvantages of much shorter life cycles (Rawson, 1988). This interaction is of great significance in relation to the future productivity of tropical agriculture, particularly if mean temperatures rise still further.

Thus, relations between yield and temperature like that illustrated for beans in Figure 1.6 may be modified not only by selection but also by atmospheric CO_2 level, irradiance and agronomic conditions. Moreover, certain stages in the life cycles of many crops are particularly sensitive to environmental conditions, such as flowering and the period just before it. Adverse conditions then, whether of temperature, water supply or irradiance, may greatly reduce yields even when conditions have otherwise been

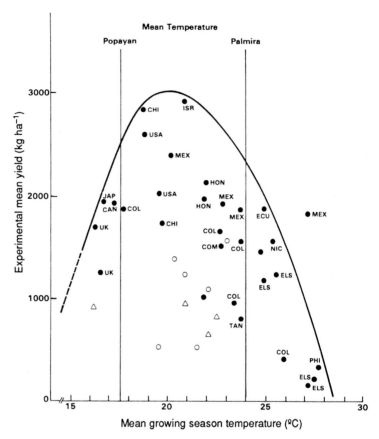

Figure 1.6.   The mean yield of 20 varieties of beans (*Phaseolus vulgaris*) as related to the mean temperature of the growing season at 41 locations throughout the world (CIAT, 1979). The significance of Popayan and Palmira is discussed later (cf. Figure 4.9).

favourable. Much plant breeding and agronomic effort has focused on these critical stages, and on reducing the vulnerability of crops during them.

If climate were wholly dependable, the close local adaptation of crops would ensure predictable food supplies. It is the variability of weather from day to day and year to year that makes crop yields so uncertain, and which challenges plant breeders to produce adaptable, reliable varieties. Adaptability is a characteristic more often praised than defined or understood. It may well have contributed to some crops becoming staple foods while others have declined, and to some varieties being more widely used than others. Presumably those that have been grown over vast and diverse areas – such as

IR36 rice and the progeny of the CIMMYT wheat cross 8156, which each occupied more than 11 million hectares at their peak – are adaptable in the sense of yielding well in many different environments. In earlier times such an ability was less important than the ability to yield well in the area where the 'land race' was used and to which it was closely adapted. But with the establishment of the international agricultural research centres such as CIMMYT and IRRI, and of globally oriented commercial plant breeding programmes, a higher priority is being given to broad adaptability than to close local adaptation.

Although adaptability is important to the plant breeder and seed merchant, of more significance to the individual farmer, especially the subsistence farmer, is the characteristic of stability or reliability of performance from one year to the next. It has been argued that variation from year to year at one site is analogous to variation from site to site in one year, i.e. that reliability and adaptability should be the same, but this proposition needs further analysis.

In a world in which global food stocks commonly represent only 2–3 months' food consumption, the reliability and predictability of food production in the face of seasonal and weather variations are clearly important features. Whether varieties can be responsive to favourable conditions and high input agronomy as well as reliable under poor conditions is a question that deserves closer analysis than it has received so far.

The physiological basis of crop improvement

Yield is the ultimate outcome of all the processes involved at all stages in the growth and development of a crop, any one of which may limit the yield of a particular crop. No one process holds the key to greater yields, but in Chapter 5 attention is focused on photosynthesis as the dominant source of crop biomass and on how that biomass is partitioned between the various organs of the plant, especially to those that are harvested. In doing so this chapter neglects many other important yield-determining processes and these omissions can be justified only on the grounds that the central concern of the chapter is with the nature of past increases in yield potential and with their implications for agronomy and plant breeding in the future. Crop physiology is essentially an exercise in comparative physiology, so we begin by examining the pitfalls of the various kinds of comparison that can be made, of cultivars with related wild species, of new with old varieties, between isogenic lines, or between different species and types of crop plant.

By focusing on photosynthesis and the partitioning of its products – on source and sink – the discussion runs the risk of being simplistically polarized. Although the leaves are commonly the main source of assimilates, photosynthesis by other organs – including the storage organs themselves – is often significant, so sinks may also be sources. Moreover, all source organs have been sinks before they become exporters of assimilate. Any growing or storing organ is a sink for assimilates, and they are of many different kinds with different mechanisms operative in the competition between them for assimilates. Even when the source and sink are quite separate, they are not independent. The supply of photosynthate at one stage often determines the potential size of the sink organs, while the demand for assimilates by the growing or storing sink organs can modulate the rate of photosynthesis. Indeed, the evidence suggests that in high-yielding crops both source and sink may be limiting, and that they are likely to be more or less in balance in the environments to which the variety is adapted.

The main source of greater yield potential so far has been increased partitioning to the harvested organs, reflected in a rise in harvest index. In raising this, plant breeders have been changing the basis of selection but not trying to improve on natural selection, whereas attempts to improve the efficiency of photosynthesis mean doing better than prolonged and intense natural selection, and will require a deeper understanding of the subtleties and trade-offs in the overall process.

It is a common fallacy to refer to photosynthesis as inefficient, given that only a small proportion of the solar radiation to which a crop is exposed may be captured in its ultimate biomass. In fact, under certain conditions the photosynthetic process rivals the most efficient photoelectric devices known, but it must cope with continually changing and often stressful conditions, as well as with continual adaptation and renewal. Apparently wasteful aspects, such as photorespiration, may have adaptive significance and may, indeed, be quite efficient. The key photosynthetic enzyme ribulose-1,5-bisphosphate carboxylase–oxygenase, with the acronym of 'rubisco', is the most abundant enzyme in the world. As a large molecule with low specific activity it may appear to be inefficient but it remains to be seen whether it can be improved upon, because it has been strongly conserved in the course of evolution. On the other hand, the success of crops and weeds with the C_4 pathway of photosynthesis nourishes the hope that other ways of enhancing photosynthesis may be found.

Because the enhancement of photosynthetic capacity may become a prerequisite of greater yield potential in the future, Chapter 5 devotes a lot of attention to the evidence for changes in photosynthetic rate achieved so far, and the answer is clear. For crop after crop there has been no increase in the maximum CO_2 exchange rate per unit leaf area (CER) with increase in yield potential. Indeed, for several crop plants the highest rate is found among the wild relatives and there is a negative relation between CER and yield. Moreover, selection for high CER has tended to reduce rather than raise yield.

This presents us with an important paradox. Increased yield frequently results when the photosynthetic rate is raised by environmental conditions such as higher irradiance or CO_2 level but not, so far, when it is raised by genetic improvement. If selection for higher CER is to contribute to increased yield potential in the future, this 'photosynthetic paradox' must be resolved. Consideration of the trade-offs between maximum CER and leaf size goes part of the way.

The absence of any genetic improvement in CER is unlikely to have impeded the rise in crop yields so far because the greater use of nitrogenous fertilizers has acted as a surrogate in raising CER through a rise in the percentage of nitrogen in leaves and increases in rubisco and other photosynthetic enzymes. Nitrogenous fertilizers also influence crop photosynthesis in other ways, e.g. by increasing the leaf area and delaying the mobilization of N from the leaves late in the crop life cycle. The survey of photosynthesis among crops suggests that domestication and improvement have usually been accompanied by greater leaf area and duration rather than enhanced CER. Relative growth rates in the seedling stage have not risen, so it is a matter of faster leaf area expansion leading to greater ground cover and interception of light and more effective exclusion of competitors. Although such a strategy is advantageous in sparse crops or in the early stages of dense crops, it may not be so once the canopy is closed in high input crops, and a shift in selection criteria may be needed. Optimization of canopy structure, through variations in leaf size, shape, spacing and inclination, is also likely to assume greater significance in high-input, high-density crops. To date, canopy structure has been discussed mainly in terms of light interception for crop photosynthesis, but many other consequences of it should also be considered.

When we turn from photosynthesis to the partitioning of its products, we are immediately confronted by processes far less well recognized, let alone understood. The mechanism of translocation is still subject to debate, and the factors regulating it and the partitioning of assimilates between the various sites of growth or storage are only dimly perceived. The geometry of the competing sinks – i.e. their relative size and distance from and vascular connection to the sources – is highly significant and has probably played an important role in crop evolution. Condensed inflorescences, for example, are not only more convenient and quicker to harvest, but possibly also more effective in competing for assimilates against other parts of the plant. Besides the geometry, the metabolic compartmentation and hormonal activity of the growing organs are also likely to be important determinants of their sink strength.

Plant breeders have not been constrained by this lack of understanding, because both the change in form and the increase in the harvest index of many crops has been quite striking. For both Darwin and de Candolle, the changes in form were epitomized by those among the brassicas (Figure 1.7). The rise in

Figure 1.7. Variations in form among wild *Brassica oleracea* and some of its domesticated derivatives, from Schwanitz (1967): clockwise from the upper left, the wild progenitor, curly kale, Savoy cabbage, white cabbage, cauliflower, Brussels sprouts, marrow stemmed kale, and kohlrabi.

harvest index of the cereals has been the major component in the enhancement of their yield potential in recent years. The ways in which this has been achieved, and the contribution to it by other organs (stems, roots, branches), by reserves, or by changes in the life cycle, vary greatly from crop to crop and change as agronomy develops. In fact, agronomic improvement is the key to major shifts in partitioning. Stems can be made shorter, thereby freeing assimilates for investment in the harvested organs, only to the extent that weed control is improved, especially in the tropics. Savings from root systems can be made only if the supply of water and nutrients is guaranteed. Reserves can be drawn down only if pests, diseases and other stresses are effectively controlled. Higher harvest index is made possible by greater agronomic support for the crop. The other side of the coin is that selection for yield under conditions of strong agronomic support may be a poor guide to yield in the absence of that support, a conclusion with which many plant breeders disagree.

Thus, although our focus on photosynthesis and partitioning in Chapter 5 is narrower than it should be, it nevertheless encapsulates the major physiological changes in crop improvement and raises several important issues in relation to plant breeding strategies, their relation to agronomic advance, the relative effectiveness of empirical vis-à-vis ideological selection, and the role of crop physiology.

Yields past, present and potential

Among the many kinds of yield watchers, the most shrill and happily seized on by the media are the plateau spotters. A pause in yield increase, associated with a few poor years or changed socio-economic policies, is soon translated into the limit to crop yield having been reached and doom being nigh. But the rise in yield eventually resumes, the noise dies down, as like as not there will be a glut for a while, and research for higher yields will be discouraged until the next round begins. Not only the public but even agriculturists themselves are confused as to what is happening. Chapter 6 considers yields past, present and potential. Like every other analysis of the subject it is coloured by the innate optimism or pessimism of its author, so let me declare myself as an optimist, more arithmetic than exponential.

Optimists have been described as pessimists without the facts, but on the subject of maximum crop yields many generations of pessimists have already fallen by the wayside. There have been those who were convinced the world was already over-populated and that famine was at hand, and those who thought there was no way forward because they couldn't see one. It pays to remember that Lord Kelvin made calculations from first principles which seemed to prove that continents could not drift and that the world could not be remotely as old as Darwin said it must be. Kelvin was not misled by the processes he knew about, but by those not yet recognized. Although some of the processes involved in yield determination are well understood, many are not and no doubt some are still unrecognized. Thus, we cannot predict the ultimate limits to crop yield with any real confidence.

Adding to the problems of prediction is the all-too-evident irregularity and variety of past trends in yield, making extrapolation hazardous. Cycles in weather or socio-economic conditions, exhaustion of soil fertility, erosion or salinization, price incentives, extension or contraction of cultivation, genetic or agronomic innovations and many other factors can lead to take-off, stasis or decline in yields, and progress can veer from apparently exponential to linear or declining.

Given the complex of influences at work, we shall not assume that yield improvement should follow any particular form. Likewise, associating a particular burst in a yield curve with a particular innovation can be a misleading game, although widely played in the scramble for research funds. For one thing it tends to conceal a vital aspect of advance in yield, that it often comes from synergistic interactions between breeding and agronomy, innovations in each successively opening up new opportunities for the other. For example, energetically more efficient ways of making nitrogenous fertilizers created a need for more radical shortening of cereal culms than had been achieved by polygenic selection. The introduction of the reduced height genes in wheat made this possible, and was in turn made feasible by improved herbicides, without which the more extreme dwarfs would have been disadvantaged by weed competition. Dwarf stature allowed still heavier applications of nitrogenous fertilizers to be profitable, creating a need to use up the residual fertilizer left in the soil at the end of the growing season, both for reasons of efficiency and in order to minimize ground water contamination. Minimum tillage techniques, permitting faster crop turn-around, allow the next crop to be sown sooner before winter and to take up the residual fertilizer, which it will presumably do even better with varieties selected under such regimes. And so this opportunistic reciprocation between breeding and agronomy will go on.

The few reports of cereal yields two thousand or more years ago suggest that, with good husbandry, they could be as high as those in Europe in the Middle Ages. Thereafter there was a slow improvement in yields, probably associated with improved farming practice. The pace picked up in the eighteenth century as individuals – such as Vilmorin in France and Knight in England – and private companies began selection and then breeding and as the use of other beneficial farm practices spread.

Yield take-offs have occurred with many crops in many countries over the past century. For rice they range in time from about 1880 in Japan (Figure 1.1) to one hundred years later in Myanmar (Burma) (Figure 1.8). They have been initiated at different yield levels, followed by different rates and durations of gain. National yields depend on the stage of economic development, on climate, and on the extent of fertilizer and other input use. They vary from crop to crop, being generally higher in the cereals than in pulses.

Crop yields at experiment stations have been taken as an indicator of what levels could be reached, so it is important to consider the causes of the difference between experiment station and national average yields, the yield gap as it is commonly called. In fact, yield gaps frequently reflect differences in environmental conditions, soil quality, water supply, access to credit, availability of labour and attitudes to sustainability and risk as much as in the varieties and practices used. There are situations where the yield gap is now small or narrowing, giving some observers cause for concern that a ceiling yield is being approached, whether for maize in Iowa (Thompson, 1975) or rice in Java.

Penalties to yield take many forms. Besides the indirect penalty from the need to focus attention on many other characteristics besides yield in most plant breeding programmes, some characteristics may exact a more direct penalty. Requirements for high protein content impose several constraints on the improvement of yield, as in the case of Canadian wheat for many years. So

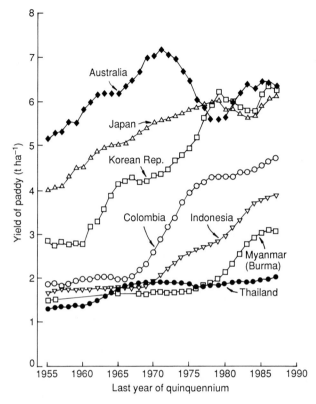

Figure 1.8. Five-yearly running average trends in the yield of paddy rice in several countries (data from FAO Production Yearbooks).

may other 'quality' requirements as in the cases of barley for brewing and rice in Thailand.

Record yields of crops hold considerable fascination and some now figure in the *Guinness Book of Records*, but they often arise from highly unusual conditions and may be of limited value as a guide to potential yields, a point which needs to be more widely recognized. Boyer (1983) has used them to assess the losses caused by 'stress', after subtracting estimates of losses caused by pests, diseases and weeds from the difference between record and average yields. He ascribes the residue to environmental stress, whereas much of it would often be due to socio-economic constraints.

Because breeding and agronomy interact so profoundly in the improvement of crop yield, the partitioning of that improvement between them is an arbitrary procedure. Nevertheless, it is worth trying to estimate what change in yield potential has been achieved by plant breeding so far. To this end yield potential may be operationally defined as the yield of a cultivar when grown in environments to which it is adapted, with nutrients and water non-limiting, and with pests, diseases, weeds, lodging and other stresses effectively controlled. These are almost hypothetical conditions but only with an approach to them can we assess progress in the assembling of productivity genes as distinct from genes for adaptation to environmental stresses, adaptedness to modern agronomy, resistance to pests and diseases, and that complex but important characteristic, quality.

Both direct and indirect estimates of the increase in genetic yield potential of many crops have now been made, and these are reviewed in Chapter 6. Overall they indicate sustained, even accelerating, progress in many breeding programmes, with increase in yield potential often accounting for a half or more of the advance in yield. In several cases yield potential has advanced substantially even though yields have not. With cotton in the USA, where there have been pressures to reduce the use of pesticides and other inputs in recent years, yield has declined in spite of a rise in yield potential.

As crop yields rise a great deal of research is needed simply to maintain past yield gains while insect pests, diseases and weeds find new niches or vary in the struggle to preserve old ones. This is what leads to 'the varietal relay race', as Plucknett & Smith (1986) have referred to it, i.e. to the continuing succession of new varieties whose most important attribute may be resistance to a new pest or disease biotype rather than greater yield potential. Such problems tend to become more varied and acute as cropping is intensified, with the result that defensive or maintenance research will probably occupy more and more of the plant breeders' and agronomists' time in future. Preventing a decline in yield may be more urgent, just as difficult and quite as creative a task as raising yields still further, although apparently with less appeal to those funding research. Famines have ensued (cf. Sen, 1981) and civilizations have declined (Jacobsen & Adams, 1958; Brewbaker, 1979; Tainter, 1988) when yield levels could not be maintained.

Inputs, energy use and synergistic interactions

Agronomic inputs have had a major impact on the raising of crop yields, both directly and also indirectly by making selection for greater yield potential more effective. Without them, it would not be possible to support the present world population (Buringh & van Heemst, 1977). Yet the most strident criticisms of modern agriculture are for its dependence on inputs and for the problems this causes, of environmental pollution, resource depletion and dependence on imported proprietary compounds. These are important problems, but the focus of this book is on the yield of crops, and Chapter 7 therefore concentrates on the relation between the use of inputs and yield.

The epigraph to the chapter, from Linnaeus, echoes a frequently heard criticism of agriculture today, namely that modern varieties are dependent on inputs, on daily sacrifices to them as the great botanist put it. The physiological analysis in Chapter 5 indicates that plant breeders have not yet succeeded in improving on natural selection for photosynthesis and growth rate, so improvement in yield potential has come from changes in the partitioning of biomass. Such changes depend on the extent to which agronomic support allows the crop to invest less of its biomass in stems, leaves, branches, roots or reserves and more in the harvested organs. Thus varietal improvement in yield potential depends on agronomic advance, but it also acts as a Trojan horse for better agronomy by making greater use of inputs more profitable (Hayami & Ruttan, 1971).

In this sense, modern cultivars are 'dependent' on agronomic inputs, and Linnaeus and critics of the Green Revolution are right to a degree. High yields can be sustained only with high inputs, because they require heavier crops as their foundation and that in turn requires more water, more nutrients such as nitrogen, more protection from pests, diseases and weeds, and a greater input of energy. Inevitably it is a case of 'potatoes partly made of oil', as Odum (1971) puts it, of prehistoric photosynthesis subsidizing current photosynthesis.

The ways in which the various agronomic inputs influence yield vary greatly and, importantly, complement one another. Nitrogenous fertilizers, for example, get the crop (even legume crops) away to a faster start, buying time that may be valuable in allowing the crop to avoid an early or late period of stress. In this sense fertilizers are a surrogate for faster development. They enhance leaf area growth, with the result that the crop canopy intercepts more light sooner, a surrogate for greater irradiance and also for more water by reducing evaporative loss from the soil. They raise the nitrogen content of the leaves and the photosynthetic rate, acting as a surrogate for selection for higher CER. They reduce the need for mobilization of nitrogen out of the leaves into the storage organs, especially in high protein crops, thereby extending the duration of photosynthetic activity and making it possible to select for a more prolonged storage phase. They have many other effects, but this list suffices to indicate the manifold ways in which this one input enhances yield, permits selection for greater yield potential, interacts with environmental conditions, allows climatic stress to be minimized, and augments other inputs. Of course, not all the effects of nitrogenous fertilizers are desirable. With more protein in the leaves the larvae of insect pests grow faster. With denser, lusher crop canopies, diseases may flourish. Residual nitrates may accumulate in the soil or in ground water, and so on, but these are not insoluble problems, they are challenges to devise better management techniques.

Of the protective inputs for the control of diseases, pests and weeds, fungicides are the oldest but herbicides are now used in the greatest and fastest-growing amounts. Protective chemicals also open up new opportunities for agronomy and breeding. Dwarfing in maize led to a higher optimum plant density (Johnson *et al.*, 1986) and associated agronomic changes. Effective weed control was a prerequisite to the use of major dwarfing genes in many breeding programmes. Without herbicides, yield losses due to weed competition are much greater with semi-dwarf than with tall cereals. A dwarf wheat, Piper's Thickset, was long ago recommended for use on rich soils by Roberts (1847) and in the 1920s the early-flowering, upright-leaved, short-statured and high yielding Ramai rice was another variety ahead of its time (Evans *et al.*, 1984). Herbicides also made the development of minimum tillage techniques possible, thereby reducing soil erosion and degradation, allowing steeper slopes to be safely cropped, and enhancing the timeliness and flexibility of sowing.

These few examples underline a crucial characteristic of agronomic inputs, namely the great variety of their synergistic interactions, which may be used to create new plant breeding opportunities, shift the timing of crop development to minimize climatic stresses, and develop more comprehensive and flexible procedures for pest and disease control. Integrated pest management, for example, aims at more sparing and effective use not only of pesticides but also of resistance genes, thereby extending the useful life of both.

Genes for resistance to pest and disease organisms may be overcome as rapidly as resistance to new protective chemicals develops. Some, such as resistance to grassy stunt virus in rice derived from the wild *Oryza nivara*, may last for a long time while others are quickly broken down. As successful modern varieties are grown over increasingly wide areas, thereby offering their pest and disease organisms greater opportunities to overcome varietal resistance to them, the greater is the need for understanding and ingenuity in devising the most effective combinations of breeding, pesticide and management for their control. Research has become a crucial input into crop production and yield.

Among other things, it has led to a remarkable reduction in the effective application rates for all three classes of protective chemicals, more than one hundredfold over the past 50 years. More effective forms of fertilizer and of their application have been developed. There has also been a major improvement in the efficiency of manufacture of the nitrogenous fertilizers that loom so large in the energy budget of modern crops.

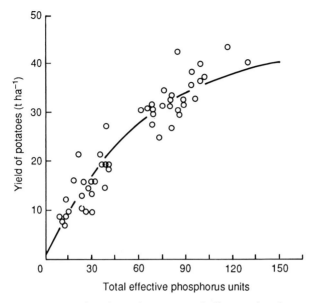

Figure 1.9. The relation between total effective phosphorus units and the yield of potatoes (adapted from Payton *et al.*, 1989).

The most impressive outcome of all these improvements has been that, at least for productive cropping systems such as maize in the USA, rice in Asia and wheat in Europe, there has been no decline in the increment of grain yield for successive increments of input energy in support of the crop. Economists may view such avoidance of diminishing returns as merely a shift in the response surface, but it is important that we understand how this has been achieved. For any one kind of input, the yield response soon approaches a plateau (Figure 1.9), following the law of diminishing returns first formulated for agriculture by Malthus. However, a succession of new inputs can keep rescuing yield from the plateau. A striking example of this process is the energy reached by particle accelerators (Figure 1.10). First presented as a joke by Fermi in 1954, this figure became, as Price (1963) noted, less and less humorous as it went on faithfully predicting when further innovations would be made. Although the gains in agriculture are less striking, a continuing series of new inputs together with the synergistic interactions between them have helped to keep the crop yield plateau at bay. In this latter respect, the advance of agriculture resembles the evolution of evolution (cf. Vermeij, 1987).

In charting our agricultural progress, the question is what unified index of inputs to use. Even with only the major fertilizers to consider, this posed a

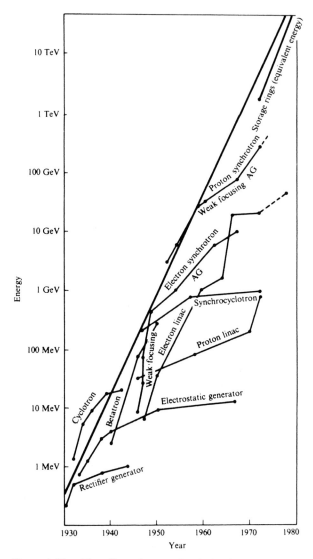

Figure 1.10.   The effect of successive design innovations in sustaining the exponential increase in accelerator beam energy (Ziman, 1976).

problem which was unsatisfactorily resolved into kilograms of $N + P_2O_5 + K_2O$ per hectare, as used in FAO statistics. But with the inclusion of trace nutrients, pesticides, larger farm machinery etc., we are left with a horizontal scale of either total energy or dollar input. Energy budgetting of agriculture has burgeoned in recent years and has made us look at some agricultural operations more critically, but it tends to over-emphasize output:input ratios. These make modern maize production seem inefficient by comparison with hunter-gathering or early agriculture, and they conceal the quite remarkable feat of having so far avoided a decline in the output:input ratio of successive increments to yield. In fact, compared with food processing beyond the farm gate and many other aspects of our lives, modern crop production is energy-efficient and merits better understanding of its achievements and constraints.

Looking ahead

Winston Churchill once said 'It is always wise to look ahead, but difficult to look further than you can see'. His remark applies to agriculture as much as to any other human activity. Those who have peered into its future have either seen the present writ larger, or they have wanted to abandon modern agriculture for its more traditional forms on the one hand or for aquaculture, algae, synthetic foods and new crops on the other. Some elements of each of these scenarios are to be found in Chapter 8: the present writ larger in the sense of still greater dependence on the staple crops, greater use of even more inputs, and higher cropping intensity; more traditional in the sense of a return to the more complete utilization of crops and a more determined conservation of resources; and more radical in the transformation of agriculture by ideas not yet conceived or whose time has not yet come. We can be certain that agriculture will be transformed by novel elements even though we cannot predict what form these will take. Many futures for agriculture await invention, and our responsibility is not to foreclose the options and opportunities.

Yields per crop will go on rising but, at least in the tropics, there may be more emphasis on greater yields per day. The scale of farming is almost bound to increase, however much we may regret it. Management will become more complex and also more comprehensive in terms of landscapes and environment. Research will become even more crucial to the future of yield, and probably more globally integrated. Merely because we cannot envisage solutions to present problems is no reason for pessimism. As Francis Bacon put it: 'They are ill discoverers that think there is no land, when they can see nothing but sea'.

CHAPTER 2

Crop yields and world food supply

'No man qualifies as a statesman who is entirely ignorant of the problems of wheat'.   Socrates

Introduction

Statesmen today can no more ignore the problems of food production and distribution than when Socrates lived 2400 years ago. In some countries 'the problems of wheat' are now those of over-supply, but in many others they are still those of uncertain or inadequate supply. Consequently, one significant dimension of that over-worked phrase 'the world food problem' is the imperfect and inequitable distribution of food both between and within countries. Use of the phrase is, however, often meant to conjure up another dimension, of imminent disaster as we approach the limit of the world's capacity to feed itself.

Is there really a world food problem in this latter sense, as against the inevitable imperfections of production and distribution in the face of variable weather and shifting policies? And is it a problem of the supply of food or of the means to obtain it?

For many centuries there have been people concerned that the end of the food was nigh, but since 1798 they have had some elementary and emotively powerful mathematics to support them. Five years earlier William Godwin had written:

> 'Three-quarters of the habitable globe are now uncultivated. The parts already cultivated are capable of immeasurable improvement. Myriads of centuries of still increasing population may probably pass away, and the earth be still found sufficient for the subsistence of its inhabitants.'

Exasperated by such optimism, Thomas Malthus put forward a contrary view, published in 1798 as *An Essay on the Principle of Population*.

So influential have this essay and his subsequent writings been for those

32

concerned with 'the world food problem' – indeed, they are often labelled Malthusians – and so frequently have his views been misrepresented (Flew, 1970), that a few comments are needed on his striking contrast between the 'natural inequality of the two powers, of population, and of production in the earth, and that great law of our nature which must constantly keep their effects equal...'. Before we look at the famous inequality of those two powers, note that Malthus envisaged not a disastrous imbalance between them but a constant trimming. It was 'that great law of our nature' which caught Charles Darwin's eye on September 28, 1838.

On the power of population, Malthus wrote: 'It may be safely asserted therefore, that population, when unchecked, increases in a geometrical progression of such a nature as to double itself every twenty-five years'. Mark that important qualification 'when unchecked', so often ignored by his critics despite Malthus' insistence near the beginning of his *Essay* that 'in no state that we have yet known, has the power of population been left to exert itself with perfect freedom'.

By contrast with the capacity of populations to increase geometrically, Malthus concluded that 'the means of subsistence, under circumstances most favourable to human industry, could not possibly be made to increase faster than in an arithmetical ratio'. Here Malthus was on shakier ground, and he knew it. Perhaps the elegant contrast between these geometric and arithmetic powers was, as Le Gros Clark (1951) suggests, his concession to Cambridge mathematics. More likely, Malthus recognized its persuasive force. The case he put for an arithmetic limit, based on British agriculture, was perfectly plausible at the time:

> 'If I allow that by the best possible policy, by breaking up more land, and by great encouragements to agriculture, the produce of this Island may be doubled in the first twenty-five years, I think it will be allowing as much as any person can well demand... The very utmost we can conceive is that the increase in the second twenty-five years might equal the present produce. Let us then take this for our rule, though certainly far beyond the truth; and allow that by great exertion, the whole produce of the Island might be increased every twenty-five years, by a quantity of subsistence equal to what it at present produces ... this ratio of increase is evidently arithmetical'.

As we have seen from Figure 1.1, the rise in British wheat yields eventually surpassed this limit. In fact, rather ironically, comprehensive but unpublished surveys of the yield of wheat in England from 1809 to 1859 suggest that it may have doubled within the 25 year period between 1835 and 1859, soon after Malthus published his *Summary View of the Principle of Population* (cf. Healy & Jones, 1962). Even for the world as a whole, food production can increase geometrically, as Figure 2.1 shows. It was Engels who put his finger on two of the reasons why, as early as 1844:

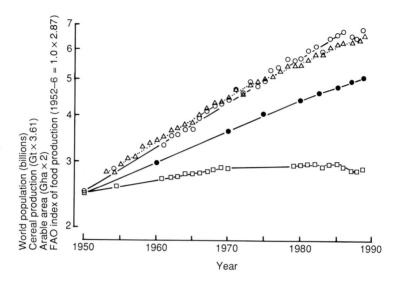

Figure 2.1. Increases since 1950 (on a logarithmic vertical scale) in world population (●), total cereal production (○), arable area (□) and the FAO index of total food production (△), all scaled to equality in 1948–52 (data from FAO Production Yearbooks).

'... has it been proved that the productivity of the land increases in an arithmetical progression? The extent of land is limited – that is perfectly true. But the labour power to be employed on this area increases along with the population; and ... there still remains a third element – which the economists, however, never consider as important – namely, science ... Science advances in proportion to the knowledge bequeathed to it by the previous generation, and thus under the most ordinary conditions it also grows in geometrical progression – and what is impossible for science?' (quoted by Flew, 1970).

Science has, in the event, proved Malthus wrong. The world's population has grown geometrically, and its food supply has more than kept pace (Figure 2.1). The rate of population increase has been less than its potential and, as Malthus put it: 'The great question, then, ... is the manner in which this constant and necessary check upon population practically operates'. In much of his later work he focused on the ways in which population growth is constrained; by implication therefore, the Malthusian view of the population – food equilibrium is that population growth is the dependent variable, and increases only to the extent that food production can be increased. This is not entirely fair to Malthus, who wrote of the limitation at the end of his *Essay*: 'We are not patiently to submit to it, but to exert ourselves to avoid it'.

This latter view is close to that put forward by Boserup (1965), although it is contrasted with the Malthusian approach by herself and others: 'It is based throughout upon the assumption – which the author believes to be the more realistic and fruitful one – that the main line of causation is in the opposite direction: population growth is here regarded as the independent variable which in its turn is a major factor determining agricultural developments'. Like Engels, she considers that the output from a given area of land can respond more generously to additional inputs of labour than had been assumed, especially when combined with other inputs.

The deliberate efforts made by agricultural scientists and policy makers over many years to increase output because of population pressure support her thesis. On the other hand, their frustration at seeing the rapid gains in food production in developing countries swallowed up by further population growth illustrates the Malthusian view that population growth can also be the dependent variable. As Borlaug (1971) emphasized in his Nobel Prize acceptance speech '... man is using his powers for increasing the rate and amount of food production. But he is not yet using adequately his potential for decreasing the rate of human reproduction'.

The point of this introductory digression is that it is important to recognize at the outset that increases in crop production and yield both determine and are determined by population growth. Moreover, it is no longer just the subsistence of populations but their rising expectations for standards of health, education and leisure to enjoy the other fruits of the earth in this and future generations that modulates both population growth and increase in crop yields.

An upper limit?

Given the mutual adjustments between population and food supply along the way, is there a limit to their increase which can be specified? Malthus did not discuss the matter, but since his time there has been a succession of people concerned that the upper limit was at hand and that another doubling of population could not occur. In 1898, one hundred years after the first publication of Malthus' *Essay*, Crookes doubted whether much more wheat could be produced. Yet the current world production is over ten times greater than it was then.

Reaching a world population of two billions set off another round of Malthusian pessimism. The opening words of a book by a group of knowledgeable and concerned scientists were: 'Within the lifetime of some of our children the world's population may be expected to reach 4000 millions. It stands at present at about 2,300 millions... How shall we work the miracle of feeding the 4,000 millions?' (Le Gros Clark & Pirie, 1951). Yet many of the authors were themselves still alive when the world population reached four billions in 1975, and that population was better fed than when they wrote: the FAO index of food production *per capita* was 23% higher in 1975 than in 1950.

The approach to a world population of four billions stimulated a further round of Malthusian concern that the upper limit was at hand (e.g. Paddock & Paddock, 1968; Brown, 1974, 1975). No doubt there will be further rounds, and each time there will be optimists to confront the pessimists with an opposing set of estimations which get more sophisticated with each round.

Buringh *et al.* (1975) compute that 25% of the land area of the world could be cultivated (compared with 11% at present), and that the absolute maximum grain production would be 49.8 billion tonnes, almost 40 times greater than that in 1975. The assumptions for their estimates are rather optimistic, but the history of such efforts suggests that further scientific advances will make them look conservative in future years. Even the FAO has described its own estimates (Higgins *et al.*, 1982) as encouraging. To quote from its fifth World Food Survey (FAO, 1987) the 117 developing countries 'could, as a group, feed almost twice their populations projected to the year 2025, even under the United Nations 'high-population' variant, by producing at the intermediate of the three levels of technology and input use envisaged in the study. The same input level would also suffice for the highest numbers . . . at which their populations may be expected to be stabilized toward the end of the next century'. Agriculture is not failing us as Calder (1967) suggested; in confounding Malthus with a geometric increase in production, it has provided one of the outstanding but unsung examples of the achievements of science.

Food production

The FAO index for world food production has increased steadily since 1950, exactly doubling in the 25 years to 1975, as may be seen from Figure 2.1. There were pauses in 1979 and 1983, but only in 1972 and 1987 did production actually fall. Since the early 1970s there has been some deceleration in the logarithmic growth of global food production, reflecting the slower rate of expansion of the area of arable land (Figure 2.1).

The change in *per capita* food supplies in recent years is indicated in Figure 2.2. Note that the FAO indexes for the two groups of countries, developed and developing, are each relative to a figure of 100 for the 1961–5 period, and therefore indicate only the relative change in their *per capita* food supplies. In absolute terms the supply of calories per person in the developing countries was only 63% of that in developed countries in 1961–5 compared with 72% in 1983–5. Food supply per person rose steadily in the developed countries until the late 1970s, but the really encouraging feature is that, after a long period of

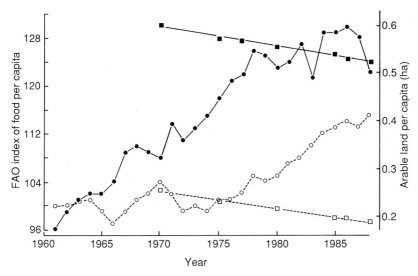

Figure 2.2. Trends since 1960 in arable land per capita (squares) and in the FAO index of food *per capita* (circles) in the developed (solid symbols) and developing (open symbols) countries of the world. The food indices are based on 1961–5 = 100, and do not expose the greater *per capita* supplies of food in the developed countries (data from FAO Production Yearbooks).

stasis, it has risen by 15% in the developing countries as a whole since 1975 in spite of a 15% fall in arable area per person (Figure 2.2). Of course, this average gain conceals considerable variation in the food supply situation from one region to another but that is not our subject here (cf. Pierce, 1990).

Food stocks and idled cropland

One of the grounds for pessimism in the early 1970s, as it had been for Crookes in 1898, was an apparent run-down in world food stocks. Brown (1975) wrote: 'Within a span of a few years the world's surplus stocks and excess production capacity have largely disappeared. Today the entire world is living hand to mouth, trying to make it from one harvest to the next'.

A favourite illustration at that time, guaranteed to ring the Malthusian alarms, is indicated in the middle section of Figure 2.3. Its implication then was that the world's reserves of food and idled cropland would be exhausted by 1975–7. The actual sequel is shown on the right hand side of the figure, and highlights the perils of extrapolating from short runs of data. Brown (1985) subsequently wrote: 'Except during the 1972–75 period, these two reserves together (i.e. carry-over stocks and idled land) have maintained a remarkable

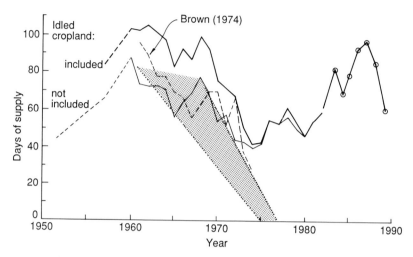

Figure 2.3. Trends in carryover stocks of grain in the world, expressed as days of utilization. Adapted from Brown (1974) and Simon & Kahn (1984) by the addition of earlier extrapolations (hatched area) by R. Bryson (Gribbin, 1974), as well as of recent data (circles).

stability. . .' Now that reserves have again fallen, a new round of pessimism has set in and Brown is claiming that the high food stocks of the 1980s were a temporary aberration (Mackenzie, 1990).

What happened, in fact, was that stocks of grain in the early 1960s were held above the usual financially profitable levels mainly because the major exporters (the USA, Canada and Australia) used them to support producers' incomes before shifting to other mechanisms (Morrow, 1980). The greater concern when stocks fall should be not that the world is running out of grain but that the poor suffer most from the rise in prices.

There are major differences between countries, regions and crops as well as between years in the proportion of annual production that is carried over as stocks. In the mid-1980s, total cereal production in the developing countries was about the same as that in the developed ones, yet the stocks in the former represented only 13.7% of production in 1986, whereas they were 31% in the latter (FAO, 1986). The world's cereal stocks are mainly held in the developed countries, and the variation from year to year in the proportion of cereals in stock is much less in developing countries, fluctuating only between 13.2 and 14.4% in the 1980s whereas in the developed countries the proportion doubled to reach more than one third of production in 1987. Among the cereals the proportion in stock is lowest and least variable for rice (10–11%) and most variable for the coarse grains (10–25%) and for wheat, which constitutes one third to one half of the total cereal stocks. For sugar the stocks currently hover around 100% of annual production.

Reliability of production data

The published data on stocks, like those on crop and animal production, must be treated with some reservation, especially those from some developing countries. Under-development is reflected at least as much in the capacity for data gathering and analysis as in the economy at large.

At the farm level, production may be under-declared for a variety of reasons, such as household use, stockpiling, black markets or the avoidance of taxes or requisitions. Farmer (1969) quotes the example of rice yields in Sri Lanka, which doubled in 1951–2 when random field sampling was introduced. Crop areas are often over-declared, thereby exaggerating the under-estimation of yield. Then there are the problems of estimating production from small subsistence holdings, mixed crops, repeatedly harvested ones such as cowpeas, sporadically harvested ones such as cassava and yams, slash and burn clearings, home gardens and inaccessible swamp rice fields (cf. Sarma, 1987). On top of these there are the losses from storage and along the way through holes, both real and metaphorical, as Farmer (1969) put it.

At the national level, the data may be over- or under-estimated for a variety of reasons. The flow of aid may be helped by under-estimation, while over-estimation in one year may compel further over-estimations or even the cessation of official publication of national data until real production catches up, as happened in China after the 'Great Leap Forward' (Evans, 1978*b*).

Such problems extend to the global level, particularly for FAO which normally publishes the official statistics of its member countries. Thus, there can be problems when countries belong to FAO as well as when they do not, particularly with large nations such as China and the (former) USSR. These problems are partly circumvented by the US Department of Agriculture data system, and since the late 1960s FAO has also made wider use of its own estimates. Discrepancies between these two major data systems are discussed by Paulino & Tseng (1980). Some of these arise because the two agencies define their reference periods in different ways. There is fairly close agreement between them for the area and production data on the major staples, wheat, rice and maize, but Paulino & Tseng found up to 10% difference for barley, 20–40% for millet and sorghum and 19% for yams.

The aggregate data for the developing countries can be greatly influenced by what happens in China or India. For example, the rapid rise in food production by developing countries evident in Figure 2.4 is less impressive when the data for China are excluded (Paulino, 1986). Indeed, aggregation of

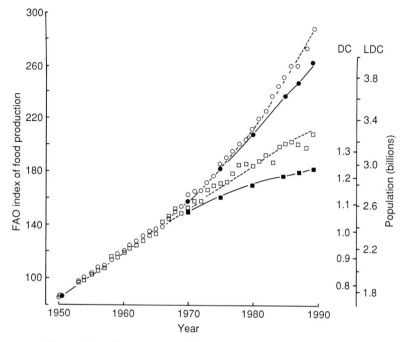

Figure 2.4. Changes in population (solid symbols) and total food production (FAO index, 1952–6 = 100, open symbols) of the developed (squares) and developing (circles) countries since 1950 (data from FAO Production Yearbooks).

data is a problem at all levels, often concealing both regression and progress, but here we can focus only on the higher levels of aggregation, on global figures and on some of the differences between the developed and the developing countries.

The importance of crops to world food supplies

There is no wholly satisfactory way of comparing the global importance of the various crops and animals as sources of food. Minor crops on the world scene may be of major significance in some countries, as tef is in Ethiopia. Even within countries there may be considerable variation from one region to another. Although rice is the staple food for India as a whole, dietary surveys have shown that to be so for only half of the population: millets and sorghum are the staples for one quarter of the people and wheat for almost as many (FAO, 1987). Most diets include many crops and animal products, although their proportions differ greatly both within and between countries.

Our purpose here is to assess the continuing significance of the various crops in providing the world's food, first by comparing their production and then by their importance in the FAO's food surveys. In Table 2.1 the main crops and animal sources are compared in terms of their production of edible dry matter and protein, while recognizing that any estimate of the percentage of dry weight and protein in the global harvests is bound to be uncertain.

In terms of edible dry matter production, over two thirds continues to be provided by the cereals, the three major staples (wheat, rice and maize) accounting for 54% of the total. Roots and starchy foods (including bananas) contribute about 8% of the edible dry matter, rather less than the legumes and oilseeds (about 10%) but rather more than sugar (5%). The estimates for fruits and vegetables are among the most uncertain, at all stages of their derivation, but altogether they probably account for only about 4%. At the global level, animal products (including fish) contribute only 6.5% of edible dry matter and, as we shall see, a large fraction of that is based on feed crops. Thus, crops still provide the great bulk of the world's food supply. Fish, on which Calder (1967) pinned his hopes, still provide less than 1% of edible dry weight (and only 4% of protein).

The relative importance of the main crops has already been indicated in Figure 1.3, which illustrates the dominance of the three main cereals in terms of production. Since much of the maize, barley and sorghum production is currently used for feed, the overwhelming significance of wheat and rice as staple foods, followed by maize, sugar cane, potato and cassava, is clear. The cereals are also the major suppliers of protein, providing about 54% of the total, followed by the legumes and oilseeds (21%), animal products (18%) and fruits and vegetables (4%).

This overall picture is much the same as that in earlier versions of the table based on 1968 and 1979 data (Evans, 1975a; Hanson *et al.*, 1982). Between 1976 and 1986, production has increased most rapidly for vegetable oils (4.3% per year), eggs, meat, and cereals (2.4%), and most slowly for the root crops (0.3%) (FAO, 1986).

Use for animal feed

The rapid growth in animal products is largely based on an escalating use of crop products for animal feed (Norse, 1976; FAO, 1983; Sarma, 1986). At the height of the enclosure movement in England, the dispossessed commoners used the slogan 'Sheep eat men'. Today, cattle, pigs, poultry and even fish eat a growing proportion of the staple crops to provide a small proportion of highly cherished foods. In 1968, 350 million tonnes of grain were used for feed, about a quarter of the total production but by 1980 twice that amount of cereal grain, 44% of the total production, was used for

Table 2.1. *World production of edible dry matter and protein, 1986*

Commodity	Production (Mt)	Edible DM (Mt)	Protein (Mt)
Cereals			
Wheat	536	472	71
Rice	476	414	41
Maize	481	423	51
Barley	180	159	18
Sorghum	71	64	7
Oats	48	43	7
Rye	32	28	4
Millets	31	27	3
Subtotal	1867	1643	204
Legumes and oilseeds			
Soybean	96	85	38
Groundnut	22	14	7
Peas	14	13	3
Beans	15	13	4
Other pulses	26	23	7
Other oilseeds	96	85	19
Subtotal	268	234	78
Roots and starchy foods			
Potato	309	68	6
Sweet potato	110	33	2
Cassava	137	51	1
Banana and plantain	69	23	1
Subtotal	661	184	11
Sugar crops			
Cane	930	86	
Beet	287	29	
Subtotal	—	115	0
Vegetables	414	54	15
Fruit	257	31	2
Animal products			
Milk and cheese	490	70	20
Meat	155	59	28
Eggs	31	8	4
Fish	86	22	15
Subtotal	762	159	67
Total	—	2420	377

Source: Based on data from FAO *Production Yearbook* 1987 and FAO *State of Food and Agriculture*, 1964 and 1986 editions. (Mt = million tonnes)

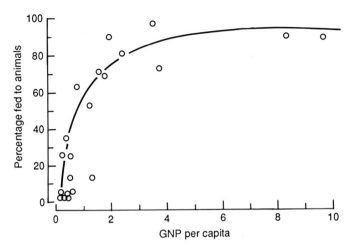

Figure 2.5. Percentage of maize used for animal feed in countries with different levels of *per capita* income (redrawn from CIMMYT, 1981). GNP is expressed as thousands of US dollars *per capita* in 1978.

feed. That proportion is bound to rise even higher in the future, given that the use of cereal grain for feed in developing countries is increasing by 4.6% per annum (Sarma, 1986).

The greater use of cereals for feed rather than food is reflected in Figure 2.1 by the faster growth in the production of cereals than in that of the FAO index for food production in recent years. In turn, this reflects the inability of forage production to keep pace with the demand for animal foods, with the result that it must be supplemented by crops to a greater extent.

For the world as a whole, forages still provide about 75% of the energy in livestock diets, compared with 17% from grains and 8% from other crop sources (Sarma, 1986). But for developed countries, FAO (1983) estimates that 40–45% of the energy consumed by livestock now comes from grains, compared with only 15% in developing countries. As gross national product (GNP) per person rises, so does the proportion of cereal grain used for feed, reaching over 90% for maize in the USA (Figure 2.5). For cereals as a whole, the relationship with GNP is similar but not so close, with the highest proportion for feed being 80% (FAO, 1979). For the developed countries as a whole, about 60% of total cereal production is now being used for feed compared with only 16% in the developing countries, but this latter proportion is rising rapidly (FAO, 1983).

The proportion used for feed differs greatly from crop to crop. For the cereals as a whole it is about half of the world production compared with over two thirds for the oilseeds and pulses and 70% for the coarse grains (Sarma,

1986). Among these latter the proportion in 1980 was 87% for barley, 68% for maize and 50% for sorghum and millet. For wheat the proportion has grown from 10% in 1962–4 to about 22% in 1986, according to FAO figures. However, Sarma (1986) estimates that when the by-products of wheat processing, such as bran, are also included the proportion of world wheat production used for feed in 1986 was 32%. The use of rice for feed is mainly as by-products of processing, and constitutes only about 8% of production. Of the root crops, about 12% of cassava is used for feed, while the proportion of potatoes so used is now quite small compared with that before World War II (Horton, 1987).

The significance of local food production

Until the late 1930s, the developing countries of Latin America, Asia and Africa were, along with North America and Australia, exporting significant amounts of grain to the industrializing countries of Europe. Since then the tide of grain has turned, and the concern now is whether the developed countries can meet the rapidly growing import requirements of the developing countries, quite apart from the question of whether the latter can afford to buy them (Paulino, 1986). Moreover, the grain surpluses currently generated by the price-support policies of the European Community and the USA could quickly be reduced if those policies were changed.

Even with the present situation, however, only a very small proportion of the world production of many crops enters international trade. For rice it is about 3%, for potatoes only 1.5%, and indeed for most crops consumption is largely dependent on local production, at least within the country and often within much smaller regions. With the feed grains, a higher proportion is traded internationally (around 15%), and the proportions are still higher for wheat (around 20%) and sugar. For these last two commodities, consumption may be less dependent on local production, but by and large food supplies reflect relatively local conditions in the developing countries. As Figure 1.2 makes clear, a sufficiency of food at the global level is no guarantee of such at the national level. Nor is national or even local sufficiency a guarantee of local availability when purchasing power is inadequate. Sen (1981) has shown that during the Irish potato famine of the 1840s, the Bengal famine of 1943 and the Ethiopian famine of 1973 there was significant export of food *out* of the famine areas. As he puts it: 'starvation is the characteristic of some people not having enough food to eat. It is not the characteristic of there being not enough food to eat.'

Although imports of cereal grains, especially wheat, by the developing countries have increased greatly since the 1960s, and are expected to reach 175 million tonnes by the year 2000 (FAO, 1981), enhanced local production and purchasing power would be the best guarantees of adequate food supplies.

The global diet

One of the first tasks undertaken by the FAO was the preparation of a world food survey, in 1946. Subsequent surveys were made in 1952, 1963, 1977 and 1987, and data from the most recent one are presented in Table 2.2. For the world as a whole, 83.7% of digestible energy and 65% of dietary protein came *directly* from crops, and much of the remainder came *indirectly* through crop-fed animals in the developed countries. The cereals directly contributed just over half of global dietary energy, a proportion that has remained fairly constant since 1962–3. By contrast, the proportion contributed by pulses and nuts has fallen progressively, as has that by roots and tubers, whereas those from sugar and from meat have risen. Fish still provide less than 1%. The composition of the world's diet revealed by these surveys agrees reasonably well with the estimate of edible dry matter in Table 2.1 after allowance is made for the proportions of the various crops used for feed, seed and industrial purposes.

The global figures conceal large differences between regions. For example, as direct sources of digestible energy the cereals contribute 67% in the Far East, 47% in Africa, 40% in Latin America and only 26% in the developed countries, whereas the respective percentages for animal products are 6, 7, 17 and 32%. A comparison of the diets in developed with those in developing countries, in conjunction with a consideration of the trends for the latter since 1961–3, should give some indication of likely future directions.

The much greater dependence in developing countries on the direct use of crops for food is apparent in Table 2.2, 91.4% of their dietary energy still coming directly from crop plants, and almost 60% from cereals, a proportion that has declined only slightly since the 1961–3 survey. The fall in the contribution by pulses and nuts, from 8.1% in 1961–3 to 5.6% in 1979–81 for developing countries as a whole, is apparent in all regions. In most it has been counterbalanced by a rise in the contributions of sugar and vegetable oils, and by a smaller rise in that of animal products. Comparison with the diets of developed countries suggests that the proportions of dietary energy from sugar, vegetable oils and animal products, especially meat and milk, are likely to rise further in the developing countries, and perhaps those from alcoholic beverages also, whereas those from pulses and root crops are likely to decrease still further. Although the cereals will eventually decline as a *direct* source of dietary energy and protein, their *indirect* contribution may increase substantially because of their increasing use for feed, like that of the oilseeds.

Table 2.2. Sources of digestible energy and protein in the world's diet, 1979–81, as percentages of total diet and in those of developed and developing market economies

Source	Digestible energy (%)			Protein		
	world	developed	developing	world	developed	developing
Plant products						
Cereals	50.2	26.4	57.7			
Pulses and nuts	3.9	2.4	5.6			
Roots and tubers	6.9	3.7	6.2			
Sugar	9.1	13.0	9.7			
Vegetable oils and fats	6.6	11.3	6.3			
Vegetables and fruit	4.0	4.8	4.4			
Stimulants and alcohol	3.0	6.7	1.5			
Subtotal	83.7	68.3	91.4	65	44	79
Animal products						
Meat	7.6	15.3	3.1			
Milk and cheese	4.4	8.5	3.1			
Animal oils and fats	2.5	4.7	1.3			
Eggs	0.9	1.6	0.4			
Fish	0.9	1.6	0.7			
Subtotal	16.3	31.7	8.6	35	56	21
Total (kcals person^{-1} d^{-1})	2630	3390	2350			
Total (MJ person^{-1} d^{-1})	11.00	14.18	9.83			
Total protein (g person^{-1} d^{-1})				68	99	57

Source: Data from the Fifth World Food Survey (FAO, 1987). (Apparent discrepancies between world and component figures are due to the omission here of data for the centrally planned economies.)

Changes in *per capita* production of the various crops in the developed and developing worlds between 1969–71 and 1986 provide a further indication of trends. In the developed countries the fastest growth over the period has been in the vegetable oil crops, such as rapeseed and soybeans, followed by maize, wheat and barley. *Per capita* production of rice and sugar has remained fairly stable, whereas that of the root crops, i.e. potatoes, has declined. In the developing countries, as in the developed, the oilseeds, wheat and maize have increased most rapidly, rice, sugar and cassava have changed little on a *per capita* basis, while sorghum, millet, sweet potatoes, groundnuts, bananas and especially the pulse crops have declined.

The incidence of under-nutrition

Before examining the estimates of how many people in the world are under-nourished, and whether the number or proportion of them is diminishing, we should first consider to what extent under-nourishment is due to lack of protein or digestible energy or purchasing power.

In 1936 a League of Nations technical commission advocated a daily intake of 1 g protein per kilogram of adult body weight, and this was subsequently adopted as a requirement by FAO. It was lowered somewhat by an FAO committee in 1957 but then raised again by an FAO/WHO joint group in 1965. Comparisons were then made between protein production and these requirements on a world scale, and the conclusion announced in 1967 that more than one third of the global population was not getting enough dietary protein. The 'world protein gap' became the favorite catch-cry of Malthusians for highlighting not only the inadequacies of the world's food supply but also one of the supposed dangers of the 'Green Revolution'. The higher-yielding cereal varieties not only had a lower protein content themselves, supposedly, but were also displacing higher protein pulse crops. Legume crops and fish production received greater emphasis.

However, the protein gap, if not the denigration of cereals, virtually disappeared when the next joint FAO/WHO expert committee on energy and protein requirements issued its report (1973). Energy needs remained much the same as before but protein requirements were reduced by about 40%. Instead of a deficit of protein there now seemed to be a surplus for most groups, even in Africa, leading McLaren (1974) to refer to 'the great protein fiasco'. The important change, in the context of this book, was that cereals and most other crops were once more regarded as providing a safe level of protein intake – i.e. with more than 4.8% of the total energy intake coming from protein (Waterlow & Payne, 1975) – provided the digestible energy supply was sufficient. In some cases this might not be so for young children, pregnant or lactating mothers or for those who obtain most of their food energy from sago, cassava or plantains (Payne, 1978; FAO/WHO/UNU, 1985). More protein may be wanted, but in most cases it is not needed, and the world's greatest dietary inadequacy is not a protein problem but a calorie and poverty problem.

However, the dietary energy standards also have their uncertainties. Between 1955 and 1971 the FAO/WHO (1973) requirement for the average moderately active reference man (65 kg) at 10 °C was reduced from 3200 to 3000 kcals (from 13.4 to 12.6 MJ) per day. That requirement has subsequently been reduced by almost 20% in estimates by the U.S. Department of Health, Education and Welfare, and even these standards are now considered too high, as are the estimates of additional requirements for pregnancy (Lipton, 1983). Turning to the developing countries, there are complex questions of what further reductions in energy requirements are associated with warmer temperatures, different body fat requirements, lower body weights, adaptation to different diets, and for individual variation in intake. These are reviewed by Lipton (1983), who suggests that those getting less than 80% of the 1973 FAO/WHO energy requirements are the 'ultra-poor', i.e. 'at income-induced nutritional risk to health or performance', and usually spending about 80% of the total household income on food. In some circumstances, however, an even lower intake of energy may suffice, as shown by Miller *et al.* (1976) for the Ethiopian highlands, where consistent intakes of 70% or less of the 1973 FAO/WHO requirements were not associated with decline of body weight or loss of fitness. Clearly, there remain considerable uncertainties in estimations of the number of undernourished people in the world.

Lipton's (1983) estimate implies that the ranks of those at nutritional risk are fewer than is generally believed, and he suggests that poverty programs should focus on them. In its most recent World Food Survey, FAO (1987) presents two estimates of the incidence of under-nutrition, based on the numbers of those whose food supply is only 20% or 40% more than the average basal metabolic rate. For the 1978–81 data these estimates are 335 and 494 million people respectively, 15 and 23% of the population of the developing countries. The comparable proportions for the 1969–71 survey were 19 and 28%, so it is possible that the proportion of under-nourished people in the developing world is at last beginning to fall. The proportion used to be highest in the Far East and lowest in Latin America, but it is now highest in Africa and lowest in the Near East. However, the extent of under-nutrition is still unacceptably high in the poorest countries of the world and, as Lipton (1988) puts it: 'handling this problem efficiently, in a time of financially induced "public squalor" alongside fashionable state minimalism, will strain to the utmost the intellectual resources of such nutritionists, political scientists, and public-finance economists as are concerned with the sustainable food security of the extremely poor'.

The components of greater crop production

The four main components of increased crop production are:

(1) increase in area of land under cultivation;
(2) increase in yield per hectare per crop;
(3) increase in the number of crops per hectare per year;
(4) displacement of lower yielding crops by higher yielding ones.

The reduction of losses from pests and diseases before harvest is encompassed under (2) above, but net crop production could also be raised by reducing post-harvest losses. Pimentel (1976) estimates these losses to be about 20% on a world scale, ranging from 9% in the USA to 40–50% in some developing countries. However, Greeley (1982*a*,*b*) argues that the post-harvest losses of cereals under traditional systems in developing countries have been exaggerated, creating what he refers to as 'the myth of the soft third option' for solving their food supply problems. His surveys showed that losses of paddy rice were in all cases less than 6% in Indian and less than 9% in Bangladesh villages, and he tabulates similar results in other tropical environments. If such low losses are indeed characteristic of traditional systems, the scope for improvement is not great, nor is there justification for removing village grain for storage in large urban silos.

Besides the four sources of greater food production listed above, two others are considered by Norse (1976). One is the replacement of cash crops by high-yielding food crops. However, cash crops currently occupy only about 10% of arable land, they are important in generating income, and the perennial ones – which occupy about 7% of the arable area – provide protection from erosion in high rainfall regions. Indeed, their relative importance may well increase in the future and it is unlikely that they will be displaced by food crops to any great extent. The other possibility Norse raises is that intensive livestock production systems could revert to extensive ones, freeing the land now used to produce feed grains for direct food production. This also seems unlikely unless the productivity of grasslands and forages can be greatly increased. In any case, although food supplies might increase, crop production would not; indeed, it might even decline if feed crops such as maize and sorghum were replaced by lower-yielding crops.

Norse (1976) combined increases in arable area with increases in crop frequency because both effectively raise the area under crops. They should be separated, however, because their dependence on agronomic inputs and plant breeding and the scope for them in various environments are quite different. For example, the extension of arable land need not depend on the breeding of shorter duration varieties and more intensive use of inputs, whereas an increase in crop frequency usually does so. Likewise, the latter is most prominent in the irrigated tropics and there is less scope for it in many temperate cropping systems where the dominant component of increased production over the past 25 years has been greater yield per crop.

Despite the different mix of components, the rise in food production in the developing countries was comparable with that in the developed ones until the late 1960s (Figure 2.4) since when they have diverged. In the developing countries total food production has increased geometrically, at a faster rate than the population, whereas in the developed countries it continues to increase arithmetically as the rate of population growth declines.

Increase in the area of cultivated land

Malthus at least foresaw the possibility of the area of cultivated land in the world continuing to increase, whereas his friend David Ricardo viewed it as virtually fixed. Ricardian prophets of doom might therefore view the slow extension in the area of arable land since 1970 as support for their concern. Instead, it probably reflects the recent and striking land-saving advances in crop yields. In 1950 there was 0.51 ha of arable land per person for the world as a whole, but only half as much by 1989.

Historical changes

The transformation of global land use over the past millennium is illustrated in Figure 2.6. The rise in the area under crops over the past century has been striking. At the global level, 1.4 billion hectares, 11% of the land not permanently covered by ice, is under crops, 24% is under grassland, and 31% is under forest, but these proportions vary greatly between regions. Almost 30% of Europe is cultivated, and more than 15% of Asia, whereas in South America and Africa the proportion is only 5–6%. In tropical Asia, more than 28% of the total land area is cultivated, this proportion rising to 55% in India and 69% in Bangladesh. Crops and grasslands increasingly replace forests and woodland, as they have done throughout history. An impression of the early spread of agriculture throughout the world can be gained from a series of maps presented by Simmons (1988).

As Boserup (1965) points out, the extension of cultivation has also been accompanied from early times by an intensification of cropping and a rise in carrying capacity. In the transition from hunting and gathering through forest fallows of 20–25 years, bush fallows of 6–10 years, pasture leys and short fallows to annual cropping and multiple cropping, the population supported per hectare has increased by three or four orders of magnitude, from 0.0001–0.01 ha^{-1} for hunter-gatherers and a maximum of 0.5 ha^{-1} for forest fallows

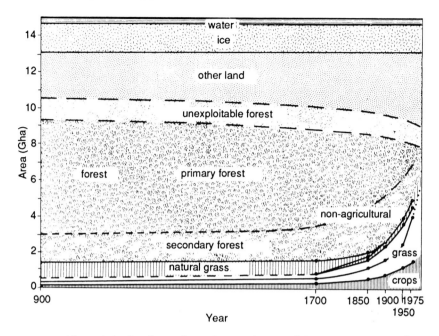

Figure 2.6. Land transformations in the period AD 900–1977 (Buringh & Dudal, 1987).

to 19.6 ha⁻¹ for cultivated land in Egypt today and about 11 ha⁻¹ in Bangladesh, China and Indonesia.

Shifting cultivation

In temperate areas, shifting cultivation may have been quickly replaced by permanent cultivation, as Rowley-Conwy (1981) argues for England, but shifting cultivation is still widely practised in Asia, Africa and Latin America. Spencer (1966) estimated that about 50 million people live by shifting agriculture in south-east Asia, cultivating about 20 million hectares each year. Although it is often viewed negatively, Spencer, Conklin (1957) and Geertz (1963), among others, regard shifting cultivation as an ecologically congruous form of agriculture. They argue that it is adaptive in the great diversity of crops grown on each swidden, in the closed ground cover it provides and in the high proportion of nutrients in the vegetation rather than the soil. But the equilibrium is delicate, and the bush fallow system of SE Asia breaks down if the population density grows to more than about 0.25–0.5 ha⁻¹ (Freeman, 1955; Conklin, 1957), with consequent shortening of the fallow period. It was the similar failure of bush fallow systems in Africa as population density rose that led Boserup (1965) to view the increase in population as the driving force towards more intensive forms of agriculture. There too, shifting cultivation has been shown to be a sustainable form of agriculture provided the duration of the bush fallow does not become too short (Nye & Greenland, 1960). Likewise, in the humid tropics of America, Sanchez & Benites (1987) suggest that the period of bush fallow should not fall below 12 years, whereas population pressure has now reduced it to an average of 4 years in some areas. The extent of shifting cultivation in South America is now recognized as a significant element in the net carbon release from tropical forests (Detwiler & Hall, 1988). Not only should the duration of bush fallow be longer than a certain period, but the duration of the crop phase must not be too long or the regrowth of trees becomes too slow (Trenbath, 1985).

In such circumstances, the progressive clearing of forests for more permanent cropping is inevitable, and with modern large scale equipment it has become rapid and relatively easy, perhaps too easy. The greater difficulty now is to develop sustainable forms of agriculture for the cleared areas (Sanchez, 1983), as well as the infrastructure to make their further intensification possible. Without such intensification, subsistence agriculture, whether shifting or permanent, inevitably results in the extension of cultivation to less and less appropriate land because of its linkage with higher rates of population growth.

Recent gains and losses

The early history of agricultural land transformation has been reviewed by Slicher van Bath (1963) and Grigg (1982). Half of the world's 1.4 billion hectares of arable land has been brought into cultivation during the last century. Richards *et al.* (cited by Buringh & Dudal, 1987) estimate that the cultivated area expanded by 0.43 billion hectares from 1860 to 1920, and by a further 0.42 billion between 1920 and 1978. In the first period most of the expansion was in the temperate regions of North America and the former USSR, whereas it was mostly in the more tropical regions of Africa, Asia and South America in the second. Expansion has slowed considerably since then, however, the area under crops rising by only 3.5% between 1970 and 1989.

Urban & Vollrath (1984) forecast that the increase in the world's arable area, about 1% per year in the 1950s and 0.3% per year in the 1970s, will continue falling to reach 0.15% per year by the end of the century. Expansion of the area results from the clearing of forests (as in the Amazon basin), from new irrigation projects in areas too arid for dryland cropping, from the drainage of wetlands (which rivals the extent of new irrigation projects in the

USSR), from mechanization (which in Britain freed about one third of the arable land which was previously required to keep the horses) and from the reclamation of saline, alkaline, toxic or eroded soils.

Offsetting these gains in arable land there are various losses, for village expansion, urban sprawl, industrial growth, roads and airports, as well as from erosion, waterlogging and salinization associated with agriculture or deforestation (Pierce, 1990). The losses for non-agricultural purposes are particularly serious not only because of their scale – about 0.5% per year in Europe and 1% per year in Egypt – but also because over 60% of them involve the most accessible and productive land. Buringh & Dudal (1987) estimate that between 1975 and the year 2000, 14% of potentially arable land will be lost in this way, including 25% of the most highly productive land. For the world as a whole, about 0.1 ha per person was required for non-agricultural uses in 1975 compared with 0.28 ha for cropping. If all the non-agricultural requirements of further population growth are met from arable land, each additional billion people would alienate about 7% of the present arable area from agricultural use.

As well as such direct losses, there are also those due to the deposition of acids and other toxic wastes in areas adjacent to cities. Then there are the losses associated with poor agricultural management, due to erosion, salinization, alkalization and desertification. Higgins *et al.* (1982) estimate that if erosion went unchecked, the potential area of rainfed crop land would be reduced by 18% and production by 29% by the year 2000, most seriously in Africa and Latin America. With good management, however, these losses could be reduced to acceptable proportions. Other estimates suggest that between 1975 and the year 2000, 50 million hectares of agricultural land could be toxified and another 50 million could become desert (Buringh & Dudal, 1987).

Potentially arable land

The President's Science Advisory Committee (1967) estimated that the potentially cultivable area of the world was 3.19 billion hectares, more than double the arable area at that time. This estimate was confirmed by Kellogg & Orvedahl (1969), raised almost 60% by Kovda (1974) and reduced to 60% by Mesarovic & Pestel (1974). Two subsequent studies based on the new soil map of the world have yielded estimates of 3.42 billion hectares, 25% of the land area of the world (Buringh *et al.*, 1975) and 3.03 billion hectares (Buringh & Dudal, 1987). Of this latter estimate, 71% is located in developing countries and 29% in the developed ones, compared with 54 and 46% respectively of the presently cultivated area. Thus, most (87%) of the potential for expansion is in the developing countries. In South America the

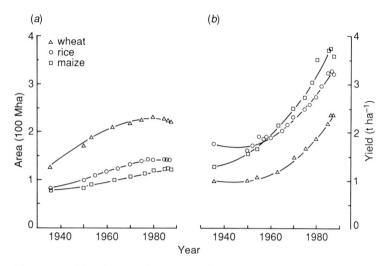

Figure 2.7.   Trends in total area (*a*) and average world yield (*b*) of the three major cereals (data from FAO Production Yearbooks, extended from Evans, 1980*a*).

arable area could increase by 560%, and in Africa by 370%, whereas in Asia it can increase by less than 10%, even on this fairly optimistic assessment. Almost 60% of the world's population lives in Asia, the region with the least scope for increase in arable area, so further increases in food production will have to come from greater yields per crop or per day.

Greater yields per crop

Environmental effects on yield are examined in Chapter 4, the physiological basis of yield in Chapter 5, the way yield has changed in various crops and countries in Chapter 6, and its dependence on inputs in Chapter 7. Here we focus on the relation between yield and the other components of increased production.

Increase in yield per crop has been going on for a long time, as illustrated in Figure 1.1, but the rapid rises are a relatively recent phenomenon, beginning in the nineteenth century for rice in Japan and wheat in several European countries, and since the 1960s for the staple crops of many developing countries (see Figure 1.8). For the world as a whole, the rise in yield has taken over from extension of the cultivated area as the major source of greater food production, as illustrated for the three major staple crops in Figure 2.7.

At the national level both the area sown and the yield of crops are sensitive to the economic as well as the physical climate. Consequently, both may

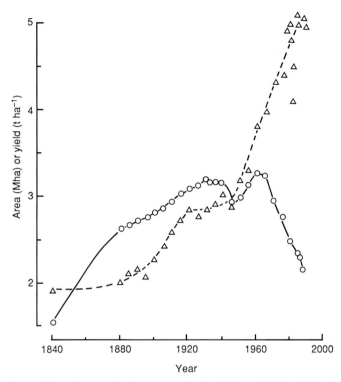

Figure 2.8. Changes in the area under rice (○) and in the yield of brown rice (△) in Japan (extended from Evans, 1978*c*).

fluctuate substantially from year to year, may follow quite irregular courses, and do not display the smooth reciprocal relationship evident at the global level of aggregation. Rice in Japan is illustrative (Figure 2.8). Before the Meiji Restoration in 1868, nearly all the increase in Japanese rice production came from extending the area under cultivation. Over the next 40 years yields rose by 40% and area by 17%. Following the Rice Revolt of 1918, both area and yield remained static, which Ohkawa & Rosovsky (1960) attribute to increasing reliance on the import of cheaper rice from Korea and Taiwan. Presumably this policy led to the lack of new technology in this period, to which Hayami & Ruttan (1971) attribute the stasis in yield. Following World War II both yield and area rose until 1960. Since then the area of rice has fallen as yields continued to rise until 1978, after which yields stabilized while area continued to fall. Clearly, the relation between crop area and yield at a national level can be quite complex.

Despite such variations at the national level, however, the increase in food production at the global level is relatively smooth, as illustrated in Figures 2.1 and 2.4, much of the variation originating in the developed countries. Variability in yield is considered later, but the coefficients of variation for world production of wheat, rice and maize have fallen from 6–8% in the 1950s to 3–4% in recent years, indicating that our greater reliance on yield enhancement has not increased the instability of food supplies at the global level.

Greater frequency of cropping

The frequency of cropping per unit land area has probably increased progressively throughout history, as Boserup (1965) argues. With shifting forest fallows, crop growth in one place for one or two years would often be followed by 20–25 years for forest regeneration. With bush fallows, a regeneration period of 6–10 years is commonly required. With grazed pasture leys for the regeneration of soil fertility, as in the Norfolk four course rotation, cropping frequency could be higher. Then, as fertilizers became available for the maintenance of fertility, annual cropping without yield decline became feasible. In areas with temperatures favourable for crop growth throughout the year, access to irrigation in dry seasons has allowed two or three crops to be grown each year.

Double cropping of rice has been known in China since Sung times (twelfth century), and has spread progressively northwards. The extent of the spread for various types of double cropping between 1940 and 1968 is illustrated in Figure 2.9. According to Henle (1974), the China-wide index of crop intensity increased from 1.15 crops ha^{-1} $year^{-1}$ in 1929–31 to 1.31 in 1952 and 1.85 by 1971, when two thirds of the cultivated area brought in more than one harvest and up to a third more than two harvests.

Similar intensification is undoubtedly occurring in many Asian countries. Cropping intensity in the Punjab now approaches 200% and continues to increase by 0.5–1% per year (Byerlee, 1990). Barker et al. (1977b) found double cropping to have contributed 1.69% per year to the overall rate of growth of rice production in the Philippines of 3.43% per year over the 1965–73 period. Increase in yield per crop contributed 2.52% per year, but was offset by a decline of 0.85% per year in the overall area on which rice was grown. Their analysis indicates the considerable contribution to rice production in the tropics that intensified cropping could make. Three rice crops grown at Mindañao within one year yielded 23.6 tonnes of paddy per hectare (De Datta, 1970), while four crops grown at Los Baños yielded 25.7 t ha^{-1} within a period of 335 days, producing 76.6 kg grain ha^{-1} day^{-1} (Yoshida et al., 1972b).

There is scope for further intensification of cropping at low latitudes, not only with successive rice crops in order to maximize yield per hectare per day,

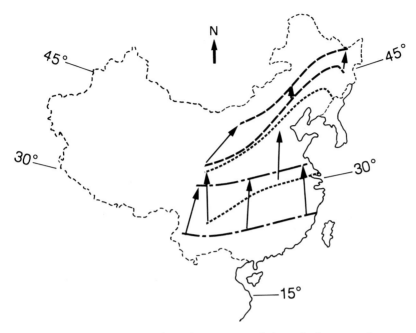

Figure 2.9. The northwards movement of rice and wheat cropping systems in China between 1940 and 1968: the arrows show the extent of the shift to higher latitudes of the northernmost boundary for double rice cropping (— - —), one rice and one other crop (- - -) and spring wheat (– – –). Adapted from Henle (1974).

but also with more diverse cropping systems such as the rice–wheat system, which now occupies 10 million hectares in South Asia. One problem with such systems is how to optimize the times of sowing and harvest for both crops. The yield of wheat following rice, cotton or soybeans is often reduced by late planting in intensive crop systems, which is reflected in yield trends. However, the breeding of shorter-duration varieties with high yield potential, the use of high inputs to accelerate crop growth, and the development of techniques for minimum tillage and fast crop turnaround, should greatly enhance the contribution to food production by multiple cropping in all its forms. In the tropics, where high temperatures tend to cut short the grain growth stage of cereals, thereby limiting yield per crop, the intensification of cropping could contribute substantially to greater food production.

Displacement of lower yielding crops

The stomach contents of the Tollund and other Iron-Age men buried in Danish bogs reveal the great diversity of plant foods in early diets (Helbaek,

1954). *Chenopodium album* was a common food plant in their day but has now become a weed, and many other formerly cherished foods, such as Job's tears in Asia and amaranths in South America, are no longer grown on a significant scale. The displacement of some crops by others has been going on for a long time, but more rapidly since the fifteenth century, when many staple crops were successfully introduced from one region of the world to others. Later, as agricultural mechanization displaced draught animals, major feed crops such as oats and buckwheat were displaced by food crops.

More recently, as the pressure on arable land has increased, lower-yielding crops have been displaced to some extent by higher-yielding ones, with the result that the world is relying to an increasing extent on relatively few staple crops. Before nitrogenous fertilizers were used on their present scale, nitrogen-fixing legume crops were widely grown in crop rotations to restore soil fertility. That role is no longer essential, nor can such crops fix sufficient nitrogen to sustain the high yields now expected of them, let alone provide enough for the following crops. Their average yields are substantially lower than those of the staple cereals and are increasing more slowly, so they tend to be displaced by the cereals. Even in India, where they are highly cherished, the area under chickpeas has fallen by 38% since 1961–5 and their production has been displaced to poorer soils. Narain (1977) has analysed the relative effects of such locational displacement and 'pure yield increase' on the productivity of Indian agriculture, and found about one third of the growth of productivity to be the result of locational shifts. Wheat particularly increased its share of the total area under crops – especially that in the more favourable, irrigated areas – and maize to a lesser extent, whereas that of the pulses and small millets declined. Although Narain emphasized the importance of such displacements prior to the 'Green Revolution', Ranade's (1986) subsequent analysis has highlighted their continuing significance.

The displacement of legumes by cereals has been said to be one of the undesirable consequences of the introduction of higher-yielding cereals into tropical agriculture, partly because of the reduced diversity of cropping and greater dependence on fertilizers, and partly from the concern of the Protein Advisory Group that it would widen 'the protein gap' still further. We have seen that this latter concern was not well founded because in most cases the primary dietary limitation was of calories rather than protein. Indeed, Ryan & Asokan (1977) subsequently showed that the small reduction in lysine intake associated with the shift to wheat was of little, if any, nutritional consequence compared with the substantial increases in the production of energy and protein resulting from the spread of high-yielding wheat varieties in India. The real price of protein in India fell by 34% between 1965 and 1977 in conjunction with the rise in wheat production over that interval. Between 1951 and 1986 the *per capita* availability of wheat in India increased from 66 to 147 g day^{-1}, and that of rice from 159 to 219, whereas that of the pulses fell from 61 to 41, and that of the other cereals from 110 to 72 g day^{-1}, indicating that the nutritional gains from greater wheat and rice production clearly outweighed the losses from reduced pulse and millet production.

Indeed, the displacement of one cereal by another may be just as significant as that of pulses by cereals. In the drier parts of Africa, and also in India as shown by Narain (1977), maize may displace sorghum, which in turn displaces millet. Cassava increasingly displaces yams, and cash crops sometimes displace food crops even in areas where food is in short supply. The more a crop succeeds, the more research attention it is likely to receive and the more its displacement of others is likely to continue, leading to ever greater dependence on fewer crops. On the other hand, as incomes rise so does the premium on dietary diversity and novelty, so that crops currently in decline may well become more prominent again in the future, although I doubt whether our diets will ever return to the diversity of that of the Tollund man in his last hours.

Green Revolution or evolution?

The phrase 'Green Revolution', encapsulating the agricultural changes that began to spread through developing countries in the mid 1960s, was introduced by USAID administrator W.S. Gaud in 1968. As Frankel (1971) put it, the phrase 'has all the qualities of a good slogan. It is catchy; it simplifies a complex reality; and, most important, it carries the conviction that fundamental problems are being tackled'. As such it raised unrealistic expectations and gave the impression of a once-off change rather than a continuing evolution. To judge by its prose, the revolution was as purple as it was green, and it attracted widespread criticism, especially from social scientists (e.g. Frankel, 1971; Poleman & Freebairn, 1973; Griffin, 1975; Hewitt de Alcantara, 1976; Farmer, 1977; Pearse, 1980; Anderson *et al.*, 1982).

In essence, most of the changes were those that had been introduced into European, Japanese and American agriculture many years earlier, but they were applied under tropical conditions and synchronously rather than successively. Hence the term 'revolution', which tended to under-estimate the skills of the traditional farmer (Schultz, 1964).

The rapid spread of the shorter-statured, potentially higher-yielding varieties of wheat and rice in developing countries has been documented by Dalrymple (1986*a*,*b*), and that of improved maize varieties by Timothy *et al.* (1988). As pointed out by Chambers (1984), too much emphasis has been given to the improved varieties *vis-à-vis* the other and equally important components. The new varieties were to some extent a Trojan horse for the more widespread use of irrigation, fertilizers, and other inputs but, as we shall see in Chapters 6 and 7, it is the interactions between all these components that were crucial.

Having criticized the Green Revolution roundly in the early 1970s, not a few social scientists then began asking whether it had really happened at all, given the evidence from here and there of its non-adoption. Since the mid-1960s, however, there has been a distinct rise in the rate of increase of food production by the developing countries as a whole, which has gone on accelerating since then (Figure 2.4). Their *per capita* food production has also risen substantially since 1976 (Figure 2.2). Given the predominance of wheat and rice in the diets of developing countries, it is reasonable to conclude that the Green Revolution has played a major role in these changes and deserves more than a question mark after it (cf. Farmer, 1977). Although damned by many social scientists in developed countries, the farmers of the developing countries have voted for it with their fields.

Like any other major and fairly rapid societal change, however, it has its costs and has generated a variety of problems, some of them generic for the improvement of agriculture anywhere (cf. Wharton, 1969; Byerlee, 1987). On the varietal front, there has been concern at the likely reduction in genetic diversity as individual modern varieties like IR-36 rice are grown over huge areas. Production could become more vulnerable to epidemics of new strains of pests and diseases. There has also been concern that the genetic resources of the staple crops would be eroded, lost, or become unavailable. Hawkes (1985) concludes, however, that the global pool of genetic resources is, in fact, more available, better documented and more widely used than ever before.

Concerns about other inputs include their effects on the environment, whether the developing countries will become too dependent on imported agrichemicals, and whether opportunities for employment for weeding and harvesting will be reduced by herbicides and mechanization. Some such displacement from the rural work force undoubtedly occurs, indeed it made the industrialization of developed countries possible, but the Green Revolution also widened the opportunities for rural employment. For example, Barker & Cordova (1978) show that the introduction of modern rice varieties increased the labour input per hectare although it decreased the labour input per tonne of rice produced.

Related to this issue is that of inequalities of impact of the Green Revolution on various social groups. Quite apart from the advantages of scale, e.g. in the use of ground water for irrigation, large farmers inevitably had better access to information, credit and influence than did small farmers and they profited more from the new agricultural techniques, partly by adopting them earlier. However, adoption by small farmers eventually became widespread (Pinstrup-Andersen & Hazell, 1985). The urban poor gained from the lower prices and greater supplies of food but the rural poor, especially the landless, have sometimes been disadvantaged. However, new agricultural technology should not be expected to stand proxy for social reform, and Lipton (1979) concludes that the technology *per se* was not to blame for the inequalities of impact; it met the criteria he would have specified for a technology to help the rural poor. As Frankel (1971) commented: 'It is precisely the social blindness of modern technology that is encouraging the most disadvantaged sections of the agricultural community'.

One disadvantaged agricultural sector which has not been encouraged by the Green Revolution is that trying to survive in poor and marginal environments. The impact of the new technology has been apparent mainly in the more favourable environments where the reliability of the rainfall or access to irrigation make the use of fertilizers, improved varieties and other inputs worthwhile. Agricultural scientists have been blamed for concentrating on these more favourable areas, and aid donors have called for new varieties to effect a 'brown revolution', failing to recognize that varietal change without greater input use can do little in such environments, even in their own countries.

So widespread in developing countries has been the spread of improved varieties and the more intensive use of fertilizers and other inputs that they have now entered the post-Green Revolution era with its different problems. These may require more emphasis on rural education and extension and an efficient and sustainable management of agricultural resources (Byerlee, 1987).

Conclusions

The major crops of the world – especially the three cereals, wheat, rice and maize – seem likely to remain the staple foods of mankind through direct consumption and indirectly through their use as feed grains. Their global production has more than kept pace with population growth through increase in the area sown (until the late 1960s), higher yields, greater cropping frequency (particularly for rice) and the displacement of other crops. However, rising yields of the staple crops in both developing and developed countries have been the key to the rise in food supplies per person over the past 20 years. They are likely to remain so over the coming decades, hence the preoccupation of this book with the sources of improvement in yield.

CHAPTER 3

The domestication of crop plants

'The difference in value, however
great, which is found among plants
already improved by culture, is less than that which exists between
cultivated plants and others completely wild'

Alphonse de Candolle (1882)

Introduction

The domestication of crop plants could hardly be called a neglected subject in recent years. Its literature is vast and wide-ranging, partly because it requires and excites the attentions of many disciplines: anthropology, archaeology, biochemistry, genetics, geography, linguistics, molecular biology, physiology, sociology and systematics among them. It is eclectic, as de Candolle (1882) insisted it must be, and advances in one discipline may suddenly create opportunities for others. Just as carbon dating and electrophoresis did this earlier, so are DNA sequencing, polymerase chain reaction amplification, ^{13}C discrimination, infra-red spectroscopy and phytolith analysis doing so now, throwing light on questions which were previously thought unanswerable. Problems with the nature of evidence for the origin and dispersal of domesticated plants have been considered by Harlan & de Wet (1973), Callen (1973), Wendorf et al. (1984) and others.

Interest in the evolution of our own species adds significance to the processes of plant domestication. Hear Isaac (1970): 'Domestication ... the invention that made possible the development of pastoral and agricultural economies and therefore made populous and complex human societies viable'. In domesticating plants and animals humankind was itself domesticated. Rindos (1984) refers to the symbiotic and coevolutionary processes between plants and people, but there are also those like Allen (1977) who argue that, although sowing preceded cities, cities preceded agriculture.

Hunter-gatherer societies can be culturally complex. They may deliberately scatter as well as gather seeds, manage their wild food plants with fire and even practise cultivation temporarily during periods of adversity. However, such sporadic husbandry would have few discernible genetic effects on the gathered plants. 'The habit of deliberately growing useful plants', which is how Bronson (1977) defines *cultivation*, was neither a unique nor a revolutionary event. For others the word cultivation implies a rather stronger involvement, such as Helbaek's (1969) 'particular and persistent interest' in a crop. This still falls short of *agriculture*, which involves a major commitment of resources to the establishment of an artificial ecosystem on which people then have to rely for their food supply.

At what point along this route domestication began depends on how it is defined. We come back to the nature of *domestication* later, but at this stage we can use Johannessen's (1982) definition of it as 'the continuing, human-controlled, evolutionary process of the modification of the genotype that has been operative since cultivation began'. Three elements of this definition should be noted. First, domestication refers to genetic changes in plants and not simply to modifying them by growth in different places and under different conditions, on which both Linnaeus and Darwin were sometimes confused. Secondly, these genetic changes are not a single event but a continuing evolutionary process, as emphasized by many biologists concerned that phrases like 'the Neolithic revolution' imply a sudden crucial change rather than the accumulation of many progressive modifications (e.g. Anderson, 1960; Johannessen *et al.*, 1970; Harlan *et al.*, 1976). Thirdly, these modifications began with cultivation. Some may have begun even before that, with gathering, as we consider below, and some changes may have had to occur before cultivation took place, as Ladizinsky (1987) has argued for seed dormancy in the pulses. Domestication involves those modifications which confer *adaptedness* to the conditions of agriculture, as distinct from *adaptation* to new environments, the subject of the next chapter, and the improvement of yield potential, the subject of Chapter 5.

Hunter-gatherers

In the Greek view, the earliest stage of human economy was that of hunter-gathering, the Golden Age. Two thousand years later, hunter-gatherers were generally viewed as ignorant, indolent savages. More recent work by anthropologists on hunter-gathering peoples such as the !Kung Bushmen and Australian aborigines has led to a more appreciative picture of their life style and of the extent of their knowledge of plants, or occasionally to a view of them as professional primitives (Wilmsen, 1989).

Because most of their archaeological remains are 'bones and stones', because most of their cave paintings are of animals rather than plants and because much of the excitement in their lives came from the predominantly male activity of hunting, the relative importance of gathering has often been under-estimated. For those living at the higher latitudes or in Europe during the Ice Ages, hunting was far more important than gathering (Bailey, 1983), but for 28 of the 33 tribes living at lower latitudes in Lee's (1968) survey, gathering provided the greater share of their food.

Their knowledge of plants

Over the silent millennia our ancestors gathered a large number of species and learned a great deal about plant life in the process. Felger (1979) records that of 2500 species of higher plants in the Sonoran Desert, about 18% were utilized for food by the various native peoples, while O'Connell *et al.* (1983) found that the Australian Alyawara recognized about 30% of the 500 species in their territory as useful. Given such comprehensive awareness of plant resources by hunter-gatherers, we should not be surprised by de Candolle's (1882) assessment that 'Men have not discovered and cultivated within the last two thousand years a single species which can rival maize, rice, the sweet potato, the potato, the bread fruit, the date, cereals, millets, sorghums, the banana, soy', nor by Anderson's (1960) comment that all sources of caffeine were known to early man: not one has been added by subsequent botanical surveys.

Gatherers had also – by a process described by Webb (1973) as 'eat, die and learn' – developed procedures for detoxifying and ameliorating many poisonous or unpleasant plant products. Cyanogenetic compounds were removed by soaking, boiling or roasting techniques that were remarkably similar among American Indians, Asians and Australian aborigines. Stands of *Macrozamia* in Australia, for example, produce high yields of large edible fruits which caused Captain Cook's sailors to be ill (both upwards and downwards as Banks put it) and poisoned many Europeans, yet provided the 'savages' of many tribes with safe and transportable food, as Grey (1841) delighted in pointing out.

The gatherers' extensive knowledge of how to prepare safe foods was probably matched by their knowledge of plant life cycles. Flannery (1968) concludes: 'We know of no human group on earth so primitive that they are ignorant of the connection between plants and the seeds from which they grow'. The role of pollination may not have been clearly understood, as Johannessen (1982) found among the Maya, but that was also true in Europe until Camerarius made his observations in 1694.

Aboriginal gatherers in Australia had an accurate knowledge of when various plants would be flowering, fruiting or ready for harvest (Palmer, 1883). Their ability to go walk-about in a hostile environment depended on it. Mitchell (1848) recorded that the aborigines would pull up *Panicum*, *Sorghum* and other wild grasses before the seeds ripened and lay them in piles so that their subsequent gathering of the seed from the ground was easier.

Sometimes they cut the stems with stone knives and beat out the seed, which they winnowed against the wind, obviating any need for resowing (Gregory, 1886). American Indians sometimes replanted wild rice (Bronson, 1977). Harlan (1975) records that 7 of 19 Nevada groups sowed the seeds of wild plants but, as Harris (1977*b*) points out, they failed to domesticate any in the sense of modifying them genetically, possibly because their harvest method of beating the seeds into a basket precluded any selection for a tough rachis.

Many tribes replanted the tops of yams and other root crops into the holes from which they had been dug, a practice reinforced by myth in many tribes. Tindale mentioned a colourful Australian variant of this to Carter (1977):

> 'women dig tubers exhorting the plant to be good, to be generous, to yield a big tuber. Once the tuber is out of the ground, no matter how large the tuber, custom decrees that the woman now complain and berate the plants: 'Oh you worthless plant, you lazy thing, you stingy plant. Go back and do better'. And so saying, she cuts the top off the tuber and puts it back in the dug hole and urinates on it'.

Even closer to cultivation is Chevalier's observation of African hunter-gatherers replanting wild yams surplus to immediate requirements close to their settlements (Coursey, 1976).

Moreover, hunter-gatherers sometimes practised elementary husbandry on plant stands which they intended to harvest. There are many reports of burning to regulate flowering time – as in *Macrozamia* (Harris, 1977*a*) – or to enhance yield, especially of root crops. Grey (1841) wrote that 'the natives must be admitted to bestow a sort of cultivation upon this root (a yam) as they frequently burn the leaves of the plant in the dry seasons, in order to improve it'. O'Connell *et al.* (1983) note an increase in the productivity of wild species of *Solanum* on recently fired sites. They also mention Tindale's observation of occasional irrigation by aboriginal tribes, a practice which has also been recorded for the Paiute Indians of Nevada (Harlan, 1975). Hunter-gatherers are, at best, reluctant weeders (Lathrap, 1968). In some cases, they recognize ownership of wild plants, as Gregory (1886) noted for fruiting *Zamia* plants and Maiden (1889) for *Araucaria* trees.

Why domestication did not always occur

Many hunter-gatherers clearly had the requisite understanding of plant life cycles and their management needed for the practice of agriculture. Plants appropriate for domestication were often available to them and they could be aware of agriculture. North Australian aborigines had, through many centuries, experienced transfers of intellectual and material culture from agricultural peoples without themselves adopting agriculture (Carter, 1977). Why?

The superficial answer is that they had no need to. As the !Kung bushman replied: 'Why should we plant when there are so many mongongo nuts in the world?' (Lee, 1968). However, his answer assumes more ironical overtones in the light of recent evidence that fully developed pastoralism and metallurgy were established in the Kalahari desert from AD 500, with extensive grain agriculture involving sorghum, pearl millet and cowpeas there by AD 800 (Denbow & Wilmsen, 1986). The reliance by the !Kung on gathering the mongongo rather than growing cereals or herding animals is more recent than has been supposed, and is not 'a window on the Pleistocene' (Wilmsen, 1989). Indeed, Lathrap (1968) has suggested that several hunter-gathering peoples are 'failed' agriculturists who have readapted to their earlier life style (cf. Levi-Strauss, 1968; Bellwood, 1985).

One of the much-touted advantages of crop domestication and the agricultural way of life was that it supposedly allowed more leisure for cultivation of the arts and sciences of civilization. Against that backdrop there was considerable surprise when the surveys of hunter-gatherers indicated that, even in the harsh environments in which they mainly live now, only a few hours each day or a few days each week of gathering were needed to secure an adequate and varied food supply. Several early Australian explorers had estimated that 2–4 hours per day sufficed, and a 1948 survey of food gathering by women in West Arnhem Land in North Australia confirmed that they could gather all they needed in a few hours and didn't harvest more than that because they preferred change in their diet and were confident that there was no need to store any surplus against hard times (McCarthy & McArthur, 1960). Sahlins (1968) also found that a few hours each day sufficed to gather food, and similar patterns of gathering were found among the !Kung. As Lee (1968) puts it, they 'eat as much vegetable food as they need, and as much meat as they can'.

Rates of gathering vary greatly, of course, but can be quite rapid. Harlan (1967), without prior practice, was able to harvest more than 2 kg of grain per hour from stands of wild wheat in Turkey. Chevalier (1932) found that Africans with a swinging basket could easily gather 10 kg of millet seed in a morning, while Harlan (1975) found he gathered enough seed of wild maize in one morning to last him eleven days. Root crop gathering also varies greatly in its yield. With productive yam stands, many kilograms can be collected in a few hours (e.g. McCarthy & McArthur, 1960).

Criteria for gathering

Of the various criteria that were important in determining which plants were most frequently gathered for food, and therefore more likely to be domesticated, ease and rate of harvest must surely have been one. Thus, greater size of the harvested organs is likely to have been favoured, as in the case of giant yams; or, if they were small like millet seeds, the ability to harvest them in large numbers from dense stands would have been important. With individually harvested inflorescences, the larger the grains, the more grains per spikelet, or the more compact the inflorescence, the more favoured the species. For these reasons, as Ladizinsky (1975) points out, wheat would have been preferred to barley, with oats a poor third. The speed of gathering reflected these differences, and may have contributed to oats being domesticated long after wheat and barley.

Seasonal distribution and predictability of harvest was another criterion of significance for nomadic gatherers. Transportability of the harvested products must often have been a consideration for gatherers, as would ease of dehusking, whereas suitability for storage would not.

Whatever their criteria for gathering, it is likely that these had no genetic impact on the gathered species, apart from influencing the choice of plants for subsequent domestication. But heavy harvesting of preferred types at particular seasons over long periods of time, especially if reinforced by elementary management or cultivation, could have modified some populations. Ladizinsky (1987) suggests that such 'domestication before cultivation' could have occurred with the wild lentil (*Lens orientalis*). As in many other wild legumes, freshly harvested seeds are highly dormant, a mechanism which helps to avoid the overcrowding of seedlings in nature but which makes the sowing of wild lentils inefficient. This germination behaviour is governed by a single major gene, and the dominant non-dormant form could have appeared in wild populations if heavy gathering of their seed reduced seedling populations and therefore the disadvantages of being non-dormant. In this way gathering could have preadapted the wild lentil, and also wild peas and chickpeas, to domestication. Similarly, by gathering wild grasses at maturity, by which time the more readily shattering forms had already begun to shed their seed on the ground, hunter-gatherers may have reduced the proportion of non-shattering plants in the wild stands while at the same time increasing it among the plants that grew in the environs of their camps.

Domestication

Why?

'Why should we?', as the !Kung bushman said to Lee. Why leave the hunter-gatherer's varied and secure life style? Cohen (1977*b*) notes that the !Kung suffered less from prolonged drought than did their agricultural neighbours, and Woodburn (1968) likewise for the Hadza. So what caused the shift to agriculture? Was it for cultural reasons, or driven by a change in climate, by population pressure or what?

Harlan (1975) gives many examples of the belief in agriculture as a gift of the Gods to mankind, whether in the Mediterranean or China or South America. Viewing agriculture as a divine gift is akin to viewing it as a discovery or sudden invention, giving rise to a revolution. However, Childe (1934) sought the origin of his 'Neolithic revolution' in the climatic changes that occurred at the end of the Pleistocene. He envisaged a progressive desiccation of the rangelands forcing people and herd animals to concentrate and intermingle at oases, where domestication and agriculture were born. Whyte (1977) thought that the combination of aridity and higher temperatures in the Neothermal led to a greater abundance of the annual cereal and legume progenitors, and hence to domestication in the Near East and in China. Palaeoecological studies suggest, however, that the climate in the Near East went from dry to moist rather than the reverse (Wright, 1977). They confirm that the cultural and climatic changes were nearly synchronous, as Childe suggested, so the possibility of environmental determinism remains, and is championed by Wright (1976, 1977) for the Near East.

Was there only one revolution or a series of local evolutions? From what we have already seen about hunter-gatherers we would expect agriculture to have evolved gradually, irregularly, and independently in different environments and forms. Adams (1966) has questioned the applicability of the concept of a revolution even for the Near East, and excavations in Mexico make it clear that the transition from hunting and gathering to agriculture was extremely gradual (e.g. MacNeish, 1958). The evolution of agriculture in Oaxaca has been described by Flannery (1968) as a series of quantitative shifts in the balance between alternative strategies for getting food rather than as a series of discoveries.

There is a great range among hunter-gatherers in the area they need for a viable economy. Lee (1968) quotes a range of 0.8–250 km² per person, but these figures apply to the more adverse environments to which hunter-gatherers are now largely restricted. In the more favourable parts of California only about 0.22 km² per person was required (Harlan, 1975), and in riverine or coastal regions abounding in fish the area may have been far less.

In some situations, increasing population pressure may have led, gradually, to more intensive and regular cultivation as envisaged by Boserup (1965), Cohen (1977a,b), Flannery (1972) and others. Binford (1968) suggests that as populations built up agriculture may have begun along the interface between the sedentary forager-fisherfolk and the nomadic hunter-gatherers. However, it remains an open question whether population pressure was the driving force towards agriculture. Sauer (1952) regards that as unlikely and argues that the founders of agriculture were sedentary folk with skills that predisposed them to agricultural experiments which 'began in wooded lands. Primitive cultivators could readily open spaces for planting by deadening trees; they could not dig in sod or eradicate vigorous stoloniferous grasses'.

When and where? The geography of domestication

If we look further back than the Neolithic for the origins of agriculture, low latitude forest regions seem the most likely cradles of domestication, and of these Sauer (1952) chose south-eastern Asia. Along with this geographic choice, he focused on clonally reproduced plants such as bananas, yams, sago palm and sugar cane. These are all good sources of carbohydrates but low in proteins and oils, which would not have mattered so long as these other dietary components were provided by fishing. These and related crops were probably used also as sources of fibres, dyes, drugs and poisons, i.e. they were multi-purpose plants and food production need not have been the primary reason for bringing them into cultivation.

This is a very different picture of the origins of agriculture from that usually presented of cereal and pulse domestication in the Fertile Crescent, and it is not without its critics (e.g. Mangelsdorf, 1952; Harlan & de Wet, 1973). Not only are the geography, the plants and the environment utterly different, but also the processes involved in domestication and the attitude to agriculture: an evolving supplement in SE Asia, a major commitment in the Fertile Crescent.

The earliest dates known for some of the main domesticated crops in various regions of the world are indicated in Table 3.1.

South-east Asia and Melanesia

If the first steps to agriculture were taken in SE Asia during the Palaeolithic, the chances of finding archaeological remains are extremely slender. Not only are clonally propagated crops in seasonally humid climates less likely to bequeath remains to posterity, but the rise in sea level of about 150 m during the retreat of the glaciers – which reduced the SE Asian land mass by about half (Bellwood, 1985) – is likely to have submerged riverine and coastal sites of agriculture.

One of the oldest cultural contexts for any food plants is in upland areas in the SE Asian region. A widespread culture called Hoabinhian has been carbon dated back to 13400 years before present (BP) in Burma, and to 11200 BP at Ongba Cave and 9200 BP at Spirit Cave in Thailand (Gorman, 1977). It continued until about 4700 BP. The earliest food plant remains found in the region so far are from Spirit Cave and are up to 9200 years old (Gorman,

Table 3.1. *Some of the crops domesticated in the various regions, with early datings and locations.*
(Brackets indicate uncertain datings, question marks uncertain domestication. From sources given in the text.)

	Years bp	Location
SOUTH-EAST ASIA & MELANESIA		
SE Asia		
Piper, Ricinus, Mangifera?	9200	Spirit Cave, Thailand
Oryza (rice)?	5500	Non Nok Tha, Thailand
Coix lachryma-jobi (Job's tears)		
Dioscorea yams		
Melanesia		
(?*Colocasia*, taro)	(9000)	Kuk, New Guinea
Saccharum officinarum (sugar cane)		
Metroxylon sagus (sago)		
Musa (fe'i bananas)		
CHINA		
North		
Setaria italica (foxtail millet)	6700	Pan p'o
Panicum miliaceum (broom corn)	7th millennium	
Cannabis sativa (hemp)	6th millennium	Yang-shao
Oryza sativa (*japonica* rice)	5300	Chien-shen yang
Glycine max (soybean)	3000	
South		
O. sativa	9000	Pengtoushan, Hunan
O. sativa (*indica*)	7000	He-mu-du, Chekiang
O. sativa (*japonica*)	7000	Luo-jia-jiao, Chekiang
SOUTH ASIA: THE INDIAN SUB-CONTINENT		
Indigenous		
O. sativa (*indica*)	8550	Mahagara, U.P.
O. sativa (*indica*)	4300	Chirand, Bihar
Gossypium arboreum (cotton)	7th millennium	Mehrgarh, Pakistan
Triticum sphaerococcum (wheat)	5500	Chirand, Bihar
Vigna radiata (mung bean)	3700	Madhya, Pradesh
From Near East		
Hordeum vulgare (2 & 6 row barley)	8th millennium	Mehrgarh
Triticum aestivum (bread wheat)	8th millennium	Mehrgarh

Phoenix dactylifera (date palm)	8th millennium	
Cicer arietinum (chickpea)	4000	Atranji Khera, U.P.
From SE Asia		
Saccharum arundinaceum (sugar cane)	4300	Harappa
From Africa		
Eleusine coracana (finger millet)	3800	Hallur, Mysore
Sorghum bicolor (sorghum)	(3700)	Ahar, Rajasthan
Pennisetum americanum (pearl millet)	3200	Saurashtra
Sesamum indicum (sesame)	4th millennium	Punjab

NEAR EAST (WEST ASIA)

Hordeum vulgare (barley)	10200	Netiv Hagdud, Israel
Triticum monococcum (einkorn)	9500	Ali Khosh, Iran
T. dicoccum (emmer wheat)	9500	Ali Khosh, Iran
Pisum sativum (pea)	9500	Cayönü, Turkey
Lens esculenta (lentil)	9500	Cayönü, Turkey
Cicer arietinum (chickpea)	9500	Cayönü, Turkey
Vicia ervilea (vetch)	9500	Cayönü, Turkey
Linum usitatissimum (flax)	9500	Cayönü, Turkey
Vicia faba (broad bean)	8500	Yiftah-el, Israel
Triticum aestivum (bread wheat)	7800	Catal Huyuk, Turkey

AFRICA

Dioscorea yams	(10000)	W. Africa
Oryza glaberrima (African rice)		W. Africa
Eleusine coracana		
Sorghum bicolor		
Pennisetum americanum		
Vigna unguiculata (cowpea)	3400	Kintampo, W. Africa
Eragrostis tef (tef)		Ethiopia
Elaeis guineensis (oil palm)		

THE AMERICAS

Central

Cucurbita pepo (squash)	10700	Oaxaca, Mexico
Lagenaria siceraria (bottle gourd)	9000	Tamaulipas, Mexico
Capsicum annuum (chile pepper)	8500	Tehuacan, Mexico
Zea mays (maize)	7700	Tehuacan, Mexico
Amaranthus cruentus (amaranth)	5500	Tehuacan, Mexico
Gossypium hirsutum (upland cotton)	5500	Tehuacan, Mexico

Table 3.1 *cont.*

	Years bp	Location
South		
Phaseolus vulgaris (common bean)	7700	Guitarrero, Peru
P. lunatus (Lima bean)	7700	Guitarrero, Peru
Solanum tuberosum (potato)	7000	Chilca Canyon, Peru
Gossypium barbadense (cotton)	4500	Coastal Peru
Ipomoea batatas (sweet potato)	4500	Ancon-Chillon, Peru
Manihot esculenta (cassava)	4500	Ancon-Chillon, Peru
Arachis hypogaea (peanut)	3800	Casma Valley, Peru
Nicotiana tabacum (tobacco)		
North		
Lagenaria siceraria	7300	Florida
Cucurbita pepo	7000	E. USA
Iva annua (sumpweed)	4000	E. USA
Helianthus annuus (sunflower)	3000	E. USA

1969). Yen (1977) has classified the botanical findings from these excavation sites into three groups. The most confident identifications include *Areca* (the betel palm), *Piper* (the betel nut), *Prunus, Ricinus, Mangifera* and *Canarium*. Less sure are *Cucumis, Lagenaria, Trapa* and what is probably a wild *Oryza* (Chang, 1976) from the Banyan Valley cave where wild rice still grows nearby. The uncertain identifications, which have been widely doubted, include *Pisum, Phaseolus* or *Glycine*, and *Vicia*. These seem improbable and must await further work. The earliest remains were presumably the results of gathering. The range of species is impressive, especially as they probably represent only those most resistant to decay. No root crop remains have been found although several species of *Dioscorea* grow abundantly nearby.

Signs of agriculture in the Hoabhinian, dating back to at least 8000 BP, are seen in the presence of associated artifacts such as ceramics and adzes. The presence of bifacially ground knives like those used for harvesting rice in Java led Gorman (1977) to conclude that by 7000–6000 BP rice was a component of agriculture along the piedmont margins of the plains. However, sickle blades imply only reaping, not necessarily cultivation. Root crops such as *Colocasia* (the taro) and *Dioscorea* yams are prominent in SE Asia, some being adapted to wet and others to dry conditions (Barrau, 1965*a*). Either taro or yams may be dominant, depending on whether the environment is a swamp, riverine plain or swidden clearing; so too for the cereals of the region. Besides rice there was another cereal, Job's tears (*Coix lachryma-jobi*), better adapted to dryland conditions.

Coix is an instructive case because its grains are large, sweet and highly nutritious. In most strains, however, the grains are enclosed by an indurated spathe which makes them difficult to husk, although attractive and widely used as ornaments. Indeed it was as such that the species was first grown in European gardens. The strains domesticated in SE Asia had a soft spathe and were easily husked. Job's tears was an adaptable, widely grown crop, but Koul (1974) suggests that it has not become a staple because of its long growing season, uneven ripening and variable yield.

Others besides the Hoabinhians left agricultural traces in the SE Asian region, if we include Melanesia. 'A claim of some audacity and considerable implications', as Golson (1982) puts it, has been made for the cultivation of wetland crops in the Kuk area of highland New Guinea about 9000 years ago (Golson, 1977, 1985). So far, the evidence is non-botanical. Increased erosion in the area at that time, possibly due to human interference, led to the deposition of a distinctive clay layer. A channel one metre deep was dug across the Kuk basin at about that time, and associated with this are pits and stake holes which could have been linked with cultivation. By 6000 BP the drainage system was much more extensive (over 200 ha) and elaborate, and pollen diagrams suggest that there had been considerable clearing of the primary forest for agriculture. Nor is the Kuk swamp the only area in which such early drainage operations have been found (Golson, 1982).

Apart from their being adapted to wetland culture, there is no indication of what plants might have been grown in the drained swamp. If the presently slender evidence that pigs were already in montane New Guinea 10 000 years ago holds up, there is no reason why the crops, as well as the pigs, may not have come from SE Asia, and the wet cultivation of *Colocasia* taro would be a likely candidate, at least in the later stages of cultivation. Less intensive cultivation of dryland crops nearby is also possible, even likely (Golson, 1982).

However, it is also possible that agriculture may have evolved in New Guinea quite independently of that in SE Asia, as Yen (1982) suggests. Quite an array of plants was probably domesticated there, including staple foods, the sugar cane (*Saccharum officinarum*) (Warner, 1962) and vegetables and fruits adapted to a wide range of conditions, both wet and dry, coastal and montane. The swamp taro (*Cyrtosperma chamissonis*) is likely to be a New Guinea or Melanesian domesticate (Yen, 1982). Although the major *Dioscorea* yams are of south-east Asian origin, several other species were probably domesticated in New Guinea or various Melanesian islands (Burkill, 1960). Of the three species of *Metroxylon* sago palms in New Guinea, two (*M. sagus* and *M. rumphii*) have domesticated forms with spineless petioles and spathes, in contrast to the wild forms. They are propagated clonally, whereas two other species adapted to a much wider range of habitats reproduce sexually. Clonal selection may also have modified the breeding system of the *Australimusa* banana, a New Guinea domesticate (Simmonds, 1962), whereas the *Eumusa* section of the genus originated in SE Asia. The breadfruit (*Artocarpus communis*) may also be a New Guinea domesticate, since it is both a wild and a cultivated plant in traditional systems. Melanesian arboriculture certainly extends back at least 3500 years (Kirch, 1989).

Clearly, the available evidence suggests that Vavilov's south-east Asian centre (cf. Figure 1.5) was a rather diffuse region in which the contributions of Melanesia and the Pacific Islands should no longer be overlooked. Independent domestications of useful local species probably took place throughout the region, in many environments, over a prolonged period. Changes in the *Dioscorea* yams, such as selection for greater size, shallower rooting, reduced spines, less bitterness and lower toxicity parallelled the series of changes in the West African yams (Coursey & Coursey, 1971). Spines have been reduced in taro and sago palm as well as in yams. Clonal selection has led to predominantly asexual reproduction in taro, yams, bananas, sago, sugar cane and even breadfruit (Yen, 1982).

Clonally propagated domesticates are prominent in the SE Asian region. Such crops establish quickly in warm, humid environments, and those with tall, climbing or dense growth habits are well adapted to compete with non-agricultural plants in such areas. They are amenable to intensive cultivation in wet beds or swidden clearings. They are also well suited to quick domestication, although their long term improvement may be more difficult unless they can be induced to reproduce sexually.

Root crop vegeculture could therefore have preceded cereal culture in SE Asia. It is suitable for small scale cultivation, and its generally low protein level need not have been disadvantageous to fisher folk or hunters. But as agriculture moved into cooler or more arid environments, to operations on a larger scale, and away from ample sources of animal protein, a shift to more cereal growing may have been necessary.

The fact that many Pacific Islands have still not moved to cereal growing has been taken as evidence that root crop vegeculture is an earlier stage of agricultural development, ignoring the fact that where cultivable land is limited, fish abundant, and rainfall high, vegeculture may be better adapted. Spencer (1966) has portrayed the eastwards retreat of both taro and yams from their areas of origin in SE Asia over the period from AD 1500 to 1950 (Figure 3.1). This has also been taken as evidence that agriculture based on rice as the staple is replacing an older form of agriculture based on tuber or root crops, fruits and nuts throughout SE Asia, a view held by many botanists

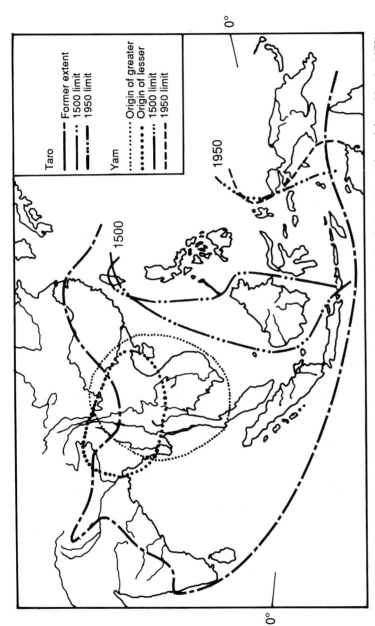

Figure 3.1. The eastwards retreat of taro and yams as staple crops in South East Asia (adapted from Harris, 1972). The former extent of taro as a staple crop is indicated by the outer envelope, and the retreat of its western limit between 1500 and 1950 is shown. The areas of origin of the greater and lesser yams are indicated by the dotted circles, and the western limits of yams as staple crops in 1500 and 1950 are shown, as portrayed by Spencer (1966).

and geographers (e.g. Haudricourt & Hedin, 1943; Burkill, 1951; Sauer, 1952; Barrau, 1965b), but recently challenged by Gorman (1977).

Whether tuberous vegeculture had priority over cereal agriculture remains to be shown. Indeed, Harlan (1975) considers that 'the cleavage between cereal agriculture and vegeculture is exaggerated and may never have been real'. For the SE Asian region as a whole, strong evidence for early domestication, as against food gathering, is still sparse.

China

In China there are many clearly identified crop plant remains, backed up by a rapidly growing number of carbon datings, begun in 1965 but held up for some years by the Cultural Revolution (Ho, 1977). There is also an exemplary written record in books and on oracle bones. Besides the earliest specifically agricultural text, *Chi-min yao-shu* from the sixth century AD, there is *Shih-ching*, '*The Book of Odes*', dating from the eleventh to the sixth century BC.

The five 'cereals' (including soybean) loom large in the *Book of Odes* and altogether it mentions about 150 different plants – compared with 83 in the Bible, 55 mentioned by the ancient Egyptians and 63 by Herodotus – together with accurate observations on their environments and management, and on wild species related to crop plants, such as wild rice. These literary records are very recent in relation to the likely times of domestication – except perhaps for the soybean – and may give biased impressions. Ho (1977) points out, for example, that most of the rice varieties referred to before AD 1000 are *keng*, i.e. *japonica* or *sinica*, whereas radio-carbon dating gives as much or more prominence to *hsien* (i.e. *indica*) types from the earliest stages of domestication.

Li (1970, 1983) has divided China into three vegetation zones each with its distinctive set of domesticates. The North China belt, roughly about latitude 34°N, is separated from the South China belt by the Quin Ling mountains, and this in turn is separated, at about 25°N, by the Nan Ling mountains from the South Asian belt. This latter region belongs with Burma, Thailand and the Indo-Chinese peninsula countries in the SE Asian region already described. The South China belt gave rise to a quite different range of domesticates from those to the north and also, with the possible exception of rice, from those in the SE Asian region. All three zones were early, independent cradles of domestication.

North China conforms to the Vavilovian concept of a well-defined centre of domestication for a great variety of cereal, legume, fibre, fruit and vegetable crops. One striking contrast between North China and South-east Asia is that whereas more humid environments and slash and burn clearings predominated in the latter, agriculture in North China began in semi-arid areas on deep, fertile, friable loess, easily worked by a digging stick, and needing to be fallowed for water rather than for nutrient supply. Domestication in North China therefore followed a different course, one better authenticated by crop remains.

Cereals were prominent from the earliest stages. The 'Lord of Millet' is the legendary founder of the Chou tribe, and three millets provided the staple foods: foxtail millet (*Setaria italica*), and both the glutinous and non-glutinous forms of broomcorn millet (*Panicum miliaceum* vars. *glutinosum* and *effusum*, respectively). Abundant remains of these have been found at Pan p'o and other excavated village sites in the north-west. An Zhimin (1989) states that foxtail millet was domesticated and widely cultivated by 8000 BP at the latest, as early as it was in Europe (Sakamoto, 1987). Carbon-13 analyses of early human skeletons from North China indicate that crops with the C_4 photosynthetic pathway provided most of the food. Domestication of the wild Eurasian grass *S. viridis* into the cultivated *S. italica* probably involved a reduction in grain shedding, increase in grain size, plant height and panicle length, more upright growth habit and longer duration (Chang, 1983; Kawase & Sakamoto, 1987).

Another early domesticate in North China was hemp (*Cannabis sativa*), cloth of which has been identified from its imprint on Neolithic Yang-shao pottery. Many fruits were also domesticated from local wild species, such as the peach, apricot, chestnut, hazelnut, mulberry and persimmon, whose wild progenitor still grows in the area. The common spiny shrub *Ziziphus spinosa* gave rise by selection to the spineless cultivated jujube (*Z. jujuba*), a change reflected visually in the ideograms for the two species (Li, 1983). Seeds of *Brassica*, possibly of the Chinese cabbage, have been found along with the millet at Pan p'o, and many other vegetables are mentioned in the *Book of Odes*.

The soybean (*Glycine max*) appears to have been a relatively late domesticate. It derived from the wild *G. soja*, still widespread on riverine lowlands in North China, and probably emerged as a domesticate during the Chou dynasty (Hymowitz, 1970). The ideogram for it can be traced back to approximately the eleventh century BC (Chang, 1983). Millet and soybean were grown in rotation from about the fourth century BC.

Asian rice (*Oryza sativa*) probably evolved from a wild progenitor – annual according to Chang (1976), perennial according to Oka (1988) – over a broad belt extending from the Ganges plains across Burma, Thailand and Laos to North Vietnam and South China (Figure 3.2). Recent evidence suggests that the area of domestication probably extended to a higher latitude in China than that indicated in Figure 3.2, and that wild rices reached as far north as latitude 38° (Chang, 1983).

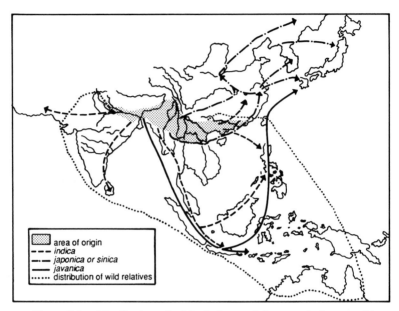

Figure 3.2. Distribution of wild relatives and the spread of geographic races of rice (*Oryza sativa*) in Asia and Oceania (Chang, 1976).

The oldest dated remains of domesticated rice in China have been found in Li's South China belt. Grains, hulls, straw and stems excavated from He-mu-du, Chekiang, have been dated to about 7000 BP. The grains are of the *indica* type and their dimensions are between those of the wild *spontanea* and modern varieties, while those found at Sung-tse, near Shanghai, dated to about 6000 BP, fall wholly within the range of modern varieties (Chang, 1983). The oldest findings of *japonica* rice are from Luo-jia-jiao, near He-mu-du, and they are also about 7000 years old (Chang, 1989). Even older remains of cultivated rice, carbon dated to 9000–7500 BP, have been found at Pengtoushan, a Neolithic mid-Yangtze site in Hunan (Yan, 1991).

The He-mu-du site in Chekiang also gives evidence of the early use of many other plants adapted to aquatic, coastal plain conditions. These are listed by Li (1983) who points out that China emerges as a region of domestication encompassing three centres giving rise, at least 7000 years ago, to quite diverse and different arrays of domesticated plants. For the two lower latitude centres, these included both cereal and tuberous crops, with no evidence of the latter preceding the former. In the course of time, many of the plants domesticated in North China diffused to lower latitudes, while rice slowly diffused to progressively higher ones (Akazawa, 1982; Chang, 1989).

The Indian subcontinent

Encompassing as it does a great range of environments and of early influences from agricultures to the east, west and south, the Indian subcontinent presents a complex picture of early crop domestication. A wide range of crops has been grown in India since early times, but apart from rice and cotton there is little evidence of very early domestication. What may be amongst the oldest dated remains of rice – possibly back to 8550 BP – come from Mahagara, UP (Chang, 1989). Wild rice still grows in various parts of India and is still gathered in areas such as the Jeypore Tract of Orissa, where modern rice growing is also practised (Vishnu-Mittre, 1974; Chang, 1989). This coexistence of gathering and agriculture is also evident in the earliest known stages of agriculture in the Indus valley, at Mehrgarh in the eighth millennium BP (Jarrige & Meadow, 1980).

Even at that early stage, the domesticates included 2-row hulled barley, 6-row barley, einkorn, emmer and bread wheat, presumably by diffusion from the Fertile Crescent of the Near East. Along with them have been found stones of the date palm (*Phoenix dactylifera*) as well as the bones of presumably domesticated cattle. By the seventh millennium BP seeds of cotton, thought to have been cultivated, were also present. The oldest known piece of cotton fabric, from Mohenjo-daro, has been dated to 4700 BP. The evidence for hunting and gathering at Mehrgarh then diminished, and by the sixth millennium BP a greater diversity was evident among the crop plants. *Triticum sphaerococcum* (Indian dwarf wheat) was a significant local domesticate by then (Vishnu-Mittre, 1974; Jarrige & Meadow, 1980).

Later, in the Indus valley, the Harappan culture – dating from about 4300 to 3500 BP – is known to have had extensive exchanges with the Near East. Wheat, barley and peas, and evidence of ploughing have been found at Harappa and Mohenjo-daro, chickpeas, lentils and the grass pea (*Lathyrus sativus*) at other sites. Local domesticates included *Vigna radiata* and *V. mungo*, the mung bean.

Whereas the more arid, northerly regions of India were well suited to many of the crops domesticated in the Near East, the wetter monsoon areas to the east were more suited to crops domesticated in SE Asia, such as sugar cane and bananas, which were already in India by the time of the earliest written records (Hutchinson, 1976).

Domesticates from Africa appear surprisingly early in the archaeological record and seem to have been particularly important to agriculture in the warmer, summer rainfall areas of peninsular India. Grains of finger millet (*Eleusine coracana*) date back to 3800 years BP, sorghum and pearl millet somewhat later, while sesame (*Sesamum indicum*) and *Dolichos lablab* also appear in the fourth millennium BP (Vishnu-Mittre, 1974).

Indian agriculture four thousand years ago already encompassed crop plants from all quarters as well as local domesticates. Impressed by the genetic diversity evident in Indian crop plants, Hutchinson (1976) compared rice, still in genetic contact with its wild and weedy relatives, with wheat and sorghum, separated geographically from theirs, and concluded that there was no evidence that rice was genetically more diverse than the other two crops in India despite its opportunities for continuing introgression.

The Near East

Accounts of plant domestication usually begin with this region, for many reasons. One is that the oldest civilizations we know much about were located there. Also, the region offers a particularly attractive and coherent model of domestication: a clearly defined centre, or at least a crescent; a wealth of archaeological sites with dated botanical remains clearly showing the transitions from wild progenitor to domesticated crop; a suite of cereals, pulses, oilseed, fibre and fruit crops undergoing concurrent domestication; a range of animals – including sheep, goats, dogs, cattle, and pigs – also being domesticated at about the same time and in the same places; and well-documented hunter-gatherer societies before agricultural domestication occurred. No wonder Childe (1934) succumbed to his concept of a Neolithic revolution, from which civilization diffused progressively beyond the Near East. And no wonder the history of this region has so powerfully shaped our views of how domestication and civilization occurred.

Consider just one aspect, the driving force towards domestication. Among the many factors that have been suggested, such as environmental change, population pressure, technological advance, social organization and the preference for a more sedentary life (cf. Redman, 1977), one is the availability of a range of potential domesticates. In this respect, the coincidence between the present distribution of the wild cereal progenitors and that of the early farming villages is striking indeed (Figure 3.3).

With the exception of Catal Hüyük and Hacilar, all the archaeological sites fall within the area where massive stands of wild barley (*Hordeum spontaneum*) occur in fairly primary habitats (Harlan & Zohary, 1966), and those two exceptions are near areas with stands of the wild einkorn wheat, the diploid *Triticum boeoticum* (Zohary *et al.*, 1969; Harlan, 1975). The location and form of the 'Fertile Crescent' suggests that the availability of potential crop plants had a major influence on the development of agriculture in this region at least, but three caveats should be borne in mind.

The first is that the present distribution of the wild progenitors, as described by Harlan & Zohary (1966) and Zohary *et al.* (1969) may not be the same as that 10000 years ago at the dawn of agriculture in the region.

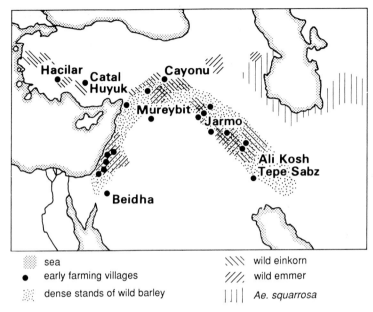

Figure 3.3. The location of early farming village sites in the Near East in relation to the recent distribution of dense stands of wild barley (*Hordeum spontaneum*) and primary habitats of wild einkorn (*Triticum boeoticum*), wild emmer wheat (*T. dicoccoides*) and *Aegilops squarrosa* (adapted from Harlan, 1971; Harlan & Zohary, 1966; Zohary *et al.*, 1969; Zohary & Hopf, 1988).

However, van Zeist & Bottema (1982) have shown by pollen analysis that the vegetation formations in which the wild progenitors now thrive were present in at least the western part of the Fertile Crescent about 12 000 years ago. Also, the fact that wild barley has been found at the lower levels of most of archaeological sites within its present range suggests that its distribution 10 000 years ago was similar to that today. Indeed, since it has also been found at Catal Hüyük, its distribution then may have been somewhat wider than now. No wild barley has been found at Hacilar, but wild einkorn occurred there at about 9000 BP, and also at Tell Mureybit and Abu Hureyra, which are currently just beyond its present range (Harlan, 1975, 1986).

Wild emmer wheat (the tetraploid *T. dicoccoides*) is more demanding in its requirements than barley or einkorn, is not a weedy plant, is never a dominant species, and now has a rather discontinuous distribution. This very discontinuity suggests that the species now occurs less widely than in earlier times. It could, for example, have been displaced downslope in the Zagros mountains, towards the south, as Harlan (1975) suggests may have occurred in the late Ice Age. The fact that wild emmer has been found at Çayönü also suggests a rather wider distribution 10 000 years ago, but taken overall the evidence indicates a close relation between the occurrence of the wild cereal progenitors and the beginnings of agriculture in the Near East (Zohary & Hopf, 1988). The same is true for legume crops such as the pea and lentil (Ladizinsky, 1989). Indeed, with the possible exception of the broad bean, the wild progenitors of all the Near Eastern crop plants have now been reliably identified within the area (Zohary, 1989).

A second caveat was raised by Harlan & Zohary (1966): 'Why should anyone cultivate a cereal where natural stands are as dense as a cultivated field?' They go on to suggest that agriculture 'may have originated in areas adjacent to, rather than in, the regions of greatest abundance of wild cereals'. Wild einkorn and emmer occur mainly on hills and plateaus flanking the piedmont areas of the Fertile Crescent where the early farming villages were located, which fits their suggestion. So does the fact that domesticated emmer appeared in Ali Kosh, Jarmo, Beidha and Çayönü before it appeared in Munhatta, the area where the wild form is most abundant today (Harlan, 1975).

On the other hand, wild barley extends further downslope than the wheats, and into the steppes and deserts along the wadis, i.e. closer to the sites where its domestication seems to have begun. Kislev et al. (1986) found both wild and domesticated barley together at Netiv Hagdud, near Jericho. In the early stages at several sites – such as Ali Kosh, Çayönü and Jarmo, about 8700–9500 years ago – there is evidence that both wild and domesticated forms of einkorn and emmer wheat were being used, i.e. there was a combination of gathering and growing (Harlan, 1977).

Long before this there are signs of human habitation in the area by Acheulean, Neanderthal and Natufian peoples (Valladas et al., 1988), indicated by the presence of many grinding slabs, hand stones, mortars and sickle blades. From her analysis of the micro-wear on flint sickle blades, Unger-Hamilton (1989) concludes that cereals were being cultivated by about 12 000 BP, but were probably harvested before they were ripe and may not have been domesticated, i.e. that cultivation preceded domestication. The great variety of plant remains at Abu Hureyra in Syria, dated to 11 500–10 100 BP, has provided no evidence of cereal domestication, despite the all-year round occupancy of the site (Hillman et al., 1989). In this instance, at least, a sedentary life style preceded plant domestication (cf. Allen, 1977).

The earliest evidence of cereal domestication comes from Netiv Hagdud, with barley dated to 10 200–9500 BP (Kislev et al., 1986), and from the Bus Mordeh phase of Tepe Ali Kosh, about 9500 BP, in which domesticated einkorn and emmer occurred along with wild barley, einkorn and emmer. The barley was classified as wild because it had a brittle rachis, but its glumes were loose and Helbaek (1966) regarded it as a 'cultivated wild barley'. No legumes were found in the early phases at Ali Kosh but peas, lentils, vetch, chickpea and *Lathyrus*, as well as flax, pistachio and almond, were found at Cayönü along with both wild and domesticated einkorn and emmer. Hexaploid bread wheat made its earliest appearance at about 7800 BP in many places (Harlan, 1975).

A third caveat that might be raised is the possibility that each of the major crops may have been domesticated more than once, as happened with cotton, peppers, and beans in the Americas. For the Near Eastern crops, however, Zohary (1989) presents evidence favouring unique domestications at least for emmer wheat, peas and lentils.

The major Near Eastern cereals and pulses all appear to have been domesticated within a period of two millennia. This was probably accomplished at different times and places for the various crop plants but frequent contact across the region, evident in the distribution of obsidian tools, presumably led to the development of a fairly homogeneous agricultural system throughout. The almost simultaneous appearance of bread wheat throughout the region from east (Tepe Sabz) to west (Knossos), following its likely origin towards the north east where the donor of the D genome (*Aegilops squarrosa*, syn. *Triticum tauschii*) occurs (Figure 3.3), may also reflect the extent of communication within the region.

Archaeological studies of the cereals give possibly too much emphasis to just one criterion of domestication, a tough rachis which ensures that the grain is retained in the ear until harvest. Given the predominance of the cereals at many early sites – wheat accounted for 83% of the seeds found at Jericho and 70% of those at Mureybit (Simmons *et al.*, 1988) – the use of a straightforward indicator of domestication has been crucial. In the pulses, this is not so easy but a smoother seed-coat and greater seed size are characteristic of several domesticated legumes (Figure 3.4). Using these features as criteria, it seems that domestication of the pea began almost as early as that of wheat and barley. The distribution of the wild species *Pisum humile* in the oak park forests of the Zagros Mountains is similar to that of the wild cereals (Ben Ze'ev & Zohary, 1973).

The same is true of *Lens culinaris* spp. *orientalis*, the wild progenitor of the lentil (Zohary, 1972), which also has much smaller seeds (Figure 3.4*b*), but the increase in seed size has been gradual and provides no clear cut criterion of domestication. Some quite large grains have been found at Tepe Sabz at 7000–7500 BP, by which time lentils were clearly domesticated. Zohary & Hopf (1973) argue for much earlier domestication on the grounds that the wild progenitor occurs only as scattered, small plants and was unlikely to be gathered in the bulk implied by its presence in most early village sites.

The chickpea is another pulse whose wild progenitor (*Cicer reticulatum*)

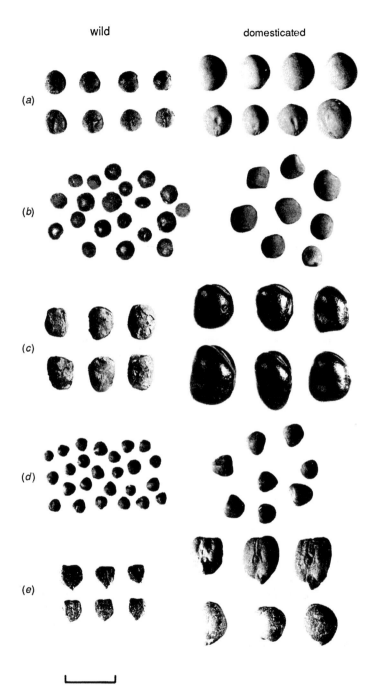

Figure 3.4. Changes in the size and other characteristics of legume seeds with domestication: (*a*) pea, (*b*) lentil, (*c*) broad bean, (*d*) bitter vetch, (*e*) chickpea (after Zohary & Hopf, 1973). (Bar = 1 cm.) (Copyright © 1989 by the AAAS.)

occurs within the Fertile Crescent (Ladizinsky & Adler, 1976; Ladizinsky, 1989). Few remains have been found at the early sites whereas fairly large seeds are common at Bronze Age sites. Another early Near Eastern domesticate was the linen flax, seeds of which have been found at Tell Mureybit and at Cayönü along with cultivated wheat and barley. Its wild progenitor (*Linum bienne*) occurs in the Fertile Crescent but also well beyond it (Zohary, 1989). The survey of the beginnings of fruit growing in the Old World by Zohary & Spiegel-Roy (1975) suggests that the cultivation of food crops such as date palm, olives, grapes and figs came well after the establishment of cereal–pulse agriculture (but cf. Barton *et al.*, 1990). Fruit remains are rare in the early Neolithic farming villages, and appear regularly only in early sixth millennium BP sites.

Africa

The contrast between the knowledge of our human as against our agricultural origins in Africa is striking. Our progenitors, in rich and well-dated variety, continue to be found in East Africa, whereas the origins of agriculture in Africa remain obscure (Harlan *et al.*, 1976; Clark & Brandt, 1984). Vavilov (1926) located one of his eight centres in Ethiopia (Figure 1.5) and Portères (1962) proposed eight African centres of domestication whereas Harlan (1971) suggested a diffuse broad belt 'north of the equator and south of the Sahara'. His belt encompasses several quite different ecological zones, forest, savanna and highlands. Forest–savanna domesticates include the oil palm (*Elaeis guineensis*), cowpea (*Vigna unguiculata*) and Guinea millet (*Brachiaria deflexa*) (see Figure 3.5). From the savanna came sorghum (*Sorghum bicolor*) and pearl millet (*Pennisetum americanum*), and from the East African highlands tef and finger millet.

Grinding slabs and handstones, probably used for the preparation of wild cereal seed, are known from several Nubian sites dated to around 15 000 years BP (Wendorf & Schild, 1976), and sickles appeared in Egypt well before they appeared in Palestine (Cohen, 1977*a*). There is then something of a gap in the Nile record until domesticated wheat, barley and flax appear at Fayum in the seventh millennium BP, and in Nubia in the fifth millennium, presumably reflecting diffusion from the Near East (Shaw, 1976).

Fertile Crescent agriculture was probably taken also to Ethiopia at an early stage, possibly by Cushitic peoples before the fifth millennium BP. Whether they encountered already domesticated local plants such as tef, ensete (*Musa ensete*) and finger millet when they arrived – as Harlan (1975) considers likely – or whether they then domesticated suitable local cereals, being accustomed to wheat and barley – as Doggett (1965) suggests – remains an open question.

However, independent local domestication is likely for many of the crops originating further west, such as pearl millet. This is the one species for which

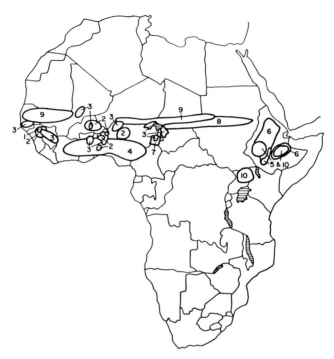

Figure 3.5. Probable areas of domestication of some African crops: 1, *Brachiaria deflexa*; 2, *Digitaria exilis* and *D. iburua*; 3, *Oryza glaberrima*; 4, *Dioscorea rotundata*; 5, *Musa ensete* and *Guizotia abyssinica*; 6, *Eragrostis tef*; 7, *Voandzeia* and *Kerstingiella*; 8, *Sorghum bicolor*; 9, *Pennisetum americanum*; 10, *Eleusine coracana* (Harlan, 1971). (Copyright © 1971 by the AAAS.)

there may be archaeological evidence of transition to its cultivation, as described by Munson (1976). However, the rapid transition around 3000 BP from the gathering of wild grasses to the dominance of domesticated *Pennisetum* almost certainly represents the introduction of previously – domesticated pearl millet into the area. As already indicated, domesticated *Pennisetum* had been introduced into India from Africa by 3200 BP, and other African domesticates such as sorghum and finger millet had reached India by 3800 BP.

Wet *et al.* (1976) concluded that domesticated sorghum probably derived from the wild races *aethiopicum* or *verticilliflorum*, mainly by selection for tough racemes and larger grains, in the region indicated in Figure 3.5. As the domesticate diffused westwards it gave rise to the *guinea* race, adapted to high rainfall conditions in West Africa, while the *kafir* race evolved from its migration southwards. Eastwards migration led to the *durra* race either before or after it reached South Asia in the fourth millennium BP, while the *caudatum* race evolved in the region of initial domestication (Harlan & Stemler, 1976). The earliest dated remains of sorghum in Africa are little more than a thousand years old (Shaw, 1976).

The origins of agriculture in the more humid forest–savanna zones of Africa, and particularly in West Africa, are still not clear. Coursey (1976) reviews several earlier suggestions that the use and domestication of the indigenous African yams, *Dioscorea cayensis* and *D. rotundata*, followed the introduction of Asian or American forms, but it seems more likely, as Harris (1976) argues, that their domestication was much earlier and may even go back to Palaeolithic times or to the Microlithic culture of about 10 000 BP. Indeed, Coursey (1976) suggests that yam digging may go back to the Sangoan and Lupemban cultures dating from 40 000 BP.

Closer to the interior, in the semi-arid zone, agriculture based on the cultivation of indigenous cereals and pulses may have replaced hunter-gathering quite early, although the earliest dates for cowpeas at Kintampo are only about 3400 BP (Flight, 1976). African rice could have been domesticated under simple flood water cultivation, near the inland delta of the Niger where the complex *crue* and *décrue* techniques of cultivation have long been practised (Harlan & Pasquereau, 1969). Although the combination of yam and cereal culture in Africa is usually regarded as a secondary development, combining the vegecultural complex from the forest with the seed agricultural complex from the savanna, Flight (1976) suggests that the combination may be a primary one, which solved the problems associated with the frequent occurrence of too much rain for the cereals alone and too little for the yams alone.

The Americas

A substantial fraction of the world's most important crop plants was domesticated in the Americas: the list includes maize, potatoes, sweet potatoes, manioc (cassava), beans, peanuts, sunflower, tomato, upland cotton and tobacco. By the criterion of carbon dating they are also among the oldest crops as well as being among the most widely used and productive. Their wild progenitors are in many cases known. Yet the origins of agriculture in the Americas are still subject to quite varied interpretations.

Vavilov allocated two of his centres of origin to Mexico and South America, subsequently adding two subcentres to the latter (Figure 1.5). He regarded these as independent centres, whereas Harlan (1971) envisaged a Meso-American centre linked with a South American non-centre in which many plants were domesticated over a wide area, not necessarily because the concept of agriculture had diffused from the Meso-American centre. Lathrap (1977) goes further in an attempt to link together the agricultural beginnings in the region. He envisages fisherfolk communities on the north coast of Brazil domesticating cassava and other crops and developing a highly productive agriculture as they spread up the Amazon to the foot of the Andes and thence in various directions, particularly northwards. These people, with their lowlands agriculture, then gave rise eventually to the Olmec and all the Meso-American cultures derived from them, and to the Chavin and the Andean cultures derived from them.

Lathrap (1977) brings together a great range of linguistic, anthropological and other evidence to support this unification, but problems remain. The very early carbon datings of crop remains in both Mexico and the Andes pose problems in relation to his suggested chronology, as does the very different nature of Mexican and Andean agricultures from that of the humid lowlands. Like all proposals giving primacy to humid zone root culture, it runs into immediate difficulties with the absence of archaeological remains, and with the extremely low probability of early root crops being preserved and found in such environments. Although cassava was almost certainly domesticated in the lowland tropics of Brazil, the earliest unambiguous remains of it have been found in the dry coastal areas of Peru, dated to 4500 BP (Hawkes, 1989).

Sauer (1952) contrasted the Meso-American dependence on seed crops with the South American dependence on root crops, but the many root crops of the Andean highlands were combined with a seed agriculture based on maize, beans, squash, quinoa (*Chenopodium quinoa*) and amaranths (Pickersgill & Heiser, 1977). In both Mexico and Peru, non-food crops appear early in the record along with the major food crops. Squash (*Cucurbita pepo*) has been dated back to 10700–9800 BP, and the bottle gourd (*Lagenaria siceraria*) to 9000 BP. Remains of upland cotton (*Gossypium hirsutum*) occur from 5500 BP in the Tehuacan Valley and from 4500 BP on the coast of Peru.

For many of the American crops, several different species were domesticated in different parts of the region, as in the case of *Phaseolus, Amaranthus, Chenopodium, Cucurbita* and *Capsicum*. Four species of *Phaseolus* were domesticated: *P. coccineus*, the runner bean, in Mexico; *P. vulgaris*, the common bean, in both Meso- and South America, probably from at least two independent domestications (Gentry, 1969; Gepts, 1990); lima bean, *P. lunatus*, which also appears to have been independently domesticated in Mexico and Peru (Kaplan, 1965; Gepts, 1990) and *P. acutifolius* var. *latifolius*, the tepary bean, in Mexico. The wild progenitors of all four domesticates are known, and their remains have been dated back to 10700 BP. Their pods dehisce along both sutures, and their seeds are smaller and more slowly permeable than those of the domesticates (Kaplan, 1965). Yet no intermediate stages between wild and cultivated forms have been found, and the earliest seeds of tepary and lima beans are as large as present-day forms.

Three species of *Amaranthus* were domesticated, *A. hypochondriacus* in central and northern Mexico, *A. cruentus* in Mexico and Guatemala, and *A. caudatus* in the Andes. Sauer (1967) found these to be closely related to three wide-ranging weedy species, *A. powellii*, *A. hybridus* and *A. quitensis*, from which they were presumably domesticated independently in different areas. Alternatively, they might all derive from *A. cruentus*, which has been found dating back to 5500 BP in the Tehuacan Valley whereas the earliest occurrence of *A. hypochondriacus* there is 2200 BP (Sauer, 1969). Plants of the domesticates are much larger than those of the wild species, with more inflorescences per plant and more highly compound inflorescences, and it is these features, rather than larger seed size, that result in their higher yield (Sauer, 1950).

Maize and the potato are the most characteristic domesticates of the Mexican and Andean regions of domestication. Both are highly productive and adaptable crops, but in their archaeological records they are quite unequal. Like most root and tuber crops the potato has left few traces. Gathered specimens of wild potatoes (*S. maglia*?) have been carbon dated to 13 000 BP in Chile (Ugent *et al.*, 1987) and there is evidence of potato gathering back to 10 000 BP in several Andean caves (Grun, 1990). However, the oldest remains of domesticated potatoes found so far come from the Chilca Canyon in Peru, dated to 7000 BP (Hawkes, 1990) and from the Casma Valley, dated to 4000 BP (Ugent *et al.*, 1982). The diploid cultigen *Solanum stenotomum* crossed with an unidentified wild diploid to give the cultivated tetraploid *S. tuberosum* ssp. *andigena*, of Andean origin, which then crossed with another species to give rise to the subspecies *tuberosum* of southern Chile (Grun, 1990; Hosaka & Hanneman, 1988).

The origin of maize is still being debated (cf. MacNeish, 1985; Galinat, 1988; Goodman, 1988); depending on the view espoused, the archaeological record either does or does not reveal the transition from wild progenitor to early domesticate. According to the tripartite hypothesis of Mangelsdorf & Reeves (1938), maize arose from a wild race of pod corn, now extinct, but to be seen among the earliest Mexican remains of maize. However, many botanists prefer the alternative view that the Mexican grass teosinte is the ancestor of maize. Beadle (1977), Galinat (1988) and Doebley (1990) review the evidence in support of this hypothesis.

The present distribution of teosinte populations in relation to major archaeological sites is shown in Figure 3.6. The oldest remains of maize (7700 BP) have been found at Tehuacan but there is palynological evidence of maize cultivation at Oaxaca at least since 10 000 BP (Galinat, 1988). Isozyme and chloroplast DNA evidence strongly supports the view that maize was domesticated from the *Zea parviglumis* subspecies of teosinte, possibly from populations similar to those now found near the Balsas river south-west of Mexico City (Doebley, 1990).

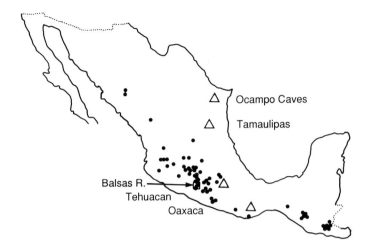

Figure 3.6. The distribution of teosinte populations (dots) in Mexico and Guatemala in relation to early archaeological sites (triangles) (adapted from Wilkes, 1972) and the Balsas River (square) (Doebley, 1990).

Pickersgill & Heiser (1977) point out that neither in Mexico nor in Peru is there an archaeological sequence demonstrating the stages of domestication of any important crop plant. Maize, bean and amaranth appear in the earliest agricultural levels at Tehuacan as fully-fledged domesticates. The slow transition observed by MacNeish (1958, 1964) was more a shift in the extent of dependence on agriculture than an illustration of the domestication process.

Maize from the Mexican centre of domestication had spread to South America by at least 3800 BP, whereas cultivated potatoes had not spread north of Colombia by the time of the Spanish conquest. Exchanges between the two American centres appear to have been quite limited, which is surprising in view of their ecological similarities as dry highlands and the fact that their domesticates have been readily exchanged within historic times (Pickersgill & Heiser, 1977).

The cultivation of maize and associated crops, especially beans and squash, spread northwards and eastwards into North America and also down to the surrounding lowlands. The analysis of human skeletons for their $\delta^{13}C$ values suggests that plants with the C_4-pathway of photosynthesis, which pass their low $\delta^{13}C$ values on to collagen in the bones of people eating them, were of overwhelming importance in Tehuacan valley diets by 6000 BP (van de Merwe, 1982), presumably reflecting dependence on maize and possibly also

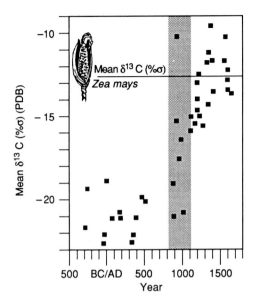

Figure 3.7. Values of $\delta^{13}C$ for human bones of various ages from mid-western US archaeological sites, as evidence of a shift to maize-based agriculture 500–1000 years ago (Smith, 1989). (Copyright © 1989 by the AAAS.)

on *Amaranthus*. High $\delta^{13}C$ values for skeletons in the Orinoco River lowlands suggest that maize had not reached that area by 2800 BP, whereas it clearly had done so by 1600 BP, when it probably made up about 80% of the diet (van der Merwe, 1982). The same technique, applied to skeletons in mid-western North America (Figure 3.7), indicates the probable shift to maize growing in those states about 500–800 years ago.

The sunflower (*Helianthus annuus*) is the major domesticate from North America, wild sunflowers having been widely used by Native Americans in the west. In the east many species were gathered and some were cultivated, including both the squash and the bottle gourd. The earliest archaeological evidence of domestication is for sumpweed (*Iva annua*) at about 4000 BP and sunflower at 3000 BP (Watson, 1989).

Centres and non-centres

Beginning with Schiemann in 1939, many crop evolutionists have pointed out that Vavilov's centres of origin are really centres of diversity. Other lines of evidence, especially archaeological remains and the distribution of wild relatives, as well as linguistic clues, have to be taken into account in determining the place of domestication of a crop. Harlan (1971) highlights the inadequacy of the Vavilovian method with the observation that barley is not particularly variable within the Fertile Crescent where it was domesticated (Figure 3.3). There is far more variation in barley and emmer in Ethiopia than there is in their crescent of origin. Vavilov's centres are sometimes regions of relict genetic diversity ('museums'), zones of more recent adaptive radiation or hybrid 'contact' zones (Bretting, 1990). Variations on the Vavilov theme, such as Zhukovsky's (1968) twelve mega-centres, have not advanced the concept; in essence they suggest that there are no centres.

Harlan (1971) considers that there are three convincing centres, in the Near East, North China and Meso-America. In the sense that there is evidence of early, more or less contemporaneous domestication of a suite of crop plants for various purposes within a well-defined and restricted area, these are good examples. Harlan points out that his three centres are more temperate than the other regions of domestication and are focused on grain crops. The other, more tropical and diffuse regions of domestication he refers to as non-centres. He links each of the three with one of his centres and suggests that each centre and its non-centre formed an interactive system.

Harlan's non-centres are really tantamount to an admission that people have domesticated plants wherever and whenever there was a need and an opportunity to do so, as may be seen by comparing Figure 3.8 with Figure 1.5. Species of *Dioscorea* yam have been domesticated in many regions, including Melanesia, the Pacific and the Caribbean. At least 24 species have been cultivated for food, another 12 have been used for drugs, and a further 26 have been gathered (Coursey, 1967). *Phaseolus vulgaris* has been independently domesticated at least twice (Brücher, 1969; Gentry, 1969; Gepts, 1990) and, as such, symbolizes one theme of this chapter, hence its use as the vignette at its head. Relatively diffuse origins are to be expected: it may be the well-defined centres that require further explanation.

What? The taxonomy of domestication

Because there are many plants hovering on the margins of domestication, estimates of the number of domesticated species vary greatly. The 2489 species listed by Zeven & de Wet (1982) are distributed among 173 families, of which by far the most important are the Gramineae with 379 and the Leguminosae with 337, these two families accounting for almost a third of all domesticates. The next most favoured families are the Rosaceae (158) Solanaceae (115), Compositae (86), Myrtaceae (73), Malvaceae (70) and Cucurbitaceae (53). These eight relatively advanced families have contributed over half of all the domesticates.

The proportion of species that has been domesticated varies greatly

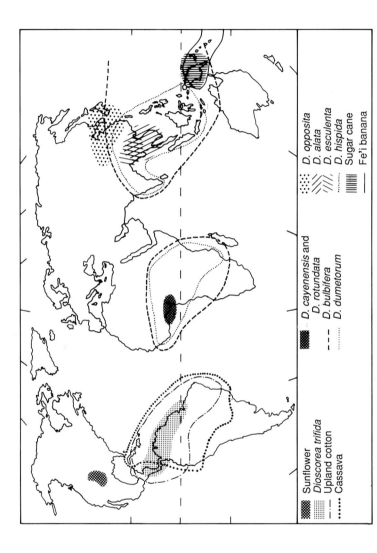

Figure 3.8. Probable areas of domestication of various species of *Dioscorea* yams and several other crops. Adapted from Harris (1972) with additions from Zeven & de Wet (1982).

Sunflower

Dioscorea trifida

Upland cotton

Cassava

D. cayenensis and
D. rotundata

D. bulbifera

D. dumetorum

D. opposita

D. alata

D. esculenta

D. hispida

Sugar cane

Fe'i banana

between families, from 0% for many to 100% for the Simmondsiaceae, jojoba (*Simmondsia chinensis*) being the only species in that family. About 3.8% of the species of Gramineae have been domesticated, 2.8% of the Leguminosae, 3.9% of the Dioscoreaceae, 5.7% of Solanaceae, 8.3% of the Cucurbitaceae and 31% of the Musaceae, on the basis of Zeven & de Wet's (1982) compilation. Their table also indicates the considerable differences between regions, not only in the total number of their domesticates but also in their taxonomic composition. The SE Asian region has contributed the greatest number (311), closely followed by China–Japan (295), Africa (292) and South America (292), these four regions accounting for almost half of the total. Africa has contributed a high proportion of the domesticated Gramineae (19%), and South and Meso-America high proportions of the domesticated Solanaceae, whereas the legumes are fairly evenly distributed across regions.

However, these numbers presuppose rather more taxonomic certainty about the classification of cultivated plants than is the case. The taxonomic problems associated with cultivated plants have been considered by Harlan & de Wet (1971), especially the relations between the various gene pools, which can be so diverse among domesticates.

How? Changes with domestication

One theme that emerges from the earlier discussion of plant domestication in the various regions is the parallelism in the changes that converted wild to domesticated plants both among the various crops and in the various regions. Likewise, similar characteristics in the wild progenitors predisposed them to domestication in the various regions. Charles Darwin's 'analogous parallel variation' and Vavilov's 'homologous series' recognized these common features both in the kinds of plants domesticated and in the changes with domestication.

Some of the changes may have been brought about by conscious selection, but in many cases selection was unconscious and indirect, as Darwin (1868) recognized. Harlan *et al.* (1973) emphasize that indirect selection pressures are multiple and interlocking, and Rindos (1984) devotes a whole book to that subject.

One caveat should be mentioned before we consider the differences between domesticates and their wild progenitors. In many instances the connecting links between them have been lost and, as Vavilov (1926) wrote, 'it must be left to the imagination to restore all the historical links in a more or less plausible form'. In the years since he wrote that, many of the missing links have been found and new ways of assessing their relationships have been developed. Nevertheless, the identified wild progenitors remain putative. I shall refer to them as wild relatives when the relation is insecure, and as progenitors when it is probable.

Seed retention

Non-shattering inflorescences in cereals and non-shattering pods in many grain legumes and other crops constitute one of the earliest and most consistent differences between domesticates and their wild relatives. Indeed, a tough rachis is often the diagnostic character used to differentiate wild from domesticated cereals in archaeological studies, although it is not altogether clear when the shift occurred in relation to domestication.

More than one mechanism enhances seed retention in the grasses and cereals, but only a single or a few genes are usually involved, with the domesticated phenotype often being recessive (Ladizinsky, 1985; Kadkol *et al.*, 1989). For example, in wild rice (*Zizania palustris*), which is currently being domesticated, non-shattering forms have been found under the control of two recessive genes (Elliott & Perlinger, 1977; Hayes *et al.*, 1989).

In some legumes the reduction of natural seed dissemination is accomplished by the strengthening of the sutures along the dorsal and ventral seams of the pods, and/or a reduction in the torsional forces developing in the pod walls as they dry out. Such mutants were found in field populations of *Lupinus luteus* by von Sengbusch & Zimmermann (1937), and of *L. angustifolius* and *L. digitatus* by Gladstones (1967), and used to breed non-shattering varieties. In soybeans, the dorsal suture is not changed but the angle of orientation of the wall fibres is less acute in non-shattering lines, the difference being controlled by one or two pairs of genes (Tsuchiya, 1987). By contrast, no change in wall fibre orientation was found in brassicas (Kadkol *et al.*, 1986).

In the cowpea there is little difference between wild and cultivated lines in the structure of the sutures, but the pod walls of the indehiscent domesticated lines have fewer spirally thickened fibres and the spirals are less tightly coiled (Lush & Evans, 1981). The difference is controlled by one gene, but its effect varies between accessions. Pods of accessions from the most humid regions dehisced when the vapour pressure deficit fell to 8–9 mbar whereas those from seasonally arid zones did so only when the deficit exceeded 19 mbar. (Lush *et al.*, 1980). Thus, a wild line from the arid zone might not dehisce in the humid zone and would be a potential domesticate there.

Ladizinsky (1987) argues that in the lentil the beneficial effect of low saturation deficits and the traditional method of harvesting combined to make pod indehiscence less necessary than non-shattering is in the cereals. However, loss of the natural means of dissemination must have been a crucial step in the domestication of most seed crops, whereas in many root crops and green vegetables, seed retention was irrelevant to domestication and their pods still shatter.

Modification of glumes and spines

Glumes, lemmas and paleas enclose the grains of most cereals and protect them from damage in many wild grasses. The awns attached to the glumes are also protective, especially against birds, and are often well developed in wild species, where they may also play a role in seed distribution and burial. The glumes may also delay seed germination. Glumes and awns have been reduced in many domesticated cereals, and they have also become more loosely articulated to give free-threshing or naked grains (cf. Kadkol *et al.*, 1989). These appeared early in the archaeological records of cereals in the Near East, and Cohen (1977a) suggests that free-threshing may have been more important than grain retention in the domestication of wheat.

Protective spines are common among crop progenitors and have often been reduced or eliminated with domestication. Several examples have already been mentioned, such as yams (whose tubers are protected from rooting by pigs and other animals), sago palm, taro, the jujube and chickpea.

Reduced seed coat thickness

Thick, hard, smooth seed coats afford protection against insect predation in wild species, but there has been a reduction in the thickness of the seed coats of many grain crops, presumably as an indirect effect of selection for faster germination, easier and quicker food preparation or better digestibility. Lush & Evans (1980a) found the testas of domesticates to be thinner and of lower specific weight than those of wild relatives among cowpeas, lupins, peas and soybeans. But even among domesticates there can be a wide range in testa thickness, as between the smooth and the rough-seeded cultivars of cowpeas, or between lupins on the one hand and groundnuts on the other.

Greater size of the harvested organs

Malthus (1798) commented that increased size was one of the most obvious features of plant improvement. Of the various differences between wild and cultivated plants, Schwanitz (1966, 1967) gives greatest emphasis to the 'gigantism' of many domesticates. This is particularly evident in vegetable and fruit crops such as sorrel (*Rumex acetosa*), tomato, peppers and raspberries, and among clonally reproduced crops such as yams. In fruits such

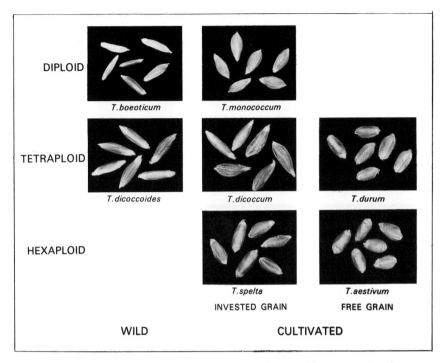

| | **WILD** | **CULTIVATED** | |

Figure 3.9. Representative grains of wild and cultivated species of *Triticum* (Dunstone & Evans, 1974).

as apples, pears and tomato it is the fleshy mesocarp rather than the seeds which are enlarged, as also in the case of the *pisifera* mutant of the oil palm, and in the olive and date, but not in the grape (Zohary & Spiegel-Roy, 1975).

The size of cotyledons in the archaeological remains of avocado (*Persea americana*) at Tehuacán increased progressively from 9500 BP until 1600–500 BP, when they reached their present size (Smith, 1966). Increase in seed size in the early stages of domestication occurred in many grain legumes, as shown by Kaplan (1965) and Zohary & Hopf (1973) (cf. Figure 3.4). The 5–10-fold increases in seed size on domestication of the cowpea or from wild *Glycine soja* to the cultivated soybean are comparable with those in the Near Eastern legumes (Lush & Evans, 1980*a*).

Among the cereals, although increase in grain size has occurred, as in wheat (Figure 3.9), it has been rather less pronounced than in the legumes. Grains of the wild *Hordeum spontaneum* may be larger than those of some modern barley cultivars. Grains of sorghum have not increased in size in India over 4000 years. In sowi millet (*Panicum sonorum*), on the other hand, grain size in the domesticates is almost twice as large as in the wild relatives although the crop has never lost its natural seed dispersal mechanism (Gentry, 1942). In another Mexican cereal, *Setaria geniculata*, grain size increased progressively throughout a period of 1500 years in the archaeological record (Callen, 1967). In several crops, such as linen flax and chickpea, Vavilov (1951) noted that seed size increased in Mediterranean cultivars far more than in Asian ones. Thus, greater seed size may reflect environmental adaptation as well as domestication *per se*.

Factors contributing to increases in organ size are considered later, but some of the consequences of greater seed size are relevant here. Larger seeds, like larger yams or fruits, would be attractive to harvesters in terms of economy of effort. Recovery of the heavier seeds might also have been favoured by the winnowing used since the days of the hunter-gatherers. Larger seeds may also have an advantage at sowing, in giving rise to larger, more competitive seedlings. Among the various wild and cultivated wheat species, for example, seedling size after four weeks was closely related to the weight of the sown kernels (Evans & Dunstone, 1970). A similar relation has been found among tropical legume species (Whiteman, 1968) and in sugar beet (Scott *et al.*, 1974).

Seedling emergence from deeper sowings is greater the larger the seed, and plants from larger seeds may have more seminal roots (Mackey, 1979*a*). Greater seed size tends to reduce the seed coat percentage (Evans, 1976), but in cereals it also tends to be associated with a reduction in the percentage of protein, as among the wild and cultivated wheats (Dunstone & Evans, 1974; Sharma *et al.*, 1981).

Correlative changes in size

As the epigraph to Chapter 5 indicates, Darwin recognized that increase in size with domestication was often greatest in those organs of interest to mankind, e.g. among fruit and root crops. Nevertheless, there are also many cases where the size of non-harvested organs has also increased with domestication, as illustrated by Schwanitz (1966, 1967).

Other examples include the parallelism between stem and leaf size in sugar cane (Bull, 1965); between kernel weight and area of the largest leaf among the wild and cultivated wheats (Figure 3.10); between kernel weight and plant height in barley (Riggs & Hayter, 1975; Hanson *et al.*, 1981); and between leaf area and pod size among cowpeas (Lush & Evans, 1981). In the first two of these cases there is a common basis for the parallelism in variations in cell size, associated to a large extent with differences in ploidy. Among the wheats, however, the parallelism in cell size between the aleurone layer of the grain and the mesophyll of the leaf did not extend to the endosperm of the grain (Dunstone & Evans, 1974). In the cowpea and in other crops such as beans

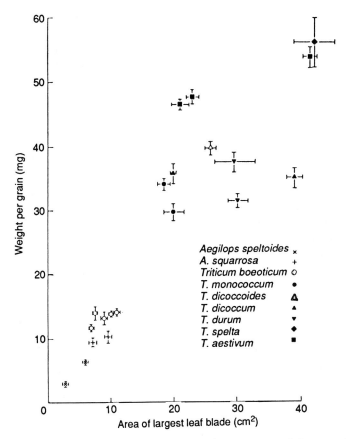

Figure 3.10. The relation between weight per grain and the area of the largest leaf among the wild and cultivated wheats (Evans & Dunstone, 1970).

(White & Gonzalez, 1990) cell size and seed size can be closely related even when differences in seed size are not due to differences in ploidy.

Schwanitz (1966, 1967) emphasizes the importance of increases in cell size in contributing to the gigantism of many domesticates. Increase in cell size apparently contributes more to storage capacity, sugar concentration and yield in beet than does greater cell number (Milford, 1976; Rapoport & Loomis, 1985, 1986), and is strongly influenced by ploidy (Butterfass, 1964). But there are also many cases where both cell size and cell number are important, as in stem diameter of sugar cane (Bull, 1965) and pea seed weight (Davies, 1975). Richards & Passioura (1981) provide an instructive example with the main xylem vessel of wheat seedlings. The diameter of this cell has a pronounced effect on the axial resistance to water flow, and has presumably been under strong selection pressure, resulting in adaptive differences between land races, with high heritability but with little difference between ploidy levels. Thus differences in cell size not associated with ploidy can match those associated with it when selection pressures are sufficiently intense.

Polyploidy and DNA content

Polyploidy is common among domesticated plants. Being larger and sometimes more adaptable, polyploids may have invited more frequent domestication, but in some cases the polyploidy has been accentuated following domestication, as in the displacement of einkorn by bread wheat. Successful autopolyploids include sugar beet and sweet potato. Successful allopolyploids include durum and bread wheat, tobacco (*Nicotiana tabacum*), sugar cane, oats and the New World cottons. These latter constitute something of a problem because of the extreme allopatry of the extant A and D genomes, which occur on opposite sides of the world. Hutchinson *et al.* (1947) attempted to resolve this by suggesting that the tetraploids arose after introduction of the A genome to South America by man. However, Wendel (1989) has shown, by mitochondrial base sequence analysis, that the maternal parent of the tetraploids was the A genome and that they probably originated 1–2 million years ago.

Most of the major cereals are diploid, including rice, maize, barley, rye (*Secale cereale*) and pearl millet. Sorghum is tetraploid; both wheat and oats range from diploid to hexaploid. The major grain legumes are also mainly diploid with only the groundnut, *Lupinus alba*, and soybean (cf. Smartt, 1980) being tetraploid, while pigeon peas (*Cajanus cajan*) are diploid, triploid and tetraploid. Thus, success as a grain crop has not hinged on polyploidy, but the picture is very different among other crops.

Both of the major sugar crops are polyploid, sugar beet ranging from diploid to pentaploid and sugar cane from 8 to 12-ploid. The success of monogerm sugar beet has led to a recent shift away from triploid or tetraploid cultivars to diploids. Although fibre crops come from many families, only four of the fourteen major species are diploid, namely hemp, jute (*Corchorus olitorius*), Manila hemp (*Musa textilis*) and possibly flax. The cottons, kenaf, henequen, sisal, Sida hemp, kapok and silk cotton are mostly tetraploid or more. Likewise, of the ten major root crops, from six different families, only *Xanthosoma sagittifolium* is wholly diploid. Most of the *Dioscorea* yams are highly polyploid, the Chinese yam (*D. opposita*) being 14-ploid, the greater yam (*D. alata*) up to octoploid and *D. esculenta* up to decaploid. The sweet potato is hexaploid, cassava (*Manihot esculenta*) is tetraploid and potatoes range from triploid to pentaploid. Some crops grown for their leaves, such as tobacco and forage plants, and many of the species grown for their decorative flowers, are also polyploid.

For sugar, fibre, root and decorative crops, therefore, whether tropical or temperate, polyploidy either made them more attractive propositions for domestication or subsequently enhanced their value as domesticates.

Chromosome number and size are not always related to the amount of DNA per gametic genome (the 1C-value), which varies more than 2500-fold in angiosperms (Bennett, 1987). Indeed polyploids may have DNA C-values similar to or even lower than diploids (Bennett, 1972, 1987), and it may well be the C-values that are associated with domestication and performance. They are closely related to chromosome volume, nuclear volume, pollen volume, cell size and both mitotic and meiotic cycle times, and are less in annuals than in perennial species. Seed weights in both *Vicia* and *Allium* species are positively correlated with their DNA C-values, as are fruit and flower sizes (Bennett, 1972). Earlier generalizations that both polyploidy and chromosome size increased with latitude have been found to have significant exceptions (Bennett, 1976).

In Figure 3.11 the DNA 4C-values of domesticates are compared with those of their likely progenitors, using the data collected by Bennett & Smith (1976) and Bennett *et al.* (1982). Over a 20-fold range in DNA per cell, the cultivars all lie close to the 1:1 slope, indicating that there has been no increase in DNA content since domestication, with the possible exception of the peppers.

Change of form: allometry and condensation

The classic example of change in form with domestication, one which greatly impressed both de Candolle and Darwin, is the group of vegetables derived from the wild kale (*Brassica oleracea*), namely the many forms of cabbage, kale, kohlrabi, Brussels sprouts, broccoli and cauliflower. Within this one species either leaves, stems, roots, axillary buds or inflorescences have been enhanced by selection to give rise to quite different forms (Figure 1.7). However, as is often the case with classical examples, the interpretation may have been oversimplified; Snogerup (1980) reviews the research suggesting that several different wild species gave rise to the several domesticated forms. The more modest array of crops derived from the wild sea-beet (*Beta maritima*) may be a more representative example of changes in form with domestication (Figure 3.12).

Of course, as Wallace perceived sooner than Darwin, wild species often encompass a considerable variation in form, some of which carries over into the domesticates. The variety of tuber shape in the clonally reproduced yam *Dioscorea alata* (Coursey, 1967) and in many other root crops such as cassava

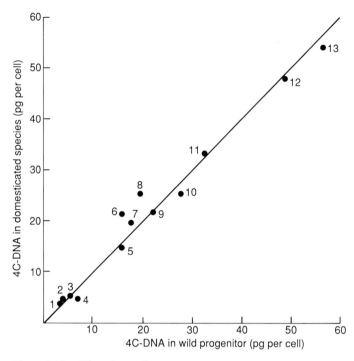

Figure 3.11.   The relation between 4C-DNA content per cell (pg) of the wild progenitors and their corresponding domesticates in several crops: 1, *Brassica campestris/oleracea*; 2, *Lycopersicum pimpinellifolium/esculentum*; 3, *Beta maritima/vulgaris*; 4, *Daucus carota gummifer/sativus*; 5, *Nicotiana sylvestris + N. tomentosiformis/tabacum*; 6, *Capsicum annuum*, wild/cv; 7, *Sorghum virgatum/bicolor*; 8, *Capsicum baccatum*, wild/cv; 9, *Hordeum spontaneum/vulgare*; 10, *Triticum boeoticum/monococcum*; 11, *Secale montanum/cereale*; 12, *Triticum dicoccoides/dicoccum*; 13, *Avena sterilis/sativa* (from data of Bennett & Smith, 1976).

and potato exemplifies this. Deep-rooting forms, which are less prone to injury in the wild, e.g. by pigs, are also less convenient for harvest; more compact, shallower-rooting clones have been selected for domestication. Some wild potatoes produce tubers on long rhizomes, distant from the parent plant. This aids dispersal in the wild, but domesticates were selected for shorter rhizomes and more tightly clustered tubers (Schwanitz, 1966).

The selective enhancement of particular organs often results from a changed allometry, with greater partitioning of assimilates to those organs from an early stage of development, a process which can be self-reinforcing.

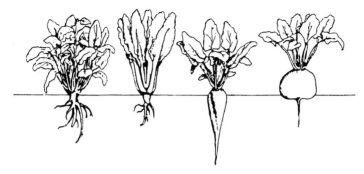

Figure 3.12. Changes in form among domesticates of the wild sea beet (*Beta maritima*), left. To the right are three forms of *Beta vulgaris*, common beet, sugar beet and mangel-wurzel (Schwanitz, 1966). (Reprinted by permission of the publishers. Copyright © 1966 by the President and Fellows of Harvard College.)

In sugar beet, for example, the hypocotyl diameter of young seedlings is a good predictor of ultimate 'root' size and yield (Doney & Theurer, 1976), and differences in ear:stem ratio among old and modern wheats are apparent from an early stage of development (Bush & Evans, 1988; Siddique *et al.*, 1989*b*).

Harlan *et al.* (1973) also refer to the process of 'condensation', whereby a shortening of branches and internodes brings many small scattered organs – e.g. florets in a cauliflower or seeds in a pearl millet inflorescence – into a more compact structure. For example, the open, branched and spreading panicles of wild sorghum have been changed by selection into the compact, single-stemmed, dense terminal inflorescence characteristic of modern sorghums. Perhaps the extreme examples of condensation are to be seen in the maize cob and the single terminal inflorescence of sunflowers, the wild forms of both crops having many small inflorescences scattered over many branches.

Given the frequency of such 'condensation', it is likely to have played a significant role in domestication. Plants with more compact inflorescences may have been preferentially gathered, because less effort was required. Galinat (1974) mentions that while collecting seeds of teosinte he must have unconsciously favoured those plants with more tightly condensed spikes. Unconscious selection for more uniform ripening may have reinforced this tendency, as may physiological processes contributing to yield (cf. p. 235).

More rapid and uniform germination

Delayed germination, due to both dormancy and hard-seededness, is a common and adaptive attribute of wild plants and many annual pasture species. In crops, however, it is undesirable, resulting in uneven stands, greater weed competition, the setting aside of more seed for sowing, and the infestation of subsequent crops. More rapid and uniform germination is likely to have been an early and important change with domestication. Indeed, Ladizinsky (1987) argues that selection for early germination was almost a prerequisite to domestication of lentils because the sown crops would otherwise have been so sparse as to be scarcely worthwhile. He suggests that the dominant allele eliminating dormancy could have established itself in wild populations in places where most of their seeds were gathered and seedling competition therefore reduced.

On the other hand, delays in germination may sometimes be adaptive in agriculture. When cereals ripen in humid conditions, a modest degree of seed dormancy is required to prevent the grains from germinating in the ear before harvest. Germination may also be modified in other ways. The seeds of many domesticates do not require exposure to light for germination to occur, whereas those of many wild progenitors do. This change is advantageous, indeed essential, when deeper sowing accompanies greater seed size, whereas many small seeded crops like lettuce and tobacco are still photoblastic.

Synchronization of flowering and maturation

Whereas maturation over a long period of time may be advantageous in wild plants, it is disadvantageous in cultivated ones except in cases where multiple harvests are wanted, e.g. for West African cowpeas in which young green pods are picked over a long period. For most agricultural crops a maximum return from one harvest is sought, and indirect selection probably enhanced uniformity of maturation, a trend that still continues (Hay & Kirby, 1991). Nabhan & de Wet (1984) took greater synchrony as evidence of domestication in *Panicum sonorum*. In the domestication of American wild rice there has been deliberate selection for synchrony in heading date (Hayes & Stucker, 1987), and crops in which synchronization has proved difficult, such as *Coix* and *Chenopodium pallidicaule*, have remained incompletely domesticated.

Closer synchronization of flower initiation on tillers and branches is one contributory element. Alternatively, there may be faster development of the later-initiated inflorescences or flowers, e.g. among the tillers of wheat (Stern & Kirby, 1979). The last grains to be set may also develop faster or for a shorter period, as in wheat. For crops which are sensitive to daylength, such as cowpeas, Bunting (1975) proposed a quite different mechanism, namely that they initiate flowers successively but that these await a particular daylength before they all develop further in synchrony. Lush & Evans (1980b) found no evidence of this in cowpeas, and suggested that the faster development of later formed flowers and pods was due to the cumulative photoperiodic induction of the plants.

Life cycle and breeding system

It is often stated that the life cycle of many crops has been shortened from perennial to annual on domestication, but this generalization has lost some of its force as more and more annuals have been identified as the likely wild progenitors, e.g. of rice, rye and soybeans. Perhaps the most convincing example is cotton. *Gossypium* is a genus of long-lived perennial shrubs; in all four cultivated species, annual types have been developed under domestication (Hutchinson, 1976), allowing cotton to be grown beyond the frost-free zone to which it was previously restricted.

Even when the wild progenitor was an annual, the life cycle has sometimes been shortened, as in beans (Brücher, 1969). While some phases of the life cycle are shortened, others may be lengthened. For example, the period from ear or panicle emergence to anthesis has been abbreviated in wheat – where it is quite prolonged in the wild relatives (Evans & Dunstone, 1970) – and almost eliminated in some barley and rice cultivars in which anthesis occurs in the sheath, whereas the duration of grain growth has commonly increased. In maize the interval between anthesis and silking has been greatly reduced (Edmeades & Tollenaar, 1990).

Another common generalization is that with domestication there has been a shift from out-breeding to self-fertilization, although maize, rye, pearl millet, sorghum, amaranths, brassicas, sugar beet, sweet potato and many fruit trees and pasture plants are still predominantly out-breeding, not to mention dioecious crops such as hops, hemp and papaya. In some instances, an increase in the rate of inbreeding has clearly accompanied domestication, as in cowpea (Lush, 1979), tomato (Rick & Dempsey, 1969; Rick, 1988), sorghum, *Capsicum*, flax, the brassicas and the egg plant (*Solanum melongena*). On the other hand, the retention of some capacity for outbreeding following domestication has played an important part in the evolution of several crops by allowing continued introgression from their wild relatives, as in the case of sunflower, sorghum, pearl millet, maize, tomatoes, potatoes and rice. These genetic exchanges work both ways; Harlan *et al.* (1973) have recorded several instances of crop characteristics appearing in the associated wild or weedy populations: of long spikes in wild pearl millet, of maize characteristics in teosinte, and of purple marker genes in weedy rice populations. Such a possibility has important implications for genetic engineers, e.g. for proposals to incorporate genes for herbicide resistance into crops whose relatives are among their most serious weeds.

A related change at domestication, and a more consistent one, has been a reduction in sterile florets and an increase in fertility and seed setting among grain crops, although the reverse is the case among root crops. Wheat has progressed from one grain per spikelet in einkorn through two in *Triticum dicoccum* to an increasing number in modern bread wheats. Wild barleys are 'two-rowed', their lateral spikelets being male or sterile, whereas these are fertile in the cultivated 6-row barleys, as the result of a single recessive mutation. Likewise in maize, an early step in its domestication involved the pedicillate spikelets becoming fertile, to give four rows of grains in two-ranked ears and eight rows in four-ranked ears, again the result of single mutations.

By contrast, among crops selected for vegetative performance domestication has often led to reduced flowering and even sterility. This is particularly noticeable among the root crops, e.g. in many species of *Dioscorea*, and may reflect the greater diversion of assimilates into the dominant storage sinks. Such sterility has not mattered so long as improvement has focused on clonal selection. However, with greater emphasis now on hybridization programmes, techniques for the induction of flowering in such crops as the yam are needed.

Loss of bitter and toxic substances, enhancement of others

Although most wild cereal and pulse grains could be eaten without special preparation, many of the other plant foods gathered by our ancestors were so unpleasant or toxic that they required treatment of various kinds before they could be eaten. About a third of the plants gathered by Australian aborigines needed pretreatment. Despite this disadvantage, many of these plants were otherwise attractive candidates for domestication. Indeed, their toxicity (and their spines) presumably evolved as protection for large nutritious storage organs, and may even have been enhanced by unconscious selection in some crops, such as cassava in South America where the tubers are stored in the ground (Lathrap, 1977).

Although selection for the loss of bitter or toxic compounds must have been an early step in the domestication of plants such as the potato, it has not always been necessary. Ricinine in castor bean seeds is not extracted with the oil and has not been bred out. Sugar beet still contains saponins and betaine, whereas these compounds have been selected out of beets for direct human or livestock consumption (Schwanitz, 1966). Cyanogenic glycosides have been reduced in cassava, alkaloids in yams and lupins, polyphenolics in safflower, S-compounds in various brassicas, trypsin inhibitors and haemagglutinins in grain legumes, and gossypol in cotton seeds. Proteinase inhibitors, which discourage predation by insect pests and small animals, still accumulate in

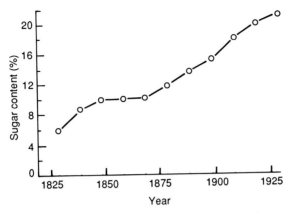

Figure 3.13. Early increase in the sugar content of sugar beet by selection (after Schwanitz, 1966).

potato tubers up to a level of 15% of the soluble proteins. In the wild tomato, *Lycopersicum peruvianum*, they account for up to 40% of the proteins in the fruit during its development, whereas only trace amounts are found in modern cultivars (Ryan, 1988).

Just as unwanted constituents have been selected out, so have wanted ones been enhanced in content. The classic case, which greatly impressed Darwin, was the creation of a new crop plant, the sugar beet, by deliberate selection for sugar content (Figure 3.13). The progressive increase in sucrose concentration was sustained for about 90 years, but there has been little advance since then. Indeed, the sucrose concentration of present-day crops is somewhat lower, partly because they suffer less from nitrogen and water stress (Ulrich, 1955; Loomis & Worker 1963). Sugar yield per hectare has continued to increase in spite of this, because of a continuing rise in harvested root weight. Sugar cane presents a similar picture, with little recent increase in the percentage of sucrose but substantial increase in the weight of cane.

Other constituents have also been enhanced or reduced by sustained selection. Perhaps the best known example among crop plants is the experiment on divergent selection for oil and protein in maize, initiated in 1896. After 70 generations of selection the high- and low-protein lines had reached 215% and 23% respectively of the original level of 10.9%, while the high- and low-oil lines had reached 341 and 14% respectively of the original level of 4.7% (Figure 3.14). Grain yields were adversely affected, especially by selection for high protein content, but these changes are quite remarkable when one considers that these are major constituents of the seeds. Where minor compounds are selected for, greater relative changes can be effected.

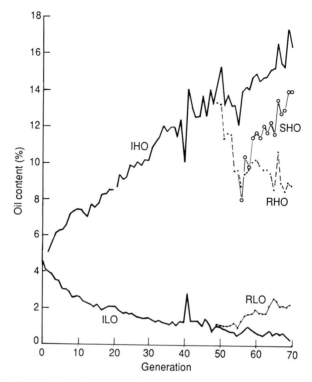

Figure 3.14.   The effect of 70 generations of selection for high (IHO) or low (ILO) oil content in maize, of reverse selections (RHO, RLO) begun after 48 generations and of a second reversal to high oil (SHO) begun after a further 7 generations (Dudley *et al.*, 1974).

Adaptedness to cultivation: crops and weeds

In the early stages of domestication of most crops, their adaptedness to disturbed soil conditions and their ability to compete effectively with weeds would have been crucial. Rapid germination of large seeds, giving large seedlings, would have been advantageous, as would the ability to compete for light above ground and for nutrients and water below. An aggressive growth habit would have been favoured at the outset whereas now, especially with good weed control, communalism rather than competitiveness may be more desirable (Donald, 1981).

Although many wild relatives of crop plants were well adapted to colonizing disturbed ground, the importance of this characteristic for crop plants may have been somewhat over-emphasized. Engelbrecht suggested that potential crops plants were also favoured by the higher soil fertility around stock camps and human habitations, as 'camp followers' and refuse heap plants (Zeven, 1973). The cucurbits, hemp, castor bean, amaranthus and some other crops fit this mould, but many others do not, and there is little evidence that domesticated plants require (as against respond to) higher soil fertility. Among *Hordeum* species, for example, phosphorus deficiency reduced the relative growth rates of the wild progenitor more than it reduced those of the cultivar or weed (Chapin *et al.*, 1989). Among wild and cultivated rice species, the wild progenitors and primitive cultivars were at least as susceptible to nutrient stress or imbalance as the modern cultivars (Cook & Evans, 1983*a*).

However, the fact that there are closely related weeds of many crop plants does suggest that at least these domesticates were naturally adapted to cultivated soil conditions, particularly if Harlan & de Wet's (1965) definition of a weed as an organism adapted to human disturbance is accepted. Many of the major cereals have closely related weedy relatives and, as already noted, introgression, gene exchange and disruptive selection is common amongst them (e.g. Harlan & de Wet, 1965; Harlan, 1969; Brunken *et al.*, 1977; Oka, 1988). In fact, the weedy relatives are among the most troublesome weeds of modern cereal crops for the very reason that their genetic and physiological similarities rule out differential herbicide action (e.g. Cavers & Bough, 1985). Weedy relatives are found not only among the cereals but in many other crops, such as chickpea, cowpea, sunflower, safflower, amaranth, chenopods, castor bean, sugar beet, water melon, pepper, tomato, potato, cassava, radish, lettuce and carrot.

Weeds may be promoted to crop plants when they spread to environments where they are better adapted than the original crop, as in the case of oats and rye as cereal culture spread into some parts of northern Europe. On the other hand, crops may become weeds when they fall from favour, such as mallow, smartweed, a lettuce (*L. denticulata*) and cocklebur in China (Li, 1983).

To what extent domestication requires genetic modifications beyond those considered in the preceding pages can be assessed from experience with more recently domesticated crops. For sugar beet in the nineteenth century and several lupins in Europe in the 1920s, and in Australia in the 1960s (Gladstones, 1970), domestication was rapid whereas adaptation and improvement still continue. Guayule (*Parthenium argentatum*) was partly domesticated as an emergency source of rubber under the exigencies of the World War II. Jojoba, another Sonoran desert shrub, has been a target for domestication in recent years because lipids with properties similar to sperm whale oil constitute about half of the dry weight of its seeds. As it is a long-lived, dioecious shrub, its initial domestication has been mainly by clonal selection (Benzioni & Dunstone, 1986). At the outset, jojoba was viewed as a crop particularly well suited to low-input, drought-stress agriculture, but it is already being grown most profitably under irrigated and fertilized conditions.

Not all plants adapt so readily to agriculture. Among the many plants evaluated agronomically by Earle & Jones (1962) and White *et al.* (1971, 1976) as potential sources of epoxy, crepenynic, erucic and other fatty acids, galactolipids, gums or steroids, quite a few have poor flowering and seed production, heavy loss of seed and conspicuous injury by pests and diseases.

A notable feature of many old domesticates is the variety of end uses to which they have been put. *Hevea brasiliensis*, originally an edible nut, is now a source of rubber. Safflower (*Carthamus tinctorius*), originally a source of dye, has become an oilseed crop, and maize a major source of sweeteners. Such examples, however, illustrate the possibilities of using old crops for new products rather than any need for prolonged domestication.

The continuing use of wild relatives

In 1880 an epidemic of sereh disease of sugar cane in Java led to a search for resistance to the virus. This was found in the wild progenitor, *Saccharum spontaneum*, and by crossing this species with the 'noble' canes (*S. officinarum*) resistant commercial varieties were bred. Most of the world's sugar cane varieties now include germplasm from *S. spontaneum*, which has contributed resistance to several pathogens, and the wild species has been an integral part of sugar cane breeding since then. Indeed, wild relatives have become an integral part of the breeding programs for many crops (Harlan, 1976; Stalker, 1980; Plucknett *et al.*, 1987). Genetic diversity is increasing within the breeding pools of some crops, such as winter wheat (Cox *et al.*, 1986), barley (Peeters, 1988) and tomato (Rick, 1988).

Plant breeders are often reluctant to include wild relatives or even land races or old cultivars in a breeding programme, as Duvick (1984*a*) documents. Considerable time may be required to break the linkage between desired characteristics and undesirable ones. With tomato, for example, it took 14 years of back-crossing to break the linkage between resistance to the root knot nematode, found in the wild *Lycopersicum peruvianum*, and undesirable fruiting characteristics (Rick, 1988). Little use has been made of tropical maize to broaden the genetic base of US maize because of such problems (Holley & Goodman, 1988). It is understandable that plant breeders prefer to cross élite lines whenever possible, especially if these encompass sufficient variation for their purposes, but for many forage, tree and fruit crops there is often no alternative to using wild relatives. All the cards in the deck are wild, as Harlan (1976) puts it, tomato providing an excellent illustration (Rick, 1988). As the techniques for making wide crosses and for transferring single genes are improved, the wild relatives are bound to be used more and more frequently.

So far they have been used most widely as sources of genes for resistance to a great range of pests and diseases, in many crops. As such, they have had a major impact on the maintenance of high yields in many crops. In some cases the needed resistance genes may be quite rare, and their use must be optimized. Resistance to grassy stunt virus in rice, for example, has been found only in one accession of the wild *Oryza nivara*, and that has now been overcome (Cabauatan *et al.*, 1985).

Wild species are increasingly being assessed as potential sources of new cytoplasmic male sterility and restorer systems in wheat, cotton, rice, sugar beet and many other crops in which hybrid vigour may help to raise yield levels still further. They are also being explored for genes that modify the breeding system in other ways, for example by rendering it apomictic and able to fix heterosis. Genes for diploidization have been used to improve seed setting in sorghum.

Various quality characteristics have also been transferred from wild relatives, such as higher protein content in cassava, or more or less of other constituents such as gossypol in cotton, nicotine in tobacco, oil in oats, etc. A wild cotton has contributed greater fibre strength to upland cotton, and many other attributes of quality have been bred into cultivars from wild relatives.

Greater tolerance of environmental stresses has also been transferred, e.g. of potatoes to cold from the wild *Solanum demissum*, *S. acaule* and *S. megistacrolobum*. Several wild relatives have also contributed to the enhancement of heat and salinity tolerance in tomato, and to comparable changes in many other crops, thereby enhancing the breadth of adaptation of modern cultivars.

All these genetic transfers from wild relatives can influence yield but to what extent yield potential can also be enhanced by wide crosses remains an open question. Where genes for short stature have been introduced from wild species, they may have enhanced yield potential, and there are many other ways in which they might do so, depending on the physiological trade-offs involved.

Conclusion

This chapter has been concerned with only the first step in crop improvement, namely domestication from wild to cultivated plants. The next two chapters deal with their adaptation and spread to new environments and with increase in their yield potential. But if we confine ourselves to the domestication process, although it was a large step for mankind it was only a relatively small and simple step at the level of the plants themselves. We have seen that our hunter-gathering ancestors took that step somewhat reluctantly whenever they were forced to and with whatever suitable wild plants were at hand.

Given their knowledge of the plants they gathered, it is not surprising that most of our food, feed, fibre and pharmaceutical crops were domesticated long ago. Only when occasional cultivation was superseded by a more sustained commitment to agriculture could the genetic changes involved in domestication begin. The earliest signs of domestication found so far in the archaeological record appear nine to ten thousand years ago in Central America, the Near East and South China, involving quite different crops and clearly independent cradles of domestication. Other regions were not far behind. If domestication began even earlier in the lowland tropics with vegetatively propagated crops, both the subsequent rise in sea level and the nature of the crops makes the chances of finding traces of them rather slender.

The archaeological evidence presents a picture of many crops undergoing domestication in many places, but at least three well-defined centres are apparent: in Mexico, the Fertile Crescent of the Near East, and North China. Although a diverse portfolio of crops was domesticated in all these centres, each was founded on cereal staples whose reliability, yield and suitability for storage may have been crucial to early commitment to an agricultural way of life and the domestication of other plants.

The genetic changes required for domestication to occur were relatively straightforward and rapid, most of them arising from unconscious selection. To what extent adaptedness to agriculture, e.g. to monoculture, had to be modified is unclear, but recent domestications suggest that this was not a major factor. Taken overall, the evidence in this chapter suggests that de Candolle's opinion, given at the head of this chapter, may have rather over-emphasized the step from wild species to domesticate.

Does not address domestication in terms of env. change @ end Pleistcene

CHAPTER 4

Adaptation and the ecology of yield

'The greatest service which can be
rendered any country is to add an
useful plant to its culture; especially a bread grain ...'

Thomas Jefferson (1821)

Introduction

In the preceding chapter we considered the domestication of crops, straying occasionally into the question of their movement beyond the region of their origin. These movements, the adaptations to new environments which made them possible, and the influence of environmental factors on yield are the subjects of this chapter. Domestication often led to redistribution, and transport away from the cradle has been seen as an essential element of domestication by some. Burkill (1952) regarded this as the factor enforcing conscious sowing or planting. Movement to different environments sometimes freed crops from their usual pests and diseases. It exposed genetic variation not apparent at their cradle, which Stebbins (1950), Darlington (1963) and others regard as an important component of domestication. Transport to new environments also exposed differences in adaptability between crops. Those which, for various reasons, proved adaptable to a wide range of environmental conditions have spread widely to become the staple foods of the world whereas others, like tef, have remained endemic to their areas of origin.

Crop redistribution

The best documented case of the early diffusion of domesticated crops from their original cradle is the gradual progress through Europe from the Near East of many of the plants domesticated there. Early accounts like those of Clark (1965) and Waterbolk (1968) emphasized the episodic nature of the north-westwards penetration, particularly along the major rivers. However, the analysis of the diffusion of wheat and barley by Ammerman &

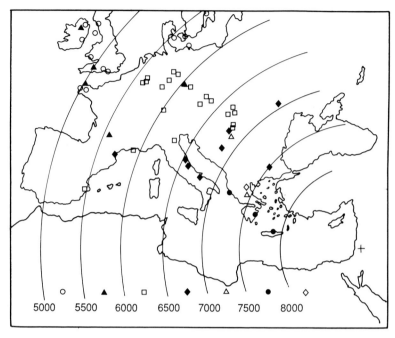

Figure 4.1. The diffusion of wheat and barley into Europe. The symbols indicate the oldest remains (within a range of 500 years BP) found at each site. Adapted from Ammerman & Cavalli-Sforza (1971).

Cavalli-Sforza (1971) implies a slow, relatively uniform spread, averaging about 0.75 km per year over a period of three millennia (Figure 4.1). This was slower than the subsequent spread of the plough across Europe or of the use of copper or iron, leading Cavalli-Sforza & Feldman (1981) to suggest that the movement was not so much of the concept of agriculture as a wave of people, the farmers themselves. Sokal *et al.* (1991) present genetic evidence for such 'demic diffusion'. Renfrew (1988, 1989) argues that the diffusion and diversification of the Indo-European languages was coupled to and made possible by the spread of agriculture. By the mid-Neolithic, the spread of a second agricultural revolution involving the plough, animal traction, slash and burn clearing of forests, and swidden culture may have been driven by population pressure (Howell, 1987).

Hubbard's (1980) analysis of the spread of Near Eastern crops through Europe illustrates not only their north-westerly progress but also the discontinuities and change along the way. Emmer was the predominant wheat throughout the Neolithic dispersal across Europe, einkorn being a minor and often weedy companion, which declined progressively in importance. The hexaploid and free-threshing wheats, by contrast, were a more prominent component of prehistoric agriculture in Europe than in the Near East, especially north of the Alps (Harlan, 1981). Rye and oats also became more prominent with movement to the north-west, whereas the importance of peas and lentils declined. Engelbrecht's (1917) suggestion that rye entered cultivation as a successful weed of wheat crops as they spread northwards is supported by the archaeological data, but occurred four millennia earlier than he supposed. In its eastwards movement wheat had reached Pakistan by the eighth millennium BP (Jarrige & Meadow, 1980) and China before 3300 BP (Ho, 1969).

After AD 1492, many crop plants crossed the Atlantic in the wake of Columbus. The sweet potato (*Ipomoea batatas*) became widely cultivated in the Iberian peninsula, was taken by sea to West Africa and on to the Far East by 1521, and soon after that from Mexico to the Philippines. Burkill (1951) comments that 'possibly no plant has ever girdled the world in cultivation more rapidly than this novelty to Europe', whereas Mangelsdorf (1986) regards maize as having spread the fastest, along with syphilis and tobacco. Clearly, the crops of American origin quickly circled the globe, and many soon began to be grown at higher latitudes.

An indication of their spread can be gauged from the dates of earliest use of their names as recorded in the Oxford English Dictionary. Wheat, barley, oats, rye and peas, among others, are words of Old English origin, while millet, cabbage, kale, and cotton are Middle English. Then come kidney bean (1548), cassava (1555), maize (1565), tobacco (1577), potato (1597), tomato (1604) and eventually, in 1835, peanut. During this period coconut (1555), yam (1588) and sorghum (1597), among others, were also added. Throughout the sixteenth and seventeenth centuries there was great interest in the introduction to Europe of newly discovered crops.

The early tropical botanic gardens played a major role in the distribution and establishment of such crops in tropical countries, as Purseglove (1957) and Plucknett *et al.* (1987) have shown. European botanic gardens, such as Kew, were also actively involved in the redistribution of crop plants such as cinchona and rubber from one region to another, an activity that has been subjected to some rather doctrinaire criticism (e.g. Brockway, 1979). The 'imperial' role played by the Botanic Gardens at Kew, Leyden and Paris in the nineteenth century has since been taken over largely by national Departments of Agriculture and more recently by the international centres of agricultural research. Plant introduction – commended so strongly by Thomas Jefferson in the epigraph to this chapter – has become a well organized and truly international activity.

The modification of environmental responses

As crop plants were taken to higher latitudes or to places with greater seasonal fluctuations in temperature or water availability they were, often unwittingly, selected for better adaptation to their new environments. Greater tolerance of more extreme temperatures, low or high, or of water stress at particular seasons, would often have been a prerequisite for their continued use as staple foods. Otherwise they might be replaced by other crops better adapted to the new conditions, as wheat was replaced by rye and oats in its northwards diffusion into Europe.

In many cases, the adaptation of crops to harsher environments has depended more on changes in the length and timing of their life cycles – allowing them to escape the most adverse conditions – than on changes in ability to tolerate such environments. As the most regular and predictable component of climate, daylength is the most potent and universal controlling element in the timing of life cycles of both wild and cultivated plants, and the modification of their responses to daylength has been a major factor in the spread and adaptation of many crop plants.

The readier change in controls on reproductive development than in tolerance of cool temperatures for growth may be one reason for the success of so many Andean domesticates in spreading to higher latitudes, where they required adjustment mainly to the longer days of northern summers. However, adaptation to cooler temperatures has been significant in the northwards spread of such crops as rice and barnyard millet. Further adaptation of many crops to higher latitudes will no doubt continue, especially as global temperatures rise. However, the adaptation of crops of temperate origin to lowland tropical conditions, to which relatively few staple food crops are well adapted, may be a greater challenge. Potatoes, maize, wheat and other crops are being explored for improved performance under lowland tropical conditions (e.g. Mendoza & Estrada, 1979; Fischer, 1985*b*; Villareal *et al.*, 1985), in a reversal of the adaptive changes that have dominated their history to date.

Growing season, crop duration and yield

Water supply and temperature, sometimes aided by the build-up of pests or diseases, determine the maximum length of the growing season, the range of crops that can be grown, their relative importance, their optimum duration and their likely yield (e.g. Dennett *et al.*, 1981*a*). In Greek mythology Procrustes was the brigand who either stretched or trimmed his victims to match the length of his bed, and there is a strong Procrustean element in the adaptation and improvement of crops (Bunting, 1975). This is apparent in the

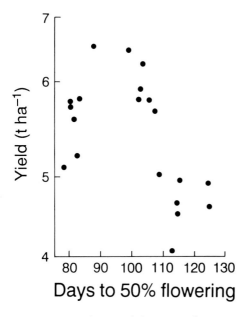

Figure 4.2. Influence of the time to flowering on the yield of rice cultivars of various duration at Hyderabad. Adapted from Venkateswarlu *et al.* (1969).

optimum durations for yield found in many crops. An example with rainfed rice in India is illustrated in Figure 4.2; many others could be cited, for example for wheat (Woodruff & Tonks, 1983; Stapper & Fischer, 1990), sorghum (Done & Muchow, 1988) and pigeon pea (Saxena, 1987).

The longer a crop is able to grow, the greater is its biomass, as illustrated by comparisons between different crops grown in one environment (Figure 4.3) and between different varieties of one crop in one environment (Figure 4.4) (cf. also Dingkuhn *et al.*, 1991). This increase in biomass with longer duration reflects not only the opportunity for more prolonged interception of photosynthetically active radiation by the crop, but also the greater opportunity for uptake of N and other nutrients, especially under low input conditions (e.g. Vergara *et al.*, 1964; Wada & Cruz, 1989).

The relation between crop duration and economic yield may be similar to that with biomass, as in the case of sorghum in West Africa (Kassam & Andrews, 1975) and many root crops. Often, however, it is quite different, as illustrated in Figures 4.3 and 4.4. In the comparison between rice varieties and hybrids of differing duration, for example, grain yield reached a plateau at the longer durations in spite of their greater biomass. Grain yield often increases with crop duration up to a certain point but what happens beyond that point

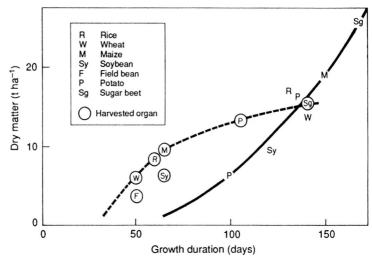

Figure 4.3.   The relation between growth duration and the dry weight of harvested organs (broken line) or total biomass (solid line) among several crops grown in the field at Hokkaido (Tanaka, 1983).

depends on environmental and agronomic conditions. With pearl millet, for example, Bidinger *et al.* (1987) found yield to increase progressively with crop duration in years with only mid-season drought stress, but to decrease progressively as the time to flowering increased beyond 40 days in years when there was a terminal drought stress. Even with wet season or irrigated crops like rice, grain yields may decline as duration increases depending on how the supply of N is managed (Vergara *et al.*, 1966; Wada & Cruz, 1989) and whether long durations are associated with a build-up of pest or disease incidence.

  Whether or not there is a decline in yield per crop with long duration, grain yield per day is usually higher in earlier-maturing varieties and this is of great significance in the irrigated or wet tropics where multiple cropping is possible. Selection under these conditions may therefore be focused on raising the yield potential of earlier flowering varieties rather than on raising yield per crop. The increase in yield per day among tropical rice varieties in recent years has been more than two-fold (Evans *et al.*, 1984; Khush, 1987). The relative rates of change in yield as duration increases can be of considerable significance in mixed cropping systems, e.g. wheat–rice, in determining which crop is better penalised by late sowing.

  Thus, control of the times to flowering and to maturity are important for the yield of both dryland and irrigated crops. The yield of cotton crops grown in the Sudan on a limited amount of stored water rose progressively from 0.6

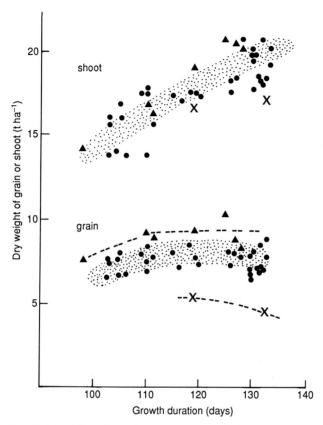

Figure 4.4. The relation between the growth duration of traditional varieties (**X**), modern varieties (**●**) and hybrids (**▲**) of rice and their yields of biomass and grain in the field (adapted from Akita, 1989).

to 2.2 tonnes seed ha⁻¹ as the time to first flowering was reduced from 75 to 43 days (Hearn & Constable, 1984). With hybrid sorghums in Israel, the time to maturity has to be earlier the lower the rainfall on which the crop is grown (Blum, 1979).

Selection for earliness in maize synthetics grown in Minnesota increased yield progressively, by 79% overall for an advance in earliness of 12 days (Troyer & Larkins, 1985). Selection for earliness over a period of 11 years in this Missouri-adapted material was sufficient to shift the area of adaptation northwards by more than 4 degrees latitude, about half the depth occupied by the US Corn Belt. Nevertheless, even at high latitudes the yield of maize may rise as the duration of the vegetative phase within the life cycle is increased (Corke & Kannenberg, 1989), more so than with increase in the duration of grain growth. Thus, not only the total duration of the life cycle but also the optimum balance between its phases varies from one environment to another. Clearly, the successful redistribution of crop plants from their original cradles of domestication would have required substantial shifts in the processes controlling the length and timing of their reproductive and storage cycles, to optimize these at each step along the way. Some of the evidence for such reproductive adaptation will now be considered.

Adaptation to daylength

Reproductive development in plants is controlled more by regulatory than by assimilatory processes. Two important consequences of this state of affairs are that development can proceed relatively independently of growth, and that it can be modified by selection more readily than growth. This is apparent in wild plants, as in the case of kangaroo grass (*Themeda australis*) in Australia: northern populations are short or intermediate-day plants; further south they are long-day plants in the higher-rainfall areas, and even couple that with a requirement for vernalization in the coldest areas; in the arid interior, where rainfall is sparse and erratic, the populations are indifferent to daylength and flower after rain at any season (Evans & Knox, 1969).

No domesticated plant displays such a wide range of response as this but many of them reveal extensive adaptive changes in their response to daylength. Through most of their history as domesticates such changes would have been selected for unwittingly and indirectly, through their effects on survival and productivity. However, since the discovery of photoperiodism, which was made with four crop plants (hops, hemp, soybeans and tobacco), selection has been more deliberate.

Soybean

Because the distribution of the wild progenitor of the soybean in China corresponded with two of the traditional Chinese areas of production (Hymowitz, 1970), no great change in daylength response was initially required. Furthermore, because the latitudes of northern China and the northern USA are similar, some of the varieties introduced from the one to the other in the nineteenth and early twentieth centuries were reasonably well adapted to their new environment (Hartwig, 1973). However, relatively few of the many varieties introduced from China, Korea and Japan have contributed to the modern gene pool (Delannay *et al.*, 1983), those of more northerly origin flowering too quickly in the southern USA.

As a result of the experiments by Garner & Allard (1920, 1930), the varietal differences in sensitivity to daylength for flower initiation were recognized as being crucial to latitudinal adaptation of flowering time and yield. Eventually this led to the subdivision of soybean varieties into 12 maturity groups best adapted to the various zones (Figure 4.5).

Since then several cycles of hybridization and varietal selection have progressively improved adaptation within these zones. Selection for earliness and/or indifference to daylength has proceeded beyond the 00 maturity group, enabling the northwards extension of soybean growing, while increasing interest in the crop by many tropical countries is leading to the better adaptation of group X varieties. Just how different varieties can be in their response to sowing date, reflecting daylength, is indicated in Figure 4.6. As a prime example of adaptive change enhanced by the power of research, the soybean is used as the vignette for this chapter.

Nevertheless, there is still much to be learned about the subtleties of its adaptation. There are several components to the differences between maturity groups. Sensitivity to daylength is particularly marked in the lower latitude groups, but some relatively insensitive varieties may even be found in Group X (Shanmugasundaram, 1981). On the other hand, although complete neutrality to daylength, most valuable at high latitudes, has been found by Polson (1972) and Criswell & Hume (1972) in their extensive screenings of groups 0 and 00, not all varieties in those groups are completely insensitive. Likewise, Nissly *et al.* (1981) found considerable variation in response to daylength among five hundred group III lines, suggesting that much more than daylength response contributes to the maturity group ratings.

Garner & Allard (1930) noted that greater sensitivity to daylength appeared to be associated with later flowering even under the most favourable daylengths, as is apparent in Figure 4.6. However, Wilkerson *et al.* (1989) have found considerable variation in the duration of juvenile insensitivity to daylength among photoperiod-sensitive lines, and inherent earliness is not entirely associated with photoperiodic insensitivity. Of the genes influencing maturity in soybean, E_1 and E_3 control both inherent earliness and response to daylength, whereas E_2 controls only earliness. Group I genotypes are e_1, e_2, e_3 whereas Groups V and above have E_1, E_2, E_3 genotypes (Wang *et al.*, 1987).

Moreover, response to daylength is not confined to flower initiation and open flowering (Johnson *et al.*, 1960; Thomas & Raper, 1976). Exposure of field crops of 8 varieties to short days hastened their flowering by 5 days but their maturity by 17 (Schweitzer & Harper, 1985). Likewise, Nissly *et al.* (1981) found long days to delay maturity much more than flowering was delayed in over five hundred Group III lines. In fact, when the later stages of development are included, few genotypes are completely insensitive to daylength (Shanmugasundaram, 1979), most requiring progressively shorter days at later reproductive stages (Morandi *et al.*, 1988). Although daylength is often the controlling factor in soybean flowering, temperature response also

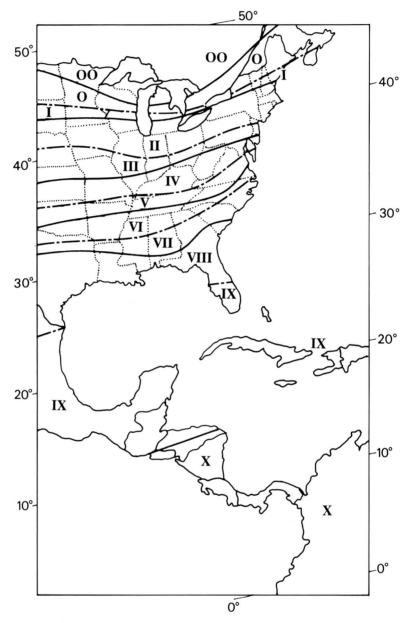

Figure 4.5. Zones of best adaptation for cultivars of Soybean Maturity Groups 00 to X (redrawn from Whigham & Minor, 1978).

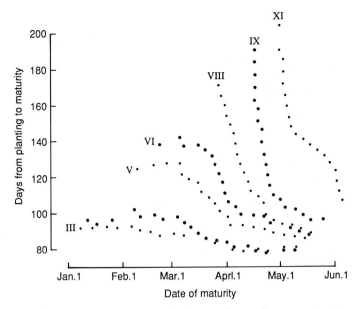

Figure 4.6.   Growth duration of soybean cultivars from six Maturity Groups planted at weekly intervals at Redland Bay, Queensland. Adapted from Lawn & Byth (1973).

plays an important role in the fine-tuning of flowering time (Hadley *et al.*, 1984).

Some other legumes

Although the photoperiodic adaptation of varieties has been most fully explored in soybeans, it has also played an important role in other legume crops (Evans & King, 1975; Summerfield & Roberts, 1985). The mung bean is a short-day plant in which higher-latitude varieties are much less sensitive to daylength (MacKenzie *et al.*, 1975). Photoperiodic adaptation is also apparent in the cowpea, in that both the wild relatives and the domesticates from seasonally arid zones are short-day plants, whereas both wild and domesticated cowpeas from the humid West African tropics are day-neutral and late-flowering. Domesticates from regions with very short and unreliable growing seasons are day-neutral and early flowering (Lush *et al.*, 1980; Lush & Evans, 1981). Daylength control of flowering in the cowpea also extends beyond initiation to the development of open flowers (Lush & Evans, 1980*b*). These effects of daylength are, of course, modified by temperature (Hadley *et al.*, 1983). For example, some apparently day-neutral varieties require short days to flower at very high temperatures, whereas others remain indifferent to daylength (Dow el-madina & Hall, 1986).

As with soybean, all stages of the life cycle of the common bean are sensitive to daylength. Varietal growth habit depends to some extent on daylength and interacts with it in determining the adaptive optima. Water stress soon after the beginning of flowering is far more injurious to the yield of determinate than of indeterminate types, which can produce later flowers and which are generally less sensitive to daylength (Coyne, 1967; Laing et al., 1984). Bean varieties are either day-neutral or short-day plants in which the photoperiod sensitivity for flower development tends to be greater than that for flower initiation (Ojehomon et al., 1968, 1973). Of more than 4000 accessions examined at CIAT in Colombia, about 60% were insensitive to daylength and many others were only slightly sensitive (White & Laing, 1989), including most high latitude varieties, which would otherwise flower too late. These latter come mostly from cooler regions in the tropics and yield poorly in the warm tropics. Daylength response in beans has been modified in ways that are highly dependent on growing-season temperature, as well as being strongly associated with growth habit and seed size (White & Laing, 1989).

Although the groundnut has been considered insensitive to daylength, it is now known to respond to short days during the peg and pod development stages (Bagnall & King, 1991). Lentils provide an example of photoperiodic adaptation among the grain legumes responsive to long days. In this case varietal sensitivity to daylength increases with latitude of origin, and wild forms respond to daylength like cultivars from the same environment (Erskine et al., 1990).

Rice

Most of the wild relatives of rice are responsive to daylength. In Katayama's (1977) survey of over 600 lines, 81% of the *minuta* group and 99% of the *perennis* group were photoperiodically sensitive. Daylength response is clearly important in the adaptation of the wild relatives of rice, and has remained so following domestication: 77% of the 227 old cultivars tested by Katayama were also sensitive to daylength. But in its spread from the low latitudes and altitudes of its origin to occupy areas ranging from 53°N to 35°S, from sea level to 2000 m altitude, and from dryland to deep-water habitats, cultivated rice has been selected for a wide range of responses to daylength.

In deep-water environments, for example, close control of flowering is essential to ensure that it does not occur before the flood level is about to fall, which varies in time from site to site but within narrow limits at each (Figure 4.7). Deep-water varieties are highly sensitive to daylength (Vergara, 1985), as are many traditional varieties from other environments (Figure 4.8).

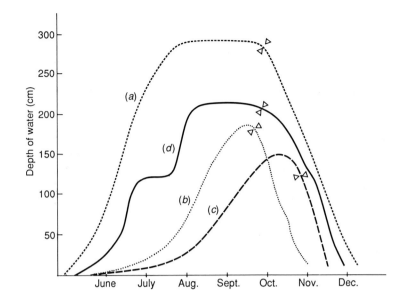

Figure 4.7.   Time of flowering (triangles) in relation to water depth changes for varieties of rice adapted to deep water conditions at (*a*) Bangladesh; (*b*) Illushi, Niger; (*c*) Huntra, Thailand; (*d*) other deep water areas (adapted from Vergara, 1985).

Both wild and cultivated *Oryza* species are short-day plants in which sensitivity to daylength generally increases towards the equator (Vergara & Chang, 1985; Katayama, 1977), although local variations from that pattern are known. Such a response is similar to that of soybeans and other short-day grain legumes except that in rice the control by daylength seems to be exerted almost entirely on the time to inflorescence initiation. Later stages are less sensitive in rice, whereas they are more so in the legumes; but like the legumes, adaptation of rice to higher latitudes has been largely associated with the loss of sensitivity to photoperiod.

In the tropics, response to daylength not only optimized the time of flowering in relation to water supply and irradiance during grain growth, but also made possible long periods of growth before flowering, allowing time for the crop to accumulate nutrients from low-fertility soils. With irrigated, high-input agriculture, however, faster crop turnaround at all seasons is more important, and day-neutral varieties are advantageous for this. Much emphasis has therefore been placed in recent years on the breeding of varieties less sensitive to daylength and more suited to out-of-season irrigation and multiple cropping. A survey of response to daylength among varieties used in the Philippines at various times this century found that there have always been some insensitive varieties but their representation has increased greatly since

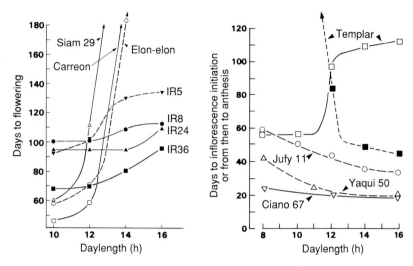

Figure 4.8. The effect of daylength on (left) time to flowering in several rice cultivars (data from Vergara & Chang, 1985); (right) time to inflorescence initiation (open symbols) or from then to anthesis (solid squares) at 21/16 °C in several wheat cultivars (data from Bush & Evans, 1988; Evans, 1987*b*).

the 1950s (Evans *et al.*, 1984). For some rice environments, such as unfavourable rainfed or upland, sensitivity to daylength is still advantageous through its effect on yield stability under adverse conditions (Poonyarit *et al.*, 1989).

Whereas in soybeans the reduction in daylength sensitivity is associated with greater inherent earliness, in rice it was initially associated with a longer vegetative phase, as in IR8 (Figure 4.8), probably as a result of indirect selection for high yield potential. Recent breeding efforts have focused on raising the yield potential of earlier lines, such as IR36 (Khush, 1987).

Other short-day cereals

Indigenous varieties of sorghum in West Africa yield best in their own locality, where their time of flowering coincides with the average date on which the rains end. At localities where the rains end at different times they perform less well (Curtis, 1968). Presumably the flowering time of these traditional varieties, all short-day plants, has been optimized through differences in their response to daylength.

With spread to higher latitudes, however, more profound modifications were required. The early introductions to the USA grew excessively tall (hence the name of Giant Milo); they flowered too late and were not adapted to cool temperatures. Dwarf and early mutants soon appeared in farmers' fields – being selected and saved in the days of hand harvesting in a way which is unlikely today – and sorghum is now well adapted to higher latitudes. Four maturity genes control the response to daylength in sorghum, tropical varieties having dominants at all four loci. A recessive at any one locus reduces the delay in long days sufficiently for temperate adaptation (Quinby, 1974).

Pearl millet is another cereal in which the short-day response so adaptive in tropical Africa has been muted in the course of spreading to temperate zones. Most pearl millets initiate inflorescences much faster in short days (Begg & Burton, 1971) but there is considerable variation in sensitivity to daylength among them, which is exposed by serial plantings at high latitudes (Ferraris *et al.*, 1973). Even at low latitudes, differences in daylength sensitivity played an important role in adaptation, the Maiwa types of Nigeria being particularly sensitive, whereas only 40 out of 290 accessions from Upper Volta required short days (Burton & Powell, 1968). Although insensitivity to daylength is of value in adaptation to extremely short or unpredictable growing seasons and to high latitudes, sensitivity can provide a mechanism for escaping the coincidence of mid-season water stress with the most vulnerable stages of inflorescence development (Mahalakshmi & Bidinger, 1985), as well as being of value for forage millets by delaying reproductive development. Thus, both sensitivity and insensitivity still have a role in the adaptation of this and many other crops.

Foxtail millet also encompasses several kinds of developmental response according to Takei & Sakamoto (1989). Low latitude varieties need short days for inflorescence initiation after a fairly long period of juvenile insensitivity. Higher-latitude varieties in Asia have a shorter juvenile period but still need short days for inflorescence initiation, though fewer of them. In European varieties, by contrast, the length of the vegetative phase depends on the rate of growth, and short days are not required for initiation although they accelerate development of the panicle.

Maize has been shown by many investigators, beginning with Garner & Allard, to be a short-day plant, like teosinte, with varying degrees of sensitivity to daylength. Francis (1972) concluded that the critical daylength for maize is about 14.5 h, and therefore that daylength played no role in the adaptation of maize within the tropics, temperature being the effective factor. Working with 16 races of tropical origin, Stevenson & Goodman (1972) found that all of them – whether from low or high altitudes – were substantially delayed by 12 h compared with 9 h days. Others (e.g. Hunter *et al.*, 1974) have found the time to inflorescence initiation or emergence to be influenced by daylengths within the tropical range. On the other hand, some early lines adapted to high latitudes are indifferent to daylength (Rood & Major, 1980).

As in the case of soybeans and rice, some of the genes for early flowering in maize may be separate from those controlling the response to daylength (Russell & Stuber, 1983). When grown in the tropics, lines adapted to high latitudes may flower too early to give an acceptable yield (Aitken, 1977), whereas tropical lines flower or mature too late to be useful at high latitudes (Holley & Goodman, 1988). Lines adapted to high latitudes are less responsive to daylength but more responsive than tropical lines to temperature in their flowering behaviour (Hunter et al., 1974).

Wheat

The wild progenitors of the three diploid genomes in wheat are all of middle latitude origin, flowering faster in long days and often responsive to vernalization. Durum and bread wheats have similar responses, but varieties adapted to high latitudes commonly show a more pronounced long-day requirement whereas those adapted to low latitudes, such as Yaqui 50 and Ciano 67 in Figure 4.8, have a reduced response to both long days and vernalization (e.g. Syme, 1968; Wall & Cartwright 1974; Hoogendoorn, 1985b). Although long days accelerate both the initiation and the development of inflorescences in most wheat varieties, there are some winter wheats from high latitudes, such as cv. Templar in Figure 4.8, in which the initiation of inflorescences is much faster in short days, even though their subsequent development requires long days (Evans, 1987b), making them short-long-day plants.

The requirements for vernalization and/or long days serve to delay inflorescence initiation until the risk of frost injury to the young inflorescence has passed. However, many Arctic grasses initiate their inflorescences before winter, as may the short-long-day varieties of winter wheat mentioned above. Presumably it is not the newly initiated inflorescence that is most susceptible to winter freezing but a later stage when the inflorescence is raised above ground level (Single, 1961).

Thus, there is a need to avoid too-early flower initiation but this is often coupled with a need also to avoid late flowering because of subsequent drought or heat stress, as in Mediterranean climates. For many wheat-growing environments, therefore, there is a well-defined optimum date of anthesis. Nevertheless, improvements in agronomy and early crop growth have fostered a trend for the date of anthesis to be earlier in more recently bred varieties, as found by Austin et al. (1980a, 1989) for British winter wheats, and by Syme & Thompson (1981), Siddique et al. (1989a), Loss et al. (1989) and Richards (1991) for Australian and Mexican cultivars. The shorter time from sowing to ear emergence has been coupled with a longer interval from ear emergence to maturity, and therefore for grain growth.

Even in environments without severe late stresses, however, the range in flowering times is quite small (Austin *et al.*, 1989) – as also among US barley varieties (Wych & Rasmusson, 1983) – so presumably the optimum balance between the pre- and post-anthesis stages is quite narrow and selection for yield may have led indirectly to particular combinations of vernalization and daylength responses. Several studies have compared the yield of nearly isogenic lines of wheat differing in their sensitivity to daylength (e.g. Lebsock *et al.*, 1973; Busch & Chamberlain, 1981; Busch *et al.*, 1984; Knott, 1986), and have found that greater sensitivity tends to be associated with higher yield, particularly at higher latitudes (but cf. Marshall *et al.*, 1989). When the sensitivity to daylength is finely tuned to local conditions, the yield advantage is likely to be even greater.

The response to vernalization by low temperatures is also adaptive, as suggested by the finding of Rao & Whitcombe (1977) that the vernalization requirements of Nepalese wheats and barleys are closely related to the altitude of the village of origin. A vernalization response appears to be most characteristic of wheat varieties from middle latitudes (Syme, 1968; Wall & Cartwright, 1974; Hoogendoorn, 1985*b*), where it may play a more significant role than response to daylength in regulating the time of flowering (Syme, 1973), although Kato & Yamashita (1991) found inherent earliness and daylength response to be more important among varieties of middle latitude origin. The ability to acclimate to cold is often associated with a vernalization response (Gott, 1961) but not all winter wheats have a requirement for vernalization (e.g. Gotoh, 1976), indeed, some spring wheat varieties have a more pronounced response to vernalization than do some winter wheats. Varietal differences in inherent earliness may be just as important in distinguishing spring from winter wheats, as first suggested by van Dobben (1965) and more recently by Hunt (1979), Ford *et al.* (1981) and Masle *et al.* (1989). Hoogendoorn (1985*a*) has implicated chromosomes 3A, 4B, 4D, 6B and 7B in inherent earliness, whereas chromosomes 2B and 2D are most involved in the daylength response and 5A, 5B and 5D in vernalization. Low-latitude wheats are not only less sensitive to daylength but also have more genes for inherent earliness (Hoogendoorn, 1985*b*). However, the breeding of varieties more suited to the lowland tropics may well require the incorporation of daylength or juvenility responses which have the effect of delaying their reproductive development (Midmore *et al.*, 1982).

Some other crops

The previous examples suffice to illustrate the crucial role of changes in response to daylength or vernalization in the adaptation of flowering time to different environments. However, there are other crops in which the selection for daylength response is focused on the yield of roots or tubers rather than on flowering.

Among the *phureja* and *andigena* potatoes, the initiation of tuberization is greatly delayed by long days (Mendoza & Estrada, 1979). The first potatoes taken to Europe were of the subspecies *andigena* and matured too late. As a result of local selection, particularly following the introduction of ssp. *tuberosum* lines, cultivars better adapted to long days were developed (Salaman, 1949; Hawkes, 1967). Although responding to short days for tuber initiation, potato varieties such as those from India may nevertheless show greater tuber growth and faster flowering in long days (Purohit, 1970). As with the initiation of flowers, that of tubers may also be dependent on temperature for its relation to daylength, tolerance of long days being greatly reduced at higher temperatures (Went, 1959).

Cassava varieties also differ in the effect of daylength on the yield of roots (Cock, 1984). The number and size of storage roots is greatest in short days (Hunt *et al.*, 1977), whereas top growth and forking is greatest in long days. To what extent the adverse effect of long days on root growth and harvest index merely reflects their promotive effect on top growth, and to what extent the differences between cultivars are adaptive is still not known.

Clones of *Saccharum spontaneum*, the wild progenitor of sugar cane, like collections of the wild progenitor of rice, require progressively shorter days for flowering the nearer their origin is to the equator (Moore, 1987). Sugar cane initiates inflorescences only during a 3 week period near the autumn equinox when the daylength is about 12–12.5 h. The later stages of inflorescence development are equally sensitive but have progressively shorter optimum daylengths (Midmore, 1980). As with sugar beet, flowering in sugar cane crops is unwanted because it reduces the yield of sugar, partly by bringing leaf appearance to an end and partly by remobilizing stored sugars into the inflorescence. Losses in yield of up to 45% in the main crop and up to 26% in ratoon crops have been recorded (Rao, 1977), but are usually less. Selection for non-flowering has therefore been prominent in most breeding programmes and has now reached the point where flowering is so rare and recalcitrant that further breeding is difficult. In crops harvested for their leaves, such as tobacco and many vegetables, the daylength requirements have often been enhanced so as to delay or prevent flowering and sustain leaf growth.

Conclusion

The examples considered above surely indicate the flexibility of adaptive response in the processes controlling the reproductive cycles of crop plants. Sensitivity to daylength has been muted in some and enhanced in others. Although flower initiation is often the most sensitive control point, all stages of the reproductive cycle may be responsive to daylength, and in several grain legumes the later stages may be the most sensitive. The response to daylength may even be reversed from one developmental stage to the next, as in some winter wheats. Although there has been much recent emphasis on the advantages of daylength neutrality, photoperiod sensitivity still has an important role to play in many environments, especially in those with fairly predictable adverse seasons.

The optimization of flowering time in temperate environments involves a complex and not wholly understood interplay between inherent earliness, response to vernalization and sensitivity to daylength. In some crops, such as soybean, maize and cotton, earliness has been associated with relative indifference to daylength, but this is not the case in rice and cowpea. Inherent earliness and photoperiod sensitivity can be independently inherited. As agronomic advance makes faster crop development possible, increasing the scope for multiple cropping in the tropics and for extending single cropping to higher latitudes, it is likely that selection for inherent earliness, which has already played a significant part in the adaptation of soybean, rice, maize, wheat, cotton (Wells & Meredith, 1984a) and other crops, will become more prominent in the future.

Adaptation to temperature

At the end of his first outline of what was to become *The Origin of Species*, Darwin listed 'non-acclimatization of plants' as one of the difficulties for his theory. Subsequently he sought, through the medium of *The Gardener's Chronicle*, to obtain evidence of the adaptation of the common bean to the cooler temperatures of Europe. His efforts came to nought. Alphonse de Candolle (1882) subsequently wrote:

> 'I have observed not the slightest indication of an adaptation to cold. When the cultivation of a species advances towards the north (maize, flax, tobacco etc) it is explained by the production of early varieties, which can ripen before the cold season, or by the custom of cultivating in the north in summer species which in the south are sown in the winter'.

This judicious summary underlines the greater role of changes in earliness in escape from low temperatures but it under-estimates the agricultural significance of even small changes in low temperature tolerance.

Consider the beans for which Darwin sought evidence. The optimum temperature for photosynthesis in the high altitude cultivar Diacol Andino is 10 °C lower than in the mid-altitude cultivar Diacol Calima, and it performs better at cool but less well at high temperatures than the mid-altitude cultivar (Laing *et al.*, 1984). That such adaptation extends through growth to yield is suggested by the results in Figure 4.9a from an experiment in which 250 accessions and advanced lines were grown at each of five altitudes with

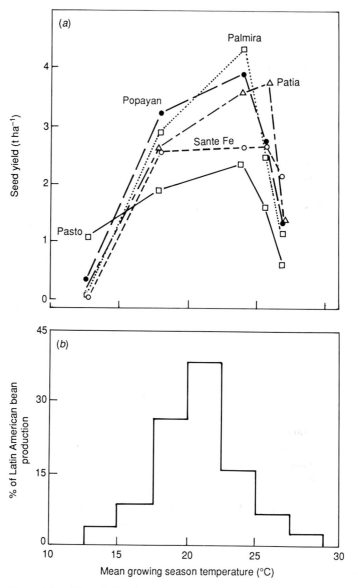

Figure 4.9.   The relation between mean growing season temperature and the yield and distribution of beans (*Phaseolus vulgaris*). (*a*) Effect on the average yield of the five highest-yielding lines from each of the five locations/altitudes. (*b*) Effect on the distribution of bean production in Latin America. (Adapted from Laing *et al.* (1983).)

irrigation, fertilizer and crop protection. The effect of location mean temperature on the mean yield of the five most productive lines at each altitude indicates that the genotypes best adapted at Popayan (1850 m) yielded well at Palmira (1000 m) and vice versa, producing 3–4 t ha^{-1}. However, the lines best adapted to Pasto (2710 m) gave low yields at higher temperatures, while those best adapted to Santa Fe (350 m) performed poorly at cool temperatures. Adaptation to both cooler and warmer temperatures has occurred, but it is limited; the cultivation of beans in Latin America is largely confined to fairly optimal temperature regimes (Figure 4.9b) where yields are highest (see Figure 1.6).

De Candolle was significantly, yet not entirely, wrong. It really depends, as so often in debates about adaptation, on the scale of change under consideration. On the grand view of human omnipotence, we have not made radical changes to the temperature adaptation of the common bean. The four domesticated species of *Phaseolus* still occupy their distinctive niches: *P. coccineus* in cooler, moister conditions than *P. vulgaris*; *P. lunatus* in more tropical habitats; *P. acutifolius* in drier ones. Smartt (1976) concludes that 'domestication has not really changed their climatic preferences or tolerance.' On a more modest view of the possibilities, however, even small adaptive changes in temperature tolerance can have considerable agricultural significance, as shown in Figure 4.9a, and it seems likely that local varieties are often quite closely adapted to the climatic conditions for which they were bred. For example, the analyses by Thompson (1969, 1970) of the effects of Corn Belt weather on the yield of both maize and soybeans suggest that the mid-season temperature optima for both crops lie close to the average Corn Belt conditions despite their disparate origins.

The temperature range for germination in cowpea cultivars is broader than in its wild relatives (Lush *et al.*, 1980), and there have been strong selection pressures for greater cool tolerance of both germination and early growth in crops such as soybean, cotton, rice, maize and sorghum, to get them away to an earlier start. Although the optimum temperature for germination is higher in soybeans than in several of its wild relatives, such as *Glycine argyrea* and *G. clandestina*, germination and seedling growth at low temperatures are nevertheless faster in some cultivars than in the wild species (Kokobun & Wardlaw, 1988).

Selection for improved germination at cool temperatures – often associated with a higher content of unsaturated fatty acids, as in sunflower, cotton and tomato (e.g. Patterson, 1988) – has been effective in many crops. In some, such as pearl millet, germination and seedling growth rate are correlated (Mohamed *et al.*, 1988) but in others seedling growth at cool temperatures has not been improved even when germination has been (Blum, 1985). Early growth in maize hybrids at cool temperatures has been improved (Hardacre

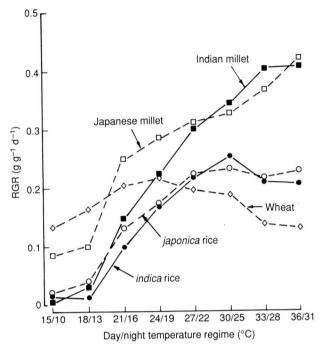

Figure 4.10. The effect of temperature regime, under high natural irradiance, on the relative growth rate of several C_3 and C_4 cereals between dry weights of 80 and 300 mg per plant (adapted from Evans & Bush, 1985).

& Eagles, 1986; Derieux *et al.*, 1987), as has the adaptation of photosynthesis and growth to cool nights at later stages (Dwyer & Tollenaar, 1989).

Evidence of adaptation of growth rate to cool temperatures in two other cereals, barnyard millet and rice, is illustrated in Figure 4.10. The superiority of the C_4 millets and of rice over wheat at high temperatures is apparent, but growth of these tropically adapted cereals was slow by comparison with wheat at the two lowest temperature regimes. Nevertheless, Japanese millet and *japonica* rice were better adapted to growth at low temperatures than their Indian counterparts. Although the differences in RGR may seem small, they have probably been of considerable significance in extending the spread of these two cereals to higher latitudes.

There are many differences between *japonica* and *indica* rice but, given the wide range of environments now occupied by each of them, clear distinctions in their responses to temperature are not always apparent. For example, some *indica* rices can germinate at low temperatures as well as or better than *japonica* varieties (Ikehashi, 1973). Adapted rates of photosynthesis in *japonica* may be somewhat higher than those in *indica* at cool temperatures and lower at warm ones (Evans & Bush, 1985). Tillering is less inhibited by high temperatures in *indica* (Oka, 1955), and grain ripening may be better adapted to high temperatures in *indica* and to low temperatures in *japonica* (Yoshida, 1981). Susceptibility to injury by cool temperatures at meiosis and anthesis is more prevalent among *indica* varieties.

Rice growing was extended northwards in Japan to the Hokkaido district not much more than 100 years ago. In a quarter of those years low temperature injury has substantially reduced yields (Satake, 1976) and selection for greater tolerance to several types of low-temperature injury has contributed significantly to yield improvement in the area (Tanaka *et al.*, 1968). Yet, paradoxically, because seedlings are now grown in protected nursery beds, traditional varieties may germinate better than recent ones at low temperatures (Toriyama, 1962).

Cool tolerance has also been improved in *indica* rice, but low temperatures still limit the use of modern varieties over large areas of S and SE Asia. In some places they are most adverse during the seedling or tillering stages, in others during flowering or grain ripening. Thus they may limit many different, independently inherited processes (Kaneda & Beachell, 1974; Munakata, 1976), and determining which limitation is operative can be a complex task. For example, although Huang *et al.* (1989) found cool temperatures during grain growth to increase the light-dependent injury to photosynthesis in rice, yield reduction was due more to feed-back inhibition of photosynthesis following impaired translocation. Different varieties may be adapted to different times of stress, and hybrids can display considerable advantage (Kaushik & Sharma, 1986). In cotton, likewise, modern varieties yield relatively better at cool temperatures (Thomson, 1986).

Adaptation to more extreme cold, and to freezing, has also occurred during the movement to higher latitudes of various crops, such as wheat, particularly following the introduction of the D genome into bread wheat. The process continues, as in potatoes (Mendoza & Estrada, 1979) and many other crops (Blum, 1985).

Improved adaptation to high temperatures is also apparent in many crops. Wild progenitors and land races from warmer environments tend to be less adversely affected by high temperatures – as shown by Smillie *et al.* (1983) among potatoes collected at various altitudes in the Andes – and are being used in attempts to breed potatoes that are better adapted to lowland tropical conditions (Mendoza & Estrada, 1979).

Better adaptation to hot conditions is also being sought in wheat. Varietal differences in heat avoidance through control of flowering time, glaucousness and leaf inclination contribute, as well as differences in heat tolerance.

Varietal tolerance of high temperatures may vary from one reproductive stage to the next; some of the wheat varieties most sensitive during the booting stage are least sensitive during grain growth (Wardlaw *et al.*, 1989*a*). Although European varieties were the most adversely affected by high temperatures, and Philippine varieties the least, both heat-susceptible and heat-tolerant varieties were found in most wheat regions (Wardlaw *et al.*, 1989*b*).

Tolerance of high temperatures involves many different processes, as has been shown for fruiting in the tomato (Opeña, 1985). Less than 1% of 4000 accessions of *Lycopersicon* tolerated high temperatures at the fruit setting stage (Villareal *et al.*, 1978), but these sufficed for the breeding of a heat-tolerant variety, Hotset.

Adaptation to more extreme temperatures may shift the optimum balance among yield components. Vavilov (1951) suggested that smaller-seeded forms might be better adapted to hot conditions and larger-seeded forms to cool ones, as in the case of the small-seeded *desi* and the large-seeded *kabuli* chickpeas. The extremely large-seeded Cuzco maize, adapted to the high Andes, may be another example, as is the common bean in which large-seeded types are best adapted to cool conditions and small-seeded ones to high temperatures (Laing *et al.*, 1983). However, there are many paths to high yield in any one environment, and Takeda *et al.* (1987) found rice varieties with larger grains to be superior in a warmer environment and inferior in a cooler one, running counter to the examples given above.

Adaptation to irradiance

Tolerance of shade plays an important adaptive role in natural plant communities, but the evidence for varietal adaptation to high or low irradiance is not strong. *Sorghum arundinaceum* from low-irradiance equatorial forest habitats has photosynthetic characteristics that confer better adaptation to low light than those of other sorghums (Downes, 1971). Among rice varieties, some are better adapted to low irradiance than others (Oldeman *et al.*, 1987; Murty & Sahu, 1987; Dash & Rao, 1990).

Greater shade tolerance may be an important component of the better performance in high density plantings being selected for in maize (e.g. Duvick, 1984*b*; Russell, 1984) and other crops, but greater tolerance of competition for water and nutrients may also be involved. Tollenaar (1991) found the superiority of the more recently released maize hybrids in Ontario to be associated with delayed leaf senescence and greater biomass production after anthesis at high planting densities.

To the extent that it leads to better performance at high planting densities, shade tolerance may become more significant in the future adaptation of crops. Another matter likely to attract more attention in the future is whether varieties have been unconsciously selected for particular seasonal sequences of irradiance, e.g. for a certain balance between vegetative and reproductive growth.

Adaptation to water stress

In approaching the subject of adaptation to water stress, we should recognize that one important role for many of the foods now harvested by mankind – such as seeds, fruits, tubers, storage roots and other reserves – was to tide the wild progenitors over seasonal periods of stress. Their life cycles were preadapted to escape from prolonged periods of water stress, and in many cases human selection has reinforced natural selection.

Escape from water stress through shorter or better-timed life cycles remains the most effective form of adaptation (e.g. Boyer & McPherson, 1975; Turner, 1979; O'Toole & Chang, 1979; Blum, 1985; Derieux *et al.*, 1987; Rao *et al.*, 1989; Siddique *et al.*, 1990*b*; Richards, 1991). Among the cereals, drought stress often has its greatest impact on yield when it occurs during meiosis and anthesis, as indicated for rice in Figure 4.11. Comparable results have been obtained with wheat (Fischer, 1973) and maize (Shaw, 1977; Bruyn & de Jager, 1978). Adjustment of flowering time to minimize the chances of drought stress at that stage has been an important element in the adaptation of many cereals, e.g. pearl millet (Mahalakshmi & Bidinger, 1985).

Avoidance of dehydration derives from the ability to maintain high turgor during conditions of water stress by either an increased water potential or a reduced osmotic potential associated with osmotic adjustment. Varieties may differ greatly in their leaf water potential under particular stress conditions, e.g. in wheat, rice, millet, sorghum and soybeans (Blum, 1985), owing to differences in such characteristics as root growth patterns, hydraulic conductivity, glaucousness and inclination or movement of leaves.

O'Toole & Bland (1987) review the many variations in root systems that have been found in plant breeding programmes for water-stress environments. Greater or deeper root growth can help to maintain leaf water potential during water stress (O'Toole, 1982; Yoshida & Hasegawa, 1982) and has been associated with adaptation and improvement in upland rice and other crops. Among old and new soybean varieties, Boyer *et al.* (1980) found greater root density and higher leaf water potential in the more recent ones.

Bean varieties adapted to dry conditions avoid stress compared with less adapted ones, their smaller reductions in yield being associated with smaller rises in leaf temperature (Laing *et al.*, 1984). Water loss and/or leaf temperature can be reduced in many ways e.g. by greater cuticular resistance

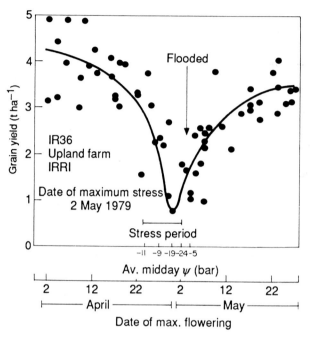

Figure 4.11. Effect on the yield of IR36 rice of the timing of a period of water stress in relation to the time of flowering as shifted by planting date (O'Toole, 1982). Crops which flowered on 2 April were near to maturity during the stress period, whereas those flowering on 22 May were near to panicle initiation during the stress period.

(Yoshida & Reyes, 1976), paraheliotropic leaf movements (Begg & Torssell, 1974), leaf rolling (O'Toole *et al.*, 1979) or the development of epicuticular waxes. Glaucousness in barley and wheat has been found to increase grain yield by 7–16% in dryland environments (Baenziger *et al.*, 1983; Johnson, D.A. *et al.*, 1983), associated with greater water use efficiency (WUE) and lower leaf temperatures (Richards *et al.*, 1986).

The trade-offs between leaf area and stomatal conductance complicate comparisons of WUE as much as comparisons of CER, because the water use term in WUE is sometimes restricted to loss by transpiration but usually includes both transpiration and the loss of water by evaporation from the soil surface. The first usage equates to the transpiration efficiency (Tanner & Sinclair, 1983), whereas the second is also greatly influenced by leaf area growth and the extent to which it reduces evaporation from the soil. Just as the ranking of varieties for canopy photosynthesis can be very different from those for CER, so WUE rankings can be very different from those for transpiration efficiency because the evaporation term can account for 40–75% of the total water loss (Richards, 1991). Unfortunately, the confusion of usage persists, and infects the following discussion in which WUE is used in the various ways defined (or not) by the authors.

Many years ago Briggs & Shantz (1913) noted the differences between crops in WUE and suggested selection for more efficient water use. While there is no doubt about the greater WUE of C_4 compared with C_3 plants, Ludlow & Muchow (1990) query whether there are any valid examples of differences in WUE between cultivars in the field. One factor inhibiting the search for such differences has been the lack of a readily usable selection criterion until Farquhar & Richards (1984) showed that WUE in wheat varieties is negatively correlated with their discrimination against ^{13}C in photosynthesis, as also among varieties of groundnut (Hubick *et al.*, 1986), barley (Hubick & Farquhar, 1989), tomato (Martin & Thorstenson, 1988) and in several cool-season forage grasses (Johnson & Bassett, 1991). Carbon-13 discrimination can therefore be used to compare cultivars for WUE. Grain yield has been positively correlated with varietal WUE in a few cases, e.g. sorghum (Hubick *et al.*, 1990) and in barley under irrigation (Craufurd *et al.*, 1991). In others, however, the relation is tenuous, as with groundnuts (Hubick *et al.*, 1988; Wright *et al.*, 1988) or even apparently negative, as with wheat (Condon *et al.*, 1987; Ehdaie *et al.*, 1991), rice and cowpea (Austin *et al.*, 1990) and dryland barley (Craufurd *et al.*, 1991). Sorghum hybrids differed in WUE under well-watered conditions but not at low stomatal conductances (Kidambi *et al.*, 1990). Thus, although greater transpiration efficiency may be adaptive in terms of biomass production under dry conditions, it may not always be so in terms of the yield of grain.

Counter-intuitive associations between seemingly desirable characteristics and crop yield have confounded others in pursuit of drought avoidance, as H.G. Jones (1977) found with selection for low stomatal density in barley. Passioura's (1972) suggestion that selection for smaller xylem vessel diameter in the seminal root, thereby increasing their hydraulic resistance, might be of value in drought-prone crops seems more hopeful. Wheat lines with xylem vessel diameters reduced from about 65 μm to less than 55 μm have been found to yield 3–11% more grain than unselected controls in the driest environments (Richards & Passioura, 1989). The improvement of WUE in crops remains an important objective for many reasons, and the negative associations between WUE and yield merit closer analysis, for which Passioura (1977) has formulated a fruitful approach.

Osmotic adjustment may also enhance the ability of certain crops or varieties to avoid drought stress, and to do so at little metabolic cost while also promoting recovery, at least in sorghum (McCree *et al.*, 1984). Morgan (1977) showed that varieties of wheat differ substantially in their capacity for osmotic adjustment as flag leaf water potential decreases. Varietal differences in this capacity were positively correlated with grain yield of wheat over a range of environments (Morgan, 1983; Morgan *et al.*, 1986), associated with greater water use particularly at depth (Morgan & Condon, 1986). Screening for osmotic adjustment can be done at the seedling stage in wheat (Morgan, 1988), and it is apparently controlled by a single gene (Morgan, 1991). Santamaria *et al.* (1990) & Ludlow *et al.* (1990) found late-flowering sorghum lines with high osmotic adjustment to have a considerable yield advantage with both pre- and post-anthesis water stress, whereas Flower *et al.* (1990) see little value in selecting for greater osmotic adjustment capacity in sorghum in severely stressed environments. Among lines of peas, yield was closely related to the extent of osmotic adjustment in two dry years but not in one wet one (Rodriguez-Maribona *et al.*, 1992).

Both *escape* from and *avoidance* of water stress involve many components and have contributed a great deal to the adaptation of crops to dry environments. It is less clear to what extent greater *tolerance* of dehydration, i.e. the ability to sustain less injury when turgor is lost, has also contributed to adaptation and been manipulated by plant breeding. Greater tolerance of the higher tissue temperatures associated with water stress is likely to be an important component of dehydration tolerance. Blum *et al.* (1989) found a close relation among wheat varieties between canopy temperature at mid-day and susceptibility to loss of yield under water stress. This suggests that varietal differences in the tolerance of high temperatures may be small, although marked differences between varieties of sorghum have been measured (Sullivan, 1972). Differences in membrane stability, proline or betaine accumulation and many other characteristics may be significant – and the list of hopeful candidates is expanding rapidly in this heyday of molecular plant breeding – but it is unlikely that any single factor will provide the master key. The most assured progress may well continue to come from empirical selection for yield in stressed environments given that the integrated and balanced action of many genes is involved in adaptation to drought stress.

Productivity beyond the centres of origin

Comparisons of the yield of crops within and beyond their centres of origin – by Purseglove (1963, 1968), Jennings & Cock (1977), Kawano & Jennings (1983) and Haws *et al.* (1983) – have led to the generalization that the average yield of most crops is higher away from the area of their domestication than within it, as indicated in Table 4.1. The table has not been updated because of uncertainties in defining the centre of origin in the same way.

With the exception of rice and cowpeas, most of the crops listed in Table 4.1 are grown on a far greater area outside their centre of origin. Yields are substantially more outside the centre, owing to several factors, listed below.

Table 4.1. *Yields of various crops within and outside their centres of origin*

Crop	Centre of origin	Within centre of origin		Outside centre of origin	
		Area planted (kha)	Yield (t ha^{-1})	Area planted (kha)	Yield[a] (t ha^{-1})
Wheat	West Asia	26966	1.41	209605	1.82(129)
Rice	South Asia	84199	1.85	58909	3.31(179)
Maize	Mexico through Andean region of Latin America	10241	1.22	106662	3.02(248)
Barley	West Asia	6592	1.33	87383	2.05(154)
Sorghum	North-east Africa	5841	0.70	45894	1.33(190)
Potato	Andean region of Latin America	615	8.20	17869	14.41(176)
Cassava	Northern South America	306	8.48	12216	9.01(106)
Groundnut	South-east South America	124	0.94	17872	0.96(104)
Soybean	China	14236	0.87	30388	1.67(192)
Field bean	Andean region of Latin America	6079	0.50	21821	0.60(120)
Cowpea	Africa	5035	0.20	135	0.70(350)
Sugar cane	South Asia	4923	47.62	7784	58.74(123)
Banana	South Asia	846	11.37	2069	13.23(116)
Coffee	North-east Africa	656	0.27	7266	0.48(178)

Note:
[a] Numbers in parentheses express yield outside the centre of origin as a percentage of that within.
Source: Kawano & Jennings (1983).

Economic development

The areas where most crops originated are developing countries whereas the yields outside the centre of origin are in many cases heavily influenced by the yields in developed countries, where agronomy and crop improvement are often at a much higher level. When the yield comparisons are made between the centre of origin and other developing countries, the difference is greatly reduced, for example when maize yields for Mexico or soybean yields for China are compared with those of all developing countries. A difference still remains, as may be seen for sugar cane, coffee, banana and cassava, which are grown almost entirely in developing countries (92% for cane, 97% for bananas and 100% for coffee and cassava). But the difference is small for cassava, and in the case of groundnuts, crops in the Latin American cradle outyield those in other developing regions.

Environmental conditions

Most crops are now grown at higher latitudes than they were domesticated in; as we shall see for maize, yields are often higher at higher latitudes, reflecting greater economic development and use of inputs but also longer days, greater radiation receipt per day during the growing season, cooler temperatures and more prolonged crop life cycles.

Reduced pest and disease losses

Centres of origin are characterized by diversity not only of the crop plant but also of its pests and pathogens, whose depredations may reduce yields there considerably. Anyone who has compared eucalypt trees in Australia with those elsewhere will recognize how important this factor can be, and both Purseglove (1963) and Jennings & Cock (1977) accord it great significance. Many pests and diseases have now spread beyond the centre of origin, but some regions are still free of those that constrain yields in their original home. Losses from some pests and diseases may actually be greater away from the centre of origin, when the pests are freed from the organisms that control their abundance or where conditions favour them. Epidemics of coffee rust in Sri Lanka, Panama disease of bananas in Central America, stem rust of wheat in North America, late blight of potato in Europe, *Sogatodes* and hoja blanca virus on rice in tropical America, southern corn leaf blight of maize in the USA and downy mildew of maize in SE Asia and Africa have all caused more serious losses of yield than in the region where the crops originated (Kawano & Jennings, 1983).

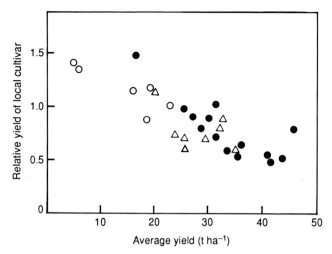

Figure 4.12. The relative yield of traditional local cassava varieties in relation to the average yield of many varieties in experiments at three sites in Colombia, Carimagua (○), Caribia (△) and CIAT (●). Redrawn from Kawano & Jennings (1983).

Vegetative vigour

Jennings & Cock (1977) argue that the survival and success of grain crops in the more tropical centres of origin, where soils are infertile, pests abundant and weed competition severe, depend on vegetative vigour and competitive ability. Beyond the centres of origin, where these adverse factors are less severe, less competitive varieties which invest more of their biomass in storage organs have been selected, contributing to the higher yields. We return to this theme later because of its important implications, one of which is that traditional varieties may out-yield modern ones in poor environments whereas the reverse may be true in more favourable conditions, as illustrated for cassava (Figure 4.12) and for rice by Kawano & Jennings (1983).

Clearly, several distinct elements contribute to the differences in the yield of crops within and beyond their region of origin, sometimes one, sometimes another being the dominant factor.

The ecology of yield

Despite the great significance of the subject and much recent attention from crop modellers, the ecology of yield remains a dark continent on agroclimatic maps, for many reasons. One is that the climatic factors with a major influence on yield, such as radiation, temperature and daylength, often co-vary to such an extent that it is difficult to unravel the effects of each. They also interact in ways which vary from one stage of crop development to another.

The effects of individual climatic factors and their interactions can be analysed to some extent with mini-crop canopies in phytotrons, but comparisons in the field are also needed. Where these involve a run of years, the effects of changing varieties and agronomy must be removed before the climatic analysis can begin. This involves assumptions about the nature of the technology trend, whether linear as in Thompson's (1969, 1970) analyses of Corn Belt yields, exponential as in the analysis by Parthasarathy et al. (1988) of Indian grain production, or stepwise as assumed by Kulkarni & Pandit (1988) for sorghum yields in India.

Data aggregated at a national or regional level are often the most readily available over a period of years but they may encompass a wide range of both climatic and agricultural conditions. On the other hand, where a fairly homogeneous region is considered, the progressively closer adaptation of varieties to that region may result in their apparent optima being close to the regional averages, as Thompson (1969, 1970) found with maize and soybean in the Corn Belt.

Statistical, physiological and modelling methods are all needed in this field, and complement one another, but our understanding remains far less than is needed for the purposes of effective crop management, international trade and global forecasting.

Latitude

The relation between national average yields of maize and latitude, as illustrated in Figure 4.13, highlights some of the complexities in the ecology of yield, in that it reflects both environmental and socio-economic factors. Chang (1981) argues that the four-fold difference in national average yields between the high latitudes and the tropics is much greater than the 2.5-fold difference between them in experimentally attainable yields because it reflects 'the technological inferiority of tropical farming'. He suggests that the difference in experimental yields is largely caused by climatic factors, yields in the tropics being limited by short days and by high night temperatures which shorten all stages of the life cycle. Chang found virtually no correlation between these maize yields and solar radiation and considered the latter a minor factor in explaining yield differences. However, the change in photothermal quotient with latitude, indicated by the sloping lines I have

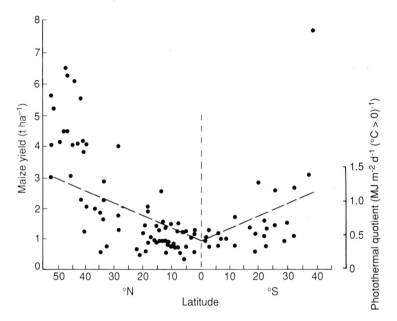

Figure 4.13. National average maize yields (1975–77) as a function of latitude as plotted by Chang (1981) but with country names omitted. The sloping lines have been added to indicate the linear change in photothermal quotient with latitude (Rawson, 1988; cf. Figure 4.21).

added to his figure, suggests that much of the variation in national maize yields is determined by solar radiation and the duration of its interception as influenced by temperature.

The sharp rise in grain yield between latitudes 25 and 35°N for maize in Figure 4.13 is apparent also for wheat crops in the International Rice-Wheat Integrated Trials (Aggarwal *et al.*, 1987). Yields of more than 1 t ha^{-1} can be obtained in the lowland tropics with good management, and potential yields of 3 t ha^{-1} may be obtainable there (Aggarwal & Penning de Vries, 1989).

These examples of grain yields rising with latitude in both C_3 and C_4 cereals are in striking contrast with the relation between annual biomass and latitude, compiled by Loomis & Gerakis (1975) and illustrated in Figure 4.14. At high latitudes the annual accumulation of biomass is, of course, limited by increasingly severe winters, but in the 25–35° latitude range neither maize (points 1–4) nor wheat (points 26–8) nor other crops display a sharp increase in biomass with increase in latitude as their grain yields do. Thus, their low yields in the tropics presumably reflect abbreviated grain growth and lower harvest indices.

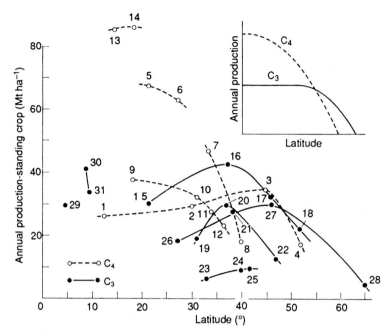

Figure 4.14. Relation between annual biomass production and latitude for various C₃ and C₄ crops, including maize (1–4), sugar cane (5–6), sorghum (7–8), sugar beet (15–18), lucerne (19–22), soybean (23–5), wheat (26–8) and cassava (30–1) (Loomis & Gerakis, 1975).

Solar radiation

As we have just seen, the yield of crops may bear little relation to their growth. Abundant growth is a necessary but not a sufficient condition for high yield. Crop growth may bear a close relation to the amount of radiation intercepted but its yield may not do so.

Although photosynthesis by single leaves is not proportional to irradiance because of light saturation, canopy photosynthesis and dry weight accumulation often are (e.g. Kiniry *et al.*, 1989). Monteith (1977) has illustrated the relation between intercepted radiation and dry matter production by crops of barley, potatoes, sugar beet and apples in the UK (Figure 4.15). The slope of the overall relation for these well fertilized crops without much water stress is equivalent to 1.4 g dry weight per megajoule of total solar energy intercepted, a conversion efficiency of about 2.4%, or 5.3% for the photosynthetically active radiation (PAR). These are all crops with the C₃ pathway of photosynthesis, growing under favourable and fairly similar conditions, and differing mainly in the extent and duration of their interception of solar

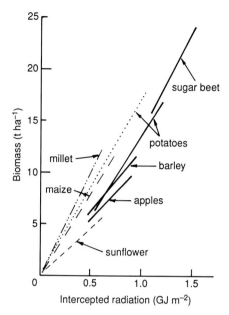

Figure 4.15. The relation between dry biomass at harvest and radiation intercepted throughout the growing season for several crops. Adapted from Monteith (1977, the solid lines) with additional regressions for millet (Craufurd & Bidinger, 1988), maize and sunflower (Kiniry *et al.*, 1989) and potatoes (Jefferies & MacKerron, 1989).

radiation. Added to Monteith's original figure are comparable slopes for potatoes in Scotland, sunflowers in Texas and for two C$_4$ cereals, maize in Texas and pearl millet in India. Linear relations between absorbed PAR and shoot weight (Figure 4.15) or crop growth rate (CGR, Figure 4.16) have been found for many crops (e.g. Fischer, 1983; Gallagher & Biscoe, 1978; Kiniry *et al.*, 1989; Sinclair & Horie, 1989; Muchow, 1989a). The differences between crops in radiation use efficiency when growing in environments to which they are adapted may not be great. Potatoes in Scotland (Jefferies & MacKerron, 1989) or sunflowers in France (Kiniry *et al.*, 1989) may have radiation use efficiencies as high as those of C$_4$ crops. Among three cultivars of cotton, radiation use efficiency did not differ significantly until they began to form bolls (Rosenthal & Gerik, 1991).

Although canopy photosynthesis, CGR and accumulated shoot weight frequently display a linear relation with radiation intercepted by the crop, this is less commonly so for harvested yield. Irrigated rice crops are not subject to the water stress so often associated with high irradiance on dryland crops. In Japan, they have long been grown with high levels of fertilizer and control of

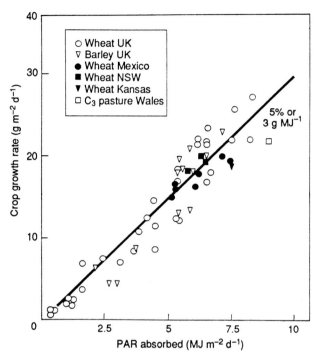

Figure 4.16. The relation between crop growth rate and absorbed radiation for wheat and barley crops and grass pastures in a range of environments (Fischer, 1983).

pests, diseases and weeds, and early investigators found that whereas yield in the northern Prefectures was positively correlated with mean temperature during grain growth, it was often more closely related to radiation during grain growth in the warmer southern Prefectures (Murata, 1975). In summarizing the International Biological Program data for rice in Japan, Kudo (1975) found a high correlation between crop yield and an index combining shoot weight at heading and radiation during the grain growth period, while Nishiyama (1985) obtained even higher correlations between yield and radiation during grain growth.

In the tropics, with improved rice varieties and good agronomy, grain yield was also found to correlate well with radiation during grain growth (De Datta & Zarate, 1970). However, shading experiments by Yoshida & Parao (1976) indicated that radiation during the middle, reproductive stage of the rice crop could be at least as important as during grain growth (Figure 4.17a). In serial monthly rice plantings over a period of ten years, radiation during the reproductive stage had the greatest influence on yield (Evans & De Datta,

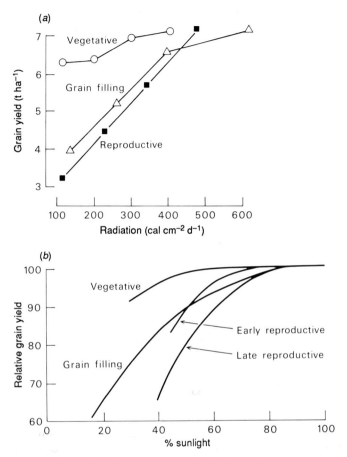

Figure 4.17. The effect on grain yield of solar radiation during three successive stages in the development of crops of (*a*) rice in the Philippines (data of Yoshida & Parao, 1976) and (*b*) wheat in Mexico (data of Fischer, 1975). (Evans & Wardlaw, 1976).

1979). With some times of planting yield was also highly correlated with radiation during the grain growth stage, and even with that during early vegetative growth. Because the correlations also varied with fertilizer level and variety, as well as with time of planting at this one site, they are likely to be less clear cut in comparisons across many sites, conditions, varieties and seasonal sequences in the tropics. This was found by Seshu & Cady (1984) in trials in 40 environments over a period of 5 years, for which the correlation coefficient between yield and radiation during grain growth was only 0.27. This low correlation reflected in part the favourable effect of lower minimum temperatures on yield, higher yields being possible when low irradiance is combined with cool nights (cf. Islam & Morison, 1992). At higher latitudes, on the other hand, cool nights can be the major cause of lower yields (Angus & Lewin, 1991).

With wheat it has proved difficult to establish a strong effect of radiation on yield by using regression or path coefficient analyses (e.g. Puri et al., 1985). However, shading experiments in the field (e.g. those of Welbank et al., 1968), phytotron experiments at different seasons (Evans, 1978b) and crop growth under a range of artificial irradiances (Bugbee & Salisbury, 1988) have all revealed a strong dependence of grain yield on the level of radiation, even up to an irradiance of 2000 μmol PAR m^{-2} s^{-1}, well beyond those experienced by wheat crops in the field.

Shading experiments by Willey & Holliday (1971b) and Fischer (1975, 1985a) show that grain yield in wheat is most sensitive to reduced radiation just before anthesis (Figure 4.17b). However, the overall effect of lower radiation on yield was relatively small; Fischer (1985a) found that, although crop growth rate was reduced in direct proportion to the reduction in radiation, spike growth rate was reduced much less. A pronounced effect of seasonal differences in natural irradiance not confounded by differences in temperature or daylength was found when mini-crops of wheat were grown in a phytotron (Evans, 1978b). In these experiments, grain yield was most highly correlated with radiation during the reproductive phase, ear number with that during the early reproductive phase, grain number per ear with that in the late reproductive phase, and kernel weight with that during grain growth.

Maize yield is also highly sensitive to irradiance. The yield of hybrids planted monthly over a period of four years in Hawaii bore a close relation to average radiation during the growing period, as may be seen in Figure 4.18. As with rice, wheat and sorghum (Fischer & Wilson, 1975b), the yield of maize is particularly sensitive to radiation just before anthesis (Figure 4.19), and grain number per plant is more or less proportional to radiation during the reproductive stage (Fischer & Palmer, 1983, 1984).

A linear relation between absorbed radiation and grain yield has also been found with groundnut crops (Waggoner & Berger, 1987). As with cereal crops, the sensitivity of groundnuts to reduced radiation varied with the stage of the crop, the podding stage being most sensitive, followed by the period of seed filling (Hang et al., 1984). The yield of potato crops has also been shown to be closely related to radiation during the growing season (Sale, 1973; Waggoner & Berger, 1987).

Thus there are quite a few instances where the yield of cereal, pulse and tuber crops has been highly responsive and more or less linearly related to radiation receipt, especially during the reproductive and storage phases of the life cycle. These instances have often come from shading treatments in the

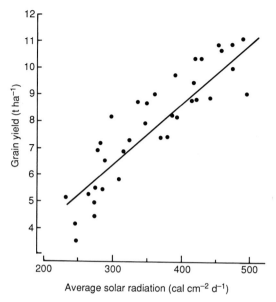

Figure 4.18. Relation between grain yield and average solar radiation in monthly plantings of maize in Hawaii (after Jong *et al.*, 1982).

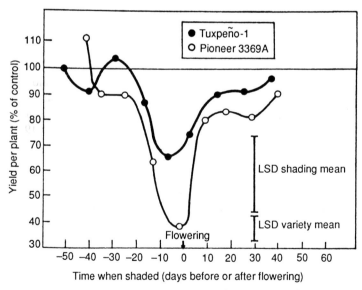

Figure 4.19. Effect on grain yield of single periods of 22 days with 54% shading at different stages of development, for two maize varieties grown at Tlaltizapan, Mexico (Fischer & Palmer, 1983). Least significant differences (at $p \leq 0.05$) of means are shown (bars).

field, phytotron experiments where confounding factors are avoided, or serial plantings in the tropics where associated changes in temperature and daylength are small. At higher latitudes or with crop data from many different environments, multiple regression analysis has often failed to reveal a marked effect of radiation on yield, for example with maize (Chang, 1981) and rice (Seshu & Cady, 1984). In such cases, the correlated variations in other factors such as daylength and temperature and their interactions with radiation may mute or over-ride the effects of radiation on yield.

Temperature

Many physiological processes influence yield and are themselves influenced by temperature. This is not the place to consider the vast literature on the effects of temperature on crop plants, but to focus on the ultimate integration of all these effects in terms of crop yield. At this level the data are sparse. Covariance of other influential components of the environment is one reason, making regression analysis inconclusive in many cases. Irradiance and temperature are often positively correlated but occasionally negatively, as in the wheat fields of north-west Mexico (Fischer, 1985a). Water stress often goes with high temperature; where it does not, pests and diseases may do so.

Serial plantings in the tropics, so effective for examining the relation between irradiance and yield, are less satisfactory for temperature because its seasonal variation is often quite small, the very reason why the influence of radiation can be so clear. Altitudinal transects have been used to eliminate differences in daylength, e.g. in the work on beans by Laing et al. (1983, 1984) and on wheat by Midmore et al. (1982), but differences in temperature may still be confounded with those in radiation, pests and diseases, etc. Enclosure of crop segments to modify day or night temperatures has been used to only a limited extent so far – for example with maize (Peters et al., 1971), wheat (Fischer & Maurer, 1976) and groundnuts (Dreyer et al., 1981) – but this approach is likely to be used more frequently in future.

Temperature influences yield through its effects on crop growth, on the number of inflorescences, seeds or storage organs initiated, and on the rate and duration of their development. In many instances the effects on the storage phase predominate. Dobben (1962) pointed out that high temperatures often shorten the period of development without sufficient compensation by way of faster growth. The effect of temperature on grain growth rate and duration has been examined in wheat (e.g. Sofield et al., 1977; Wiegand & Cuellar, 1981; Sanford, 1985), rice (Yoshida & Hara, 1977), pearl millet (Fussell et al., 1980), sorghum (Muchow & Carberry, 1990) and other crops. The fall in the duration of grain growth as temperature rises, illustrated for several cereals in Figure 4.20, reflects a relatively constant

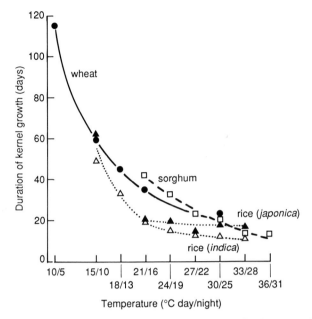

Figure 4.20. Effect of temperature regime on the duration of grain growth in several cereals (Chowdhury & Wardlaw, 1978).

thermal time for grain growth, as Muchow & Carberry (1990) have shown for sorghum. The fall in duration at higher temperature is compensated to some extent by associated increases in the rate of grain growth, but not for wheat and barley at temperatures above 15/10 °C, nor for rice above 24/19 °C, nor for sorghum above 27/22 °C in the experiments of Chowdhury & Wardlaw (1978). In those of Fischer & Palmer (1983), the optimum compromise between rate and duration of maize grain growth was at 24.8 °C, but even at temperatures up to 32 °C the compensation between rate and duration may be effective in preventing a fall in yield when irradiance is high, e.g. in maize and sorghum (Muchow, 1990*a*,*b*).

Thus, although in many temperate crops, and even rice, the faster rates of development and storage at high temperatures do not compensate for the shorter durations, they can do so at least in the C_4 cereals under high irradiance.

Rawson (1988) argues that tropical temperatures *per se* do not reduce yield, even of temperate crops such as wheat, but rather that reduced interception of radiation owing to the shorter life cycles at high temperatures is the limiting factor. Evidence in support of this view is presented in Figure 4.21, in which wheat crops grown at both moderate and high temperatures displayed the

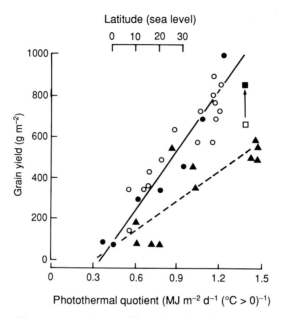

Figure 4.21. Grain yield as related to photothermal quotient in wheat crops from four environments. The triangles represent hot sites in Mexico (Rawson, 1988).

same relation between grain yield and photothermal quotient, i.e. radiation per unit area per day degree. Rawson & Bagga (1979) and Fischer (1985a) likewise found a close relation between grain number per square metre and the photothermal quotient for wheat crops ranging from cool to hot environments. Rawson (1988) suggests, therefore, that in areas or seasons where high temperatures are combined with high radiation, high yields per crop are feasible provided the supplies of water and nutrients are not limiting for growth and development, an approach which has important implications for the future of tropical agriculture.

A few examples of the relation between yield and temperature will now be considered.

Bean

The relation between temperature and the mean yield of many varieties of beans (Figure 1.6) and of those best adapted to high or low temperatures (Figure 4.9a) plays a major part in determining where beans are grown (Figure 4.9b). Yields are highest in environments which combine warm days with cool nights, high radiation and freedom from disease but as in all such international trials, many factors besides temperature influence yield (CIAT, 1979).

Maize

In serial sowing studies in the tropics, Jong *et al.* (1982) found a weaker positive correlation of yield with temperature than with radiation. Both crop enclosure and regression analyses suggest that yield falls as night temperature rises (Peters *et al.*, 1971; Chang, 1981). In Japan the highest correlations with maize yield were for temperature during the early period of growth, which were much higher than those with temperature later in the life cycle or with rainfall or irradiance (Kudo, 1975). An expression for mean daily temperature in July combined with radiation in August accounted for more than 90% of the variation in grain yield.

With crops in the US Corn Belt, Thompson (1969, 1986) found temperatures above normal in June, July or August all to be associated with a fall in yield (Figure 4.22), whereas slightly lower than normal temperatures in July and August were favourable, probably owing to their association with additional rainfall. Thus, temperatures close to normal Corn Belt temperatures were most favourable for yield.

Rice

Working in tropical environments, De Datta & Zarate (1970) found a negative correlation between yield and the minimum temperature for 30 days after transplanting, and Yoshida & Parao (1976) a negative relation between yield and daily mean temperature for the 25 days before flowering. Seshu & Cady (1984) found a negative relation between yield and the average minimum temperature during grain growth. In general, the yield of rice tends to be negatively related to temperature in the tropics (except at high altitudes), whereas it may be positively related to temperature at higher latitudes (cf. Murata, 1975; Munakata, 1976). Night temperatures below 20 °C at the stages of microsporogenesis and anthesis can be a major source of yield loss (Satake, 1976). Nishiyama (1985) found no correlation between rice yield and average temperature during the ripening stage in Japan. Although both Yoshida & Parao (1976) and Seshu & Cady (1984) imply that there is no interaction between temperature and radiation in the determination of rice yields, this seems unlikely in that at least the adverse effects of higher temperatures may be substantially lessened when irradiance is high.

Like Thompson's (1969, 1986) analysis of maize in the US Corn Belt, that of Uchijima (1981) for rice in Japan also emphasizes the relatively close varietal adaptation to the average temperatures at each locality, the optimum

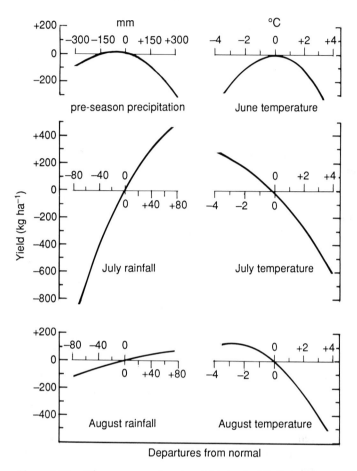

Figure 4.22. The response of maize yields in the US Corn Belt states to weather variables (Thompson, 1986). (American Society of Agronomy, Inc. Reprinted with permission.)

temperature in July–August ranging from 19 °C for varieties grown at Hokkaido to 25.5 °C for those at Hokuriku.

Wheat

The effects of temperature on grain number, weight and yield have been analysed under controlled environmental conditions more thoroughly with wheat than with any other crop. Temperature during the period of vegetative growth is less significant for yield than that in the later stages, but no stage between inflorescence initiation and flag leaf emergence is particularly sensitive to high temperatures (Rawson & Bagga, 1979). High temperatures during the reproductive phase greatly reduce grain number (e.g. Warrington *et al.*, 1977; Bagga & Rawson, 1977; Sofield *et al.*, 1977; Kolderup, 1979), whereas those during grain growth reduce kernel weight (Sofield *et al.*, 1977; Wiegand & Cuellar, 1981; Wiegand *et al.*, 1981). Some varieties are much more sensitive than others (Wardlaw *et al.*, 1989a,b).

The effects of higher temperatures on wheat crops in the field have been examined by Fischer & Maurer (1976), using enclosures which varied the mean temperature from 2 degrees below to 7 degrees above ambient. Higher temperatures were most deleterious when given during the reproductive period, causing a 4% loss of yield per 1 degree rise, close to the estimate of Wardlaw *et al.* (1989a).

The adverse effects of high temperatures on the yield of wheat have usually been associated with too rapid development. Using an altitudinal transect in Mexico, Midmore *et al.* (1982) found the days to flowering, leaf number, tiller number and potential spikelet number all to decrease in early flowering varieties as mean temperature rose. In varieties with some response to vernalization and daylength, however, the early stages of development were delayed at high temperatures, but yield components were not improved. Rawson (1986) has obtained high yields of wheat in a high temperature glasshouse regime (30/25 °C) with high radiation, long days and near-optimal water and nutrient availability. Under these conditions there were at least six ears per plant and little evidence of floret sterility, so high temperatures *per se* may be less deleterious than the inadequacy of water and nutrient supplies, the high pest and disease pressures, and the limiting irradiance with which they are often associated. The dependence of wheat yield on radiation is greater at high temperatures (Bush & Evans, 1988; Wardlaw *et al.*, 1989a).

Thus, as the atmospheric CO_2 level continues to rise, and as irrigation and fertilizer management improve, it is likely that the adverse effects of high temperatures on the yield of many crops will be ameliorated, especially when varietal differences in heat tolerance, such as those found by Bagga & Rawson (1977) and Wardlaw *et al.* (1989a,b), have been used in breeding programmes.

Rainfall

In considering the relations between climate, weather and yield, Watson (1963) noted that in those situations where one climatic factor is predominant in its effect on yield, rainfall or the lack of it is likely to be the one. A classical example is R.A. Fisher's (1924) analysis of wheat yields at Rothamsted. Using his technique, Gangopadhyaya & Sarker (1965) accounted for three quarters of the total variation in the yield of wheat in

India in terms of rainfall distribution, above-average rain before or during germination being beneficial, whereas during the tillering phase it is detrimental to yield. Excessive rainfall during the vegetative stage of the crop may be detrimental for several reasons, such as poor aeration, greater loss of nitrogen by leaching or enhanced pest and disease problems (Gales, 1983).

Excessive rainfall is a problem for some, but inadequate rainfall is a far more widespread and dominant one. In the drier regions of the world, yield may be proportional to the amount of precipitation or of available water in the soil, whereas yield in moister climates may depend heavily on moisture availability during critical stages of development, such as just before anthesis in the cereals.

Linear relations between yield and crop water use have often been found, e.g. for wheat in the Punjab (Arora & Prihar, 1983) and in South Australia (French & Schultz, 1984). In the Punjab, the total water supply (i.e. that stored in the soil at seeding plus growing season rainfall) was a better predictor of yield than seasonal water use by the crop, and the predictive power of the regressions was improved by taking account of the stages most sensitive to water supply. With both sorghum and wheat crops in Queensland, Nix & Fitzpatrick (1969) found the relation between yield and the ratio of actual:potential evapotranspiration to be much closer during the heading and early grain growth stage than either before or after. Thus, seasonal or year to year differences in the yield of wheat and sorghum varieties in Queensland were largely a function of anthesis date in relation to the environmental water stress prevailing at that time.

Likewise, Runge's (1968) analysis of the effects of interactions between rainfall and temperature on maize yields highlights the period one week each side of anthesis as being most sensitive to the adverse effects of low rainfall and/or high temperature. Thompson (1986) also identifies July rainfall as having the greatest effect on maize yields in the US Corn Belt (Figure 4.22). It also has a large effect on the relation between yield and July temperature, warmer conditions being deleterious when the rainfall is low but favourable when it is high.

For crops that are grown in semi-arid areas, the relation between yield and rainfall or soil moisture tends to be simpler, as in the case of the close linear relation between sorghum or millet yields and growing season rainfall in several West African countries (Dennett et al., 1981b), or to soil moisture at flowering in India (Seetharama et al., 1984). In the semi-arid regions of the US Great Plains, soil moisture status accounts for most of the variation in yield of winter wheat crops, whereas high temperatures are a more important variable for spring wheats (Feyerherm & Paulsen, 1981).

Rainfall interacts not only with temperature in its effects on crop yield, but also with other climatic factors, with the incidence of pests and diseases, and with the response to nitrogenous fertilizers (Young *et al.*, 1967; Russell, 1968; Feyerherm & Paulsen, 1981).

Assessment

This brief survey of the ecology of yield has been confined to only three components of climate (solar radiation, temperature and rainfall); it has dealt with these only in terms of yield and its main components and it has not tried to integrate these effects as crop simulation models do. Other books cover that ground. Having already looked at the extent to which crop plants have been adapted to climatic conditions quite different from those in which they were domesticated, we needed to assess how responsive their yield still is to variations in weather and shifts in locale.

Certain stages in the life cycle, such as the approach to flowering in cereal and pulse crops, are particularly sensitive to radiation, temperature and water stress. Variations in these climatic components and in the interactions between them during the grain growth stage also have a strong impact on yield. For example, the effect of irradiance on yield is more pronounced at high temperatures, and the effect of July temperature on maize yield in the Corn Belt depends greatly on the rainfall (Thompson, 1966).

The annual, seasonal and daily departures of the weather from climatic averages accounts for much of the variation in yield from long term trends, as in the case of Corn Belt maize and soybean crops (Thompson, 1969, 1970) and of rice in Japan (Uchijima, 1981). Nearly all departures from normal temperatures in June, July and August are associated with reduced yields. An implication of such analyses, and of the optimality of the mean temperatures where beans are grown in South America (cf. Figures 1.6 and 4.9*b*), is that modern varieties are often closely adapted to the local temperatures where they are grown.

To what extent are crops grown where the climate makes them most productive rather than having been adapted by plant breeding to local climates? There is a circularity of logic in some of the evidence for close adaptation to local conditions, but the 6.5 deg range in optimum July–August temperatures estimated by Uchijima (1981) for rice varieties adapted to four locations in Japan implies considerable selective change. With maize in monsoon India, on the other hand, above average rainfall during emergence or above-average maximum temperatures before flowering increase yield (Huda *et al.*, 1976), and there are many other examples of yields being raised by departures from average conditions, not only for rainfall but also for radiation and temperature. More important than close adaptation to average conditions may be the capacity to take advantage of departures from them.

Adaptability and reliability

Adaptability is characteristic, almost a *sine qua non*, of all the major crops. How else could they have spread so widely, changed so much and been put to so many uses? However, some crops and varieties are more adaptable than others.

Traditional agriculture was characterized by great spatial diversity of varieties adapted to local tastes as well as to local conditions. Location × variety interactions for oats in Iowa were progressively reduced as that state was divided into smaller zones, reflecting local adaptation (Horner & Frey, 1957). On the other hand, although the Thai rice variety Khao Dawk Mali originated in a deep water area of the Central Plain, it is also one of the most reliable yielders in the dry conditions of the north-east and is preferred for its quality in both the deep water and the drier areas. Thus, local adaptation and adaptability need not be mutually exclusive.

As the earlier spatial diversity of land races gives way to the temporal diversity associated with varietal improvement, greater emphasis is now being given – especially in international plant breeding programmes – to varieties suitable for use over wide areas. The products of CIMMYT's wheat cross 8156, IR36 rice, and the wheat varieties Bezostaya-1 and Mironovskaya-808 in the (former) USSR, have all been sown on more than 11 million hectares in one year. All are widely adaptable varieties in a geographic sense and possibly also in an environmental one, i.e. able to yield well not only at many locations but also across a wide range of environments. Despite the emphasis on adaptability, however, considerable local diversity remains. For example, across the USA in 1984, at least 429 varieties of wheat were grown, 16% more than five years earlier (Siegenthaler *et al.*, 1986). The most widely grown variety occupied only 6.1%, and the top ten only 39.2%, of the total area under wheat.

Another desirable characteristic is the ability to yield well over successive years, best referred to as *reliability* although often called *stability*, a term that has other genetic and varietal registration implications. Insofar as different years at one location may provide different environments, adaptability and reliability are related although different characteristics. In some situations reliability of yield from year to year depends more on the incorporation of multiple resistance to pests and diseases than on varietal response to climatic conditions. Khush (1987) has illustrated this for rice yields by comparing those of IR8, which has relatively few resistance genes, with those of IR36 and IR42, which have multiple resistance (Figure 4.23). In other situations, as in the case of maize hybrids in France, reliability of yield depends more on earliness of flowering (Derieux *et al.*, 1987).

The emphasis given to adaptability in modern plant breeding has led to

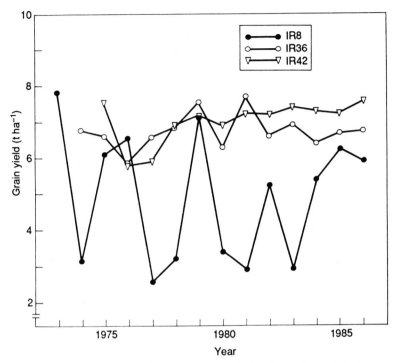

Figure 4.23. Variation from year to year in the dry season yields of three rice varieties at IRRI (modified from Khush, 1987).

widespread use of the regression technique, introduced by Yates & Cochran (1938) and developed by Finlay & Wilkinson (1963) and Eberhart & Russell (1966), of using the mean yield of all varieties as the environmental index for each site. Yates & Cochran saw this as a way of comparing varietal responses to fertilizer, whereas Finlay & Wilkinson recognized the usefulness of the regression coefficient for comparing the adaptability of many varieties and lines across a wide range of environments.

Figure 4.24 presents some of the results obtained by Finlay & Wilkinson (1963) for barley varieties grown in South Australia. This indicates the wide range in regression slope, from about 0.1 for the unresponsive Bankuti Korai to 2.1 for the highly responsive Provost, while Atlas had an approximately average value of 0.9. The differences in slope are such that there are crossovers in performance, some varieties being better suited to the 1959 conditions, others to the 1958 ones. Such crossovers had already been found by Yates & Cochran (1938).

The most significant climatic factor varying between years and sites in

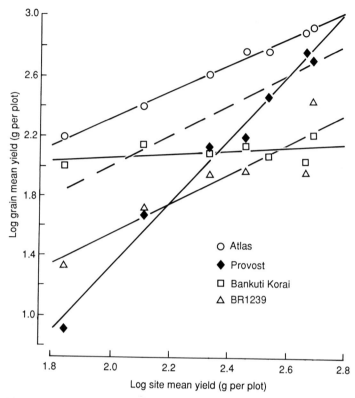

Figure 4.24. Regressions of the relation between the yields of four individual barley varieties and the mean yield of 277 varieties (broken line) grown at various sites and seasons in South Australia (Finlay & Wilkinson, 1963).

South Australia was rainfall. In this example, therefore, the varietal differences in 'adaptability' were largely dependent on their maturity times in relation to the length of the growing season as determined by moisture availability. Bankuti Korai matured early and therefore yielded quite well in the short dry season of 1959 but no better in the long wet season of 1958. The later-maturing Provost, by contrast, was cut short in 1959 but could take full advantage of the longer growing season in 1958. Although the mean yields of Provost and Bankuti Korai were similar, their regression coefficients were very different, and their impact on a subsistence farmer would be very different.

Soil moisture availability in relation to maturity was also the predominant factor in the maize trials analysed by Eberhart & Russell (1966) and Russell

(1974). In many international yield trials, however, there may be differences between sites not only in the length of the growing season but also in temperature, radiation, daylength, soil fertility and pest and disease pressures, with the result that the regressions may be less significant.

Some limitations of regression analysis

Regression diagrams like that in Figure 4.24 have become the plant breeder's icons, ubiquitous but with a variety of styles to support a variety of dogmas. We should therefore be aware of their weaknesses, as well as of what they reveal.

(1) The use of site mean yield as a biological measure of the environment gives no insight into which factors affect performance. The environmental effects are often assumed to be too complex for analysis. Thomson (1986) has shown that the relative performance of cotton varieties can be closely related to individual climatic factors such as temperature, and in some regional comparisons one factor, such as water supply, may be obviously dominant.

(2) When one factor does predominate in the differences between sites and/ or years, the regression coefficient indicates adaptation to that one factor rather than a more general adaptability, and the limits to its relevance for other environments must be recognized.

(3) Low site mean yields are just that, but they are often assumed to refer to low input or adverse conditions. When the trials are carried out on experimental farms, the low yielding sites may still be much more favourable than those encountered by many farmers. Consequently, the regressions may not extend to really adverse conditions, and may over-emphasize performance in more favourable ones. Finlay & Wilkinson (1963) write: 'Plant breeders are inclined to ignore the results obtained in low-yielding environments (e.g. drought years), on the basis that the yields are too low and are therefore not very useful for sorting out the differences between selections. This is a serious error, because high yielding selections under favourable conditions may show relatively greater failure under adverse conditions'.

(4) These days, the regression lines are often presented without the data points, which may obscure individual site responses, especially at the low yield sites where even small differences may be highly significant to a subsistence farmer. In fact, the linear regressions often account for a relatively small proportion of the genotype × environment interactions, except when single factors predominate. They may also not be linear, having significant quadratic curvatures (Pederson & Rathjen, 1981).

(5) The regression coefficients are in no sense absolute values but depend on the particular set of varieties and sites as Knight (1970), Lin *et al.* (1986) and Crossa (1988) have shown. For example, the regression coefficients for

both Kharkov and Scout 66 wheats in yield trials for the Great Plains have fallen progressively over the years because they are estimated in relation to the average for all the varieties in each trial, and presumably the responsiveness of the new entries has risen progressively (Peterson *et al.*, 1989).

(6) The exclusion of extreme points, particularly at the low yield end, can have a drastic effect on the regression coefficients, and therefore on varietal evaluation, as Crossa (1988) has shown.

(7) To the extent that data from both different sites and different years are used, as in Figure 4.24, adaptability and reliability may be confounded in the regression coefficient. In cases where one environmental factor predominates across both sites and years that may not matter, but in other situations the factors determining adaptability and reliability may be quite different. Using two different sets of rice yield trials, Evenson *et al.* (1978) found adaptability and stability (i.e. reliability) to be positively correlated in the Indian trials, but not in the wider-ranging international yield nursery. Flinn & Garrity (1986) found adaptability and reliability to be positively correlated in both irrigated and upland rice yield trials. However, one of the problems encountered in these analyses is that few varieties have been retained in the trials for long enough to make reliable estimates of reliability.

The relation deserves further analysis, because although the plant breeder may want wide adaptability, that is of little or no concern to the subsistence farmer who wants reliability of performance, at as high a yield level as possible. Kawano & Jennings (1983) take issue with the belief that wide adaptability confers reliability, and are supported by the finding of Crossa *et al.* (1988) that maize populations selected on the basis of multilocational performance are no more stable (i.e. reliable) than those assessed at a single location.

Crossovers in performance: how common and significant are they?

In recent years, and especially in international plant breeding programmes such as those for rice at IRRI and for wheat at CIMMYT, much emphasis and publicity has been given to varieties with superior performance across a wide range of environments. Indeed, the breeding of such varieties consistutes a major *raison d'être* for international agricultural research centres. Their improvement continues, as indicated by the flow of regression diagrams illustrating the superiority of new lines, such as the 'Veery' wheats, over the previous best varieties across a range of environments (e.g. Ceccarelli, 1989).

Such comparisons have powerful implications: first, that genetic improvement in yield can continue independently of any advances in agronomy; secondly, that yields under adverse conditions will improve along with those under more favourable conditions, bringing betterment to all; thirdly, that plant breeders can select under favourable conditions, in which varietal differences are magnified, for improved performance under less favourable conditions (e.g. Roy & Murty, 1970; Mederski & Jeffers, 1973; Blum, 1985; Johnson & Geadelmann, 1989); fourthly, that breeding programmes which generate such adaptable varieties can be genuinely international in their impact. Given the long history of such portrayals, aid agencies might well wonder why yields under adverse conditions, and for poor farmers without inputs, have not been substantially raised. Indeed, they may even be led to believe that agricultural scientists are neglecting poor farmers and poor conditions when the means for progress are already at hand.

Two points deserve emphasis, therefore. First, the performance of the older standard varieties is likely to be declining as new pest and disease biotypes evolve, i.e. continued breeding effort is needed simply to avoid declines. Secondly, the regression lines from such trials frequently do not extend to really adverse conditions and they may conceal a lot of variation from site to site, especially at the lower end, reflecting significant differences in local adaptation.

Crossovers in relative performance, reflecting differential adaptation, often occur, as in rice (Kawano & Jennings, 1983), barley (Figure 4.24) (Schaller *et al.*, 1972; Ceccarelli, 1989), wheat (Laing & Fischer, 1977; Fischer, 1981; Blum, 1982; Carver *et al.*, 1987; Bruckner & Frohberg, 1987), sorghum (Seetharama *et al.*, 1982; Francis *et al.*, 1984), maize (Russell, 1974; Fischer *et al.*, 1982, 1989), broad bean (Dantuma *et al.*, 1983), soybean (Cooper, 1985), chickpea (Saxena, 1987), bean (Figure 4.9*a*), cotton (Thomson, 1986) and other crops. An example with cassava is illustrated in Figure 4.12, in which the improved varieties substantially outyielded the traditional varieties in favourable conditions with good cultural practices at CIAT and at Caribia, whereas in the low fertility conditions at Carimagua the reverse was the case.

In the breeding of dwarf soybeans, it was recognized at the outset that such varieties would be specifically adapted to high yield environments (Cooper, 1985). The analysis by Ceccarelli (1989) suggests that although selection under optimal conditions may be relatively efficient for environments where yields are above 3–4 t ha^{-1}, it is not efficient for those in which yields are only 0.5–2 t ha^{-1}. In tropical maize trials, stresses must reduce yield by at least 50% before significant changes in varietal ranking become apparent (Edmeades & Tollenaar, 1990). However, two other considerations can amplify the significance of crossovers in grain yield. Traditional varieties often have cherished qualities of taste or aroma which win a higher price, so that the cross-over point for economic returns can be higher than that for yield. Secondly, the longer, heavier straw of the old varieties may add even more to their value, raising the cross-over point for returns still higher.

The differential adaptation of varieties, as indicated by crossovers in their

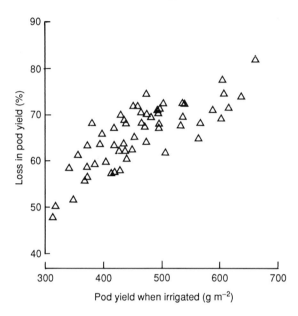

Figure 4.25. The relation between percentage yield loss due to end of season drought and pod yield potential among 64 peanut genotypes (Rao *et al.*, 1989). (American Society of Agronomy, Inc. Reprinted with permission.)

relative yield, has often been found when differing water supply is the dominant variable, whether via drought stress (e.g. Fischer & Wood, 1979; Blum, 1982; Cooper, 1985; Austin, 1990) or the extent or timing of irrigation (e.g. Innes & Blackwell, 1983*b*). Differing temperature regimes may also result in yield crossovers, as with beans (Figure 4.9*a*), wheat (Bagga & Rawson, 1977), cassava (Cock, 1983) and cotton (Thomson, 1986). Crossovers with differing irradiance have been found in rice (Dash & Rao, 1990). Crossovers might also be expected with differences in soil fertility and nutrient supply but were not apparent among old and new British wheat varieties (Austin *et al.*, 1980*a*) nor among maize hybrids (Crossa, 1988). However, maize synthetics based on selections at low soil N levels have been found by Muruli & Paulsen (1981) to yield more under those conditions than synthetics based on selections at high N levels.

Yield under particular adverse conditions may be raised by selection for better adaptation to those conditions and sometimes by selection for greater yield potential. In the latter case, selection under favourable conditions might improve performance in adversity as rapidly as selection under adversity itself, as Johnson & Geadelmann (1989) found in the selection of maize for water-stressed environments. With 64 genotypes of peanuts, however, Rao *et al.* (1989) found that the higher the yield potential under irrigation, the greater was the percentage yield reduction due to drought stress during grain filling (Figure 4.25). They concluded that 'it is unlikely, therefore, that breeders will be able to combine high yield potential with low sensitivity to drought spanning the seed-filling phase.' With barley, Ceccarelli & Grando (1991) found a substantially higher risk of crop failure among genotypes selected under favourable conditions than among those selected under low yield conditions.

The most effective breeding strategy for adverse conditions will depend on the net effect of two quite separate components, on the one hand the selection differentials and heritability of yield in the favoured and stressed environments, on the other the extent to which selection for yield potential conflicts with adaptation to the particular stress. It also depends on the extent of replication, which must be more extensive under adverse conditions (Atlin & Frey, 1990).

Conclusion

Selection for high performance across a wide range of environments can be effective (cf. Hughes *et al.*, 1987), and is desirable from the point of view of increasing the useful range of new varieties from both international and commercial breeding programmes. Both economies of scale and the increasing commercialization of plant breeding are likely to lead to even greater emphasis on varietal adaptability in future.

Nevertheless, the limits to the useful breadth of adaptation should be more openly acknowledged, certainly for environments with severe water stress and probably for those with poor soils and low nutrient levels, just as they are for those with disease and pest problems. Varieties with a greater degree of local adaptation may be required to maximize yields in difficult environments. For the subsistence farmers who so frequently have to cope with these conditions, reliability of yield is of much greater significance than broad adaptability. To the extent that reliability and adaptability are positively correlated, adaptable varieties may be of value to them, but this correlation has yet to be firmly established.

Selecting for improved performance and reliability under adverse, variable conditions where, at best, progress will be slow, uncertain and location-specific is not an attractive proposition for plant breeders. Moreover, government policies for varietal testing and release may discourage such work, as may policies limiting the allocation of fertilizers to such areas. It is easy to see why so much emphasis, and hope, is placed on breeding for high average yields and broad adaptability. It may be the only practicable approach in many cases. But in recognizing that, we should not deny the important role that local adaptation has played in the improvement of crop plants, and could still do so. The wealth of genetic variation in adaptive responses to soil and climatic conditions conserved in the world's gene banks is little known and less used relative to that in resistance to pests and diseases, but it may yet prove to be the most important genetic resource of all.

CHAPTER 5

Physiological aspects of crop improvement

'We may conclude that whatever part
or character is most valued – whether
the leaves, stems, tubers, bulbs, flowers, fruit or seed of plants ... that
character will almost invariably be found to present the greatest amount
of difference both in kind and degree.' Charles Darwin (1868)

Introduction

The power of sustained empirical selection is such that improvement
of the yield, quality and adaptation of many crop plants has been striking even
when our understanding of their genetic and physiological basis has been
minimal. Further progress is likely, therefore, even in the absence of further
understanding.

As Hutchinson (1965) points out, however, plant breeding has been more
successful when its objectives have been more clearly defined, as in breeding
for resistance to pests and diseases. These are readily observable characters,
as are many of those which played a role in the domestication of crop plants.
By contrast, the physiological characteristics that determine yield potential
are not readily observable (Frankel, 1947); indeed, they are scarcely known
and certainly not agreed upon.

In one sense they never will be, because crop yield is the ultimate outcome
of the whole life cycle of the crop, and of the rates, durations and interlinkages
of many processes at all stages of development. No one process provides the
master key to greater yield potential. Quite different processes may limit the
yield of different cultivars at one site and season, or of one cultivar at different
sites or seasons (Crosbie & Mock, 1981; Willman *et al.*, 1987). Yield potential
is an extremely complex character, a matter for recognition rather than
delight.

Nevertheless, there are strong reasons for trying to identify the physiological characteristics which have contributed most to past increases in yield
potential or which could be exploited in the future. In the first place, the major
pathway to greater yielding ability so far, namely the investment of a greater

169

proportion of crop biomass in the harvested organs, may be approaching its limit in some crops, and alternative paths must be explored. Secondly, the physiological nature of the greater yield potential has major implications for both plant breeding and agronomy, which need to be more widely understood. Thirdly, the promise of molecular plant breeding lies in the specificity with which it may be able to transpose and modify genes, and to realize the full value of these techniques we should be able to specify the processes most limiting to yield potential.

The comparative element in crop physiology

History has been called an examination of the present in the light of the past for the purposes of the future. Crop physiology, likewise, compares present high-yielding varieties with their cultivated and wild progenitors to see which processes have been altered and to consider how much more they might be modified or what other processes might be changed. Thus crop physiology inevitably has a strong retrospective element, but this perspective on past changes is valuable because it is all too easy to design physiological masterpieces when not constrained by the trade-offs and counter-intuitive associations between characters and by the imperatives of environmental adaptation.

There are equal dangers, however, in focusing our comparisons only on crop plants and their current range of characteristics. The physiological attributes and adaptations of wild plants may enlarge our vision of what is possible, especially as recombinant DNA techniques develop, because in principle the genetic resources of all plants, indeed of the whole biosphere, may become available for crop improvement. Modern varieties should be compared not only with their progenitors but also with other crop species and with the full range of plant behaviour.

Physiological comparisons between older and newer, higher-yielding or better-adapted varieties have often been used to identify characteristics that may have contributed to crop improvement. Varieties usually differ in many features, however, and rarely if ever have varietal comparisons established that a particular physiological characteristic was the cause of yield enhancement rather than merely being correlated with it. There are many paths to success in crop plants.

The inclusion of related wild species and progenitors in our physiological comparisons offers a wider perspective on the processes contributing to domestication and improvement than can be got from purely varietal comparisons. This has proved helpful in relation to the role of photosynthetic rate and duration, and the rise in harvest index becomes even more striking with the inclusion of wild species.

Major genes have frequently been manipulated in plant improvement, particularly in breeding for resistance to pests and diseases, quality characteristics and environmental adaptation, as we saw in Chapter 4. The fact that they have not often been used in breeding for yield potential is due mainly to the fact that relatively few major genes suitable for such an approach have been identified. The height of cereals had been reduced by polygenic selection for many years, but when the advent of cheaper nitrogenous fertilizers made it necessary to reduce height more drastically, the major genes for height were used with great effect. Now the development of recombinant DNA techniques adds urgency to the task of analysing the physiological effects of major genes that influence yield, and comparisons of near-isogenic lines can be helpful in this context.

Pleiotropic effects of major genes may complicate the analysis of their influence on yield, as in the case of erect leaf inclination in barley in which the more erect isolines also had less culm flex, more erect spikes, fewer ears, later maturity and greater lodging resistance, but no net effect on yield (Tungland *et al.*, 1987). The major dwarfing genes have a variety of effects, e.g. in wheat (King *et al.*, 1983; Gale & Youssefian, 1983) and influence yield in several ways (Fischer & Stockman, 1986; Bush & Evans, 1988).

The genes for cereal awns also have a variety of effects (Rasmusson, 1987). Awns may enhance ear photosynthesis and provide a source of assimilates of particular significance for the more distal spikelets (Evans *et al.*, 1972), especially under dry conditions. On the other hand, developing awns may compete for assimilates with developing florets (Rasmusson & Crookston, 1977). Grain number may be fewer but kernel weight greater in the presence of awns (Patterson *et al.*, 1962; Patterson & Ohm, 1975; Qualset *et al.*, 1965). Awned crop canopies may be cooler (Ferguson, 1974) and less susceptible to lodging, but on the other hand they may trap more water and thereby accentuate disease problems or sprouting in the ear (King & Chadim, 1983). Another illustrative example is the gene that reduces the capacity of barley to produce abscisic acid in response to water stress. It also reduces nitrate reductase activity and increases molybdenum dependency (Walker-Simmons *et al.*, 1989), making its net effect on yield difficult to analyse.

Added to such complications are the interactions between major genes and the genetic background into which they are introduced. For example, the gene for sub-okra leaf shape in cotton increases yield on some backgrounds but not on others (Meredith & Wells, 1987). The effects of cereal awns on yield also vary with genetic background in both barley (Scharen *et al.*, 1983) and wheat (Olugbemi & Bush, 1987), as do those of the dwarfing genes in wheat (Allan, 1986; Bush & Evans, 1988), barley (Ali *et al.*, 1978) and oats (Meyers *et al.*, 1985). The Maryland Mammoth gene, which confers a requirement for short days for flowering to occur in *Nicotiana tabacum*, is expressed as a response to low temperatures in *N. rustica* (Steinberg, 1953). Similarly, whereas the expression of one gene for male sterility in rice is controlled by daylength in some genetic backgrounds, in others it is controlled by temperature.

The effects of some major genes on yield also depend on environmental conditions, as in the case of those for awns (e.g. Schaller *et al.*, 1972; Evans *et al.*, 1972) or stem height in wheat (Bush & Evans, 1988). Such environmental dependence is likely to be found for many of the major genes influencing yield. Consequently, even when truly isogenic lines, differing only in one gene, displace those currently available from back-crossing programmes, crop physiologists may still get ambiguous answers as to the effects of particular genes on yield. This situation may be frustrating to those in search of unqualified recommendations, but the integrating, all-encompassing nature of yield makes it inevitable.

Of sources and sinks

As an example of the complexities of the physiology of yield, let us consider the question of whether the capacity to generate assimilates or that to store them is more limiting to further advances in crop yield, i.e. whether they are source- or sink-limited. Source or sink? is too simple and polarised a question, not least because there are no agreed definitions of the terms. For most crops there are several different sources of assimilates, and several competing and quite different kinds of sink, each with different mechanisms for attracting assimilates. Those in the cell division phase may be scattered and small in total yet crucial to crop development. Nearby competitors will be sinks in the cell elongation phase, larger in their demands and involving different endogenous growth regulators. Then there are the many kinds of storage sinks, those which directly determine yield, and those which provide temporary storage along the way, some for sugars, others for polysaccharides such as starch. The various kinds of sinks are likely to differ in the mechanisms by which they attract and retain assimilates.

Sinks may become sources, as when young leaves cease elongation. Since this is progressive, the distal parts of a leaf may become a source while the proximal parts are still sinks. Cessation of import does not depend on the attainment of export, as Turgeon (1984) has shown with tobacco leaves.

Crops differ greatly in the timing of competition between their various sinks. By and large the cereals establish their complete assimilatory structure, both leaves and roots, and then enter the final storage phase with their grains as the predominant sink. In axillary flowering crops, such as many grain legumes, vegetative growth and reproductive storage may overlap and compete for some time. With plants such as sugar beet and cane, such competition between growth and storage continues throughout the life of the crop, so that renewal of the source is as important a sink as the harvested storage organs. Finally, with forages and vegetable crops like lettuce the source and the sinks that renew it all contribute to the harvested yield. Thus, the relations between source and sink differ greatly between crops.

They also differ with environment. For many crops the source may be less sensitive to temperature than the sink organs, but more sensitive to irradiance or water stress. Thus, even in the environment to which it is adapted, a variety may veer from source to sink limitation as the weather changes, as well as diurnally between day and night. Moreover, the sequence of source and sink activities in different varieties may be adapted to different seasonal sequences of radiation, temperature or daylength and may differ in the way in which photosynthetic capacity before flowering feeds forward to define the storage capacity after flowering.

Such feed-forward effects of source on sink indicate that source and sink are not independent entities. Feed-back effects of storage capacity on photosynthetic rate and duration also highlight their mutual dependence.

Regulation of photosynthesis by sink activity

Feed-back effects of sink on source activity have been reviewed by Neales & Incoll (1968), Herold (1980), Guinn & Mauney (1980), Foyer (1987), Geiger (1987) and others, so only the aspects most relevant to crop yield need be considered here. The subject deserves our attention, however, because of its relevance not only to determinations of source–sink limitations to yield but also to comparisons of photosynthetic rates between cultivars and crops. A great variety of treatments altering the balance between source and sink has been applied to many crops to examine the effect of demand on leaf photosynthetic rate (Table 5.1).

Pretreatments with high or low irradiance or CO_2 have been used to increase or reduce the levels of sugars and starch in the leaves of soybean, cotton, sunflower and sorghum (Hofstra & Hesketh, 1975; Nafziger & Koller, 1976; Mauney et al., 1979; Sawada et al., 1986). Leaf carbohydrate levels have also been raised by exposing leaf discs to sugar solutions in barley and beet (Natr & Ludlow, 1970; Austin, 1972), or by preventing translocation from the leaves of sugar cane, beet, soybean and *Amaranthus* (Hartt, 1963; Waldron et al., 1967; Hall & Loomis, 1972; Setter et al., 1980; Blechschmidt-Schneider et al., 1989).

A reduction in sink demand has led to a fall in CER in many experimental systems (Table 5.2). This has been found in at least four cultivars of both wheat and soybean, and has been compared in several hybrids and inbreds of maize (Barnett & Pearce, 1983; Crafts-Brandner & Poneleit, 1987a,b; Connell et al., 1987). In maize it appears that although the CER can be raised by

Table 5.1. *Source–sink manipulations to influence photosynthetic rate*

(A) *For reduced demand from leaves*		
Prevention of fertilization	maize	Moss, 1962; Ceppi *et al.*, 1987
Removal of ears	wheat	King *et al.*, 1967
Removal of grains	wheat	Rawson *et al.*, 1976; Atsmon *et al.*, 1986
Removal of flowers	soybean	Heitholt & Egli, 1985
Removal of pods	soybean	Setter *et al.*, 1980; Wittenbach, 1983
Excision of tubers	potato	Burt, 1964
Tuber exposure to light	sweet potato	Tsuno & Fujise, 1965
Cooling of tubers	potato	Burt, 1966
Cooling of roots	sugar beet	Geiger, 1966; Hall & Loomis, 1972
Cooling of rest of plant	peanut	Bagnall *et al.*, 1988
Physical restraint	soybean	Crafts-Brandner & Egli, 1987
Physical restraint	cucumber	Robbins & Pharr, 1988
(B) *For enhanced demand from leaves*		
Ear photosynthesis inhibited	wheat	King *et al.*, 1967
Removal or shading of leaves	wheat	King *et al.*, 1967
Removal or shading of leaves	oats	Criswell & Shibles, 1972
Removal or shading of leaves	rice	Palit *et al.*, 1979*b*
Removal or shading of leaves	cucumber	Mayoral *et al.*, 1985
Pod shading	soybean	Thorne & Koller, 1974
Pod warming	soybean	Seddigh & Jolliff, 1984
Crop thinning	soybean	Lauer & Shibles, 1987; Diethelm & Shibles, 1989
Rooting of leaves	bean	Humphries & Thorne, 1964
Rooting of leaves	sweet potato	Spence & Humphries, 1972
Rooting of leaves	soybean	Sawada *et al.*, 1986, 1990
Grafting	spinach/sugar beet	Thorne & Evans, 1964; Rapoport & Loomis, 1985
Grafting	sweet potato	Hahn, 1977; Hozyo, 1977
Seedling growth in long days	sugar beet	Humphries & French, 1969
Gibberellin treatment	sugar cane	Nickell, 1982

Table 5.2. *Some responses of photosynthetic rate to changed demand*

(A) Fall in CER with reduced demand	
wheat	Birecka & Dakic-Wlodkowska, 1963; King *et al.*, 1967; Rawson *et al.*, 1976; Atsmon *et al.*, 1986
maize	Moss, 1962; Christensen *et al.*, 1981; Tollenaar & Daynard, 1982; Pearson *et al.*, 1984; Connell *et al.*, 1987
pea	Flinn, 1974
soybean	Thorne & Koller, 1974; Nafziger & Koller, 1976; Mondal *et al.*, 1978; Setter *et al.*, 1980; Wittenbach, 1983
peanut	Bagnall *et al.*, 1988
sugar beet	Thorne & Evans, 1964; Habeshaw, 1973
potato	Burt, 1964, 1966; Nosberger & Humphries, 1965
sweet potato	Spence & Humphries, 1972
cassava	Ramanujam, 1987
ryegrass	Gifford & Marshall, 1973; Deinum, 1976
egg-plant	Claussen & Biller, 1977
capsicum	Hall & Milthorpe, 1978
cucumber	Robbins & Pharr, 1988
mulberry	Satoh & Hazama, 1971
apple	Fujii & Kennedy, 1985
(B) Rise in CER with increased demand	
wheat	King *et al.*, 1967
oats	Criswell & Shibles, 1972
maize	Barnett & Pearce, 1983; Pearson *et al.*, 1984
soybean	Thorne & Koller, 1974; Sawada *et al.*, 1986; Lauer & Shibles, 1987; Diethelm & Shibles, 1989; Shibles *et al.*, 1989
peanut	Bagnall *et al.*, 1988
pea	Nath & Bhardwaj, 1987
ryegrass	Gifford & Marshall, 1973; Deinum, 1976
mulberry	Satoh & Hazama, 1971

increased demand in high-yielding genotypes, and falls more rapidly in them following ear removal, the differences between hybrids in the rates at which nitrogen is mobilized from leaves and CER falls are maintained even after ear removal, i.e. demand is not the only internal factor modulating the photosynthetic rate.

Removal of the major sink has not always led to a fall in CER, e.g. in wheat (Apel *et al.*, 1973; Austin & Edrich, 1975), oats (Criswell & Shibles, 1972), cotton (Nagarajah, 1975) and soybean (Heitholt & Egli, 1985). Active alternative sinks were present in all these cases. Rawson *et al.* (1976) found grain removal from wheat to cause a rapid fall in flag leaf CER only when tillers were also removed, not when they were present as a major alternative sink. Geiger (1976) found no effect on CER of the prevention of assimilate export from bean leaves, but their CER was already so low that it would take a long time for inhibitory amounts of carbohydrate to accumulate in them. Likewise, Borchers-Zampini *et al.* (1980) found no rise in bean leaf CER when the other source leaf was removed, but since ^{14}C translocation from the remaining leaf doubled, no rise in photosynthetic rate was needed in the short term. Ryle & Powell (1975) record a similar case with uniculm barley. Thus, these negative results do not invalidate the conclusion that the photosynthetic rate in many crop plants is highly responsive to demand.

Many of the experiments have demonstrated only a fall in CER when demand is reduced. However, a rise in CER when demand is increased has been found in many cases (Table 5.2B). Increased demand often appears to delay or retard senescence and therefore increases CER at later stages even when the peak rate is not increased, as in wheat, soybean and ryegrass. However, increased demand may have conflicting effects on leaf senescence. When the demand is mainly for carbohydrates, the duration of photosynthetic activity of the upper leaves is commonly enhanced, but when there is a high demand for remobilized nitrogen, the photosynthetic life of the leaves may be shortened, as in many grain legumes (Sinclair & de Wit, 1975).

It has been suggested that the fairly drastic treatments used to expose the effect of demand on CER are irrelevant to field conditions, and that such responses are unlikely to be of significance in field crops (cf. Geiger, 1976, 1987; Evans, 1991). Several kinds of evidence suggest that feed-back control of CER does occur in the field.

(1) Leaves supplying fruits commonly have a higher CER than those further removed from them, as in apple trees (Hansen, 1970).

(2) Ontogenetic trends in CER may reflect the changing pattern of demand on the leaves. One of the clearest examples is in peach trees, in which CER increased, decreased and then increased again in parallel with the rate of fruit growth, before decreasing again after the fruit was harvested and assimilates accumulated in the leaves. These trends were most marked in mid-canopy where most fruit was present (Chalmers *et al.*, 1975, 1983). Similar results have been reported with apple trees (Fujii & Kennedy, 1985). Likewise in pea and soybean, CER peaked first during rapid fruit elongation and again at pod inflation and early seed growth (Flinn, 1974; Shibles *et al.*, 1989). In wheat, CER may fall during the period just before active grain growth begins, when alternative sinks are also inactive, and then rise again as grain growth gets under way (Birecka & Dakic-Wlodkowska, 1966; Rawson & Hofstra, 1969; Evans & Rawson, 1970; Rawson & Evans, 1971). In potatoes, CER reaches a minimum just before tuberization begins, and then rises markedly (Moll & Henniger, 1978).

(3) With several crops there is also evidence that CER of leaves on intact plants may begin to fall after several hours under high irradiance, associated

with the accumulation of carbohydrates, for example in wheat (Azcon-Bieto, 1983), soybean (Upmeyer & Koller, 1973), egg plant (Claussen, 1977), bean (Moldau & Karolin, 1977), lucerne (Chatterton, 1973) and sugar cane (Hartt, 1963). Likewise, although the CER may rise initially in reponse to higher atmospheric CO_2, eventually it declines towards its initial rate and rubisco activity may be substantially decreased, as in tomatoes (Yelle *et al.*, 1989*a,b*).

(4) With soybean in the field, warmer night temperatures, causing faster pod growth, led to substantial increases in leaf CER (Seddigh & Jolliff, 1984), as did shading of the pods which can occur in natural crop canopies (Thorne & Koller, 1974). Cool nights, on the other hand, may slow down grain growth and assimilate utilization, leading to a fall in CER, as in rice (Huang *et al.*, 1989).

(5) Taking insensitivity of CER to oxygen level as an indicator of feed-back limitations on photosynthesis, Sage & Sharkey (1987) conclude that it is quite common in field crops (but cf. Bagnall *et al.*, 1988).

(6) In comparisons of CER, varietal differences may not be apparent until the grain growth stage, when they probably reflect differences in demand as in wheat (Fischer *et al.*, 1981), rice (Lafitte & Travis, 1984), soybean (Buttery *et al.*, 1981; Wells *et al.*, 1982) and hybrid sorghum (Khanna-Chopra, 1982).

Thus, differences between varieties in photosynthesis may reflect rather than cause the differences in yield. The mechanisms involved in such photosynthetic regulation are diverse. The response is sometimes rapid – within one hour in maize according to Pearson *et al.* (1984), – but often requires a day or longer before it becomes apparent. Rapid responses may be associated with changes in stomatal conductance, as in soybean (Setter *et al.*, 1980) and maize. In both maize, where it is rapid, and wheat, where it is slower, the changes in stomatal conductance are largely confined to the abaxial surface of the leaves (Pearson *et al.*, 1984; Rawson *et al.*, 1976). Changes in mesophyll conductance tend to be slower, e.g. in soybean (Thorne & Koller, 1974; Nafziger & Koller, 1976). They often occur in parallel with changes in stomatal conductance, e.g. in barley, bean and capsicum (Natr & Ludlow, 1970; Moldau & Karolin, 1977; Hall & Milthorpe, 1978).

Leaf starch levels beyond a certain threshold are often negatively related to photosynthetic rate, e.g. in soybean (Nafziger & Koller, 1976; Thorne & Koller, 1974; Hofstra & Hesketh, 1975), cotton (Mauney *et al.*, 1979), capsicum (Hall & Milthorpe, 1978) and sugar beet (Milford & Pearman, 1975), but not always, e.g. in egg plant (Claussen & Biller, 1977), sunflower and soybean (Potter & Breen, 1980). In some cases, both starch and sucrose levels in the leaves are negatively related to CER (Sawada *et al.*, 1986), and in others either starch or sucrose (Claussen & Biller, 1977). In *Amaranthus* the fall in CER was associated with a rise in sucrose but not in starch

(Blechschmidt-Schneider *et al.*, 1989), while in wheat the depression of CER is closely related to the total carbohydrate concentration in the leaves (Azcon-Bieto, 1983).

The fact that CER in soybean has been found by different investigators to be negatively related to both starch and sugar accumulation in the leaves, or to starch alone, or to be relatively unaffected by high starch levels, suggests that the extent of feed-back inhibition varies with conditions and cultivars, and may involve an overflow loop rather than a simple feed-back loop (Foyer, 1987).

Yield limitations by source and sink

The evidence presented above indicates that source and sink are not independent entities. The source at one time feeds forward to determine the later sink, while sink activity feeds back to modulate the photosynthetic source. For these and other reasons considered above there might seem little point in further discussion of whether source or sink most limits yield. However, much research is still being done within this context (Wardlaw, 1990) and plant breeders want help in determining which selection criteria to focus on, so the kinds of evidence need assessment.

Evidence that source is limiting

Reduced source   If source and sink in modern varieties are more or less in balance, it is not surprising that partial defoliation of a crop reduces yield. Gifford's (1974*b*) analysis of the data of Boonstra (1936) and Stoy (1965) on partly defoliated wheat crops suggests almost complete source limitation of yield. Willey & Holliday (1971*b*) obtained similar results with wheat defoliation treatments, yet the same authors (1971*a*) found shading during grain growth to have little effect on yields in barley.

Normal source   A common argument for source limitation of yield is that many crops (e.g. wheat, soybean, and cotton) produce far more florets than they use to generate yield, so how can sink be limiting? Quite apart from the fact that many of these florets may not be competent to develop further, there is also the argument that apparent overproduction is needed to ensure sufficient survival of the hazards of frost, drought, pests, disease and failure to be pollinated effectively. The counterpart argument is that crops often develop apparently excessive leaf area at anthesis, but an excess then may be needed to ensure sufficient to the end of grain filling.

Crop growth is often closely related to the total intercepted radiation (e.g. Figure 4.15). Yields may also be highly correlated with radiation (Figures

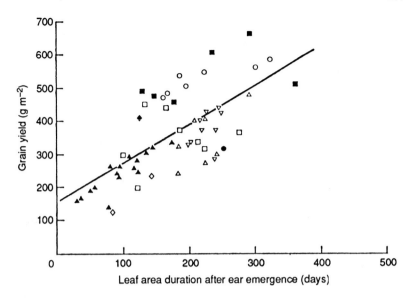

Figure 5.1. Grain yield of wheat in relation to leaf area duration after ear emergence for nine sets of wheat crops grown in a range of environments (Evans *et al.*, 1975).

4.17, 4.18). However, such relations do not prove that yield is source-limited because temperature and other climatic factors often co-vary with irradiance.

Another kind of evidence adduced that yield is source-limited is the frequently close relation between the leaf area index (LAI, the total leaf area per unit ground area) and its integral over time, the leaf area duration (LAD), with yield. Figure 5.1 illustrates the relation with LAD across varieties and environments for one crop, and Figure 5.2 that across different grain legume crops. The relation for wheat was close in the experiments of Fischer & Kohn (1966, solid triangles in Figure 5.1), in which the LAI began falling before anthesis so that light interception was incomplete and likely to be limiting to yield. With greater LAD values, as in the experiments of Welbank *et al.* (1966, 1968), the relation was less close. In any case, the relation between LAD and yield may reflect the strong influence of duration after anthesis on yield and of LAI and shoot weight at anthesis on potential storage (i.e. sink) capacity.

Yield compensation   Another line of evidence used to argue for source limitation of yield is the well known phenomenon of compensation between the various components of yield. As long ago as 1861 Hallett observed that a 21-fold increase in the sowing rate of wheat resulted in only a 7% increase in the number of ears per square metre. If the components change

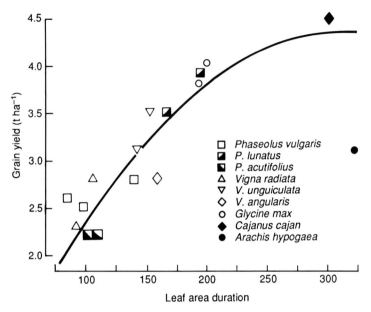

Figure 5.2. Seed yield in relation to leaf area duration from emergence to maturity in legume crops grown at Palmira, Colombia. Adapted from Laing *et al.* (1983).

but the yield remains the same, it could be that the supply side is limiting (e.g. Adams, 1967). A striking example is the experiment of Kirby (1969) in which both grain and total dry matter yields were almost perfectly compensated in barley over a range of sowing densities from 50 to 800 plants m^{-2} (Figure 5.3). Yield compensation was achieved at the lowest planting density by an 8.2-fold increase in ears per plant, a 1.6-fold increase in grains per ear, and a 1.2-fold increase in weight per grain, giving a 15.7-fold increase in grain yield per plant when sown at one sixteenth of the plant density.

The compensation between yield components is usually less perfect (e.g. Darwinkel, 1978; Tungland *et al.*, 1987). Reduction by two thirds in the density of maize plants in a high-yielding crop several weeks before silk emergence led to a halving of yield despite increases in grain number per plant and kernel weight of 28% and 9%, respectively (Schoper *et al.*, 1982). With sorghum, a loss of up to 20% of the spikelets could be almost fully compensated when they came from the base of the panicle, the last to flower, but there was virtually no compensation when they came from the top, where flowering begins and where the heaviest grains occur (Hamilton *et al.*, 1982). The ability to compensate for the loss of grains in wheat also varies. In some

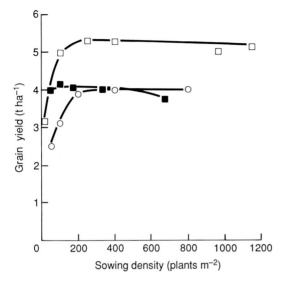

Figure 5.3. Yield compensation in barley and wheat crops over a wide range of sowing densities, from data of Kirby (1969) (■) and Willey & Holliday (1971*a*) (□) for barley and of Anderson (1986) (○) for wheat.

cases the removal of grains has little effect on the weight of the remaining ones (e.g. Buttrose, 1962*a*), suggesting that there was no spare storage capacity. In other experiments (e.g. Stoy, 1965; Bingham, 1967), the kernel weight of the remaining grains increased progressively as more grains were removed, although not enough to compensate for the reduced grain number. Early grain removal has even led, through compensatory grain setting and heavier kernels, to a significant increase in grain yield per ear (Rawson & Evans, 1970; Aufhammer & Zinsmaier, 1982), suggesting that yield was not source-limited.

Experiments with soybeans by McAlister & Krober (1958) and Adams (1967) showed that drastic reductions in pod number could be wholly compensated by increases in seeds per pod and weight per seed. However, Egli (1988) found that while this can happen, especially with indeterminate varieties at fairly high planting densities, at less than 6 plants m^{-2} yield is proportional to density of planting. With broad bean crops, likewise, yield compensation was incomplete even when the population was only halved (Ishag, 1973).

Such manipulations allow us to see whether the yield components determined later in the life cycle can compensate for deficiencies in the components determined earlier. The very fact that such compensation can occur demonstrates that there is spare storage capacity that can be called on if required. However, the sorghum, wheat and legume experiments indicate not only that there may be narrow limits to such compensation, but also that regulatory processes other than assimilate availability may control it.

Brocklehurst (1977) found, for example, that endosperm cell number in the remaining grains was increased by partial sterilization of wheat ears well before their faster grain growth became apparent. Busch & Kofoid (1982) found that selection for larger kernel size in wheat was associated with correlative reductions in the numbers of spikelets per ear and of kernels per spikelet, even though these characteristics were determined long before kernel size.

Enhanced source   The clearest evidence that the source is limiting comes from experiments in which yield is increased when crop photosynthesis is enhanced with raised CO_2 level or irradiance.

CO_2 enrichment raises the yield of many crops (e.g. Kimball, 1983), but in relation to the source-sink debate the significant question is *when* the rise in CO_2 level has its greatest effect on yield. If it is before flowering, storage capacity as determined by photosynthesis then is the more limiting; if after, then the source for grain growth is more limiting. True to crop physiological form, we have answers in favour of the source (e.g. Hardman & Brun, 1971; Ackerson *et al.*, 1984), the sink (Fischer & Aguilar, 1976) or both (Cock & Yoshida, 1973; Gifford *et al.*, 1973; Krenzer & Moss, 1975; Fischer *et al.*, 1977). Once again, the source–sink polarity is too simplistic an approach. Both may be limiting, and the measured response depends on seasonal conditions (e.g. Fischer & Aguilar, 1976), including temperature as well as irradiance. The yield of wheat grown at low temperatures did not respond to increased CO_2 at any stage of the life cycle although it did so at a higher temperature (Krenzer & Moss, 1975).

Enhanced irradiance at various stages has also been used to determine whether source is limiting to yield. Schoper *et al.* (1982) used reflectors to increase irradiance in the upper canopy of a high-yielding maize crop. Treatments before and during silking raised yield more than those during grain growth, mainly due to increase in kernel number and in the weight of apical grains.

The thinning of dense crops at various stages has also been used to improve the light environment of the remaining plants. With maize, Schoper *et al.* (1982) obtained results similar to those with reflectors, namely that thinning before silking increased yield per plant much more than thinning after silking. With wheat crops, likewise, earlier thinning was more beneficial (Willey & Holliday, 1971b; Fischer & Laing, 1976), as is to be expected when yield per plant is the measure.

Evidence that sink is limiting

Reduced sink   Just as reduction of the source below normal levels would be expected to reduce yield, so would reductions of the sink by partial removal of florets or grains. However, compensatory increases in later-determined yield components may occur, and the interpretation of sink removal treatments is also complicated by feed-back effects leading to reduced photosynthesis or earlier senescence by the leaves.

Normal balance   One argument that source is not entirely limiting to yield is the presence of sometimes quite substantial carbohydrate reserves in crops at maturity. In maize, for example, stem sugars may remain high, especially in inbreds (Johnson & Tanner, 1972) and at lower latitudes. Such stem reserves may be important, however, in reducing the incidence of stalk rot in maize and sorghum (Dodd, 1980; Colbert *et al.*, 1984), so their presence may reflect empirical selection pressures rather than sink limitation.

Another line of evidence is that the leaves of a crop may still be photosynthetically active when grain growth ceases, e.g. in wheat (Jenner & Rathjen, 1975) and maize (Crafts-Brandner & Poneleit, 1987a). Feed-back inhibition of photosynthesis normally precludes such behaviour, but its occurrence suggests that the duration of sink activity limits yields in some crops at least. Similarly, the fact that crop photosynthesis often increases when the storage organs begin to grow suggests that the source could have been more active earlier had there been a greater demand.

Enhanced sink activity   This is not so readily arranged as is enhancement of photosynthesis by raising the CO_2 level. As we have seen, however, additional CO_2 or radiation may enhance yield most strongly when given before flowering, during the period when the potential storage capacity of the crop is being determined. In these cases later storage capacity is determined by earlier photosynthesis, but it may also be determined by genotype. The influence of genotype is most clearly seen in grafting experiments, particularly those with root crops, to the extent that such grafting has been used to enhance cassava production with the Mukibat system practised in Indonesia (Bruijn & Dharmaputra, 1974). The work of Borger *et al.* (1956) with potatoes suggested that tuber size and starch content were largely independent of the characteristics of the above ground parts. Thus yield could be strongly sink-determined, as also in the case of reciprocal grafts between sugar and spinach beets (Thorne & Evans, 1964; Rapoport & Loomis, 1985), and between high- and low-yielding sweet potatoes (Hozyo & Park, 1971; Hozyo & Kato, 1976; Hozyo, 1977) and cassava varieties (Ramanujam & Ghosh, 1990).

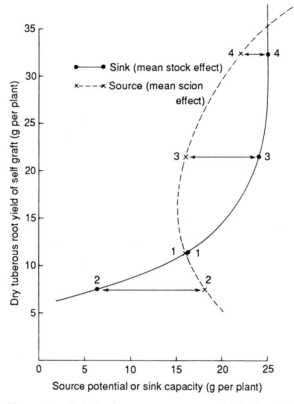

Figure 5.4. Relation between source potential (broken line), sink capacity (solid line) and yield in four varieties of sweet potato, as revealed by cross and self-grafts (Hahn, 1977). (Crop Science Society of America, Inc. Reprinted with permission.)

Grafting experiments with sweet potato varieties by Hahn (1977, 1982) have revealed a wide range in both source and sink activity (Figure 5.4), some varieties having high source but low sink potential, others the reverse. Yield was more closely related to sink than to source activity, but high yields required both to be active. Those varieties with largest sink capacity showed the greatest response of sink to source in a diallel graft, while those with the largest source potential showed the greatest response of source to sink.

Assessment

Because the source activity feeds forward to determine the sink capacity while the sink activity feeds back to modulate the source, they are not independent determinants of yield and may both be limiting to it.

Gifford (1974*b*) has presented a conceptual model indicating how both source and sink may limit yield in a cereal crop after anthesis, by assuming that a common pool of sugars controls both the rate of photosynthesis by the leaf and the rate of grain growth. If the rate of grain growth increases progressively with the concentration of sugars in the pool (e.g. Gifford & Bremner, 1981; Lichtner & Spanswick, 1981; J.H. Thorne, 1982) whereas the photosynthetic rate declines, the crop cannot operate at both full source and full sink capacities because the sugar concentration that maximizes one rate minimizes the other. However, as we have seen above, the level of sugars in the leaves may not be the factor regulating photosynthetic rate, and because loading into the phloem is an active process (Wardlaw, 1990), low sugar levels in the photosynthetic tissues can be associated with much higher sugar levels at the storage organs. The assumption of a common pool of sugars for both source and sink is inadequate.

Thus, the explanation of why source and sink can be simultaneously limiting to yield, as they appear to be in barley (Gifford *et al.*, 1973), must be sought elsewhere. It may be simply that the main limitation oscillates diurnally from day to night – as suggested by the occurrence of growth shells in starch grains (e.g. Buttrose, 1962*b*) – and possibly even from hour to hour as radiation and temperature change, radiation having more effect on the source, temperature more on the sink. In modern varieties, growing under the conditions to which they are adapted, the source and sink are likely to be reasonably balanced except to the extent that the seasonal weather sequence departs from the norm, or that source or sink are reduced by pests and diseases or enhanced by additional CO_2 at one stage or another. Clear evidence of marked imbalance between source and sink under conditions to which modern crops are adapted has not been found. Some spare capacity on both sides, for which there is evidence, is surely needed if yields are not to be too sensitive to environmental conditions. Indeed, it may well have been selected for empirically by plant breeders.

On this view, further increases in yield potential will depend on coordinated increases in both source and sink capacities. Changes in source capacity will be considered first.

Crop photosynthesis

The preceding discussion has emphasized the need to enhance crop photosynthesis, by both genetic and agronomic means, while recognizing that this may not on its own be sufficient to raise yields.

It is a common fallacy to assume that crop photosynthesis is inefficient because only a small proportion of the incoming radiation is converted to crop yield, or because there are parts of the overall process, such as photorespiration, which appear to be wasteful. In fact, at its maximum efficiency of energy capture, photosynthesis is comparable with the best of photoelectric devices, and its apparently wasteful aspects may be of adaptive value. The photosynthetic process has been subject to intense and prolonged natural selection and plant breeders have yet to improve on it. The enzymes, the structures and the regulatory processes involved in photosynthesis may be more efficient than we realize, like many of the other basic metabolic processes (Baldwin & Krebs, 1981).

Crop yields depend on both the rate and the duration of photosynthetic activity. Considering first the rate, in the early stages of crop growth, before a closed canopy is established, crop photosynthesis is most influenced by the extent to which solar radiation is intercepted. There is a premium at this stage on leaf area expansion and on leaves that are more horizontally disposed in ways that minimize mutual shading. Once the canopy is fully intercepting, however, crop photosynthesis is more dependent on the net CO_2 exchange rate per unit leaf area (CER), and there are conditions in which more inclined or aggregated leaf postures are beneficial.

Thus, several characteristics of the photosynthetic system should be able to change during the development of the crop. In the absence of such developmental trends, the optimal solution will be a series of compromises, e.g. between rapid early leaf area expansion and high CER later, between early horizontal leaves and later inclined ones, and between early entire leaves and later dissected ones. It is these compromises that help to explain the apparent paradox that intraspecific variation in CER is known, is highly heritable and can be selected for, but seems not to have contributed to increase in yield potential so far. They must be borne in mind as we compare photosynthetic rates.

Photosynthetic rate comparisons

Many aspects of CER are relevant to crop yield. The maximum rate under high irradiance and atmospheric CO_2 level is relevant to all stages of a crop, and is the most readily compared rate. The quantum requirement under low irradiances is highly relevant to canopy photosynthesis. CER under optimal conditions provides a measure of photosynthetic capacity, but differences between cultivars under more extreme conditions of temperature, drought or nitrogen stress may be at least as important in determining yield. The general effects of various environmental and developmental factors on CER are therefore considered before we examine intra-specific differences.

Irradiance

Some indication of the overall range in the effect of irradiance on CER is given in Figure 5.5, which combines two of the earliest figures

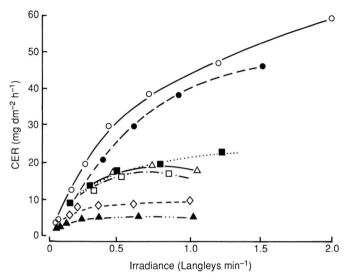

Figure 5.5. Indication of the range among crop plants in the reponse of CO_2 exchange rate to irradiance: maize ($\bigcirc$), sugar cane ($\bullet$), orchard grass ($\blacksquare$), red clover ($\triangle$), tobacco ($\square$), oak ($\diamond$) and maple ($\blacktriangle$). Combined from Hesketh (1963) and Hesketh & Moss (1963). (One Langley is equal to 1 cal cm^{-2} or 4.184 J cm^{-2}.)

highlighting the high rates at high irradiance of plants with the C_4 pathway of photosynthesis, such as maize and sugar cane.

Temperature

Species and varieties may differ greatly in the effect of temperature on CER, as is apparent in Figure 5.6, which presents the 'adapted' CER for several C_3- and C_4-pathway cereals. The higher CER of the barnyard millets (C_4) at high temperatures compared with wheat and rice (C_3) is apparent, but at temperatures below 20°C wheat has a distinct advantage. Photosynthesis in rice (C_3) is as adversely affected by cool temperatures as that in the millets (C_4), but with each of these crops small but adaptively important differences are apparent at low temperatures. The CERs of *japonica* rice and Japanese barnyard millet surpass those of the more tropically adapted *indica* rice and Indian millet below 20°C. The immediate, short-term effect of low temperatures on the CER of these tropical cereals was much less drastic, as illustrated in Figure 5.6 for Indian millet. Thus, relative CERs depend on the temperature at which they are measured and on the length of time at those temperatures.

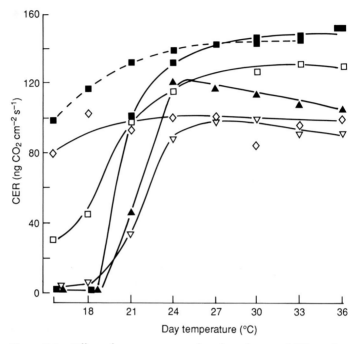

Figure 5.6. Effect of temperature on the adapted rates of CO_2 exchange in recently expanded leaves of several C_3 and C_4 cereals under a flux density of 730 μmol PAR m^{-2} s^{-1}. The adapted rates were measured at a temperature equal to the day temperature in which the plants were grown, whereas the unadapted rates for millet (broken line) were for plants grown at 36/31 °C but with the measured leaves cooled progressively. Results are shown for Indian millet (*Echinochloa frumentacea*, ■), Japanese millet (*E. utilis*, □), *indica* rice (▲), *japonica* rice (▽) and wheat (◇). (Modified from Evans & Bush, 1985.)

Daylength

The greatest influence of daylength on CER is probably indirect, through its effects on leaf size and senescence. In *Poa pratensis*, for example, the longer the day the greater the leaf blade length and area but the lower the CER, as indicated in Figure 5.7, with the result that photosynthesis *per leaf* is always greater with greater elongation.

Previous conditions

In comparisons between species or cultivars, it is not only the irradiance and temperature *during* measurement of CER that are important,

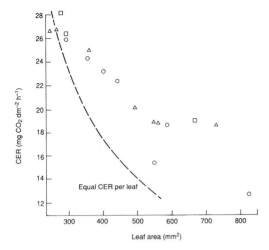

Figure 5.7. The trade-off between photosynthetic rate and leaf area as influenced by daylength in three experiments with *Poa pratensis*. Data of Heide *et al.* (1985). The dashed curve indicates one level of equal photosynthesis per leaf.

but also those under which the leaves developed (e.g. Hesketh, 1968; Bunce, 1985; Begonia *et al.*, 1988). In a comparison of wheat varieties and species, for example, the high CER of the wild diploids under high irradiance was apparent only when the plants were also grown under high irradiance prior to measurement (Dunstone *et al.*, 1973). As Bunce (1986) points out, genotypic differences in such acclimation responses can confound photosynthetic comparisons.

Leaf rank

Comparisons of CER must also take leaf rank into account. With sugar beet (Figure 5.8), Hodanova (1981) found CER to reach its highest value with leaf 20, area with leaf 15, and longevity with leaf 30. Progressive increases in CER with ontogenetic rank have also been found in soybeans (Woodward, 1976) and in wheat up to the seventh leaf (Rawson *et al.*, 1983). In wheat the maximum CER may then decline at higher positions or, if the area of the penultimate and flag leaves is reduced by competition with the developing ear, the CER may rise substantially (e.g. Dunstone *et al.*, 1973). In cotton, CER is high in the vegetative leaves, falls in those leaves that develop in competition with the bolls, but may rise again in still later leaves (Wells, 1988). In tobacco, by contrast, the CER of leaves 6–14 fitted a common curve for leaf development time (Rawson & Hackett, 1974), and rank was less important than leaf age when measuring CER for inter-varietal comparisons.

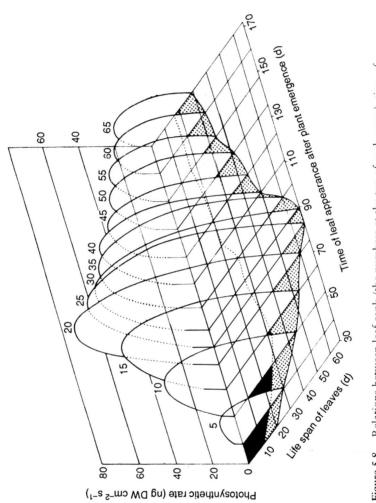

Figure 5.8. Relations between leaf rank (the numbers at the top of each curve), time of appearance, life span and photosynthetic rate in sugar beet (Hodanova, 1981). The dotted line extending the black areas in the base shows the time to full leaf expansion, the hatched areas the remainder of their life span.

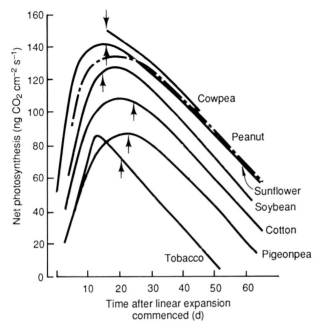

Figure 5.9. Changes in net photosynthesis by the leaves of several crops as they age. Arrows indicate the attainment of final leaf area (Rawson & Constable, 1981).

Leaf age

The effects of leaf age can be a major problem in comparisons of CER, as may be seen from Figure 5.9, which indicates how rapidly CER changes with leaf age in seven dicotyledonous crop plants. Peak rates were maintained for only a short time, especially in tobacco. The overall time course was similar in the seven crops, but peak CER was reached at rather different stages of leaf expansion, at about 37% of full expansion in tobacco (when cell division was giving way to cell elongation), close to full expansion in pigeon pea, and some time after it in soybean. The decline in CER after its peak is closely related to leaf nitrogen content per unit area in tobacco (Rawson & Hackett, 1974), maize (Crafts-Brandner & Poneleit, 1987*a*), soybean (Shibles *et al.*, 1989) and wheat (J.R. Evans, 1983). Because the rate of decline was similar for all species in Figure 5.9, peak CER was positively related to the duration of photosynthetic activity, but this is not always so. For example, the decline in CER of the wild progenitors of wheat is faster than in modern cultivars despite the lower maximum CER of the latter (Evans & Dunstone, 1970).

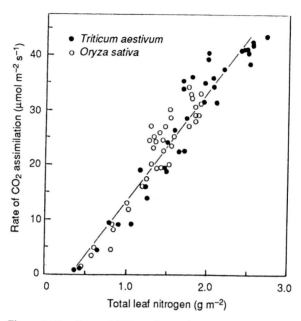

Figure 5.10. Rate of CO_2 exchange (at 1800 μmol PAR m^{-2} s^{-1}, 25 °C, 345 μbar CO_2) as influenced by leaf nitrogen content in the flag leaves of wheat and rice (Makino *et al.*, 1988).

Demand

As we have already seen, CER is greatly influenced by the demand for assimilates, the course of decline being more influenced than the peak. Thus, even when CER is measured under standard conditions of irradiance, temperature, CO_2 and leaf age, varietal differences in CER may reflect differences in sink demand or developmental stage rather than differences in photosynthetic capacity.

Nutrition

Varietal relativities in CER may be affected by nutrition through its effects on the levels of N, P, K, Mg, Fe etc. in the leaves (e.g. Natr, 1972; Terry & Ulrich, 1973; Osman *et al.*, 1977; Terry & Farquhar, 1984; Rao & Terry, 1989).

The relation between nitrogen content per unit leaf area and CER for rice and wheat is illustrated in Figure 5.10. The slope of the relation is identical to those found earlier by Cook & Evans (1983*a*) for rice and by J.R. Evans (1983) for wheat, suggesting that varietal differences in the relation are slight. That

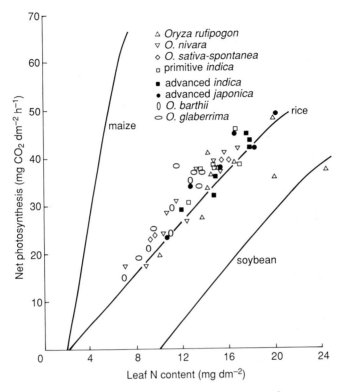

Figure 5.11. The relation between leaf N content and net photosynthesis in several species and races of rice (Cook & Evans, 1983*a*) compared with those presented for maize, rice and soybean by Sinclair & Horie (1989).

the relation may be quite different in other crops is indicated by the curves for maize and soybean in Figure 5.11, based on equations derived by Sinclair & Horie (1989) from a survey of earlier work on these crops. The curves highlight the marked differences between crops in their nitrogen use efficiency for photosynthesis. Comparable work on barley (Natr, 1975), sugar beet (Nevins & Loomis, 1970), tobacco (Rawson & Hackett, 1974) and *Brassica* (Sasahara, 1971) extends these differences. Still wider comparisons indicate a 10-fold range between species in CER for a given level of N per unit leaf area, reflecting different strategies of nitrogen partitioning, electron transport capacities per unit of chlorophyll, and specific activities of rubisco (Evans, 1989).

Thus, when different varieties are compared for CER, we need to know whether any differences between them in CER are associated with differences in nitrogen content per unit leaf area. The data given in Figure 5.11 for six species and several varieties of rice suggest that the relation between CER and N per unit leaf area was similar for all of them except O. *rufipogon*, but that they differed in their leaf N contents, the O. *sativa* cultivars tending to maintain higher values, consistent with their higher specific leaf weights (SLW, i.e. dry weight per unit leaf area). However, Takano & Tsunoda (1971) found some species and varieties of rice to have quite different relations between leaf nitrogen per unit area and CER, especially under high irradiance, the CER of O. *fatua* and of *indica* varieties being higher than those of *japonica* varieties at the same N content per unit leaf area.

In both wheat and soybean much of the variation in CER between varieties is associated with differences in leaf N (Khan & Tsunoda, 1970a; Ojima & Kawashima, 1968; Hesketh *et al.*, 1981). Thus, varietal comparisons of CER may be confounded by differences in leaf N even when plants are grown under the same nutritional conditions. At high levels of N supply the flattening response may conceal varietal differences in CER which are significant to yield at lower N levels. At these lower levels, on the other hand, the use of nitrogenous fertilizers can be a powerful surrogate for genetic improvement in CER.

With this cautionary background on the effects of various factors on CER when measured under constant conditions, we now turn to the evidence for varietal differences and improvement in CER. In the light of that background it might seem almost impossible to identify small gains in CER, yet one could scarcely hope for large gains in a process that has already been subjected to prolonged and intense natural selection.

Intra-specific differences in CO_2 exchange rate

Differences in single leaf photosynthetic rate within species have been surveyed by Evans (1976, 1984), Elmore (1980), Eagles & Wilson (1982), Zelitch (1982), Wilson (1984), Yamashita (1984), Bunce (1986), Austin (1989) and others.

Although photosynthetic rates can be usefully compared in many ways, e.g. per unit dry weight or nitrogen or water use, etc., the following discussion is mainly in terms of CER per unit leaf area because this relates most readily to canopy rates per unit ground area, to light interception and to ontogenetic changes.

Forage plants

Because yield is more directly related to total shoot growth in forage plants than in most other crops, we might expect them to provide the clearest evidence of selection, whether direct or indirect, for higher CER unless obscured by other factors such as palatability, quality, suitability for conservation or competitive ability and survival (Wilson, 1984).

Perennial ryegrass   Wilson & Cooper (1969*a,b*) found highly heritable differences among genotypes of perennial ryegrass (*Lolium perenne*) in the rates of both light-limited and light-saturated photosynthesis. The ranking of the genotypes differed in these two conditions, and yet again under high CO_2 levels. The light-saturated CER was negatively correlated with mesophyll cell size, and selection for small mesophyll cells increased net assimilation rate but reduced the leaf area ratio, the net effect being no change in relative growth rate (Wilson & Cooper, 1970). However, selection for smaller cells also resulted in heavier seeds, giving an advantage in initial growth which was reflected in final shoot weight. Among four families selected from *L. perenne* S321, Rhodes (1972) found the most productive one to have the lowest CER.

Tall fescue   Sufficient genetic variation in CER, and a high enough heritability for effective recurrent selection, have also been found in tall fescue (*Festuca arundinacea*) (Asay *et al.*, 1974). However, the variation in CER is not correlated with yield (e.g. Nelson *et al.*, 1975; Wilhelm & Nelson, 1978*a*, 1979). In one study the highest CER was found in the lowest-yielding selection (Cohen *et al.*, 1982). High yield among these genotypes has been consistently associated with rapid leaf area development (e.g. Nelson *et al.*, 1977; Wilhelm & Nelson, 1979). Randall *et al.* (1977), Byrne *et al.* (1981), Joseph *et al.* (1981) and Meyers *et al.* (1982) all found an increase in CER with ploidy which does not seem to be associated with either the specific activity of rubisco (Joseph *et al.*, 1981) or electron transport and photophosporylation capacity (Krueger & Miles, 1981).

In reed canary grass (*Phalaris arundinacea*), Topark-Ngarm *et al.* (1977) found high SLW and CER to be negatively related to early yields at low densities but positively related in later, denser stands.

Lucerne or alfalfa   With *Medicago sativa* there are substantial differences in CER between lines. These differences bear a close positive relation to SLW (Pearce *et al.*, 1969) but a negative relation to leaflet area and to yield (Delaney & Dobrenz, 1974*a*). When clones were selected for high or low SLW (and therefore CER), yield was found to be positively related to leaf area but either not or negatively related to CER (Hart *et al.*, 1978). When selected for large leaflet size, yield rose although CER fell (Leavitt *et al.*, 1979).

For all these forage plants, therefore, there appears to be a negative relation between CER and productivity.

Cereals

Maize   Selection for faster growth in young maize plants was found by Hanson (1971) to be associated with a fall in CER along with a rise in leaf area growth rate. Significant variation in CER has been found in maize by Duncan & Hesketh (1968), Heichel & Musgrave (1969), Gaskel & Pearce (1981, 1983), Crafts-Brandner & Poneleit (1987b), Rocher (1988), Rocher *et al.* (1989) and others. Comparisons between lines or hybrids are complicated by the finding of Vietor *et al.* (1977) that rankings between hybrids depended on plant age and on which leaf was compared. They were broadly indicative of canopy CER but not clearly related to grain growth (Vietor & Musgrave, 1979). Fakorede & Mock (1978) found no change in CER with recurrent selection for yield.

Crosbie *et al.* (1977) found their rankings of CER at two different stages to be consistent; their estimate of heritability was moderately high, suggesting that selection for high CER could be effective. However, CER was found not to be significantly correlated with yield traits among 64 inbred lines, which Crosbie *et al.* (1978b) concluded could be due to inadequate sink activity. Significant advance in CER under selection was obtained (Crosbie *et al.*, 1981a,b), but whereas height, time to anthesis and other characteristics were modified, grain yield was not significantly changed (Crosbie & Pearce, 1982). In fact, higher CER tended to be associated with lower yields. For maize, clearly, a positive relation between photosynthetic capacity and yield has yet to be established.

Wheat   Varietal differences in CER were first found by Stoy (1965) and Natr (1966). The inclusion of the diploid and tetraploid wild progenitors of modern wheat in the comparisons has led to a wider perspective and to the surprising finding that maximum flag leaf CER is higher in the wild progenitors than in modern cultivars (Evans & Dunstone, 1970; Khan & Tsunoda, 1970a; Austin *et al.*, 1982, 1986; Johnson *et al.*, 1987a). The higher CER in the wild progenitors, especially the diploids, is coupled with smaller flag leaf area (Figure 5.12) and faster senescence (Evans & Dunstone, 1970), and with higher contents of N per unit leaf area (Khan & Tsunoda, 1970a). Nitrogen content per unit leaf area, CER and SLW were all closely and positively related (Khan & Tsunoda, 1970c).

Dunstone *et al.* (1973) found the higher CER of the wild progenitors to be expressed most strongly in the flag leaves and when the plants were grown, as well as measured, under high irradiance. Their higher CER was associated with parallel reductions in stomatal and residual resistances, and with increased stomatal density. Across the range of genotypes from wild diploid to modern cultivar, CER fell as mesophyll cell size increased (Dunstone &

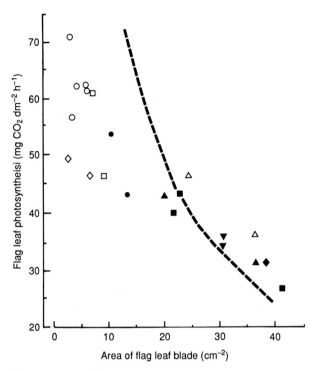

Figure 5.12. The relation between the area and the rate of photosynthesis of flag leaf blades of various wheat species and their wild progenitors (Evans & Dunstone, 1970): *Triticum boeoticum* (○), *T. monococcum* (●), *T. dicoccoides* (△), *T. dicoccum* (▲), *T. durum* (▼), *T. spelta* (◆), *T. aestivum* (■), *Aegilops speltoides* (◇) and *Ae. squarrosa* (□). The broken line depicts the relation for equal photosynthesis per leaf.

Evans, 1974). The number of chloroplasts and the amount of rubisco per cell increased almost threefold from diploid to hexaploid (Dean & Leech, 1982), the chloroplasts covering a relatively fixed proportion of the cell surface (Ellis & Leech, 1985). A strong negative relation between leaf size and CER among wheat species (Evans & Dunstone, 1970; Austin *et al.*, 1982; Johnson *et al.*, 1987*a*) has also been found among cultivars by Planchon (1969, 1979) and Gale *et al.* (1974), and a weaker one by Rawson *et al.* (1983).

According to Planchon & Fesquet (1982) the D genome, besides conferring baking quality and wide climatic adaptation on wheat, also weakened the negative relation between CER and flag leaf area. Singh & Tsunoda (1978), Kishitani & Tsunoda (1981) and Johnson *et al.* (1987*b*) suggest that the smaller, high CER leaves of the wild species are advantageous under arid conditions, but some accessions of the wild tetraploid *T. dicoccoides* combine high CER with large leaves (Carver & Nevo, 1990). Under favourable moisture conditions the gain in leaf area by modern wheats may outweigh the fall in CER, as indicated in Figure 5.12, although Austin *et al.* (1982) found no difference between the wild progenitors and modern cultivars in biomass production when crop duration was comparable.

Among hexaploid bread wheat cultivars, Shimshi & Ephrat (1975) found rankings for yield to be positively correlated with those for leaf permeability, which could reflect varietal differences in access to water. However, Fischer *et al.* (1981) found CER to be only weakly related to crop growth rate before anthesis, although significantly correlated both with CGR during grain growth and with grain yield. Because these characteristics were strongly associated with time to anthesis, they suggest that the correlations probably reflect a feed-back effect of greater sink size (i.e. more grains) with later anthesis, i.e. that the higher CER may be an effect rather than a cause. Among six cultivars and 120 random progeny, Rawson *et al.* (1983) found no correlation between the peak CER of the third and of the flag leaves, and no correlation between peak flag leaf CER and grain yield per ear. Flag leaf area was correlated positively with cumulative dry matter production and negatively with CER.

In wheat, therefore, CER has tended to fall with domestication, as leaves have become larger, and most of the evidence suggests that recent improvements in yield potential have not been associated with any increase in CER.

Rice   Many investigators have recorded varietal differences in CER in both *japonica* (e.g. Murata, 1961) and *indica* rice (e.g. Osada, 1963; Ohno, 1976). When Takano & Tsunoda (1971) compared varieties on the basis of N content per unit leaf area, CER was found to increase in all genotypes throughout the full range in N contents, the highest CER being found in a wild species of *Oryza*. At any one level of N per unit leaf area, there was no evidence that higher-yielding cultivars were superior to wild species or old cultivars. This is also evident in Figure 5.11, although there was a tendency for CER to be lower in the African and wild Asian species of *Oryza* than in *O. sativa* cultivars (Cook & Evans, 1983*b*).

When only cultivars were compared, no increase in CER was found among those grown in the Hokkaido area since 1905 (Tanaka *et al.*, 1968; Samoto, 1971), nor among 50 cultivars grown at various times this century in the Philippines (Evans *et al.*, 1984). Among Indian cultivars, also, differences in peak CER were not positively related to grain yield (Janardhan & Murty, 1978; Palit *et al.*, 1979*a*), as found also by Dey et al. (1990), whose data indicate a negative relation between CER and leaf area. Arjunan et al. (1990) found a positive correlation between CER and yield among 30 Indian varieties, but only in the final stage of crop development.

In their comparison of six varieties, Lafitte & Travis (1984) concluded that the higher CER apparent in some during grain growth was in response to greater sink:source ratios. Kuroda & Kumura (1990*a*) found the CER of modern Japanese varieties to be no higher than those of older ones at heading, but to fall less rapidly during grain growth. This may to some extent reflect greater demand, but it was also associated with more erect leaves (Saitoh et al., 1990) and with higher stomatal conductance (Kuroda & Kumura, 1990*b*) and higher leaf N content (Kuroda & Kumura, 1990*c*) which they consider as likely to reflect greater root activity during grain growth in the modern varieties.

Barley   As in several other cereals, the CER of the wild relatives can exceed that of modern cultivars (Chapin *et al.*, 1989). However, in a comparison of 115 barley cultivars, Apel & Lehmann (1969) found a relatively high CER among 7 high-yielding varieties. The heritability for varietal differences in CER is moderately high (Takeda, 1978).

Oats   Significant and consistent differences in CER among varieties of oats have been reported by Criswell & Shibles (1971) and Lawes & Treharne (1971). CER was higher in some higher-yielding isogenic lines (Brinkman & Frey, 1978).

Sorghum   In a comparison of *Sorghum* species, across three ploidy levels, Downes (1971) found higher CERs among the wild species than in the cultivars. Comparing hybrids and inbreds, Khanna-Chopra (1982) observed heterosis for CER only during grain growth and possibly, therefore, in response to greater demand. Differences in CER among 30 hybrids were most marked under favourable growing conditions (Kidambi et al., 1990). Those among 22 parental lines before anthesis were found by Peng et al. (1991) to be positively correlated with total biomass and grain yield under both well-watered and water-limited conditions.

Millets   When two cultivars of pearl millet were compared with a related wild species, *Pennisetum mollissimum*, Lavergne *et al.* (1979) found the CER to be substantially higher in the wild species. With the barnyard millets, the CER was at least as high in the wild *Echinochloa turneriana* as in cultivars of the tropical *E. frumentacea* and temperate *E. utilis* (Evans & Bush, 1985).

Grain Legumes

Soybean   As in many cereals and other crops, the CER of some wild relatives of the soybean is higher than that of the cultivars (Kokobun & Wardlaw, 1988). Differences in CER between soybean varieties have been found by many investigators (e.g. Ojima & Kawashima, 1968; Dreger *et al.*, 1969; Dornhoff & Shibles, 1970; Watanabe, 1973; Bhagsari *et al.*, 1977). Varietal differences in CER are positively correlated with chlorophyll content (Buttery & Buzzell, 1977; Hesketh *et al.*, 1981), nitrogen content (Ojima, 1972; Buttery *et al.*, 1981; Hesketh *et al.*, 1981), chloroplast number (Watanabe, 1973), & SLW (Dornhoff & Shibles, 1976; Kaplan & Koller, 1977; Hesketh *et al.*, 1981). On the other hand, CER is negatively related to leaf size, as found also by Burris *et al.* (1973) and Sasahara (1984).

Curtis *et al.* (1969) found no relation between CER and yield among 36 varieties, nor did Gay *et al.* (1980) in a comparison of two old and two new varieties. Among four varieties, plants from larger seeds had larger leaves, lower CER but greater yield (Burris *et al.*, 1973). Buttery *et al.* (1981) found r values of 0.65–0.78 between CER during pod filling (August) and grain yield among 12 varieties grown for 3 years, whereas July CER values were not correlated with yield. Such a correlation with CER only during grain filling may, as with wheat and rice, merely reflect differences in demand. Peak CER in soybean is related to crop duration, decreasing as the length of the crop life cycle is extended from Group 1 to Group IV (Ojima & Kawashima, 1968; Gordon *et al.*, 1982), whereas the duration of photosynthetic activity increases. Thus, the relative rankings for CER change in the course of development. Both the duration of maximum CER in upper leaves and the rate of decline in mid-canopy leaves appear to reflect the demands of seed growth as determined by varietal maturity (Hesketh *et al.*, 1981; Shibles *et al.*, 1987, 1989).

High CER in soybeans can be selected for (Ojima & Kawashima, 1970; Wiebold *et al.*, 1981; Secor *et al.*, 1982). Ojima (1972, 1974) obtained preliminary evidence that such selection might lead to increased yield, but Ford *et al.* (1983) found that although such selection was effective, F_7 lines with either high or low CER did not differ in either CGR or yield. Varietal comparisons of canopy photosynthesis are considered later.

Bean   Varietal differences in CER in *Phaseolus vulgaris* were shown by Izhar & Wallace (1967) to be heritable. Peet *et al.* (1977) found the variation in CER among nine bean varieties to be positively correlated with both biological and seed yield when CER was measured at early pod set. However, varietal rankings for CER changed with the stage of development, and may have reflected differences in demand (cf. Liu *et al.*, 1973). Kueneman *et al.*. (1979) found a positive correlation between CER and biomass only in one year and then only at close spacing, with no significant correlation between CER and seed yield, either per crop or per day at any spacing in either year.

Pea   CER in pea is a genetically variable character, which is reproducibly expressed from year to year and of moderately high heritability (Mahon & Hobbs, 1981; Mahon *et al.*, 1983). However, when six genotypes selected for high or low CER were compared in the field over two years, both CGR and yield were higher in the low CER genotypes in an early study (Mahon, 1982), whereas in a later one (Mahon, 1990) differences in CER were not related to LAI, and RGR was positively related to CER.

Peanut   Bhagsari & Brown (1976*a*) surveyed the variation in CER among 6 wild species of *Arachis* and 24 cultivars. The average CER for the cultivars was about 21% greater than that for the wild species, and the highest CER was found in a high-yielding cultivar. However, the correlations between CER in different years and between field and pot values were low, and the only closely related wild species, *A. monticola*, had CER values close to the average for all the cultivars. Rao & Rama Das (1981) found a two-fold range in the CER of third leaves among six peanut varieties; these CERs appeared to correlate highly with their CGRs, which were low. Data presented by Bravdo & Pallas (1982) indicate a negative correlation between CER and leaf area.

Vigna   Among accessions of green gram (*V. radiata*) there was no clear relation between CER measured before flowering, and shoot weight or yield, but positive correlations were apparent when CER was measured after flowering (Srinivasan *et al.*, 1985), as also in black gram (*V. mungo*) (Babu *et al.*, 1985).

Cowpea   Domestication and improvement of cowpea does not seem to have been associated with any change in CER, but the duration of photosynthetic activity may have been increased (Lush & Rawson, 1979).

Other crops

Sugar cane   Bull (1971) found the CER of several commercial clones of sugar cane to be no higher than those of clones of the wild *Saccharum spontaneum* or of *S. robustum*. Yield was not related to CER, but was related to leaf area. Among ten varieties of sugar cane, Irvine (1967) found substantial differences in CER, which were only weakly correlated with stalk weight. His subsequent experiments showed that CER did not increase with increasing yield, which was more strongly related to leaf area. In fact, the wild *S. spontaneum* lines had the highest average CER, 73% higher than that for the *S. officinarum* cultivars (Irvine, 1975). Rosario & Musgrave (1974) also found no significant correlation between CER and yield of either cane or sugar.

Cotton   In a survey of CER in *Gossypium*, the highest rates were found in the wild species, usually associated with smaller leaf area per plant (El-Sharkawy *et al.*, 1965). High dry matter production per plant was associated with larger seeds, cotyledons and leaves, and not with higher CER. Although Muramoto *et al.* (1965) found no differences in CER among varieties of *G. hirsutum* and *G. barbadense*, significant differences in CER among varieties of *G. hirsutum* were found by Elmore *et al.* (1967). With 14 upland cotton cultivars, Bhardwaj *et al.* (1975) found high seed yield to be associated with low CER.

Tobacco   Avratovscukova (1968) and Matsuda (1978) found consistent differences in CER between varieties of tobacco which could be used in a breeding program. Peterson & Zelitch (1982) made frequent measurements of both photosynthesis and respiration by two cultivars of tobacco in field crops. The overall mean CERs for the two varieties were similar and the differences in growth rate and yield were due to the differences in leaf area growth.

Potato   Moll & Henniger (1978) measured CER on 18 clones of potato and found a high positive correlation between CER × leaf area and tuber yield. Comparing only CER and yield in 19 clones, Dwelle *et al.* (1981) found no correlation between them in 1978 when yields were high, but a positive correlation in 1979 when yields were low. By contrast, yield was highly correlated with integrated LAI in 1978, but not in 1979. This suggests that CER may determine yield more strongly when leaf area growth is inhibited by unfavourable conditions. However, three clones that were higher-yielding than Russet Burbank were also found to have higher CER (Dwelle *et al.*, 1983).

Cassava   Among 10 cultivars of cassava and 9 wild species of *Manihot*, Mahon *et al.* (1977) found the highest CER in three of the wild species. Nevertheless, the variation in CER among the cultivars was positively related to biomass, although not to the yield of roots. Differences in CER between cultivars have also been found by Palta (1982) and El Sharkawy *et al.* (1984). In a comparison of only two cultivars, the higher-yielding one had the lower CER (Veltkamp, 1985). Across 127 cultivars El-Sharkawy *et al.* (1990) found a significant correlation between CER and root yield only among those varieties with greater top weight and interception of radiation (i.e. with greater leaf area).

Sweet potato Among 21 genotypes of sweet potato, Bhagsari & Harmon (1982) found a positive relation between CER and biomass in one year, but no significant correlation with root yield, as also found by Bhagsari (1981).

Sunflower Sobrado (1983) found the wild and cultivated forms of sunflower to have similar peak CERs. Lloyd & Canvin (1977) compared CER, photorespiration and dark respiration in 107 genotypes of sunflower, both wild and cultivated, and found the highest CERs among the wild species. Among 10 cultivars of Jerusalem artichoke (*Helianthus tuberosus*) CER was positively correlated with biomass but not with root yield (Soja & Haunold, 1991).

Brassica Differences between lines and species in CER were correlated with cell surface area, leaf N content (Sasahara, 1971) and leaf area (Kariya & Tsunoda, 1972). The wild species had higher CERs than the cultivated ones, associated with their smaller leaves and higher N per unit leaf area (Tsunoda, 1980). No evidence of higher CER being associated with yield improvement was found by Hobbs (1988).

Assessment of single leaf photosynthetic rate studies

The extensive research with many crops on varietal differences in CER has been reviewed in some detail because it is important not only in relation to understanding what changes have contributed to past crop improvement, but also for exploring the way forward. Our survey indicates that, for crop after crop, there has been no increase in maximum CER with increase in yield potential. Indeed, for quite a few of the crops – wheat, rice, sorghum, pearl millet, soybean, sugar cane, cotton, cassava, *Brassica* and sunflower – the highest CERs recorded are for the wild species, and in many instances the comparisons among cultivars revealed a negative relation between CER and yield.

Some positive relations between CER and yield have been found, but most of these were for CER measured during grain growth or pod filling rather than for the earlier maximum CER values. In these cases, therefore, the positive correlation with yield may reflect a feed-back effect of greater sink size, as seems likely in the wheat, rice, sorghum, soybean and green and black gram studies. A longer duration of leaf photosynthetic activity in higher-yielding varieties may also have contributed to these few positive relations between CER and yield. However, such explanations do not apply to the work of Peng *et al.* (1991) with parental lines of sorghum, although their finding that yield was more highly correlated with CER measured during the development of panicles than with that measured at their initiation or exsertion is again suggestive of an effect on demand.

Bunce (1986) claims that 9 of the 23 studies he reviewed showed a positive relation between CER and growth or yield, which sounds more optimistic. However, of these 9 studies, all but one of which have already been discussed, three considered growth but not yield, two revealed no clear relation between CER and yield, and three displayed a positive relation only during the grain growth or pod fill stage. The remaining one, on a grass, revealed a positive relation with forage production only in high-density stands where the usual positive relation with leaf area would not apply.

More telling is the experience with selection for high CER, which has so far not resulted in any increases in yield, and in several cases has resulted in a decrease, e.g. in lucerne, maize, and peas. Moreover, such selection may affect many other traits. In *Phalaris*, for example, selection for high SLW and CER was associated not only with greater leaf thickness but also with notably more upright leaves and heavier but fewer tillers. The net effect of these shifts in partitioning depended on sward density and growing conditions (Topark-Ngarm *et al.*, 1977).

We are, therefore, confronted with an important paradox. There is substantial variation in CER within most crops; this variation is heritable and can be readily selected. Increases in CER associated with higher irradiance or CO_2 levels generally result in increased yields, but genetic increases in CER do not. This 'photosynthetic paradox' has been misunderstood by Zelitch (1982), who suggests that it ignores the positive relation between yield and CER as determined by environmental conditions. Yet that is the very crux of the paradox, that yield is favoured by environmental improvement of CER but not, *so far*, by genetic improvement of individual leaf rates. We need to resolve this paradox if selection for higher CER is to contribute to increased yield potential in the future. Three elements in it are considered here and others (e.g. respiratory losses) later.

The relation between CER and leaf area

The preceding survey of variation in CER within crop species indicated that it is frequently, although not always, negatively related to leaf area. In their survey of this relation, Bhagsari & Brown (1986) found a consistently negative correlation in all 16 of the crops they examined, with a

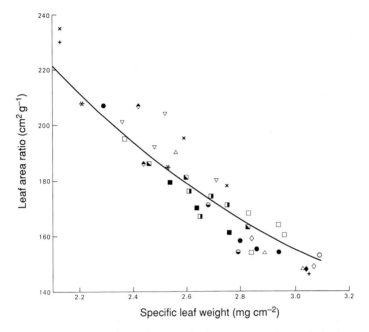

Figure 5.13. The relation between leaf area ratio and specific leaf weight among 9 species and races of *Oryza* (Cook & Evans, 1983*b*).

coefficient of up to -0.94 in soybean. Varietal differences in leaf area and CER are not independent variables in many crops, and the trade-offs between them are a major factor in the photosynthetic paradox.

In classical growth analysis, net assimilation rate (related to mean CER) and leaf area ratio were interdependent components. From their survey of many growth analyses, Heath & Gregory (1938) concluded that the net assimilation rate during the vegetative phase was approximately constant and that 'the main determining factor leading to the very great differences in dry weight accumulation in different types of plants is the rate of extension of the leaf surface ...'. Although the original basis for this conclusion could be questioned, our preceding survey provides many instances of crop biomass and yield being strongly and positively related to leaf area and weakly or negatively to CER.

There are several elements in the trade-offs between leaf area and CER, which play an important role in ecological adaptation (cf. Bloom *et al.*, 1985; Field & Mooney, 1986; Givnish, 1986). For a particular crop the proportion of plant dry weight and N which is in the leaves does not vary greatly between genotypes unless there are marked differences in stature. Figure 5.13 shows the relation between leaf area ratio and SLW for a large number of both wild and cultivated genotypes of *Oryza* from a wide range of environments. The drawn curve depicts the relation when 46% of the dry weight of the plants is in the leaves, and some variation in this proportion with stature is apparent. By and large, however, the consequence of greater leaf area is reduced dry weight and N content per unit leaf area and therefore reduced contents of rubisco and other photosynthetic proteins per unit area, and lower CER, as Oritani *et al.* (1979) have shown in comparisons of rice cultivars. Hesketh *et al.* (1981) refer to this area/rate compensation as 'the dilution hypothesis'. A clear example of it has been seen in Figure 5.7, while Figure 5.12 indicates its likely operation among genotypes of *Triticum*. There are, of course, other factors that may complicate the compensatory relations between CER and leaf area. For example, it may not be apparent when leaves of different ontogenetic rank are compared (e.g. for sugar beet in Figure 5.8), and it may not hold across genotypes when LAI or total rather than individual leaf area is considered, e.g. among pea varieties (Mahon, 1990) or maize hybrids in which high CER is associated with profuse but unproductive tillering (Gaskel & Pearce, 1981). Nor is the relation between specific leaf weight and CER always strongly positive (Bunce 1983, 1986).

The relative advantage of any particular CER/leaf area combination depends on the stage, density and environment of the crop. In the earlier stages of crop growth, at least, the gain in leaf area for the interception of light may outweigh the fall in CER. With dense canopies, on the other hand, the relation between yield and leaf area should be weaker and that with CER stronger, as found with beans and *Phalaris*. Thus an ontogenetic shift from large, low CER leaves during early growth to smaller, high-CER leaves once the canopy is closed could be advantageous. Such a shift occurs to some extent in cereals like wheat and maize, combined with a change from more horizontal to more inclined leaves, but whether the shift is more pronounced in higher-yielding varieties is not known, although implied by the results of Saitoh et al. (1990) with rice and proposed in the ideotype for maize of Mock & Pearce (1975).

Another factor that could favour larger leaves rather than higher CER in selection for greater yield potential would be any allometric relation between leaf size and fruit or grain size. Such relationships have been frequently observed (see, for example, Schwanitz, 1967), especially among polyploid crops. Among wheat cultivars and their wild relatives, for example, there is a high correlation between kernel weight and the area of the largest leaf, each over a ten-fold range (see Figure 3.10). To some extent this is associated with the shift from diploids to hexaploids, but the largest step in the range is from the wild to the cultivated diploids. Larger grains are also linked with larger leaves and lower CER in soybean (Burris *et al.*, 1973) and bean (Adams, 1975), where there is no change in ploidy.

Parallel changes in cell size could be one basis for such an allometric relationship, but among the wheats the projected area of mesophyll cells varied only two-fold, bore no relation to endosperm cell size, and was only weakly correlated with leaf blade area (Dunstone & Evans, 1974). Thus, although changes in cell size with ploidy have contributed to the increasing leaf area and falling CER among the wheats, increase in cell number has also contributed significantly to the greater leaf area (Pyke *et al.*, 1990).

The negative relation between mesophyll cell size and CER found among the wheats has also been found in other crops (El-Sharkawy & Hesketh, 1965) and has been suggested as the basis of the occasionally higher CER in leaves of the semi-dwarf wheats (Le Cain *et al.*, 1989; Morgan *et al.*, 1990). Wilson & Cooper (1969*a*) suggested that in perennial ryegrass it could be associated with the fall in surface:volume ratio of larger cells, but no fall in this ratio was found in the hexaploid wheats because of the additional lobing on their mesophyll cells (Parker & Ford, 1982).

Selection for mesophyll cell size is possible, as Wilson & Cooper (1970) showed with perennial ryegrass, where it led to compensating effects on leaf area and net assimilation rate. Stomatal size, density and arrangement also vary with cell size (El-Sharkawy & Hesketh, 1964), and selection for cell size may have unexpected effects. H.G. Jones (1977) hoped to reduce water use in barley by selecting for low stomatal density but in the event this proved to be associated, via larger cell size, with larger stomata, larger leaves and delayed senescence, all of which increased water use. Jones has done us a service in reporting this reverse, because the complexities of crop response may often lead to such counter-intuitive effects, just as selection for higher CER has resulted in reduced yield, while selection for larger leaves in perennial ryegrass reduced the rate of leaf appearance (Edwards & Cooper, 1961).

Optimum leaf size also depends on radiation and on the availability of nitrogen and water, as these influence the relative magnitudes of boundary layer and stomatal conductance (Farquhar & Sharkey, 1982), as well as on other factors affecting photosynthesis.

The relation between photosynthetic capacity and duration

The improvement of yield potential in many crops has been associated with a greater duration of photosynthetic activity, e.g. in maize (Russell, 1974, 1984; Duvick, 1977, 1984*c*; Crosbie & Mock, 1981; Pearson *et al.*, 1984; Tollenaar, 1991), wheat (Stoy, 1965; Welbank *et al.*, 1966; Evans & Dunstone, 1970), rice (Cook & Evans, 1983*b*; Kuroda & Kumura, 1990*a*), soybean (Wells *et al.*, 1982; Boerma & Ashley, 1988) and cowpea (Lush & Rawson, 1979). If this were, in turn, associated with a reduction in peak CER it would help to explain the photosynthetic paradox.

A negative relation between maximum CER and the duration of photosynthetic activity was not apparent among maize (Crafts-Brandner & Poneleit, 1987*a*) or tall fescue genotypes (Wilhelm & Nelson, 1978*b*), but was found in comparisons between the wild and cultivated wheats, in which the highest CERs are associated with reduced duration (Evans & Dunstone, 1970; Austin *et al.*, 1982). A negative relation is apparent also in comparisons between soybeans of different maturity groups (Gordon *et al.*, 1982). In both these instances, however, the longer duration of photosynthetic activity in the lower-CER varieties may reflect the more prolonged demands for grain growth (see, for example, Shibles *et al.*, 1989).

Whether there are inherent differences between genotypes in the duration of leaf photosynthetic activity as distinct from differences in the duration of grain growth and demand is not clear. Greater demand may have quite different effects on the duration of photosynthetic activity, depending on the relative rates of protein and carbohydrate storage and on the nitrogen status of the crop. When the nitrogen status is high and carbohydrate storage predominates, the photosynthetic activity of the leaf may be prolonged by demand, whereas if protein storage is high or nitrogen status is low, the photosynthetic enzymes may be broken down and the N remobilized out of the leaf and the duration shortened. This is particularly apparent in high-yielding, high-protein grain legume crops (Sinclair & de Wit, 1975, 1976).

In cases where greater or more prolonged demand in higher-yielding crops is reflected in a higher CER during the late stage of grain growth, rankings for CER at that stage may parallel those for yield, even when no positive correlation was apparent at an earlier stage when CER was higher, as found in wheat (Fischer *et al.*, 1981), rice (Arjunan *et al.*, 1990), soybean (Buttery *et al.*, 1981; Boerma & Ashley, 1988), bean (Peet *et al.*, 1977), *Vigna mungo* (Babu *et al.*, 1985) and *V. radiata* (Srinivasan *et al.*, 1985). Varietal differences in photosynthetic longevity may also reflect differences other than those in demand. In rice, for example, more prolonged photosynthesis is also associated with more prolonged root activity (Kuroda & Kumura, 1990*b*) and more upright leaves (Saitoh *et al.*, 1990).

Quite apart from the differences in duration of demand, there may well be inherent differences between varieties in the duration of their photosynthetic activity which could be independently selected. In soybeans, for example, the removal of all flowers did not affect the time when CER began to fall in different varieties, although it did enhance the rate of fall (Crafts-Brandner & Egli, 1987). With maize, the leaves of some hybrids remain green at maturity, with 20% of their maximum CER, whereas others are senescent (Crafts-Brandner & Poneleit, 1987*a*); this difference was not eliminated by early ear removal (Crafts-Brandner & Poneleit, 1987*b*; Connell *et al.*, 1987). The hybrid with delayed senescence (i.e. greater 'stay-green') in both these studies was that giving the world record yield of maize in 1985.

Genetic differences in photosynthetic longevity may be more important to yield potential than those in maximum CER if they are less encumbered by trade-offs, particularly in view of the analysis by Sinclair & Horie (1989) that further increases in CER will have relatively little effect on radiation use efficiency.

CER in optimal and sub-optimal conditions

Even though CER under optimal conditions may not have increased in the course of crop improvement, changes in the ability of photosynthesis to withstand less favourable conditions may have occurred, and may be of considerable significance to crop yield, for example for rice varietal improvement in Hokkaido (Tanaka *et al.*, 1968) and maize in Ontario (Dwyer & Tollenaar, 1989). Although Sinclair & Horie's (1989) analysis suggests little advantage in raising maximum CER further, it also highlights the major impact on radiation use efficiency of falls in CER. However, in relation to the question under consideration, namely why CER has not increased with crop improvement, the relevant issue is whether selection for greater CER at extremes is associated with a reduction in CER under favourable conditions.

For both rice and barnyard millet, the results shown in Figure 5.6 indicate a lower maximum CER in the races with greater tolerance of low temperatures, and other examples of such cross-overs in relative CER are known (see, for example, Duncan & Hesketh, 1968; Bunce, 1981). On the other hand, higher altitude varieties of bean with greater photosynthetic tolerance of cool temperatures showed no reduction in their maximum CER (Laing *et al.*, 1984). Similarly, the data of N. Norcio for several genotypes of sorghum and maize suggest that greater heat tolerance of photosynthesis need not be accompanied by any reduction in maximum CER (Eastin & Sullivan, 1984). Insofar as greater tolerance of drought is associated with smaller leaves, it can be coupled with higher, rather than lower maximum CER as in wheat (Kishitani & Tsunoda, 1981; Johnson *et al.*, 1987*b*), and *Brassica* (Tsunoda, 1980).

Thus, selection for greater tolerance of adverse conditions need not result in a lower maximum CER, so the 'photosynthetic paradox' is presumably due mainly to the frequently negative relation between leaf area and CER, or to other trade-offs involving respiratory losses or changes in partitioning associated with CER.

Some component processes

We have focused our attention so far on single leaf CER because of the large amount of comparative data available and because CER has been considered a feasible criterion for selection. But the CER is simply the net result of many leaf properties and processes which influence the photosynthetic rate either directly or indirectly. Photosynthesis was initially divided into light and dark reactions, or the Hill reactions of water photolysis and the dark reactions of CO_2 fixation, but each of these sets of reactions has been dissected into more and more steps, revealing more and more interdependencies and control points. As the analysis has progressed, so has the identification of limiting processes veered from one candidate to another. The diffusion of CO_2 within the leaves – and even to the leaves within the crop canopy – has been viewed as a major limitation, as have the light harvesting and electron transport processes. On the CO_2-fixing side, the content, activation, kinetic properties and oxygenase activity of rubisco (ribulose-1,5-bisphosphate carboxylase–oxygenase) have been the focus of much attention, to the relative neglect of many other less abundant yet important enzymes. Some attention is now being turned to the limitations imposed by triose phosphate utilization, as well as to those by rubisco and RuBP regeneration, and hence to limitations beyond the photosynthetic process *per se*, associated with the capacity of the plant to use the products of photosynthesis (Herold, 1980; Sharkey, 1985; Stitt, 1986).

Although the unravelling of the C_4-photosynthetic syndrome tended to focus too much attention on rubisco activity, compartmentation and photorespiration, the modelling of C_3 photosynthesis in single leaves (see, for example, Caemmerer & Farquhar, 1981; Farquhar & von Caemmerer, 1982) has brought a more balanced view of the limitations to photosynthetic rate.

Research on the many component processes of photosynthesis may well prove to be crucial to future advances in the yield potential of crops, but cannot be reviewed here. Only a few general comments can be made.

Adaptive changes

Under extreme conditions, photosynthesis may be limited by one particular process, but within the usual range of environmental conditions the overall photosynthetic rate is often limited by the rates of many reactions (cf. Makino *et al.*, 1988; Shibles *et al.*, 1989). With such a tightly linked sequence of adaptive reactions, the concentration of metabolites at branch points may be particularly influential. The enzymes catalysing these reactions tend to be highly regulated, especially the bisphosphatases, the energy-requiring reactions, the irreversible reactions, those involving inorganic phosphorus, and the rubisco reactions (Sharkey, 1985). Even when photosynthesis is limited by light, it may still respond to changes in CO_2 level, and under some conditions it can be light-limited throughout the range of 50–500 μbar CO_2 (Farquhar *et al.*, 1980). With sugar beet crops in the field, the balance between the electron transport and the rubisco-dominated systems is such that both could be simultaneously co-limiting (Terry & Farquhar, 1984).

The many regulatory and feed-back processes controlling photosynthesis at all structural levels from membrane organization to crop canopy, and at intervals ranging from nanoseconds to months, operate to adapt and optimize photosynthetic performance over a wide range of environmental conditions (Gifford *et al.*, 1984). Consequently, even when CER is limited by a particular process, that situation is unlikely to remain true for long, as the balance between processes adapts to current conditions (Sharkey *et al.*, 1988). Sage (1990) has indicated how the activity of rubisco and the rates of electron transport and orthophosphate regeneration by starch and sucrose synthesis in response to changes in irradiance or CO_2 are regulated and balanced.

Such adjustments have been most extensively documented for sun and shade plants under changing irradiance, but also occur with shifts in other environmental conditions such as water availability, where there is continuous control of stomatal aperture to maintain sub-stomatal CO_2 concentration and photosynthesis with minimal water loss (Cowan, 1984). The activities of the two photosystems may change in parallel, as may RuBP carboxylation and RuBP regeneration across various lines of wheat (J.R. Evans, 1986). Parallel changes in the various photosynthetic enzymes during senescence have been found in soybean (Ford & Shibles, 1988; Crafts-Brandner *et al.*, 1990), and in wheat at different levels of nitrogen supply (J.R. Evans, 1983) or irradiance (Kobza & Edwards, 1987). In maize, by contrast, greater N supply led to a pronounced fall in the content of rubisco *vis-à-vis* two other photosynthetic enzymes without altering the proportion of soluble protein allocated to these three enzymes (Sugiyama *et al.*, 1984).

Thus, species, and presumably varieties, differ in the extent to which they adjust the component processes of photosynthesis in order to optimize their performance.

Rubisco

The enzyme rubisco is a major component of leaf protein. In wheat it constitutes about 20% of total leaf nitrogen and 40% of total soluble proteins (Dean & Leech, 1982; Evans & Seemann, 1984), but these proportions vary substantially between species (Figure 5.14) and to some extent between varieties. Across species the greatest difference is between those with the C_3 and those with the C_4 photosynthetic pathway (Long, 1983), the latter having a much lower proportion of leaf N invested in rubisco. Among varieties, the proportion of soluble protein present as rubisco ranged from 48 to 57% in rice (Makino *et al.*, 1987) and from 44 to 56% in lucerne (Meyers *et al.*, 1982).

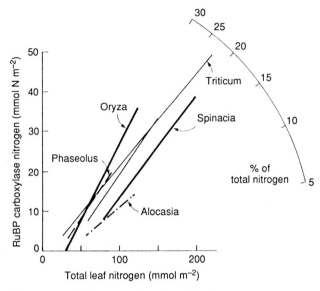

Figure 5.14. The proportion of total leaf nitrogen invested in rubisco by various C₃ crop plants (Evans, 1989).

Bjorkman (1981) found a high correlation between the light-saturated CER and the fully activated rubisco activity across many species. Within species the relation is also close, for example in rice (Makino *et al.*, 1983, 1988), wheat (J.R. Evans, 1986; Evans & Seemann, 1984; Lawlor *et al.*, 1989), soybean (Hesketh *et al.*, 1981; Seemann & Berry, 1982; Shibles *et al.*, 1989) and tomato (Besford *et al.*, 1985).

The limitation of photosynthesis by rubisco in spite of its considerable abundance, coupled with consideration of its high molecular weight and low turn-over number, has made the enzyme a prime target for 'improvement' (see, for example, Raven, 1977), on the grounds that its kinetic properties reveal it to be an anachronism under present atmospheric concentrations of CO_2 and O_2. Such a view may be simplistic, however, given the long period and intense natural selection to which it has been exposed under these conditions. One example of apparently adaptive change in rubisco is the difference in its kinetic properties between plants with the C₃ and C₄ pathways of photosynthesis, the $K_m(CO_2)$ being lower in the C₃ grasses (Yeoh *et al.*, 1980) and the specific activity higher in C₄ plants (Seemann *et al.*, 1984).

Intra-specific variation in rubisco is less clearly established. Makino *et al.* (1987) found only small differences in $K_m(CO_2)$, V_{max} and the ratio of carboxylase to soluble protein among 25 cultivars of rice but greater differences in these characteristics among related *Oryza* species with the A and C genomes. Among barley lines, Ecochard *et al.* (1991) found some differences in rubisco as a percentage of soluble protein (65–74%) and in the specific activity of rubisco. Some variation in rubisco may occur in allopolyploid series in which different cytoplasms specify the large subunit. Evans & Seemann (1984) found the specific activity of rubisco to be 30% greater in the hexaploid wheat *T. aestivum* than in the diploid *T. monococcum*. This *in vitro* difference has been associated with the influence of the B cytoplasm on the large subunit of rubisco (Evans & Austin, 1986), but did not result in a difference in turn-over number for the enzyme *in vivo* (J.R. Evans, 1986). Thus, even where differences in kinetic properties are found *in vitro*, it needs to be shown that they operate *in vivo* as well.

Makino *et al.* (1988) found wheat and rice to exhibit the same relation between CER and leaf N (Figure 5.10), and also between CER and rubisco *activity*, even though rice appeared to have 25% more rubisco per unit leaf area (Figure 5.14) and as a proportion of total N. Superficial analysis might suggest that the efficiency of rubisco in rice could be improved by at least 25%, but too little is known of the optimization and regulatory processes to be at all sure of this. Both the amount and the activation of rubisco change diurnally in response to irradiance (Kobza & Seemann, 1989; Salvucci, 1989), as well as more slowly in response to changes in source–sink balance (Sawada *et al.*, 1990) and CO_2 level (Yelle *et al.*, 1989*b*), such that the regulated level of rubisco activity is closely correlated with CER over a wide range of irradiances.

Electron transport

The conclusion reached by Terry & Farquhar (1984) for sugar beet, that photochemical capacity may be co-limiting with the capacity to fix CO_2, probably applies to many other crops, at least over the course of a day. As Sharkey (1989) puts it, 'If photosynthesis is limited solely by Rubisco in the middle hours of the day, and by electron transport capacity at the beginning and end of the day, this may be the best use of scarce resources'. Increase in the quantum efficiency of photosynthesis would be welcome and useful, but it appears that most crops are already able to use absorbed radiation with close to maximum efficiency. McCree's (1972) survey of 22 very different crop plants indicated a close similarity in their relative quantum yields at different wavelengths.

In Osborne & Garrett's (1983) examination of several diploid and polyploid crop and pasture plants, quantum yields approaching the maximum value of 0.08 mol CO_2 mol^{-1} photons were obtained with C_3 plants, but only at reduced O_2 levels. The yield was lower in C_4 plants, as found also by Ehleringer & Pearcy (1983) who measured efficiencies ranging from 0.052 to 0.069 mol CO_2 mol^{-1} photons, reflecting the higher intrinsic energy requirement of C_4 photosynthesis. However, there was little variation evident among species when photosynthetic rate was plotted against absorbed PAR, the quantum yield at atmospheric CO_2 and O_2 levels being about 0.05 for both C_3 and C_4 species. Variation in leaf absorptance rather than in quantum efficiency is therefore likely to have a greater influence on crop photosynthesis in the field.

Thus, many of the differences between crops in their photosynthetic behaviour are associated with adaptive adjustments and trade-offs between the various constituent processes of photosynthesis rather than to inherent differences in efficiency.

Canopy photosynthesis and stand structure

Measurements of CER made on single leaves when their photosynthetic rate should be close to maximal are operationally feasible as a selection criterion in a breeding program but, as we have seen, they tend to be negatively or not related to yield. Measurements of net photosynthesis by the crop canopy – which integrates the compensating effects of leaf area, presentation (inclination, shape etc) and CER – would be a better criterion, especially if integrated over the life of the crop or during the grain growth phase. Such measurements should relate closely to the crop growth rate even though they may not correlate with yield. Even when they do so, however, they may not be a practicable criterion for selection. With lines from two soybean crosses, Harrison et al. (1981) found that seed yield and the canopy photosynthetic rate per unit ground area were significantly related in both crosses. Heritability was higher for canopy photosynthesis than for yield, but the cost and time for measuring it greatly exceeded those for measuring yield. However, techniques for such measurements are improving rapidly.

Eagles & Wilson (1982) have tabulated many records of both single leaf and canopy photosynthesis measurements. The highest canopy rates for eleven C_3 species in their survey range from 31 to 126 mg CO_2 $dm^{-2} h^{-1}$, while those for the four C_4 species range from 50–144 mg CO_2 $dm^{-2} h^{-1}$. Gifford (1974a) has explained why the pronounced differences between C_3 and C_4 crops in their rates of single leaf photosynthesis are attenuated at the canopy photosynthesis level. Nevertheless, C_4 crop canopy rates can greatly exceed those for C_3 crops, particularly when irradiance is high. Kanemasu & Hiebsch (1975) compared the rates for sorghum, wheat and soybean canopies throughout a growing season. At low irradiance the rates were similar, but whereas the wheat and soybean crops did not exceed rates of 45 mg CO_2 $dm^{-2} h^{-1}$, sorghum reached 100 mg $dm^{-2} h^{-1}$ and was up to four times higher than the concurrent rate for the soybean crop despite their similar leaf area indices.

Varietal comparisons of canopy photosynthesis integrated over the whole growing season have rarely been made. In one early study comparing two wheat cultivars, Puckridge (1971) found differences in LAI to be the major determinant of canopy photosynthesis and yield. With one variety of soybean grown at four different spacings, Christy & Porter (1982) found a close relation between integrated photosynthesis and grain yield even though the main differences in photosynthesis occurred early in the growing season, associated with different rates of canopy closure. Differences in canopy rates only early in the growing season are uncommon in comparisons between varieties. In cotton they were positively related to yield (Wells *et al.*, 1988) but not in a comparison of tall and dwarf wheat varieties by Gent & Kiyomoto (1985). In this case, although the tall variety had the higher canopy rates early on, and similar rates later, its grain yield was less. Early differences in canopy rates may not, therefore, be a guide to differences in yield.

Duration of canopy photosynthesis

Most varietal comparisons of canopy rates of photosynthesis have been confined to the period of grain growth, for example in soybean. Jeffers & Shibles (1969), working with three varieties of soybean, found the canopy photosynthesis rate to be highly dependent on LAI at high irradiance, and on irradiance at high LAI, but much less so on variations in temperature and on the relatively small differences between varieties. Egli *et al.* (1970) likewise found the differences between their three varieties to be small compared with those due to irradiance or CO_2 level. Larson *et al.* (1981) compared 13 varieties from maturity groups I–IV: canopy photosynthesis was closely related to LAI during early growth, while later it was related to maturity. The maximum rates were slightly lower in the three oldest, lower-yielding varieties, but the duration of these high rates depended mainly on the earliness of flowering and maturity.

Wells *et al.* (1982) used the same 16 varieties from 4 soybean maturity groups as Bhagsari *et al.* (1977) used to compare individual leaf CERs. Comparison of these leaf rates with the canopy rates over two intervals in 1975 from the data of Wells *et al.* gave correlation coefficients of only $+0.30$ and $+0.15$, an indication of how poor a determinant and guide leaf CER is for canopy CER. Grain yield was correlated ($r = 0.64$) with canopy photosynthesis integrated over the period from full pod to maturity. Within each maturity group, the decline in canopy photosynthesis occurred five or more days later in the highest yielding variety than in the lowest one, although they did not differ consistently in their peak rates (Figure 5.15).

Boerma & Ashley (1988) found a strong association between greater canopy photosynthesis from flowering to maturity and greater duration of

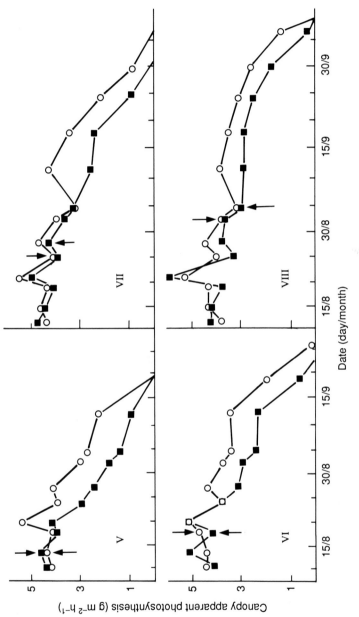

Figure 5.15. The time course of photosynthesis by crop canopies of the lowest (squares) and highest yielding (circles) varieties of soybean in Maturity Groups V–VIII (adapted from Wells *et al.*, 1982). Arrows signify the start of pod filling.

grain growth, both being positively correlated with grain yield. Increased yield was associated with delayed decline in photosynthesis, which could in turn have been due to a greater demand for photosynthates by the varieties with higher yield potential as much as to photosynthesis driving grain growth, which Sinclair & de Wit (1976) assume. The same uncertainty arises from the work of Pearson *et al.* (1984) on varietal differences in canopy photosynthesis of maize. The modern hybrid had a higher canopy photosynthesis rate only towards the end of grain filling, which could have reflected a greater demand for assimilates because leaf CER was shown to be enhanced or diminished within one hour by detopping or ear removal respectively.

Light interception and stand structure

The relation between light interception, canopy photosynthesis and yield is clear in the work by Wells *et al.* (1986) with isogenic cotton lines differing in leaf morphology. The more dissected the leaves, the lower was light interception by the crop, the lower was canopy photosynthesis and the lower was the final lint yield, especially with super-okra leaves. Lint yield correlated quite highly with integrated canopy photosynthesis during the reproductive stage.

Besides leaf shape, leaf inclination and arrangement also influence photosynthesis. Simulation models have predicted that more erect cereal leaf posture should result in higher photosynthesis at high LAI than with lax-leafed forms, but in lower photosynthesis at low LAI (see, for example, McCree & Keener, 1974). With barley, Angus *et al.* (1972) found such an interaction between leaf posture and density of planting on the yield of barley, whereas Tungland *et al.* (1987) found no yield advantage for erect leaved barley lines at any planting density. Comparing two lines of wheat, Austin *et al.* (1976) found the yield to be higher in the erect line in the year when the LAI was higher, and in the lax-leafed line in the year when the LAI was lower, although canopy photosynthesis during grain filling was somewhat higher with the erect line in both years. Innes & Blackwell (1983*a*) found a small increase in yield associated with more erect leaf habit in winter wheat, whereas the highest yielder among five winter wheat varieties compared by Green (1989) was the one with the least erect leaves. In his comparison, the varieties with higher extinction coefficients had higher radiation use efficiency after anthesis, whereas higher yield among Japanese and Philippine rice varieties is associated with more erect leaves and a lower extinction coefficient after anthesis (Saitoh *et al.*, 1990; Evans *et al.*, 1984). Recurrent selection for yield in maize also resulted in more upright leaves at the top of the canopy (Fakorede & Mock, 1978), even though Hicks & Stucker (1972) and Russell (1972) found no yield advantage with comparisons of upright and lax-leafed forms, probably because their comparisons were not done at high enough planting densities and with density-tolerant genotypes (Mock & Pearce, 1975).

Leaf inclination has many different effects and pleiotropic associations, whose net effect on yield depends greatly on variety and environmental conditions. Like many other characteristics, its value is more interactive than intrinsic.

Effect of leaf area index

We have seen that LAI has a profound effect on canopy photosynthesis. Early crop models and experiments (reviewed by Yoshida, 1972), suggested the existence of an optimum LAI for net canopy photosynthesis, on the assumption that dark respiration increased in proportion to LAI. This was found not to be so with cotton communities except at high temperatures (Ludwig et al., 1965) nor with white clover (McCree & Troughton, 1966b) nor with many other crops such as wheat, rice, maize, sunflower, soybean, lucerne and others (cf. Yoshida, 1972; Bunce, 1989). Plateaus rather than optima also characterize the relation between yield and LAI, as in rice (Yoshida & Parao, 1976).

The overall conclusions from canopy photosynthesis measurements are that:

(1) unlike single leaf CER, canopy photosynthesis often bears a close relation to yield;
(2) in the early stages of crop development, it is more closely related to leaf area growth than to single leaf CER;
(3) in the later stages of crop development, varietal differences in canopy photosynthesis may reflect differences in demand rather than be the cause of differences in yield;
(4) the influence of canopy structure on yield may sometimes be through associated effects rather than through those on canopy photosynthesis.

Respiratory losses

Respiratory losses by crops commonly account for about half of the carbon fixed in photosynthesis, according to the survey by Amthor (1989), while photorespiration reduces the amount of carbon initially fixed by C_3 crops in photosynthesis by a further 15–20%. Many attempts have been made to reduce the photorespiratory losses but far less effort has been given to the search for reductions in 'dark' respiration, possibly because of the view that, as the product of long evolution, respiration is already a highly efficient process. However, the operational distinction between growth or synthesis respiration and maintenance respiration stimulated a reassessment of the opportunities because even if there is not much scope for reducing the growth component, that for maintenance – so dominant in the later stages of the crop – may be amenable to change once it is better understood.

Photorespiration

For those geneticists and physiologists with little faith in the power of natural or empirical selection to improve efficiency, photorespiration seemed an obvious target for specific 'improvement' (Zelitch, 1975). It was an apparently wasteful process, and had been circumvented by anatomical rearrangement and biochemical compartmention in plants with the C_4 pathway of photosynthesis.

Many C_3 crops were therefore screened for genotypes with reduced photorespiration. In the event, most of these programmes found little variation in the CO_2 compensation point or in the relation between photorespiration and gross photosynthesis, for example in rice (Akita, 1980), soybean (Cannell *et al.*, 1969) and sunflower (Lloyd & Canvin, 1977). Rather greater differences between lines have been reported in tobacco (Zelitch & Day, 1973) but these may reflect anatomical or adaptive differences.

Lorimer & Andrews (1973) suggested that the light-dependent production and metabolism of phosphoglycollate is an inevitable consequence of the reaction mechanism of carboxylation and of the oxygenase function of rubisco. That being so, the recovery of three quarters of all the carbon lost to phosphoglycollate through the processes of photorespiration is highly efficient compared with excretion of glycollate by algae.

In fact, photorespiration is probably of considerable adaptive value and may play a protective role in the regulation of reductant and the avoidance of photochemical damage, especially in plants at low temperatures or under water stress (Osmond & Bjorkman, 1972; Lawlor, 1979; Evans *et al.*, 1988). When wheat was grown in low O_2 to reduce photorespiration, with the CO_2 level also adjusted so that photosynthesis was unchanged, reproductive development was slowed and seed setting inhibited (Gerbaud *et al.*, 1988). Reduced O_2 levels also arrested seed development in sorghum and soybeans (Quebedeaux & Hardy, 1973, 1975), but whether this was due to interference with photorespiration or to some other process is uncertain. If photorespiration is of adaptive significance, there will be constraints on the extent to which it can be varied in order to enhance yield.

Dark respiration

About 40–60% of the carbon fixed in photosynthesis is commonly lost through subsequent dark respiration, e.g. in maize (Loomis & Lafitte,

1987), sorghum (McCree, 1988), rice, maize and soybean (Yamaguchi, 1978), tobacco (Peterson & Zelitch, 1982), perennial ryegrass (Danckwerts & Gordon, 1987) and other crops (Amthor, 1989). The actual proportion varies with the type of crop, with environmental conditions and with stage of crop development, as in potatoes (Gawronska et al., 1984) and barley (Biscoe et al., 1975).

Given this magnitude of respiratory loss, one might have expected almost as much attention to have been paid to the efficiency of respiration as to that of photosynthesis. This has not happened, partly because respiratory loss is compounded of many processes and several pathways whose overall magnitude is not easy to assess, and partly because of the earlier conviction that respiration was already a highly efficient process tightly coupled to cellular work except in special cases of adaptive uncoupling, and governed by the pace of the processes consuming its products (Beevers, 1970).

One major problem in the assessment of respiratory loss has been uncertainty about the magnitude of 'dark' respiration in the light (Graham, 1980). Suggestions ranged from its being completely switched off in photosynthetic tissue to being enhanced because of the greater availability of assimilates for synthesis by day, as in the roots of wheat (Neales & Davies, 1966). Most estimates have taken the easiest course of assuming that it continued at the rate it would have been in darkness, which now appears to be approximately correct. This was suggested by Brown's early work (1953) with oxygen isotopes, and from the use of photosynthetic inhibitors on several crops (King et al., 1967; Downton & Tregunna, 1968; El-Sharkawy et al., 1968). More recent work by Azcon-Bieto (1986), Brooks & Farquhar (1985) and McCashin et al. (1988) suggest that respiration continues in the light at 75–80% of its dark rate.

An operational advance in how to compare respiration rates arose out of early attempts to model crop photosynthesis. There was uncertainty whether to relate respiration to biomass or LAI on the one hand, as done by Davidson & Philip (1958), or to photosynthesis on the other. One consequence of making respiration proportional to biomass was a well-defined optimum leaf area index. However, experiments with cotton, sunflower, wheat and lucerne communities (Ludwig et al., 1965; King & Evans, 1967) and with white clover swards (McCree & Troughton, 1966a,b) indicated that the respiration rate quickly adjusted to the photosynthetic rate, with the result that no optimum LAI was apparent. An outcome of this work was McCree's (1970) suggestion that crop respiration was the sum of two terms, one related to gross photosynthesis and the other to biomass, being 25% of the former and 1.5% per day of the latter in his example. Although these two components are not biochemically distinct, their functional identification as, respectively, growth respiration and maintenance respiration (Thornley, 1970) has proved to be useful. Growth respiration has been reviewed by Lambers *et al.* (1983) and maintenance respiration by Amthor (1984).

The *growth respiration* term is relatively independent of environmental conditions or cultivar differences except insofar as these influence the composition of the growing tissue. Penning de Vries (1974, 1975*a*) concludes that the synthetic processes operate at high efficiency, so that there is unlikely to be much scope for improvement by plant breeding, a conclusion supported by the elemental analyses of McDermitt & Loomis (1981), Loomis & Lafitte (1987) and Lafitte & Loomis (1988). Estimates of growth respiration losses have ranged from 14% in maize and 18% in sorghum to 34% in barley (Biscoe *et al.*, 1975).

With *maintenance respiration*, however, there may be some scope for selection, albeit possibly at the expense of adaptation to stress. Penning de Vries (1975*b*) includes within this component the CO_2 evolved in the generation of energy for (1) resynthesis of the substances undergoing renewal during metabolism, such as enzymatic proteins, ribonucleic acids and membrane lipids; (2) maintenance of the required gradients of ions and metabolites; and (3) processes involved in physiological adaptation to stressful or changing environments. Maintenance respiration is, therefore, affected by environmental conditions such as temperature and possibly by differences between cultivars in their adaptation to these. It is also, of course, greatly influenced by the stage of crop development and biomass accumulation.

Loomis & Lafitte (1987) estimate that of the total C fixed photosynthetically by a maize crop, 47% was lost by respiration, two thirds of which was maintenance respiration. With a barley crop Biscoe *et al.* (1975) estimated that over the whole growing season maintenance respiration accounted for about half of the carbon loss, rising from 18% in the early stages to 72% during grain filling when the accumulated biomass reached its maximum. However, as Stahl & McCree (1988) have shown for sorghum, the coefficients for both growth and maintenance respiration may fall as the crop ages, owing to changes in its composition such as the reduced protein content at later stages (McCree, 1988).

Besides possible differences between cultivars in protein turn-over rates or adaptive processes, there may also be differences in the extent to which cyanide-insensitive respiration occurs. Not being coupled to ATP synthesis, this alternative pathway has been considered wasteful. It is more prominent in older tissue (Day *et al.*, 1985), but may be of adaptive significance. Lambers (1982) estimates that it reduced potential grain yield of wheat by 6% in one instance. He ascribes an energy overflow function to the alternative pathway, but there was no additional diversion of C through it in wheat grown at high CO_2 (Gifford *et al.*, 1985).

Cultivar differences

Differences between cultivars in respiration rates per unit leaf area or dry weight have been reported for many forages and crops, but in most cases their relation to yield potential is not clear. Much depends on the stage of growth at which the comparisons are made. In young plants, when synthesis respiration is the dominant component, one might expect a positive relation between growth and respiration rate. At later stages, when maintenance respiration is the dominant component, the relation between growth and respiration may be negative.

Positive relations between respiration rate and growth when cultivars are compared have been found, for example, in lucerne (Delaney & Dobrenz, 1974b), tall fescue (Jones et al., 1979), sorghum (Gerik & Eastin, 1985) and with the growth of the wheat ear (Voss, 1979). However, the relation depends on the basis of comparison, the lucerne and sorghum yields being positively correlated with respiration rates expressed on a whole plant basis but not on a leaf area basis. Likewise, Volenec et al. (1984a) found the respiration rate of diverse tall fescue genotypes to be positively correlated with leaf nitrogen content but negatively related to yield per tiller, whereas Volenec et al. (1984b) found leaf respiration rates not to be consistently correlated with any characteristic measured, in spite of their fairly high heritability. In soybean a quite large difference between two cultivars in leaf respiration rates was not clearly related to yield (Mullen & Koller, 1988), nor was the extent of respiratory loss related to yield among four varieties of potato (Gawronska et al., 1984). With sorghum, Fernandez (cited by Gerik & Eastin, 1985) found no difference across cultivars in the efficiency of growth respiration, but a negative relation between yield potential and maintenance respiration.

Perennial ryegrass still offers the most hopeful example of the possibility of selecting for low respiration rates. Wilson (1975) found substantial variation in the respiration rate of fully expanded leaves of S_{23} perennial ryegrass, which was highly heritable, readily selected for, and correlated negatively with leaf expansion ($r = -0.78$) and regrowth ($r = -0.89$). The differences in growth rate, however, were not apparent until after canopy closure (Wilson, 1982), suggesting that they may reflect differences in maintenance respiration. They are not associated with differences in the cyanide-resistant pathway (Day et al., 1985) nor with a difference in protein turn-over rates (Barneix et al., 1988). Robson (1982b) also found no difference in early growth between the high and low respiration lines, and his model suggested that no differences need be assumed except in the respiration rate of mature leaves. The yield advantage of the slow respiration line increased over successive regrowth periods to 23% (Robson, 1982a), disappeared in N-deficient swards (Robson et al., 1983), but was enhanced in mixed swards (Pilbeam & Robson, 1992).

Another case where low respiration rate appears to be advantageous to yield is that of the sugar beet. Wyse *et al.* (1978) and Doney *et al.* (1985) found that hybrids with high sugar yields had lower respiration rates during storage than the inbred parents. Whether or not varieties differ significantly in the use of the CN-resistant pathway is still being debated. Musgrave *et al.* (1986*a,b*) claim that pea varieties can differ substantially in its operation, yet Obenland *et al.* (1988) failed to find such differences using the same varieties.

Thus, there are some circumstances in which low respiration rates appear to be advantageous to yield, as in pasture regrowth and sugar beet storage, but the relation between respiration rate, grain yield and its stability over a range of environmental conditions remains unclear. Selection for low respiration rate seems to be quite feasible, but its effects on yield potential and on performance under stress will require careful assessment.

Growth rates

Relative growth rate

Pronounced differences between crops or cultivars of a crop are often apparent in the size of seedlings a few weeks after sowing, and may be of crucial importance in the competition with weeds. Moreover, such differences can persist at least until canopy closure (Black, 1957), leading to the impression that some cultivars grow much faster than others. In many cases, however, these differences reflect differences in either the speed of germination or the size of the seed sown. Figure 5.16 illustrates the latter relation for a number of wild and cultivated wheats. Similar relations are apparent, for example, among varieties of rice (IRRI, 1979*a*) and cowpeas (Lush & Wien, 1980). In the common bean, RGR is negatively correlated with seed size (White *et al.*, 1992), and across tropical legume species (Whiteman, 1968). High biomass lines of barley all had large grains and rapid early growth (Hanson *et al.*, 1985).

The relative growth rate (RGR), i.e. the exponential rate of increase in the mass of plants, falls as their size and absolute growth rate increases (e.g. Bush & Evans, 1988; Poorter, 1989). Consequently, valid comparisons between crops or cultivars for RGR can be made only with plants of comparable size, as in Figure 4.10. When cowpea genotypes were compared at the same age, higher RGRs were found among the smaller seeded wild forms, but when compared across the same size range, their RGRs were similar (Lush & Wien, 1980).

Heichel (1971) compared two maize inbreds which did not differ in photosynthetic rate, and claimed that the inbred with the higher respiration rate had the lower RGR. However, it also had the heavier kernels and so, by

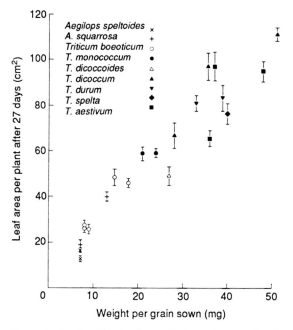

Figure 5.16. Seedling leaf area 27 days after germination as related to kernel weight in various wheats and their wild progenitors (Evans & Dunstone, 1970).

analogy with the cowpea results, its lower initial RGR was to be expected. This indicates the care needed in making comparisons of RGR. Since temperature, irradiance, daylength etc also influence RGR profoundly, comparisons must be made under the same conditions.

Differences between crops in RGR were indicated in Figure 4.10. Those between the C_3 and C_4 cereals are striking – cf. Kawanabe (1968) for grasses – especially at high and low temperatures. However, even the smaller differences between *indica* and *japonica* rice, or between the Indian and Japanese barnyard millets, may have considerable agricultural significance. Small differences between cultivars in RGR under sub-optimal conditions undoubtedly occur, and may be important, but there is no evidence that varietal improvement has, so far, been associated with an increase in maximum RGR under favourable conditions.

No consistent differences in RGR were found among the wild and cultivated wheats (Evans & Dunstone, 1970; Khan & Tsunoda, 1970*b*), nor among wheats differing in year of release (Siddique *et al.*, 1989*b*) or in the presence of the Rht dwarfing genes (Bush & Evans, 1988), nor among 22 races of maize and teosinte ranging from land races to modern hybrids (Duncan &

Hesketh, 1968). Among tomatoes, likewise, highly productive modern cultivars have RGRs no higher than their wild progenitors (Warren Wilson, 1972). Among the rices, all three African species and *Oryza nivara*, the wild annual progenitor of *O. sativa*, had somewhat higher RGRs, while the more advanced *indicas* had lower rates than the older cultivars (Cook & Evans, 1983*b*). Thus, no increase in RGR with the progress of breeding was apparent, nor was any found among Hokkaido rice varieties by Tanaka *et al.* (1968). Comparison of wild, weedy and cultivated barleys revealed higher RGRs in the weedy forms, associated with their smaller seed size (Chapin *et al.*, 1989) and among *Echinochloa* millets, the RGR of the cultivars was no higher than that of a wild relative (Evans & Bush, 1985).

Thus, while differences in RGR do occur, often associated with differences in partitioning to leaf area growth (Poorter *et al.*, 1990), there is no evidence that crop improvement has been accompanied by an increase in RGR when comparisons are made between seedlings over the same size range and under standard conditions. One consequence of this is that relative differences in seedling size associated with those in seed weight or speed of germination may persist at least until canopy closure and may therefore be reflected in substantial differences in *crop growth rate* at particular times, a possibility to be borne in mind with our next topic.

Crop growth rate

The rate of increase in dry weight of the crop per unit ground area per unit time integrates the gains by photosynthesis and the losses by respiration, the compensatory effects of leaf area and photosynthetic rate, and the trade-offs between the geometric effects of crop height, leaf inclination and shape. Yet this important attribute of the productive capacity of crops was without a name, its significance unrecognized, until Watson (1958) called it the crop growth rate (CGR). Although simple to measure, there are various problems with some published rates which have been traversed by Monteith (1978) and Loomis & Gerakis (1975).

Data quality

Only when crop canopies are fully intercepting can the CGR of different genotypes be validly compared. Otherwise, as Fischer (1983) indicates, the differences between cultivars are mainly in the proportion of photosynthetically active radiation absorbed. The canopies should also have representative planting densities and must be sampled in such a way as to avoid the edge effects which probably vitiate several of the high CGRs in the literature. Comparisons of CGRs should also be based on harvests over relatively short intervals rather than on estimates for the life cycle as a whole. It is tempting for plant breeders to use measurements of shoot weight only at flowering or maturity to estimate CGR, but besides the loss of leaves along the way, and the pronounced changes in CGR as crops develop, there are also problems posed by differences between cultivars in their time to maturity and in the extent of post-anthesis growth.

In the work of Brinkman & Frey (1977) with oat lines, the CGRs were virtually identical until flowering. Thereafter the lines diverged considerably and it was these differences during the grain growth period that gave rise to the differences in overall CGR, and were positively related to yield (Salman & Brinkman, 1992). Likewise, the shoot weights of modern maize hybrids grown in Ontario were found by Tollenaar (1991) to be similar to those of old hybrids until one week before silking but the modern hybrids added 30% more shoot weight in the following month, and 480% more in the period from 3 weeks after silking until maturity, because of their greater LAI and delayed leaf senescence.

Dependence on leaf area index

Before the crop canopy is closed, LAI has a pronounced influence on CGR, as indicated in Figure 5.17 for a variety of crops and environments. Note that the oil palm estimate is an annual average, and that the cassava values are also for a long interval during tuber growth, and rise to 20 g m^{-2} d^{-1} when leaf loss is allowed for (Hunt et al., 1977). At the highest CGR for the maize crop, Lemcoff & Loomis (1986) estimate the efficiency of conversion of PAR to have been about 8%.

Environmental dependence

At full interception, whether the CGR reaches a plateau or passes an optimum depends on the crop and on radiation and temperature. The CGR may be more or less proportional to the intercepted radiation, as with wheat (Fischer, 1983; Bugbee & Salisbury, 1988), rice (Suzuki, 1983) and maize crops (Williams et al., 1968). The effect of temperature on CGR varies between crops, and with stage of development for any one crop (e.g. Suzuki, 1983). The CGR is also highly responsive to N supply (Green & Vaidyanathan, 1986).

Photosynthetic pathway

The fact that the only two crops reaching a CGR greater than 40 g m^{-2} d^{-1} in Figure 5.17 both have the C$_4$ photosynthetic pathway is not

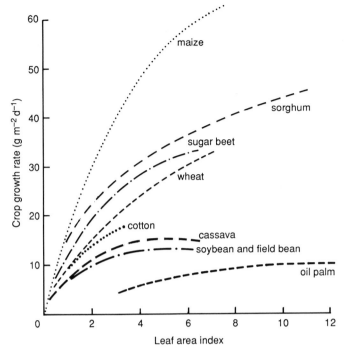

Figure 5.17. Crop growth rate as influenced by leaf area index in various crops: maize (Lemcoff & Loomis, 1986); sorghum (Eastin, 1983); sugar beet (Watson, 1958); wheat (Stoy, 1965); cotton (Hearn & Constable, 1984); cassava (Cock, 1983); soybean (Shibles & Weber, 1965); field bean (Laing *et al.*, 1983); oil palm (Corley, 1983).

surprising. The higher CER reached by many C_4 crops (Figure 5.5), and their higher canopy photosynthesis rates, lead to the expectation of higher CGR. Murata (1981) presents two figures relating single leaf CER with CGR for a range of crop and pasture plants, the first based on short-term measurements of CGR, the second on mean CGR over the whole growth period. For C_4 plants the correlation coefficient is quite high in both cases, but for C_3 plants it is much higher in the first version, illustrated in Figure 5.18. Murata (1981) suggests that when both LAI and radiation are high, the influence of CER on CGR becomes clear, as well as the superiority of C_4 plants under these conditions. However, he has excluded several of the highest CGRs recorded for C_3 plants even though he includes two of those rejected by Monteith (1978), for *Typha* and rice, thereby highlighting the problems of data selection.

In the table of maximum CGR over short periods compiled by Eagles &

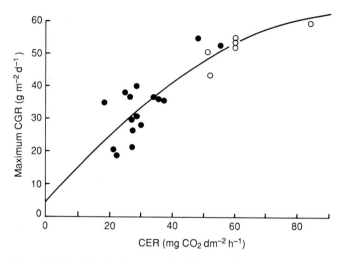

Figure 5.18. Relation between leaf photosynthetic rate and maximum crop growth rate across a range of C_3 (filled circles) and C_4 (open circles) crop and pasture species (adapted from Murata, 1981).

Wilson (1982), those for C_4 plants range from 19.3 to 78 g m^{-2} d^{-1} and those for C_3 plants from 13.4 to 90.6 g m^{-2} d^{-1}. In that of Gifford (1974a) the highest recorded CGRs for C_4 species (three of them) ranged from 51 to 78 g m^{-2} d^{-1}, and those for C_3 species (five of them) from 53 to 76 g m^{-2} d^{-1}. My previous compilation (Evans, 1975b) included the same entries, but with the addition of another C_4 forage plant, *Pennisetum purpureum*. Loomis & Gerakis (1975) give reasons for excluding the highest CGRs recorded for carrot and sunflower, and ignore other high rates for C_3 plants. Monteith (1978) gives reasons for excluding *all* the high C_3 rates tabulated by Gifford and myself.

   If one agrees with such a procedure, then of course the distinction between the C_3 and C_4 crops becomes clearer. As Monteith (1978) puts it: 'When all unreliable figures are discarded, the maximum growth rate for C_3 stands fell in the range 34–39 g m^{-2} d^{-1} compared with 50–54 g m^{-2} d^{-1} for C_4 stands'. Excessive densities and sampling problems certainly contributed to some high C_3 rates, such as that for carrots, and edge effects probably contributed to others, e.g. with *Typha*, *Phragmites* and some of the sunflower experiments. But some are well based. The CGR of 55 g m^{-2} d^{-1} for rice (Tanaka *et al.*, 1966) has been approached again recently by Shi & Akita (1988). A CGR of 60 g m^{-2} d^{-1} has been recorded for wheat at ambient atmospheric CO_2 levels by Rawson (1988) and up to 140 g m^{-2} d^{-1} with CO_2 enrichment by Bugbee & Salisbury (1988). After a critical survey of radiation use efficiency in several C_3

and C_4 crops, Kiniry *et al.* (1989) conclude that the differences between them were not 'as distinct as previously thought' (cf. also Sinclair & Horie, 1989).

My reasons for traversing this issue again are twofold. First, if the relation between the photosynthetic rate of single leaves and CGR is as clear as Murata's figure (5.18) and Monteith's assessment imply, we are faced with the problem of explaining why the many intraspecific comparisons reviewed above reveal a mainly negative relation between CER and CGR whereas the C_3–C_4 comparisons suggest a positive one. One possible contributing factor is that the trade-offs between leaf area and CER evident *within* each species are less significant in comparisons *between* C_3 and C_4 species because the anatomical and physiological compartmentation in the latter permits more efficient use of rubisco and other proteins. The C_4 pathway is also associated with faster and more complete export of assimilates from leaves (see below) and with faster rates of leaf appearance and expansion (Evans & Bush, 1985). Thus, the higher CGR of C_4 plants may reflect their different set of trade-offs between the components of growth, rather than a direct effect of the C_4 pathway.

Secondly, it is not the record CGRs that should be compared but typical rates for crops growing under characteristic conditions. The C_4 pathway is particularly adapted to conditions of high radiation and temperature, and C_4 crops may surpass C_3 crops in CGR under those conditions yet be inferior under lower radiation and cooler temperatures. Such a cross-over in performance is indicated in Figure 4.10 for RGR × temperature, and in Figure 4.14 for annual shoot growth × latitude.

Intra-specific differences

Pasture grasses are likely to have been under strong selection pressure for greater CGR yet, after quoting from a colourful account by Dickinson (1847) on the productivity of Italian ryegrass pastures kept well supplied with water and with the urine of horses, servants and maids, Alberda (1971) writes: 'the conclusion can be drawn that no substantial increase in dry herbage production has been achieved during more than one hundred years since Dickinson's results'.

Data from British trials presented by Robson (1980) indicate that modern strains of perennial ryegrass are, at most, 5% more productive than the old Aberystwyth strains. As mentioned above, selection for low respiration rate may offer limited further improvement, on the evidence of initial sward trials (Wilson & Jones, 1982), but only at high N fertilizer levels (Robson *et al.*, 1983). With lucerne, likewise, the improvement in productivity since 1956 has been only about 5%, i.e. 0.2% per year (Hill *et al.*, 1988).

Among 16 landraces, varieties, hybrids and wild and weedy relatives of pearl millet, Bramel-Cox *et al.* (1984) found the highest CGR in a wild relative followed by the weedy forms and landraces in both wet and dry seasons. Comparisons of old and new rice varieties in Japan (Tanaka *et al.*, 1968) and in the Philippines (Evans *et al.*, 1984) revealed no evidence of any improvement in CGR. Wheat varieties from the UK, USA, Mexico and Australia all displayed the same relation between CGR and absorbed PAR, of close to 5% efficiency of conversion (Fischer, 1983), which they shared with British barley crops and pastures (see Figure 4.16). No difference in CGR was found among four peanut varieties released between 1943 and 1977 (Duncan *et al.*, 1978). Varietal differences in CGR do occur, of course, especially under non-optimal conditions, e.g. with wheat crops at high temperatures (Rawson, 1986) and maize at low temperatures (Tollenaar, 1989), and these may be important for the improvement of yield.

Plant height also influences the CGR, being positively correlated with it in comparisons between varieties of wheat (Fischer, 1983) and oats (S.K. Johnson *et al.*, 1983). The differences are not great but may confound comparisons between old and new varieties. Reductions in shoot growth of modern semi-dwarf cereal varieties compared with older, taller ones, as in wheat (Sinha *et al.*, 1981) and barley (Boukkerou & Rasmusson, 1990) may reflect the effect of stature on CGR.

Where inter-varietal differences in CGR have been found, almost invariably they are positively related to leaf area and negatively to CER, as in wheat (Rawson *et al.*, 1983; Fischer *et al.*, 1981), soybean (Kaplan & Koller, 1977; Burris *et al.*, 1973), pea (Mahon, 1982), cotton (Muramoto *et al.*, 1965; Ibrahim & Buxton, 1981), perennial ryegrass (Rhodes, 1972), tall fescue (Horst *et al.*, 1978) and lucerne (Leavitt *et al.*, 1979).

The one body of data that claims that increases in growth rate can be obtained and do contribute to greater yield potential comes from K.J. Frey and his colleagues. Brinkman & Frey (1977) claimed that oat lines resistant to crown rust had higher growth rates, but their data make it clear that this is only when the grain growth stage was included in the estimation of growth rate. Differences in growth rate before anthesis were notably absent, and the fact that the resistance genes appeared to affect growth rate only among early isolines makes it possible that their effects on yield were associated with the evident differences in leaf area duration and other characteristics. Using growth rates estimated over the whole life cycle, Takeda & Frey (1976) found heterosis for growth rate in their crosses between *A. sterilis* and cultivated oats, and a positive association between growth rate and yield. Gupta *et al.* (1986*a,b*) concluded that selection for growth rate can be effective in raising the yield of oats by inter-specific crosses.

Similar work with pearl millet by Bramel-Cox *et al.* (1986) indicated that the introduction of exotic genes by crosses with wild relatives could increase growth rate, whereas crosses with land races and weedy relatives were more effective in raising yield. Sorghum and barley have also responded to this approach, according to Frey *et al.* (1984).

Shoot weight and CGR in relation to yield

Although yield in pasture grasses is fairly directly related to CGR, this need not be the case for crops. Oats selected only for high vegetative growth index may have low harvest index and yield (Takeda & Frey, 1987), while the pearl millet crosses that resulted in the fastest growth rate were not those with the highest yield (Bramel-Cox *et al.*, 1986). Although CGR in wheat increased with height, yield decreased (Fischer & Quail, 1989). Among cassava lines, yield was not highly correlated with CGR but was closely related to tuber growth rates (Ramanujam, 1985).

Thus, although inter-specific differences in CGR can be potent in determining differences in yield, we lack evidence that the improvement of yield potential in any crop has been associated with higher pre-anthesis CGR. The possibility raised by Frey and his colleagues that exotic genes may increase yields through their enhancement of growth rate merits further exploration, particularly within the context that further increases in harvest index and crop duration will be constrained.

Although the maximum CGR of crops seems not to have been increased by selection so far, growth under unfavourable conditions has been improved, and also the duration and rate of photosynthesis and growth after anthesis. Consequently, although the shorter stature of modern varieties of many crops tends to reduce the CGR, the additional growth late in the life cycle may more than compensate for this reduction, resulting in greater shoot weight at maturity in high-yielding recent varieties than in taller older ones. Such increases are apparent in recent British winter wheats (Austin *et al.*, 1980*a*, 1989), Mexican bread and durum wheats (Waddington *et al.*, 1986, 1987), barley (Riggs *et al.*, 1981; Wych & Rasmusson, 1983), oats (Brinkman & Frey, 1977; Wych & Stuthman, 1983), maize (Tollenaar, 1991) and several other crops. Gifford (1986) assesses the relative contribution to higher yield potential of increases in harvest index and shoot weight at maturity in six different crops, and concludes that 23% has come from increased crop growth. This proportion may well rise from now on as the agronomic support and protection of crops late in their life cycles improves, without requiring any enhancement of the maximum CGR.

Translocation and partitioning

The processes of photosynthesis and partitioning form a striking contrast in relation to yield enhancement. Our understanding of crop photosynthesis and our capacity to simulate it greatly exceed those for partitioning, yet it is changes in the latter rather than the former that have allowed yield potential to increase. Understanding has not been correlated with impact, and the reason is clear. In seeking to enhance crop photosynthesis we have been trying to improve on prolonged natural selection and, in the event, we have failed. Increase in partitioning to the organs for which we grow the crop, on the other hand, may not enhance fitness under natural selection, and has been encouraged by greater agronomic support for crops. It has been achieved by empirical selection for yield, and we still have only limited insight into the processes involved. We remain unsure of their limits, and even of what controls them. In some cases, at least, the translocation process seems to be less subject to stress than the preceding photosynthesis and the subsequent growth (see, for example, Wardlaw, 1967; Boyer & McPherson, 1975). But this sequence of processes is so linked by feed-back and feed-forward mechanisms that it is often difficult to identify which step in the sequence is limiting.

Export from leaves

Species differ greatly in the rate and extent of export of assimilates from their leaves. The differences are particularly striking when C_3 and C_4 plants are compared, as done by Hofstra & Nelson (1969), Lush & Evans (1974), Gallaher et al. (1975) and Stephenson et al. (1976). In fact, these comparisons indicate a close relation between the CER and the extent of early export, as illustrated in Figure 5.19, which also holds for individual C_3 and C_4 crops as irradiance changes (Wardlaw, 1976). Among species of Arachis, both wild and cultivated, a similar relation was found by Bhagsari & Brown (1976b). Differences in CER may drive the differences in export but it is also possible that faster export elicits faster photosynthesis, as Liu et al. (1973) suggest for cultivars of Phaseolus. In turn, faster export can be elicited by greater demand, as in tall fescue (Wong & Randall, 1985).

This caveat also applies to the translocation out of the leaf of those assimilates not exported immediately. Much of this remainder is exported during the night, and the relative proportions of day-time and night-time export depend on daylength and temperature, being quite tightly programmed (Lush & Evans, 1974; Chatterton & Silvius, 1979; Rocher & Prioul, 1987). Varieties also differ in the relative extents of export by day and by night, as Mullen & Koller (1988) have shown for soybean.

Phloem capacity

Among 22 wild and cultivated wheats, ranging from diploid to hexaploid, the cross-sectional areas of phloem in the culm varied over a 16-fold range in proportion to the maximum rate of assimilate import by the ears

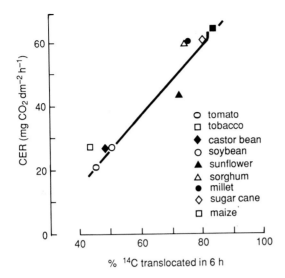

Figure 5.19. The relation between maximum CO_2 exchange rate and the initial export of ^{14}C assimilated by leaves of several crops (after Hofstra & Nelson, 1969).

(Evans *et al.*, 1970). On reasonable assumptions for the rate of translocation per unit area of phloem, the amount of phloem present coincided closely across the whole range with the amount required to sustain the maximum rate of import by the ear. So, was there just enough phloem in all genotypes, or did phloem area limit the rate of grain growth in all of them, as Canny (1973) and Patrick & Wareing (1980) suggest? If there was just enough in each case, how could that happen?

Several kinds of evidence suggest that the amount of phloem is not usually limiting. Removal of one primary leaf in beans quickly leads to a doubling of export from the other (Borchers-Zampini *et al.*, 1980). Partial severing of the vascular system of the culm leads to faster translocation and no reduction in grain growth of wheat (Wardlaw & Moncur, 1976) and to little effect on grain yield in sorghum (Fischer & Wilson, 1975a). Comparison between C_3 and C_4 grasses of the rates of mass transfer per unit area of phloem indicates that rates much higher than is usual in C_3 plants can be maintained (Lush & Evans, 1974). Moreover, translocation increases in both speed (Troughton *et al.*, 1974, 1977) and rate (Servaites & Geiger, 1974) up to the highest levels of irradiance and photosynthesis. Along with other evidence (Milthorpe & Moorby, 1969; Patrick, 1972; Passioura & Ashford, 1974; Wardlaw, 1990), these results suggest that assimilate transport capacity is not generally limiting, although it may be so in specific cases.

If the phloem area is generally sufficient, it is because there seem to be effective mechanisms for varying the number and size of vascular bundles that are differentiated. For example, when spikelet and grain numbers per ear in one cultivar of wheat were varied over a two-fold range by vernalization, the cross-sectional area of phloem in the peduncle varied in proportion to demand (Evans *et al.*, 1970). In oats, likewise, the phloem area and number of vascular bundles in the stem varied in proportion to spikelet number across varieties (Housley & Peterson, 1982) and environments (Peterson *et al.*, 1982). Likewise, the number and size of vascular bundles in leaves of millet species closely paralleled the leaf area served by them (Nakamoto & Yamazaki, 1988).

The partitioning of assimilates

The pattern of assimilate partitioning varies greatly between species and with environmental conditions (see, for example, Wardlaw, 1968, 1990; Snyder & Carlson, 1984). Differences in plant form imply strong genetic control of partitioning. Indeed a shift in partitioning has been the crux of crop improvement, as we shall see. Its genetic control can be illustrated by the experiment of Snyder & Carlson (1978) in which sugar beet seedlings were selected for either the highest or the lowest leaf:taproot ratio. After only two cycles of selection the progenies differed in their ratio by 60%.

Grafting experiments also highlight genetic differences in the capacity for growth or storage of various organs. Reciprocal grafts between the roots and tops of sugar beet and spinach beet or chard have shown that the genotype of the root determined the rate of root growth and storage (Thorne & Evans, 1964; Rapoport & Loomis, 1985). Sugar beet roots grew faster than chard roots regardless of shoot type. Sugar beet leaves were smaller regardless of root type, but were somewhat larger than usual when grafted on chard roots. Similar results have been obtained in grafts with sweet potato (Hozyo, 1977; Hozyo & Park, 1971; Hahn, 1977; see also Figure 5.4), potato (Bunemann & Grassia, 1973) and cassava (Bruijn & Dharmaputra, 1974). In this latter case the technique is used by Indonesian farmers to obtain high-yielding crops. Proportionately more ^{14}C is exported from leaves when they are grafted to faster-growing root systems, e.g. with grafts of chard/sugar beet (Wyse, 1979) and tomato/potato (Khan & Sagar, 1969). Similarly, the sugar or starch content of the root or tuber is also determined by its genotype, not by that of the shoot (e.g. Borger *et al.*, 1956).

Partitioning is a fashionable word in current research grant proposals, and is used in a wide range of contexts: between leaf area growth and higher CER; between photosynthetic assimilation and respiratory loss; between sugar and starch formation; between carbohydrates and other compounds; or between the various organs of the plant. All these partitionings are relevant to the yield of crops, but here we focus on partitioning at the organ level. Not only has this been central to crop improvement, but it also has a marked influence on partitioning at the lower levels of organization, although separate genetic controls may operate at each level. For example, genetic differences among wheat and soybean varieties in the activity of sucrose phosphate synthase influence their partitioning between sucrose and starch (Huber, 1981; Huber & Israel, 1982).

What factors influence partitioning between organs? In the competition for assimilates between similar organs, such as the many cotton bolls, soybean pods or citrus fruits on a plant, geometric factors such as their relative size, distance from the source and vascular connection to it can be crucial to survival. No doubt these geometric factors are also important in the competition between organs of different kinds, but the outcome is also influenced by other processes, still unknown but likely to be determined by interactions among the various plant hormones.

Geometric factors

The relative *size* of competing sinks was varied in experiments with wheat by varying the number of grains in two ears equidistant from the source leaf (Cook & Evans, 1978). The grains in the larger sink received more than their *pro rata* share of ^{14}C-labelled assimilates, and the bias in their favour increased with the relative size of the larger sink, as Peel & Ho (1970) had found with competing colonies of aphids. The smaller sink had to depend more on local sources and reserves. In subsequent wheat experiments the competing sinks were two spikelets with different numbers of grains, the source of ^{14}C-labelled assimilate in this case being a centrally located awn (Cook & Evans, 1983c). Here again the bias in favour of the larger sink, when they were equidistant from the source, was approximately proportional to the square of its relative size. The more grains in the spikelet, the more counts per grain. This Matthew effect – 'unto everyone that hath shall be given' (Matthew 25:29) – may have been a potent force in crop evolution, e.g. in the progressive condensation and enlargement of the harvested organs at the expense of the others. Consider the single maize cob or sunflower head which replaced the many small ones of the wild progenitors.

Although the distance *per se* between source and sink may not limit growth (Wardlaw, 1990), the *relative distance* (d) of competing sinks from the source in the wheat ear experiments had a powerful effect on the partitioning of labelled assimilates, which was proportional to $1/d^2$ (Cook & Evans, 1983c). This disadvantage in being further from the source of assimilates may be a reflection of phloem resistance (cf. Thornley, 1972). However, in our experiments the disadvantage of distance could be overcome by an increase in relative sink size, and it may be this interaction which has permitted the improvement of crop plants with sinks ranging in position from the top of the plant (e.g. sorghum and sunflower) to the root (e.g. sugar beet and cassava).

Good *vascular connections* to the source are also important in partitioning. A sink on the opposite side of a wheat ear from the source of assimilate received only 3–10% as much ^{14}C as a comparable sink on the same side as the source (Cook & Evans, 1983c). This implies a considerable variation in the phloem resistance of translocation pathways to the competing sinks, in agreement with earlier work showing that movement in the phloem of intact plants is often confined within one orthostichy, for example the movement of ^{14}C into sunflower heads (Prokofyev *et al.*, 1957) and among the leaves of tobacco (Jones *et al.*, 1959; Shiroya *et al.*, 1961) or wheat (Patrick, 1972). Vascular connections may favour spikelets in particular positions in rice panicles (Inosaka, 1957; Asada *et al.*, 1960) or on the same side of wheat ears (Evans *et al.*, 1972). However, these preferential patterns of movement can be varied, for example by partial defoliation or by grain removal in wheat (Patrick & Wardlaw, 1984). When demand is high, any canalizing effect due to the greater resistance of certain minor vascular connections can be overcome.

As in human affairs, it pays to be large, close to the source, and well connected. These geometric factors influencing the partitioning of assimilates are presumably supplemented by other mechanisms in the competition between organs of different kinds for the allocation of assimilates, especially by the intervention of the various endogenous plant growth substances. It is these other mechanisms to which the term 'sink strength' refers, but how they act and in what way they are integrated is not understood.

Hormonal effects

The 'nutrient diversion' theory of Went (1939) first introduced the concept that assimilate distribution may be controlled by endogenous plant growth substances. These could operate at the source end (e.g. on photosynthesis and phloem loading), along the transport pathway on the mechanisms of translocation, and/or at the sink end through effects on phloem unloading, growth and storage processes. In fact, hormonal effects are apparent at all these stages (Patrick & Wareing, 1980). Phloem unloading can be enhanced by several growth substances even when greater growth of the sink organ is prevented. Hormone-induced growth or storage may also have indirect effects on translocation and unloading.

There is no simple relation between sink strength and its hormone content, however. For example, although wheat grains may grow at a constant rate over a quite long period, this is in spite of large fluctuations in their content of cytokinins, gibberellins, auxins (Wheeler, 1972) and abscisic acid (King,

1976). The same is true for wheat florets before anthesis (Lee *et al.*, 1988). Moreover, the rate of growth at one time may have been determined by earlier hormonal action, as in sugar storage by cane internodes whose volume has been enhanced by earlier applications of gibberellic acid. In any case it may not be hormonal levels so much as differing sensitivities to them that govern partitioning, as with the Rht dwarfing genes in wheat which render the stem and leaf tissue insensitive to gibberellins (Gale & Youssefian, 1985). Gibberellin levels are actually higher in the Rht dwarfs than in tall isolines.

The fact that the relations between the growth rate of an organ and its inventory of hormones are complex in no way rules out their role in modulating sink strength (Patrick & Wareing, 1980), as revealed for example in the effects of localized infections by pathogens or gall insects. But how they act varies from one hormone and organ to another. Cytokinins may act more locally than auxins or gibberellins, for example, and may influence ultimate demand through effects on cell number or protein levels whereas gibberellins may do so predominantly through cell enlargement. In the competition between organs, terminal meristems would seem to be at a great disadvantage on purely geometric grounds. They are very small, often distant from the sources of assimilate and lacking in vascular tissue. Yet their survival and development is crucial, and is probably aided by their exceptionally high levels of endogenous growth substances.

Functional equilibria

Partitioning is also influenced by the constraints imposed by one organ on the functions of another, the classic example being the mutual interactions of root and shoot. These may involve hormonal interactions of the kind just considered. Besides these, however, the partitioning between root and shoot depends on the supply of water and nutrients by the former in relation to the requirements for shoot photosynthesis and growth, and on the supply of assimilates by the latter in relation to the requirements for water and nutrient uptake and other root functions. The 'functional equilibrium', as Brouwer (1962*b*) first referred to it, between these mutually dependent processes varies with the light environment of the leaves on the one hand and soil nutrient and water levels on the other, as well as with temperature. But for any one species at a particular stage of development, the functional equilibrium is a stable characteristic, quickly restored following partial defoliation or root removal, as Brouwer (1962*a*) showed with beans.

Developmental trends

Control of assimilate partitioning by functional equilibria and geometric factors alone is too conservative. Seasonal changes in the environment coupled with progressive trends in the impact of geometric influences would lead to some developmental change, and any accumulation of assimilates could have spill-over effects on branching or tillering and even on the initiation of reproductive development in some plants. But to establish new sinks such as primordial inflorescences and to ensure their survival and coordinated development in conjunction with, and often in competition with, a rapid increase in stem growth in the cereals at least, requires the intervention of additional mechanisms, almost certainly hormonal in nature.

The initiation of flowering once the daylength becomes inductive is an example of such an intervention. Our understanding is still insufficient for explanatory models to be formulated. Consequently, crop growth models without hormones, as de Wit & Penning de Vries (1983) refer to them, must be resorted to. These rely on empirically determined trends in partition coefficients such as those for rice (Monsi & Murata, 1970) or sugar beet (Patefield & Austin, 1971; Fick et al., 1975), and allow crop modellers to bypass our lack of understanding of the processes which control partitioning as plant breeders have done before them. Better understanding does not guarantee more effective progress in plant breeding, but as we approach the limits to harvest index in some crops, it is not only for intellectual satisfaction that we shall need greater insight into the factors which control partitioning.

Harvest index

Introduction

In his prize essay *On the management of wheat*, written in 1847, Roberts presented grain and straw yield data for ten varieties which showed, with one exception, grain yield rising as straw yield fell. The higher-yielding varieties had partitioned more of their shoot weight into the grain. Donald & Hamblin (1976) have reviewed the subsequent history of this proportion, referred to as the 'migration coefficient' by Beaven in 1920, as the 'coefficient of effectiveness' by Nichiporovich in 1956, and as the 'harvest index' by Donald (1962). In the same year Dobben (1962) drew attention to the crucial role of the rise in harvest index (HI) in the improvement of yield potential of wheat by showing that modern Dutch varieties produced no more shoot weight than their older, lower-yielding predecessors.

The inclusion of wild relatives in the comparisons makes this rise even more striking because their HI is often less than 0.2, as in wheat (Austin et al., 1982), rice (Cook & Evans, 1983b; Oka, 1988), American wild rice (Grava & Raisanen, 1978) and *Echinochloa* millets (Evans & Bush, 1985), whereas it may be 0.5 or more in improved crops. However, Sano & Morishima (1982) record harvest indexes of up to 0.5 in wild annual rice, and Sobrado & Turner (1986) up to 0.43 in wild sunflower. Many factors influence HI and some of these will be considered before further comparisons are made, after which we shall analyse the ways in which HI has been increased and could be further enhanced. Two aspects should be borne in mind throughout. The first is that HI is a ratio and, as such, may rise because yield has risen while shoot weight either did not increase or increased proportionally less, or because yield did not increase but shoot weight fell. In this latter situation we gain nothing unless the crop is of shorter duration. In the former case, where shoot weight does not increase – a common situation – HI is of course highly correlated with yield.

Secondly, harvest index can be unsatisfactory in focusing too strongly on only one product of the crop, such as the grain. In earlier times cereal straw was also highly valued, indeed it still is in many developing countries, and as we move towards total crop use in future the HI may lose much of its current significance. On what basis should we compute the HI of sugar cane, for example? Irvine (1983) points out that it could range from 0.19 if sugar is the only product of interest, via 0.23 if syrup and molasses are included to 0.63 if stalk fibre is counted and to 0.99 if the crop as delivered to the mill is the basis.

The scope for increase in HI and the way in which selection can enhance it depends to a considerable degree on its magnitude. Where the primary product of the crop is a high value component – such as a pharmaceutical precursor, ultra-sweetener or rare compound – which represents only a small proportion of the biomass, enhancing its content may not prejudice the growth and adaptability of the crop. But when it constitutes a major proportion, its enhancement may be at the expense of organs that determine productivity and survival, in which case it will be possible only with additional agronomic support for the crop.

Recorded HI values commonly neglect the roots, for practicality, and also the fallen leaves in many cases, thereby over-estimating the index and limiting its comparative value. Some of the high values recorded for legume crops probably reflect these limitations.

The HI usually refers only to that proportion of the total dry weight (biomass) in the harvested organs, yield being measured in the same terms. The HI for nitrogen, phosphorus or other constituents may be more relevant for some purposes. They are not considered here although they may present a quite different picture. For example, although the HI for dry matter in English winter wheats has risen from 0.36 in old varieties to 0.51 in recent ones, the HI for N has not risen at all; however, its average value fell from 0.78 to 0.71 as soil fertility improved (Austin *et al.*, 1980*a*). In soybean, likewise, the HI for N reached much higher values, with smaller differences between varieties, than the HI for dry weight (Salado-Navarro *et al.*, 1985). Among malting barleys, by contrast, the HI for N rose from 0.46 to 0.64 along with a rise in the HI for dry matter from 0.27 to 0.41 (Wych & Rasmusson, 1983).

Developmental effects

Among crops there are two quite different developmental pathways for HI, depending on whether the growth of the storage organ is predominant and terminal (as in the cereals) or continues over a long period in equilibrium with the growth of other organs (as in sugar cane and beet), with pulse crops being somewhat intermediate. Harvest index in the cereals is therefore influenced by the relative duration of the various stages of the life cycle, whereas in crops like cassava or sugar beet it depends more on the genetically and environmentally determined equilibrium between growth and storage. In the cereals and pulses, HI begins to rise more or less linearly from sometime after flowering until it reaches its varietal maximum at maturity, as in soybeans (Salado-Navarro et al., 1985). In cassava, by contrast, the HI begins to rise early in the life cycle, approaches its limit and then maintains that value – varying between 0.28 and 0.67 among six cultivars – for a prolonged period of growth and storage (Hunt et al., 1977; Ramanujam, 1985). The HI in this case reflects fairly directly the genetic differences in partitioning to storage, as proposed in Boerboom's (1978) model and supported by Cock (1983). A similar situation applies to sugar beet (Fick et al., 1975).

With the cereals and pulses, the developmental factors determining HI are more complex. Early flowering can limit inflorescence development and HI, but HI can be reduced even more drastically by late-flowering and large biomass accumulation. In some cases such negative correlations between HI and flowering time largely reflect environmental stress late in the season, as in Finlay's study with barley (Donald & Hamblin, 1976). In others, however, it reflects an imbalance between vegetative and reproductive development.

Among ten cultivars of soybean, from five maturity groups planted at various times in Missouri, both HI and yield were highest in those from Group II and lowest in those from Groups I and V (Johnson & Major, 1979). Control by daylength of the time to flower initiation and of the rate of inflorescence development can have a profound effect on HI. Exposure to shorter days more than doubled the HI of four pearl millet hybrids not only by reducing vegetative growth but also by increasing inflorescence size (Ong & Everard, 1979). In root crops, also, the HI may be profoundly influenced by daylength through its control of development: for example, in *phureja* potatoes, decreasing the daylength from 15 to 11 hours reduced biomass by only 7% but increased HI from 0.25 to 0.58 (Mendoza & Estrada, 1979).

The rise in HI of rice varieties as their growth duration is reduced is illustrated in Figure 5.20. Similar results have been obtained by Venkateswarlu *et al.* (1969), Akita (1989) and Dingkuhn *et al.* (1990) with rice as well as

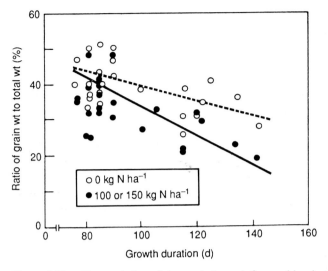

Figure 5.20. Harvest index of rice varieties as influenced by their growth duration and N fertilizer application (Yoshida, 1981).

in wheat (Siddique *et al.*, 1990*b*), maize (Edmeades & Tollenaar, 1990) and pigeon pea (Lawn & Troedson, 1990). However, the correlation is a loose one, and among maize hybrids HI was more influenced by crop density and environment than by time to maturity (De Loughery & Crookston, 1979).

Environmental effects

As the population density of a crop is raised, so is its shoot weight and its yield up to a point, whereas its HI tends to fall (Donald & Hamblin, 1976). The adverse effect of high densities on HI is exacerbated under stress and with older cultivars selected for sparser populations. Indeed, selection for reduced barrenness in dense populations has been a major route to higher yield and HI in modern maize hybrids (Fischer & Palmer, 1984).

As might be expected from the fall in HI with increasing crop density, HI tends to rise with increase in irradiance, as in wheat (Bugbee & Salisbury, 1988) and cassava (Cock, 1983). Likewise, HI can be raised by CO_2 enrichment, e.g. in rice (Cock & Yoshida, 1973) and wheat (Gifford, 1979), although apparently not in cowpea (Bhattacharya *et al.*, 1985). Irrigation increases HI (Donald & Hamblin, 1976), and the increase would surely be even more striking if root weights were included in the estimates. Water stress can lower the HI drastically, as in maize (De Loughery & Crookston, 1979) and wheat (Gifford, 1979).

Whereas provision of water usually raises the HI, provision of nitrogen fertilizer tends to lower it. Donald & Hamblin (1976) illustrate this by reference to the Broadbalk wheat experiments from 1852 to 1925, the HI of wheat over this period being 0.40 with minerals only, and 0.35 with minerals plus nitrogen. Pearman *et al.* (1978), with modern varieties grown at Rothamsted, also found the HI to be reduced by N fertilizers. The lowering of the HI at higher N levels can be more pronounced when water is in short supply and when varieties are tall or later maturing (Donald & Hamblin, 1976). Comparisons of HI across crops or cultivars need to take these effects into account.

Comparisons between different crops

Given the great range in growth habit, duration, environment and harvested product among crops, there is little point in comparing their HI values except to illustrate their diversity, which may be salutary. One might be tempted to conclude, otherwise, that the higher HI of maize and sorghum compared with pearl millet (Chaudhuri & Kanemasu, 1985; Muchow, 1989*a*) reflects their more advanced development as crops, since their total shoot weight is similar. The HI of modern wheat, maize and rice cultivars approaches 0.55 (see Figures 5.20 and 5.21). Higher values have been recorded for several pulse crops, possibly associated with a greater loss of leaves before measurement, as in the case of the HI of 0.7 recorded for beans (Peet *et al.*, 1977) and values up to 0.59 for chickpeas (Saxena *et al.*, 1983). Nevertheless, it is likely that the HI can be at least as high in pulses like soybean (Salado-Navarro *et al.*, 1985) as in cereals.

The HI of root and tuber crops tends to be even higher, as shown by comparative experiments such as those of Haws *et al.* (1983) in which potato, sweet potato and sugar beet reached HI values of 0.88, 0.60 and 0.53 respectively. Gawronska *et al.* (1984) record values of up to 0.84 for the old potato variety Russet Burbank; Ramanujam (1985) up to 0.7 for cassava; and Bhagsari & Harmon (1982) and Hahn & Hozyo (1983) up to 0.59 and 0.72, respectively, for sweet potatoes.

The figure of 0.53 given above for sugar beet refers to the roots, but it is their content of sucrose for which the crop is primarily harvested. Since this is commonly about 12–13% of root weight, the HI of sugar beet would be substantially reduced, except that the leaves and the residue of the roots are used in other ways, as also for cane whose harvest index for sugar is about 0.19 (Irvine, 1983).

With perennial crops the HI may be estimated at final harvest, such as the value of 0.64 for the boles of *Pinus sylvestris* stands at least 30 years old (Ovington, 1957). But with regularly harvested perennial crops, the HI is more appropriately estimated from the annual growth increment. On this basis Corley (1983) suggests maximum values of 0.2 for coconut, 0.31 for rubber, 0.33 for tea, 0.34 for oil palm and 0.47 for sago. Given the high calorific value of some of these products, he estimates the HI on an energy basis as reaching 0.52 for both oil palm and rubber, comparable with that for modern cereal varieties, while Chalmers & van den Ende (1975) estimate the fruit crop of peach trees as equivalent to 0.70 of the annual dry weight increment.

Against this background we can now turn to the comparison of older and newer cultivars of particular crops.

The rise in harvest index

Wheat

Waddington *et al.* (1986) complain that 'it has now nearly become accepted as dogma that wheat breeders have improved only the crop's partition efficiency to the grain without an effect on total phytomass produced'. This 'dogma', first enunciated by Dobben (1962) for wheat varieties used in the Netherlands over the period 1902–55, has continued to receive strong experimental support.

Dobben's data and others for comparable British winter wheat varieties released at various times are illustrated in Figure 5.21, along with data for barley and rice. They agree in indicating a pronounced rise in the HI, particularly among varieties released since 1950. Many comparisons of old and new wheat varieties in other environments indicate a similar rise in HI in recent varieties, such as those of Ruckenbauer (1971), Jain & Kulshrestha (1976), Sinha *et al.* (1981), Syme & Thompson (1981), Hucl & Baker (1987) and Siddique *et al.* (1989*b*), as well as Bamakhramah *et al.* (1984) with diploid, tetraploid and hexaploid wheats.

In most of these studies the increase in HI is closely related to the increase in varietal yield, as found by Syme (1972), Aguilar & Fischer (1975), Kulshrestha & Jain (1982), Loffler *et al.* (1985) and Slafer & Andrade (1989). Of course, a positive linear relation between HI and yield does not preclude a similar relation between shoot weight and yield. However, when grain weight is plotted against straw weight for varieties of differing vintage, as in Figure 5.22, the negative 1:1 slopes indicate that the gain in grain yield approximately equals the loss in straw weight, and that the rise in HI accounts for most of the rise in yield, as found also by Aguilar & Fischer (1975), Perry & D'Antuono (1989) and Siddique *et al.* (1989*b*). Note, however, that with the higher level of fertilizer application in Figure 5.22, the most recently released varieties (9–12) are above the line of average shoot weight. Gifford's (1986) analysis of these results suggests that 82% of the increase in yield potential has come from the

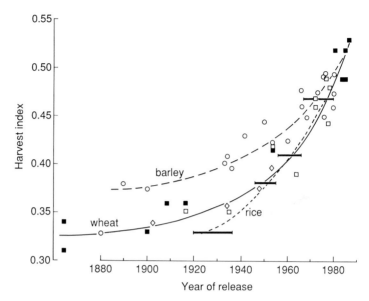

Figure 5.21.   Rise in the harvest index of wheat, barley and rice varieties with year of their release. Data for wheat from Dobben (1962, ◇), Austin *et al.* (1980*a*, □, 1989, ■); for barley from Riggs *et al.* (1981, ○) and for rice from Evans *et al.* (1984, bars).

rise in HI, and the remainder from greater shoot weight. Some rise in the shoot weight at maturity of recent varieties in spite of their reduced stature is also apparent in the comparisons by Austin *et al.* (1989), Cox *et al.* (1988) and, most strikingly, by Waddington *et al.* (1986, 1987) on varieties bred recently at CIMMYT in Mexico. With the bread wheats, HI rose from 0.33 in 1950 to 0.46 in 1970 but declined with further increase in yield potential since then. The number of grains per ear has risen more than kernel weight has declined, while ears per square metre have not changed. Thus, the rise in final shoot weight probably reflects an increase in non-ear-bearing tillers following the introduction of the Veery genes, and further selection for yield is likely to see a resumption of the rise in HI as the proportion of fertile tillers and high kernel weight are restored. With the durum wheats, Waddington *et al.* (1987) found a rise in HI from 0.23 in Tehuacan 60 to 0.5 in Mexicali 1975, but no further rise since then. With the CIMMYT durums, as with the bread wheats, recent progress has been associated with increases in shoot weight and in grains per ear and per spikelet. The large rise in grains per square metre seems to have been the driving force behind the increase in yield potential, and may have elicited the rise in shoot weight. It remains an open question, therefore,

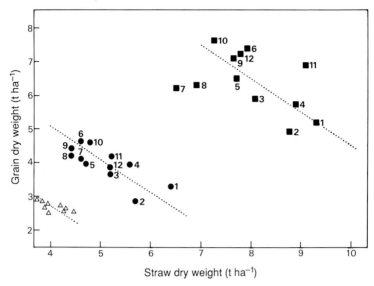

Figure 5.22. The relation between grain yield and straw weight in experiments with wheat varieties of differing vintage and height. For the data from Austin *et al.* (1980*a*), the 12 varieties (1 being the oldest and 12 the most recent) were grown at both high (■) and intermediate (●) levels of fertilizer application at Cambridge. The much older data of Roberts (1847, △) are added for comparison. Dotted lines represent the average shoot weight.

whether the CIMMYT results since the early 1970s illustrate a quite different pathway to increased yield potential, or are simply setting the stage for further increases in HI.

Rice

Comparing old and new Japanese rice varieties, both Ito (1975) and Takeda *et al.* (1984) found a rise in their HI which largely accounted for the rise in yield potential. Among Philippine varieties used in this century, the rise in HI was comparable with that in European wheat and barley varieties, as indicated in Figure 5.21. A wider comparison of rices from all stages of evolution of the crop also illustrated the rise in HI (Cook & Evans, 1983*b*). Kawano (1990) notes, however, that although grain yield may be closely related to HI in high-yield environments, it may be more closely related to shoot weight in low-yield ones.

Barley

A pronounced rise in HI from old to modern varieties of barley has been found by Aufhammer & Fischbeck (1964), Hayes (1970), Riggs et al. (1981) and Martiniello et al. (1987) (cf. Figure 5.21). Grain yield was highly correlated with HI, but was also significantly correlated with shoot weight in the comparison by Riggs et al. (1981). Gifford's (1986) analysis of their data suggests that only 55% of the increase in yield potential was attributable to the rise in HI and only 50% in the comparison by Wych & Rasmusson (1983) in which the raising of yield has been constrained by the requirements for malting and brewing quality. Boukerrou & Rasmusson (1990) found only a 10% increase in HI among American spring barleys over six decades, compared with a 25% increase in shoot weight at maturity. Barley lines with greater shoot weight have been identified (Hanson et al., 1985), but all are tall and it remains to be seen how readily high HI and greater shoot weight can be combined.

Thus, although a rise in HI has been an important component in the increase of yield potential in barley, so too has increase in shoot weight, possibly associated with the need to maintain a high HI for N in malting barleys or with the shorter life cycle and/or less drastic dwarfing of this cereal.

Oats

Comparisons of oat varieties of differing vintage have generally indicated a rise in HI with increasing yield potential, accounting for much of the rise in the latter (Sims, 1963; Aufhammer & Fischbeck, 1964; Lawes, 1977; Wych & Stuthman, 1983; Peltonen-Sainio, 1990). However, the relation between HI and yield of successive varieties of oats is not as strong as it is for wheat and rice, possibly reflecting the more limited shortening of the stems in oats (cf. Figure 5.23) and the relatively short life cycle, as suggested by Takeda & Frey (1976). Like Rosielle & Frey (1975a), these authors found grain yield in oats to be more highly correlated with shoot weight than with HI. Payne et al. (1986) found no change in HI after three cycles of selection for greater yield, possibly because any rise in HI may have been counterbalanced by a fall associated with the delay in heading and maturity (cf. Figure 5.20).

C_4 pathway cereals

Retrospective comparisons of varieties or hybrids of different vintage among the C_4 cereals have not usually included HI. In maize, one of the main trends in the course of yield improvement has been the reduction of barrenness in high-density crops (Duvick, 1984c; Edmeades & Tollenaar,

1990), which is likely to be reflected in a rise in HI, as found by Tollenaar (1989) among hybrids released in Ontario between 1959 and 1988. No rise in HI was apparent at low densities of planting whereas at the highest density the HI increased from 35 to 52%. Nevertheless, Tollenaar concludes that only 15% of the genetic gain in grain yield can be attributed to the rise in HI. Russell (1985) found HI to be the attribute most highly correlated with yield ($r = 0.83$) among 28 lines ranging in vintage from open-pollinated to 1980s hybrids. Selection for reduced plant height could raise HI substantially in maize (Johnson *et al.*, 1986) (cf. Figure 5.23) as also in sorghum (Goldsworthy, 1970). Among a large number of pearl millet hybrids Kapoor *et al.* (1982) found a positive correlation between HI and grain yield in all environments, whereas shoot weight and grain yield were correlated only in low-yielding environments. Peng *et al.* (1991) found a relatively low correlation between HI and yield among 22 lines of sorghum.

Pulse crops

Yield was correlated with both HI and shoot weight among varieties of field bean, according to the data of Peet *et al.* (1977). Shoot weight and HI in soybean are strongly influenced by maturity group and crop duration, and also by leaf drop as maturity approaches (Johnson & Major, 1979). In a comparison of old and new varieties within two maturity groups, Gay *et al.* (1980) found a pronounced rise in the HI of recent varieties in both groups. Schapaugh & Wilcox (1980) found a strong positive correlation between HI and yield in one year whereas Buzzell & Buttery (1977) found a negative one.

In a comparison of four peanut varieties released between 1943 and 1977, Duncan *et al.* (1978) found no increase in CGR, but a progressive increase in the proportion of assimilate partitioned to grain growth in the higher-yielding varieties, as found also by Wells *et al.* (1991) in a comparison of ten cultivars released between 1944 and 1990. Data for HI are not presented by Williams *et al.* (1975) in their comparison of four varieties but they also conclude that the differences in yield are not due to differences in leaf area or CGR, but in the ability to develop the reproductive sink, as determined by the timing of grain growth relative to the decline in leaf area. Kwapata & Hall (1990) found a high correlation between HI and yield among cowpea varieties.

Sugar crops

With both sugar beet and cane there is a question as to what HI refers to (cf. Irvine, 1983), but this discussion is confined to the sucrose content of the harvested organs. Following the blockade by the British navy of supplies of cane sugar from the West Indies to France, Napoleon decreed in 1811 that beet should be developed as a source of sugar. The subsequent progress by French and German plant breeders in raising the sugar content provided Darwin with a favourite example of the power of artificial selection, illustrated in Figure 3.13. However, the sucrose concentration in sugar beet roots has subsequently declined to 10–14% of the fresh weight, reflecting partly the more adequate nitrogen nutrition of the crop and partly the greater emphasis on 'yield' types (i.e. large roots with moderate sucrose levels) rather than 'sugar' types (Fick *et al.*, 1975).

A somewhat similar course has been followed in sugar cane breeding. Sucrose constitutes only about 4% of fresh weight in the wild progenitor *Saccharum spontaneum* compared with up to 17.5% in *S. officinarum* (Bull & Glasziou, 1963). In many breeding programmes, the percentage sucrose in cane was raised progressively by selection, to reach peak values in Java, Hawaii, Australia and elsewhere about 50 years ago. Sucrose content is a strongly inherited clonal character (Brown *et al.*, 1968) which is negatively correlated with dry matter yield. In recent years more emphasis has been given in many countries to selection for high cane yield, resulting in a decline in percentage sucrose. In Louisiana, on the other hand, where a shorter growing season is coupled with a substantially lower percentage sucrose, selection is still aimed at raising the sugar content (Breaux, 1984).

Other crops

Comparing obsolete and modern varieties of cotton, Meredith & Wells (1989) found a high correlation between yield and the ratio of boll weight to total dry weight, which ranged from 0.41 to 0.56. The HI of hybrid sunflowers surpasses that of older open-pollinated varieties (Amir & Khalifa, 1991). With root crops, yielding ability among varieties and clones is fairly closely related to HI in both sweet potatoes (Bhagsari & Harmon, 1982) and cassava (Williams, 1972; Ramanujam, 1985) but not, apparently, in potatoes (Gawronska *et al.*, 1984).

Conclusion

The rise in HI with increase in varietal yield potential over the years is clear for wheat, rice and feed barley (Figure 5.21), and its positive correlation with yield is high in many crops. More comparative data are needed for the pulses, as also for the root crops where losses by leaf fall and differences in duration of storage need to be taken into account. Gifford (1986) estimates that 77% of the increase in yield potential across six crops has come from the rise in HI but in some crops, such as malting barley and maize, additional shoot growth has been the more significant contributor.

The analyses by Kawano (1990) with rice and cassava and by Kapoor *et al.* (1982) with pearl millet suggest that the rise in HI has been the more important under high-yield conditions whereas greater shoot growth has been more so under low-yield conditions, but that does not appear to be the case for maize. There is still scope for further increases in HI, e.g. in wheat for which Austin *et al.* (1980*a*) suggested a possible limit of 0.62, and the components of further increases in HI are considered below.

Already, however, there are signs that increase in shoot weight – from a longer duration of active growth rather than any increase in maximum CGR – may contribute more to further increases in yield potential. The example of the two sugar crops may be instructive. Following a sustained period of increase in percentage sucrose, their equivalent of HI, this has now fallen somewhat as selection for greater growth has assumed more significance. The shift to shorter stature in many crops under intensive management has contributed greatly to the rise in HI, probably at some cost to crop growth, but in several of the comparisons of old and new varieties of wheat, barley, oats, maize and sorghum there are signs that the greater yield potential of recent varieties is associated with greater shoot growth. The further rise in HI may become less significant than it has been over the past 30 years or so.

Sources of the rise in harvest index

Given the great significance of the rise in HI for crop improvement, it is remarkable that we have so little understanding of the ways in which HI has been raised. This is doubly unfortunate not only because it limits a rational analysis of the way ahead but also because the likely mechanisms have powerful implications for the dependence of further improvement on agronomic advances, and therefore on the relation between plant breeding and agronomy.

Reduced investment in other organs

In the many cases where there has been no increase in shoot weight with crop improvement, greater yield has been achieved through reduced investment in other organs. That, in turn, may be possible only at the expense of adaptation to stress or by the provision of better agronomic support for the crop. The provision of irrigation and fertilizers may permit less partitioning to root growth; effective weed control may allow shorter plants to be selected; control of pests, diseases and other stresses may reduce the requirement for reserves to aid recovery. The operative word in all these instances is 'may' and, except with stem height, evidence that improvement in yield potential through a rise in HI has been coupled with reduced partitioning to other functions is hard to find.

Plant height and stem growth

Old European land races of the small grain cereals and traditional tropical varieties of rice and the coarse-grained cereals were tall, this being a desirable attribute where straw was useful and weeds had to be suppressed. As nitrogenous fertilizers became cheaper and more heavily used, however, tall varieties were prone to lodging, and shorter stems were selected. Substantial height reduction was achieved in many crops by empirical selection, but the use of major dwarfing genes has led to more drastic shortening since the 1960s (Athwal, 1971). In many cases these genes reduce either the endogenous level of gibberellins (as in many rice and maize dwarfs) or the capacity of the stem to respond to them (as with the commonly used Rht genes of wheat).

Stem weight is reduced in parallel with stem height in wheat (Bush & Evans, 1988) and maize (Johnson et al., 1986), and substantial savings of assimilate are freed in the dwarf plants for investment in other organs. However, only by further selection and adjustment of the genetic background can their investment in additional grains rather than additional tillering or root growth be ensured (Ali et al., 1978). Isogenic lines with dwarfing genes on genetic backgrounds from tall varieties often yield less than the comparable tall varieties, as in barley (Austin et al., 1980b), oats (Meyers et al., 1985), wheat (Bush & Evans, 1988) and pearl millet (Rai & Rao, 1991).

The negative relation between plant height and HI in five cereals is indicated in Figure 5.23. Similar relations have been found many times (e.g. Tanaka et al., 1966; Jain & Kulshrestha, 1976; Allan, 1983; Fischer & Quail, 1989). In some instances the gain in grain yield is approximately equal to the reduction in straw weight (Figure 5.22), which might suggest a direct reallocation from stem to grain growth. As plant height was progressively reduced from 2.8 to 1.8 m by selection in tropical maize, the savings in stem and sheath weight equalled 2.8 t ha^{-1} (Johnson et al., 1986). From data presented by Fischer & Palmer (1983), these savings resulted in the dry weight of female inflorescences at anthesis increasing from 0.20 to 0.32 t ha^{-1} and HI from 0.30 to 0.45 (Figure 5.23). Clearly, more than a simple reallocation is involved: throughout the 15 cycles of selection for reduced height there appeared to be a close relation between the additional investment in inflorescence differentiation, final grain number, yield and HI.

In wheat the concentration of water soluble carbohydrates in the developing ear and stem reaches a peak about two weeks before anthesis, but neither the peak concentration nor that at anthesis is any higher in dwarf than in tall lines (Fischer & Stockman, 1986; Bush & Evans, 1988) and the total sugars are less, not more. This could happen if the gain in ear growth rate matched the fall in stem growth, as suggested by the results of Brooking & Kirby (1981), but their results may merely reflect the somewhat earlier

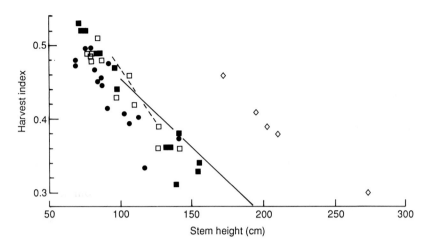

Figure 5.23. The relation between stem height and harvest index in five cereals: wheat (data of Austin *et al.* 1980*a*, □, 1989, ■); barley (Riggs *et al.*, 1981, ●) and maize (Fischer & Palmer, 1983, ◇) and regressions estimated from data of Lawes (1977) for oats (broken line) and of Tanaka *et al.* (1966) for rice (solid line).

inflorescence initiation in dwarf lines. Alternatively, the savings from reduced stem growth could be used to support the development of more ears, as often found in dwarf spring wheats (Gale & Youssefian, 1985). Evidence that the savings are conserved to support increased grain set and growth after anthesis is lacking. Indeed, Austin *et al.* (1980*b*) found the contribution of pre-anthesis reserves to be no higher in dwarf than in tall lines of barley.

Clearly, we are still far short of understanding how assimilates saved when stem growth is reduced are utilized to enhance grain number and yield in dwarf wheats, the most comprehensively studied case. Many mechanisms appear to be involved, some to do with the relative timing of development of the competing organs, and some which amplify an initially small shift in partitioning to a substantial rise in HI. Nevertheless, it is clear that reduction in plant height can eventually be translated into a rise in HI and grain yield in many cereals and other crops, provided weeds are effectively controlled. This has been the dominant component of yield enhancement over the past 20–30 years, and may have led to some over-emphasis of the significance of the rise in HI.

Root growth

Although no reduction in root growth might be expected among dryland crops, unconscious selection for it could have been possible in varieties adapted to high-input irrigated agriculture. On the other hand, the prolongation of leaf photosynthesis and grain growth in high-yielding varieties may require more prolonged root activity and growth. For example, Austin *et al.* (1977) found that the increase in shoot weight during grain growth among British winter wheat varieties was paralleled by the increase in N uptake at that stage.

Among barley and potato lines, the root:shoot ratio is highest in the wild relatives (Chapin *et al.*, 1989; Iwama & Nishibe, 1989). Siddique *et al.* (1990*a*) found the root:shoot ratio at anthesis in a range of old and modern wheats to have fallen progressively (from 0.64 to 0.53) in more recent varieties. Among wild and cultivated wheats the proportion of assimilates translocated to the roots after anthesis was lowest in the hexaploid cultivars (Evans & Dunstone, 1970). These examples suggest that savings on root growth may have occurred during crop improvement, but in rice and soybeans the change may have been in the opposite direction (Cook & Evans, 1983*b*; Kuroda & Kumura, 1990*b*; Boyer *et al.*, 1980). Both unconscious and deliberate selection are likely to have modified crop root systems in many different ways (O'Toole & Bland, 1987), some increasing the proportion of roots, others decreasing it.

Because selection for greater drought tolerance involves a relative increase in root growth, as in wheat (Hurd, 1974) and sorghum (Bhan *et al.*, 1973), it might be assumed that selection for higher yield potential under irrigated, high fertilizer conditions goes in the opposite direction, but this is not necessarily the case. For example, although the root:shoot ratios of wild, weedy and cultivated barleys diverged strikingly when phosphorus supply was low, there was little difference between them when P was supplied, nor was there any consistent difference in the phosphate absorption rate of their roots (Chapin *et al.*, 1989). Varietal differences in root:shoot ratio which are significant for performance under water stress can persist even in nutrient culture, as in sorghum (Nour & Weibel, 1978).

Differences in the spatial and temporal patterns of root development may well be of greater significance than differences in total root growth. MacKey (1979*b*) has argued that the domestication and improvement of wheat has been coupled with a shift to greater seminal and less crown root development. Along with Fritsch (1977) he has shown that the number of seminal roots has risen as kernel weight has increased, across a wide range of *Triticum* and *Aegilops* genotypes, and that this trend has been reinforced by the reduction in tillering. One consequence of this trend is a higher proportion of roots at depth. MacKey (1979*b*) suggests that such a shift in root distribution enhances the efficiency of nutrient uptake and results in a root system better adapted to the demands of grain growth. However, the optimum geometry of the root system also depends a great deal on soil and climatic conditions and on agronomic inputs.

Conflicting processes may be at work. Selection for greater leaf area may be associated with greater root length, as in sorghum (Blum *et al.*, 1977*b*). Selection for shorter stems may result in shallower root systems (MacKey, 1979*b*; Dev *et al.*, 1980; Virmani, 1971) or may not (Lupton *et al.*, 1974; Cholick *et al.*, 1977; Pepe & Welsh, 1979; Jordan & Miller, 1980). Yoshida *et al.* (1982) found no association between plant height and root depth in rice, in which Ekanayake *et al.* (1989) observed recombinants exhibiting short stature and deep root systems with moderate frequency. To the extent that it is associated with greater tillering, the dwarf habit in wheat is commonly associated also with relatively more shallow roots (Virmani, 1971; Dev *et al.*, 1980; Tandon & Saini, 1979).

Differences in the time course of root growth may also be significant to yield potential. Root systems of cereals reach their peak weight at about the time of anthesis, but varieties can differ in the rate and extent of decline in roots after anthesis, e.g. in sorghum (Blum *et al.*, 1977*a*). A reduction in root density of 1 cm cm^{-3} at that time could save about 0.8 t ha^{-1} in sorghum (Jordan & Miller, 1980), but might well result in a substantial abbreviation of grain filling. Root senescence can be a significant source of N for grain growth, e.g. in wheat (Simpson *et al.*, 1983). Since more prolonged grain growth is a common feature of higher-yielding varieties, it seems unlikely that HI could be further increased at the expense of root growth.

Branching and tillering

Prolific branching and tillering in the wild progenitors of crop plants increased competitiveness during the vegetative phase and safeguarded recovery from herbivores and pests. In earlier agriculture, also, this characteristic could be valuable in aiding recovery from biological or climatic stresses and in compensating for poor establishment. With improved agronomic control there has been selection for reduced tillering and branching in many crops. Donald's (1968) ideotype for wheat proposed its complete abolition. Some tillering capacity has been retained in modern varieties, however, both to allow the crop to compensate for injury from unfavourable conditions and to take advantage of favourable ones. Moreover, as experience with uniculm gigas wheats shows, the additional crown roots associated with tillers can be useful in reducing lodging. McDonald (1990) found tillered varieties of barley to have a higher yield than two uniculm lines over a wide range of sowing densities and growing conditions.

Most of the N taken up is recovered by the grain from tillers whether or not they bear ears (Palfi & Dezsi, 1960), but most of the C invested in the growth of non ear-bearing tillers is wasted and reduces the HI, especially with higher levels of N fertilizer (Donald & Hamblin, 1976). Consequently, emphasis has been given to selecting lines with a higher proportion of ear-bearing tillers, for example in wheat (Jain & Kulshrestha, 1976; Siddique et al., 1989a) and barley (Riggs et al., 1981). This trend has undoubtedly contributed to the rise in HI, and could enhance it still further.

Reserves

In 1920, Beaven described the last stage of cereal crops as follows: 'Some time ... before the grain is ripe the plant ceases to gain in weight of solid matter. Its last effort is to transfer its accumulated reserves into the grain. But all, or very nearly all, the dry matter of the grain is first stored up in the leaves and stems of the plant ... It is mainly ... on the extent to which this "uplift" takes place that plenty or scarcity of the staple food of man depends'. On this view of crop development, which may have been correct for the varieties and agronomic conditions at that time, the extent of reserves was a major determinant of yield. But as N fertilizer applications were raised and new varieties introduced, the dependence on reserves has diminished. Indeed, even by 1945 Archbold was suggesting that the carbon required for grain growth was derived 'principally, if not entirely, from current assimilation'. Some crop physiologists, such as A.A. Nichiporovich (1956) in his influential Timiriazev Lecture, continued to agree with Beaven that a phase of accumulation was followed by a phase of translocation, whereas others such as Bunting (1975) recognized a shift in dependence from savings to current account.

The use of ^{14}C to label assimilates at various times during the life cycle led to a reduced emphasis on reserves as a source of yield, especially in the cereals. Wardlaw & Porter (1967), for example, found that stem reserves of wheat contributed at most only 5–10% of final grain weight, while Rawson & Evans (1971) found that proportion to range from 3 to 12% among varieties of various heights. These proportions were rather higher in water-stressed plants, however, as found also by Austin et al. (1980b) and Pheloung & Siddique (1991).

With field crops of barley and wheat, Gallagher et al. (1975, 1976b) concluded that pre-anthesis reserves contributed 43% on average, and up to 74% of grain yield in crops under water stress, whereas the estimates by Bidinger et al. (1977) were only 13 and 12% for irrigated crops of wheat and barley, respectively, and 27 and 17% for droughted crops. Austin et al. (1980b) found pre-anthesis reserves to contribute only 11% of grain yield in a wet, cool year but 44% in a hot dry one. Hall et al. (1990) estimated that 15% of the C in seeds of irrigated sunflowers, and 27% of that in water-stressed ones, originated from pre-anthesis assimilation. However, respiratory losses, which can account for about half of the fall in stem weight in dwarf wheats

(Stoy, 1965; Rawson & Evans, 1971), can lead to an over-estimation of the role of reserves.

Given the problems of determining the magnitude of the contribution of pre-anthesis reserves to grain yield, it is difficult to say whether they play a greater or a smaller role in modern, high-yielding varieties. Even comparisons between crops are uncertain, although it seems that among the cereals such reserves are most important in rice (cf. Cock & Yoshida, 1972; Dingkuhn *et al.*, 1990). Only 5–18% of grain yield in maize derives from pre-anthesis assimilation (Simmons & Jones, 1985; Wardlaw, 1990), and 10–12% in sorghum (Fischer & Wilson, 1971; Chamberlin & Wilson, 1982).

Several factors, such as differences in stem height and in the extent of N remobilization from leaves, could affect inter-varietal comparisons of the contribution of pre-anthesis reserves to yield. However, no difference in that contribution was found by Rawson & Evans (1971) among tall and dwarf varieties of wheat or by Austin *et al.* (1980*b*) among barley varieties. The latter authors point out that a minimum contribution of 10% by reserves to grain yield in barley is needed merely to account for the remobilization of N into the grain.

In the maize hybrid examined by Reed *et al.* (1988) the availability of reduced N was more limiting to grain growth than the availability of C, which was not the case for that used by Swank *et al.* (1982). Thus, the contribution of reserves to grain yield depends on varietal N status, and on when the remobilization of N to the grain occurs.

Indeed, as this latter is deferred to a progressively later stage in the crop life cycle by later and greater fertilizer applications, pre-anthesis reserves may become less important than post-anthesis assimilates which are temporarily stored as fructans or other compounds in stems and leaves. Blacklow *et al.* (1984) found fructan storage in wheat stems to reach its peak 20–24 days after anthesis, and Kühbauch & Thome (1989) found these post-anthesis reserves to be twice as abundant as pre-anthesis ones. As grain growth duration is extended and leaf senescence is deferred, so is the contribution by reserves to grain yield in cereals likely to become more important again. With the highest-yielding maize hybrid, Swank *et al.* (1982) found reserves to be mobilized mainly during the final period of grain growth.

In pulse crops, in which the demand for N can be so high during seed growth that it leads to the extensive remobilization of leaf proteins, reserves almost certainly play a greater role than they do in the cereals. At the other extreme are the root and sugar crops with relatively little N in the harvested organs, so that remobilization of leaf N during storage is minor. In most of these crops, early reserves play less of a role, final storage being based mainly on the partitioning of current assimilates, with the striking exception of the Jerusalem artichoke, *Helianthus tuberosus* (Incoll & Neales, 1970).

The question at issue is whether the rise in HI of modern varieties derives to some extent from either (1) more complete use of reserves during grain growth or (2) reduced partitioning to reserves in the early stages of the life cycle. The answer is unclear.

Stoy (1965) compared three spring wheat varieties and concluded that the higher-yielding the variety the less it depended on reserves and the less completely it used them, but his data indicated a greater fall in the straw weight of the newer varieties. This latter trend was also apparent among the varieties compared by Austin *et al.* (1980*a*) who concluded that the margin between demand by the grains and supply by assimilation is smaller in the higher-yielding varieties.

From their comparison of many rice varieties, Venkateswarlu *et al.* (1969) concluded that the contribution by reserves to grain yield rose from 25% in early-flowering varieties to 45% in late ones. However, HI fell from 0.51 in the early-flowering varieties to 0.31 in the late ones, so in this case a rise in the contribution from reserves was not coupled to a rise in HI. Weng *et al.* (1982) found a higher level of pre-anthesis reserves of starch in their higher-yielding rice varieties, and that yield was always closely related to such reserves, more so than to post-anthesis growth. With rice, therefore, there is no indication of a reduced investment in reserves with crop improvement, and they may become more important as atmospheric CO_2 level rises (Rowland-Bamford *et al.*, 1990).

Rocher (1988) found the more modern hybrids of maize to partition less photosynthate into leaf reserves of starch, and Tollenaar (1991) and Gardner *et al.* (1990) found them to mobilize their reserves to a smaller extent.

Thus, there is no clear trend among the cereals that either investment in or use of reserves has changed as yield potential has been enhanced. Nor should we expect one, given the various roles of reserves and the complexities of the benefits and costs involved (cf. Bloom *et al.*, 1985). Quite apart from their role in grain growth, reserves may enhance crop survival (through osmotic adaptation, for example), recovery from environmental stresses or pest injury, and the prevention of lodging in cereals like sorghum and maize (Colbert *et al.*, 1984). On the other hand, their very existence may attract pests and there are both direct and indirect costs associated with their storage. Direct costs are associated with their transport, synthesis, turn-over, protection and ultimate mobilization. Indirect costs from not compounding the resources by investment in new growth could be considerable.

As pests, diseases and other stresses are better controlled, and as the N supply is improved with fertilizers, it should be possible to defer the breakdown of reserve proteins (Staswick, 1989), rubisco and other enzymes and to support more prolonged grain growth without loss of photosynthetic capacity.

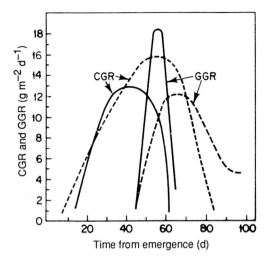

Figure 5.24. Time course of crop growth rate (CGR) and grain growth rate (GGR) in soybean (broken lines) and bean (solid lines) at Palmira, Colombia (Laing *et al.*, 1983).

The reliance of pulse crops on symbiotically fixed N precludes such an option. As illustrated in Figure 5.24, grain growth of beans begins as the CGR approaches its maximum, and rapidly escalates until the grain growth rate exceeds the CGR. Highly nitrogenous reserves must be mobilized to support this situation, which results in a rapid decline first in crop and then in grain growth rates. In soybean, the grain growth rate increases more slowly, never exceeding the CGR and thereby permitting more prolonged grain growth and a slower fall in CGR. However, soybean varieties vary in the extent to which their grain growth rate approaches or exceeds their CGR (Hanway & Weber, 1971), and in their need to mobilize N, as do other pulse crops. Egli *et al.* (1987) found that varieties with a longer grain growth duration mobilized N more slowly out of the leaves, although there was no difference between varieties when the changes were expressed on a seed development scale.

As illustrated in Figure 5.25, some pulses such as cowpea reach extremely high rates of grain growth, sustained by rapid mobilization of reserves, but these rates may last for less than 10 days. Others, such as groundnut and soybeans, grow less rapidly but over a longer period. The grain growth rate sometimes exceeds the CGR in some cereals, e.g. wheat (Evans, 1978*a*; Austin *et al.*, 1980*a*), and sorghum (Chamberlin & Wilson, 1982), indicating the use of reserves, but this need not require the mobilization of leaf N if fertilizer N is applied. Under these conditions, typical of modern cereal agronomy, grain growth can be relatively constant over quite long periods regardless of

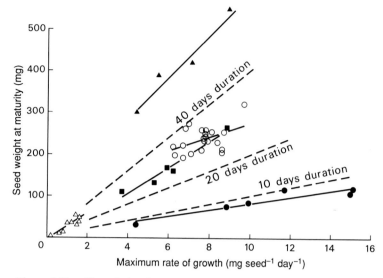

Figure 5.25. The relation between maximum seed growth rate and final seed weight in several crops: groundnut (▲), maize (○), soybeans (■), cowpeas (●) and wheat (△) (Lush & Evans, 1981; cf. also Egli, 1981).

variations in radiation (see, for example, Brenchley & Hall, 1909; Gallagher *et al.*, 1976*a*; Sofield *et al.*, 1977).

Consequently, even if selection for faster and more prolonged grain growth is accompanied by effective selection for a longer duration of leaf photosynthetic activity when supported by improved agronomy, it is likely that reserves will still be needed to sustain grain growth and protein storage through periods of low irradiance or other adversity.

Changed balance in the life cycle

Increase in HI through reduced investment in other organs or reserves seems to have been confined mainly to stem shortening, so prominent in many crops in recent years, and to the reduction of non-ear-bearing tillers. Roots and reserves have not been conspicuously reduced, and other factors capable of contributing to the rise in HI must be considered.

Reduced investment in leaf growth is a significant possibility, in two contexts. With faster early growth, aided by irrigation, fertilizer application and crop protection, the earlier initiation of flowering, after fewer leaves have been formed, reduces the investment in leaves. This could help to account for the rise in HI with earlier flowering evident in rice (Figure 5.20) and found in several other crops.

Secondly, irrigation, fertilizers and crop protection also make possible a longer leaf life, so selection for delayed senescence could also contribute to the rise in HI. As we have already seen, this appears to have occurred in wheat, rice, soybean, cowpea and maize although the causes remain unclear. In some instances the longer photosynthetic life of the leaves is due to more prolonged demands for grain growth. In others it is due to more inclined upper leaves delaying senescence in lower ones, e.g. in wheat (Austin *et al.*, 1976) and rice (Hayashi & Ito, 1962; Saitoh *et al.*, 1990), or to more prolonged root growth as in maize (Reed *et al.*, 1988), but sometimes presumably to greater inherent leaf longevity.

Further rises in yield and HI are likely to require a coordinated increase in the duration of both grain growth and leaf photosynthesis, as found by Crosbie & Mock (1981). One without the other could be ineffective, which may be why Metzger *et al.* (1984) found no effect of increased grain growth duration on yield in barley and Hartung *et al.* (1989) none in maize. Varieties can differ to a considerable extent in the duration of their grain growth, as in maize (Daynard *et al.*, 1971), wheat (Sofield *et al.*, 1977), barley (Rasmusson *et al.*, 1979), rice (Senadhira & Fu, 1989), pearl millet (Fussell & Pearson, 1978), soybean (Egli *et al.*, 1981) and bean (Izquierdo & Hosfield, 1983).

Among maize hybrids, a greater duration of grain growth is associated with greater yield potential (e.g. Daynard & Kannenberg, 1976; Fakorede & Mock, 1978; Ottaviano & Camussi, 1981; Meghji *et al.*, 1984; Cavalieri & Smith, 1985). In some comparisons, the duration of grain growth is more highly correlated with yield than its rate is (e.g. Daynard *et al.*, 1971; Cross, 1975); in others the reverse was found (Hartung *et al.*, 1989), while Crosbie & Mock (1981) found their relative importance to vary from one cross to another. In comparisons among wheat varieties, yield may be related to grain growth duration either strongly (Gebeyehou *et al.*, 1982), weakly (Austin *et al.*, 1989) or not at all (Nass & Reiser, 1975), as also in soybean (Gay *et al.*, 1980; Boerma & Ashley, 1988; Hanson, 1985). In these crops as in several others – e.g. bean (Izquierdo & Hosfield, 1983), oats (Wych *et al.*, 1982), rice (Jones *et al.*, 1979) and potatoes (Evans & Neild, 1981) – the relation between storage duration and yield is often but by no means always positive. Whether the duration or the rate of storage is more usefully enhanced probably depends on the environmental conditions, on growing season length, and on the extent to which early development of the crop can be accelerated.

With maize hybrids grown over the past 50 years in central Iowa, Cavalieri & Smith (1985) found no change in flowering time but a greater duration of grain growth in the more recent hybrids, associated with more prolonged leaf activity and later physiological maturity. With English winter wheats, on the other hand, the date of anthesis has been progressively advanced in more recent varieties (cf. also Cox *et al.*, 1988), in association with an increase in the duration of grain growth (Austin *et al.*, 1980*a*, 1989) This shift will have contributed, along with the reduction in stem height, to the increase in HI evident in Figure 5.21. With barley, likewise, the rise in HI and yield potential in British varieties has also been associated with earlier ear emergence as well as reduced plant height (Riggs *et al.*, 1981), and there is no evidence to suggest that these trends are exhausted. Nor are they likely to be confined to the cereals. Higher-yielding genotypes of sweet potato, for example, initiate storage root formation earlier than low-yielding genotypes, as well as partitioning a higher proportion of assimilates to the storage root (Bhagsari & Harmon, 1982).

The rise in HI clearly has many components and they, rather than HI *per se*, should be the focus of attention in future.

Selection for harvest index

Beaven (1920) and Engledow & Wadham (1923) long ago suggested the value of selecting for migration coefficient (i.e. HI) to enhance yield, as have Ito (1975) for rice, Rosielle & Frey (1975*a*) for oats, Fischer & Kertesz (1976) for wheat, and others. A high heritability for HI has been found in several crops (Donald & Hamblin, 1976).

Rosielle & Frey (1975*b*) found HI to be an efficient criterion for indirect selection for yield in oats, although less so than growth rate (S.K. Johnson *et al.*, 1983). Intensive selection for HI resulted in raising HI but not yield because the selected lines had poor vigour, and Takeda & Frey (1985) suggested that serial selection for HI and growth rate would be more effective. Selection for high HI in wheat resulted in reduced shoot weight, shorter plants and earlier flowering, and was a poor predictor of grain yield (Sharma & Smith, 1986). Selection for high HI reduced yield in soybean (Kenworthy & Brim, 1979). As so often with yield potential, undue attention to any one feature in selection, other than yield itself, can be counter-productive.

Yield components

For anyone to write a book on crop yield without a section on yield components would have been regarded as treasonable by Engledow & Wadham (1923) but as reasonable by some of today's crop physiologists. Individual yield components are mentioned many times in this book, e.g. in relation to domestication (Chapter 3), adaptation (Chapter 4), storage capacity and compensation (Chapter 5) and yield potential (Chapter 6). They figure prominently in analyses of yield because of their accessibility for measurement and the insights they offer. Over-emphasis of any one component can be stultifying, however. As Hallett said of wheat in 1861: 'The

Table 5.3. *Yield components of some highly yielding wheat crops*

Cultivar	Grain yield (t ha^{-1} DM)	Ears m^{-2}	Grains ear^{-1}	Weight grain^{-1} (mg)
Hobbit[a]	7.30	429	46.1	36.9
Norman[a]	7.57	366	43.6	47.5
Benoist[a]	7.23	715	27.8	36.4
Maris Fundin[b]	8.46	563	41	41.3
Maris Templar[b]	8.24	606	32	39.5
Maris Huntsman[b]	7.76	426	39	45.9
Huntsman[c]	9.56	479	39.0	51.2
Huntsman[c]	9.48	411	47.0	50.4
Huntsman[c]	9.07	448	42.3	47.9
Huntsman range[c]	3.25–9.56	330–511	17.4–47.0	33.5–55.0

Sources:
[a] Austin *et al.* (1980a); [b] Hubbard & Ross (1975); [c] Gales (1983).

tacit assumption, that improvement in the size of ears can be obtained only at a sacrifice of their number, has been a great stumbling block in the way of advancement, as it closes the only path in which we can proceed with any prospect of success'.

There are many paths to high yield, not only among different crops but also among different varieties of a crop, as is evident in Table 5.3 for the three highest-yielding winter wheats examined by Austin *et al.* (1980a) and the three highest-yielding farmer's wheat fields analysed by Hubbard & Ross (1975). The ranges of yield components for 32 crops of Huntsman winter wheat compared by Gales (1983) from several sites over several years are included in Table 5.3, along with the data for the three highest-yielding crops, to illustrate how much each component can vary for any one variety. These indicate that the highest yields are often the outcome of extreme values for one component, rather than of some golden mean or 'allometric harmony' (Grafius, 1965, 1978; Rasmusson, 1987). The balance among the yield components is an important adaptive characteristic and is likely to differ across environments (Rasmusson & Cannell, 1970), as in durum wheats (Spagnoletti-Zeuli & Qualset, 1987) and barley (Acevedo *et al.*, 1991).

Because they are determined sequentially, the yield components often behave in a compensatory manner, as we have already seen. Low values for the first components to be determined can result in almost perfect yield compensation (see, for example, Figure 5.3; Grafius *et al.*, 1976; Darwinkel, 1978; Tungland *et al.*, 1987). Such cases rather undermine the much publicized opinion of Jonathan Swift of those who could make two ears of corn grow where only one grew before, but yield compensation is often less than perfect, e.g. in sorghum (Kiniry, 1988) and rice (Sasahara & Itoh, 1989).

The correlations between individual yield components and final grain yield reflect their sequential determination in relation to the seasonal sequence of conditions rather than the greater importance of some components compared with others. For example, favourable early conditions may result in many ears per square metre, but whether or not a high yield is realized will depend largely on subsequent conditions, and the correlation between ears per square metre and yield may be modest, as in Gales' (1983) survey. At the other end of the life cycle, varietal differences in potential kernel weight may be highly significant to yield under conditions favourable for prolonged grain growth, especially when these follow earlier adversity, but irrelevant when the conditions for grain growth are poor. As a result, correlations between kernel weight and grain yield are often quite low, both for individual varieties (e.g. Gales, 1983; Evans, 1978a) and across varieties (e.g. Aggarwal & Sinha, 1987), or even negative as in the common bean (White & Gonzalez, 1990).

High grain number per square metre, whether from more ears per square metre or more grains per ear, is a pre-requisite for high yield; a high correlation between yield and grains per square metre has often been found, e.g. in wheat (Fischer et al., 1977; Evans, 1978a; Spiertz, 1978; Gales, 1983), barley (Gallagher et al., 1975; Dyson, 1977), rice (Yoshida et al., 1972a), sorghum (Krieg & Dalton, 1990), and pearl millet (Craufurd & Bidinger, 1989). Figure 5.26 is illustrative, although the highest grain numbers reached per square metre are exceptional, as may be seen by comparison with Table 5.3. However, the record Chinese wheat crop had 28 063 grains m^{-2} (Cheng et al., 1979) while the record Chinese and Egyptian rice crops had 70 000 spikelets m^{-2} (Xu et al., 1984; Tanaka et al., 1987). Even though the grain yields in Figure 5.26 increased along with grain number per square metre, kernel weight fell progressively, as the last component to be determined. This is a common occurrence in other cereals also, e.g. in pearl millet (Craufurd & Bidinger, 1989). Or, if looked at the other way round, kernel weight increased substantially to compensate for inadequacy of the earlier-determined yield components, and the extent to which it can rise is an important varietal characteristic in such conditions (Fischer & Hille Ris Lambers, 1978). Ma et al. (1990) found that the smaller the kernel weight in intact ears, the greater was the relative increase in kernel weight when grain number per ear was halved.

Optimum kernel size therefore depends on the seasonal sequence of conditions and on variety, through their effects on the number of grains per square metre. But it may also depend on pleiotropic effects of selection for grain size or quality. In wheat, for example, selection for larger kernels led also to later flowering, fewer spikelets per ear, fewer kernels per spikelet and

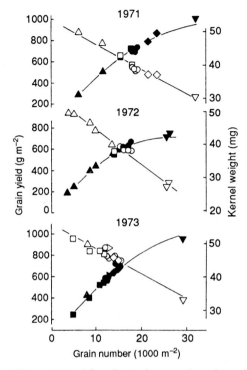

Figure 5.26. The relation between kernel weight (open symbols), grain yield (solid symbols) and grain number across various shading ($\square$), thinning ($\triangle$), crowding ($\triangledown$), CO_2 fertilization ($\diamond$) and cooling ($\triangleright$) treatments to Yecora wheat over three years (Fischer *et al.*, 1977).

almost perfect yield compensation (Busch & Kofoid, 1982). Likewise, selection for larger kernels in maize reduced the other components so that there was no net effect on yield (Odhiambo & Compton, 1987). Selection for greater kernel size can also be associated with greater leaf size, reduced photosynthetic rate and slower leaf appearance, trade-offs that have already been considered, so its effect on grain yield is likely to be complex.

Final kernel weight is the outcome of differences between varieties in both rate and duration of grain growth, as in wheat in which kernel weight was correlated more highly with the rate of grain growth than with its duration (Sofield *et al.*, 1977; Sanford, 1985). Selection for high grain growth rates rather than for long duration may be more effective in short-season environments, as suggested by Nass & Reiser (1975). Even in long-season environments, however, the ultimate outcome of selection for longer duration of grain growth might not be greater kernel size. Instead, it might well result in indirect selection for lower grain growth rates and increased grain number per ear. In this context it is notable that the improvement of yield potential in both wheat and barley has been mainly associated with greater grain number per ear, kernel weight and ear number showing only modest increases or even decreases (e.g. Austin *et al.*, 1989; Hesselbach, 1985; Martiniello *et al.*, 1987).

The associations between yield components are often subtle and counter-intuitive, and by no means always negative. It is the balance between them in relation to environment and agronomy that is important, not only for cereals but also for pulses and root crops.

Hybrid vigour

We have seen, rather painfully, that many different physiological processes have contributed to increases in yield potential, no one being predominant, so it should come as no surprise that the same is true for hybrid vigour, or heterosis as Shull first referred to it. Nor has lack of understanding of the mechanisms involved prevented heterosis from being used to good effect in plant breeding programmes, just as it has not prevented shifts in partitioning patterns from being selected.

Much of the genetic debate about hybrid vigour has centred around the relative significance of dominance on the one hand and of over-dominance arising from interactions between alleles at one locus on the other. In the latter case, hybrid enzymes may be formed, or the alleles may display complementarity of action, which might enhance stability of performance even when yield potential is not increased. Besides the complementarity between alleles in the nucleus, there is also that of complementary action between nuclear genes and those of the chloroplasts and mitochondria (cf. Srivastava, 1983; McDaniel, 1986). Although neither Price *et al.* (1986) nor Lamkey *et al.* (1987) found yield heterosis to be related to the extent of allelic differences at a number of loci in maize, both Walton & Helentjaris (1987) and Lee *et al.* (1989) have found a positive correlation between yield and genetic distance as determined by RFLP mapping of maize chromosomes.

As to which processes are most affected by heterosis, many have been identified at one time or another, as reviewed by Sinha & Khanna (1975), Frankel (1983) and others. Heterosis in stem height is often pronounced, and could indirectly influence crop photosynthesis and growth once the canopy is closed. Heterosis in root growth has often been observed, e.g. in sorghum (Blum *et al.*, 1977*b*) and rice (Ekanayake *et al.*, 1989) and may influence total nutrient uptake. Heterosis in leaf area is one of the commonest and most striking manifestations of hybrid vigour, as in beans (Duarte & Adams, 1963), rice (Akita *et al.*, 1990) and sorghum (Khanna-Chopra, 1982).

Where heterosis for leaf area is marked, heterosis for CER may not be evident or CER may even fall, as in some sorghum hybrids before anthesis. On the other hand, heterotic combinations of yield components could enhance demand and raise the CER, which could account for heterosis in CER being apparent only after anthesis in sorghum (Khanna-Chopra, 1982). Heterosis in CER has been found with maize by Fousova & Avratovscukova (1967), Heichel & Musgrave (1969), Crosbie *et al.* (1978*a*), and Monma & Tsunoda (1979). Akita *et al.* (1986) found it early in the season but not later. With cotton, on the other hand, heterosis for canopy photosynthesis was marked early in the growing season, but not for CER, reflecting the greater leaf area of the hybrids (Wells *et al.*, 1988). The lower CER sometimes found in rice hybrids (e.g. Kabaki *et al.*, 1976) or the absence of heterosis in CER (e.g. Yamauchi & Yoshida, 1985), may reflect the pronounced heterosis for leaf area (e.g. Kawano *et al.*, 1969). Where heterosis in CER has been found (e.g. Murayama *et al.*, 1982), it may have been due to the greater demand for assimilates in the hybrid.

Ashby (1932) considered that there was no heterosis in RGR of maize and that hybrid vigour 'is nothing more than the maintenance of an initial advantage in embryo size'. Since then, many have supported and others have refuted these conclusions (cf. Yamada *et al.*, 1985). Like other aspects of the physiology of heterosis and yield, it all depends on which genotypes are being compared. With rice, for example, the greater early growth of hybrids is related to greater embryo weight in *japonica* × *japonica* crosses and in *indica* hybrids, but not in *japonica* × *indica* crosses where faster leaf area expansion and respiration are the significant factors (IRRI, 1988*a*; Akita *et al.*, 1990).

At least until the canopy is closed, faster leaf growth in the hybrid is likely to be associated with greater LAI, light interception and CGR, as in cotton (Wells *et al.*, 1986, 1988) and rice in the dry season (Ponnuthurai *et al.*, 1984). In some cases, however, there is little evidence of increase in early growth (e.g. Lupton (1976) for wheat) or even in final shoot weight, as in rice (see Figure 4.4) and sunflower (Amir & Khalifa, 1991), yet grain yield can be higher. In such cases there is bound to be a higher harvest index in the hybrids although this is not always apparent (Sinha & Khanna, 1975). Heterosis in both shoot weight and HI has been found in rice (Ponnuthurai *et al.*, 1984) and wheat (Virmani & Edwards, 1983), and is implied for maize (Djisbar & Gardner, 1989).

One route to a greater HI in hybrids is a positive interaction between the yield components rather than the usual compensatory adjustment. Many examples of such behaviour in hybrids are mentioned by Sinha & Khanna (1975) and Frankel (1983). Blum (1977) has analysed a striking example in sorghum. However, there are also instances where no yield heterosis was found even though the parental lines differed reciprocally in the components of yield.

Clearly, the source of heterosis in yield is diverse, both genetically and physiologically, with both dominance and overdominance playing a part. One further aspect of the latter, i.e. of complementation, should be mentioned, however, in relation to the effects of stress on the stability of yield in hybrids. Even if there were little heterosis apparent under favourable conditions, allelic complementation could be important under stress. For example, McWilliam & Griffing (1964) found heterosis in maize to be strongly temperature-dependent, with little hybrid advantage at optimal temperatures, but with heterosis increasing at both higher and lower temperatures. This finding was confirmed with interspecific hybrids of *Phalaris* (McWilliam *et al.*, 1969). Blum (1989) also found the CER of sorghum hybrids to be more stable at high temperatures. The temperature dependence of heterosis is not always so marked (e.g. Rood *et al.*, 1988) but if such a response were general, heterosis would be greater under stressful conditions and the yield of hybrids would tend to be more stable across environments. On the other hand, yield component complementation in hybrids might be more pronounced under favourable conditions.

Yield heterosis in rice is quite pronounced under the cold conditions of the Himachal Pradesh (Kaushik & Sharma, 1986), but not under the low irradiances of the monsoon (Ponnuthurai *et al.*, 1984). Heterozygosity improved stability in maize (Schnell & Becker, 1986) whereas Carver *et al.* (1987) found wheat hybrids to be more responsive than varieties to favourable conditions.

Not only the nature of hybrid advantage but also the conditions in which it is best expressed are in need of further investigation. This microcosm of the physiology of yield is no clearer than the whole field. The widespread use of hybrid maize, sorghum, fodder, vegetable and ornamental crops may not have required a better understanding of heterosis, but certainly merits it, as does the growing use of hybrids in other crops such as wheat and rice (Virmani & Edwards, 1983).

Conclusion

Selection for greater yield potential has not, could not and never shall wait on our fuller understanding of its functional basis, despite the pleas of physiologists such as Cowan (1984), Sharkey (1985) and others. In that sense crop physiology may be retrospective, but the purpose of its backwards glance is to discern some of the ways forward and to provide at least a partial map for plant breeders. We have seen, for example, that the predominant improvements so far have not been in the efficiency of the major metabolic and assimilatory processes, but in the patterns of partitioning and the timing of development. We have also seen that there is still scope for significant further changes in these directions, but that they are likely to involve physiological trade-offs which require the amelioration of environmental stresses on crops through more intensive and diversified agronomic support.

This perception is crucial to our understanding of the nature of greater yield potential, of the inter-dependence of plant breeding and agronomy, and of the role of continuing agronomic innovation, discussed in Chapter 7. Such innovation might not be needed if plant breeding had improved the efficiency of the major metabolic and assimilatory processes, but convincing evidence of such improvements is lacking, whether for photosynthesis, respiration, translocation or growth rate. Regulatory processes, by contrast, have been profoundly modified.

As Jacob (1982) points out, the really creative part of biochemical evolution occurred very early, and highly efficient pathways like the Krebs cycle had evolved long before any plants were domesticated. The predominant evolutionary processes since then are what Jacob refers to as 'bricolage' or tinkering, the modification of pre-existing processes or parts for new purposes.

In essence, artificial selection has extended natural selection in this respect, but in new directions made possible by changing agronomic environments, as exemplified by the rise in harvest index. To date, such changes have been achieved mainly by empirical selection for yield. Many plant breeders and physiologists regard this as a slur on their disciplines, that they cannot do better by design than by empiricism. Hence the frequent attempts to improve the 'efficiency' of important processes such as photosynthesis, which have foundered on the multiple and often unrecognized trade-offs or adaptive roles of the apparently inefficient processes. Moreover, crop yield is the integrated end product of a great variety of processes, and focusing on any one of these, however important, is likely to have counter-intuitive effects, even when supported by quite comprehensive simulation models.

Focusing on several complementary characteristics rather than on a single predominant one, as in Donald's (1968) ideotype for wheat, Rasmusson's (1987) for barley or that of Mock & Pearce (1975) for maize, may be preferable yet still simplistic and counter-productive. There is no one optimum genotype (Marshall, 1991) and supposedly desirable traits may bring undesirable trade-offs or interactions with them. With the proposed uniculm habit, for example, lodging in the absence of crown roots, injury by frost or pest to the only shoot meristem, or inability to take advantage of good conditions may undermine the apparent advantages. Whether Donald's (1968) more radical concept of a weakly competitive 'communal' ideotype will fare better remains to be seen. In any case, most successful plant breeders have a comprehensive set of characteristics in mind when crossing and selecting, as Darwin recognized long ago (Evans, 1980*b*).

Does this mean, then, that there is no real role for the crop physiologist in plant breeding, but only in agronomy? Not at all. Rather, it is a plea for a less doctrinaire approach to the physiology of yield than is apparent in many projects focused on one process, one technique or one gene. Empirical selection is an extremely powerful agent of change but selection by design may yet prove to be even more powerful when our understanding of the physiology of crop yield is more comprehensive than it is at present.

CHAPTER 6

Increases in yield: trends and limits

'*With monotonous regularity, apparently competent men have laid down the law about what is technically possible or impossible – and have been proved utterly wrong, sometimes while the ink was scarcely dry from their pens. On careful analysis, it appears that these débâcles fall into two classes, which I will call Failures of Nerve and Failures of Imagination*'.
Arthur C. Clarke (1962)

Introduction

Arthur C. Clarke did not have the yields of crops in mind when he made the comment above, but they provide an excellent illustration of his theme. With increasing frequency over the last century, agriculturists have been concerned that the limit to crop yield has been reached, putting an end to further progress. Failures of nerve can be seen among those who understand the crucial role of higher yields in ensuring the adequacy of the world's food supply, and failures of imagination among those who see no way forward.

Take the case of wheat. Following the publication of *The Origin of Species*, many of Darwin's critics argued that there was a limit to the extent of variation within species. When Joseph Hooker told Darwin that Hallett believed he had reached a limit in his improvement of wheat yield, Darwin replied (7/8/1869) that 'it would, I think, be rash to assume, judging from actual experience, that a little more improvement could not be got in the course of a century...'. In 1951 another perceptive and experienced plant breeder, S.C. Harland, summed up current opinions with the conclusion that British wheat yields might possibly be raised by 20%, although many thought they could not. Yet the average wheat yield in England has more than doubled since then (see Figure 1.1) while the record yield is more than five times as high. Nevertheless, Austin (1978) estimated the maximum potential yield of wheat to be 11.4–12.9 t ha^{-1}, subsequently revised upwards to 12–14 t ha^{-1} (Austin, 1982) but still less than the current record yield of 15.7 t. ha^{-1} (see Table 6.2). As with earlier calculations of the impossibility of continents drifting, aeroplanes flying or nuclear energy being harnessed, it is not the processes that we understand so much as those we do not, or have not recognized, which constrain our imagination.

In this chapter we begin by considering how yields have changed in the past, for various crops, in different countries, under a range of environmental and socio-economic conditions. In most cases our criterion will be yield per hectare, the form in which the statistics are mostly collected these days. But just as in earlier times a different criterion was more appropriate, namely seeds harvested per seed sown, so also in future more appropriate comparisons of yields might be on other bases, such as per hectare per day (especially in the tropics), or per unit of water or energy or phosphorus used.

After examining the nature of the gaps between national yields, experimental yields and record yields, we consider the potential yields of present day crops and the extent to which these have been raised by plant breeding. This leads us to analyse the interdependence of improvements by plant breeding on the one hand and by agronomy on the other, which has important implications for agricultural development and for the stability of world food supplies.

Change in yield: general form

For any one criterion of yield, change with time is likely to be approximately S-shaped (sigmoid) in form. After an early period of slow change, there is acceleration followed by a temporary or more prolonged slowing down, or yields may even fall. Many factors operate to impose an irregular form on most yield curves, including the year-to-year variation associated with weather conditions, longer climatic cycles – some of which may be associated with sunspot activity (King et al., 1974; Hall & Unwin, 1974; Laird et al., 1990) – and socio-economic and policy shifts.

Changes in the yield of sorghum, the vignette for this chapter, appeared to be more or less sigmoid in form in the USA until about 1984 (Figure 6.1). In fact, like many other yield curves, this was a sigmoid fraud, the earlier rapid rise in yield reflecting the shift of sorghum to more favourable, irrigated conditions after the introduction of hybrids, while the subsequent stasis reflected a shift back to non-irrigated conditions as profitability declined (Jackson et al., 1980; Miller & Kebede, 1984). There is often more to a sigmoid yield curve than meets the eye, and an apparent plateau should not be taken as a sign that yields cannot increase further (cf. Brown, 1985) as is apparent for sugar beet in the UK, for which Elston et al. (1980) volunteered several regressions (Figure 6.2).

Widespread exhaustion of the accumulated fertility of New World soils late in the nineteenth century led to falling cereal yields. Similar declines have occurred throughout history, associated with soil erosion or salinization (cf. Jacobsen & Adams, 1958), epidemics of new pests and diseases (cf. Brewbaker, 1979), displacement of the crop to poorer environments or

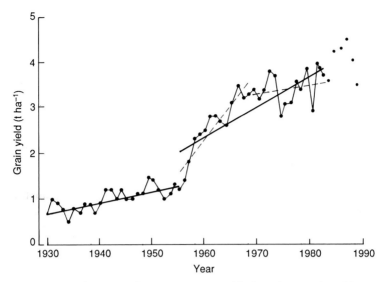

Figure 6.1. Changes in the average grain yield of sorghum in the USA since 1930 (Duvick (1984*b*), updated from FAO *Production Yearbooks*).

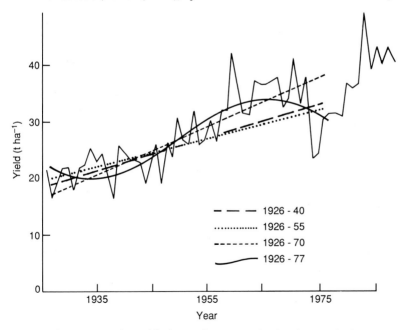

Figure 6.2. The yield of sugar beet in England and Wales with several regressions calculated by Elston *et al.* (1980) for different periods. The yields since 1977 have been added.

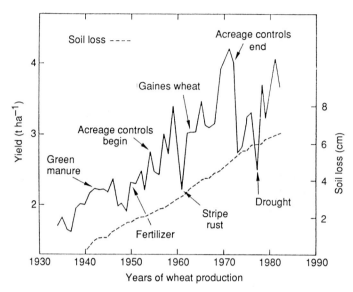

Figure 6.3. Trends in average wheat yields in Whitman County, Washington, USA, as influenced by various factors against a background of progressive loss of topsoil (Papendick *et al.*, 1985). (American Society of Agronomy, Inc., Crop Science Society of America, Inc. and Soil Science Society of America, Inc. Reprinted with permission.)

adverse socio-economic conditions. Nevertheless, yields may rise in spite of a progressive loss of topsoil (Figure 6.3). Like many other yield figures (e.g. Calder, 1953; Donald, 1965), this identifies some of the innovations and adversities contributing to the changes in yield.

While it may have been possible to single out innovations with an identifiable impact on yield in earlier years, it is more characteristic of modern agriculture that at any one time there are many innovations, both genetic and agronomic, at various stages of adoption. Stepwise changes in yields at the national level are therefore less likely than in earlier years, but can occasionally be seen, for example for sorghum in India (Kulkarni & Pandit, 1988). Moreover, synergistic interactions between the various innovations (cf. Chapter 7) may result in yields increasing exponentially in some cases. The long term changes in the yield of wheat in England (Figure 1.1) or New Zealand (Hall & Unwin, 1974), for example, are suggestive of exponential increase, although in the short term progress may appear to be linear, irregular or absent. With rice in Japan (Figure 2.8), for example, yields were relatively static from 1920 to 1940, variously attributed to lack of new technology (Hayami & Ruttan, 1971) or to the availability of cheaper rice from Taiwan and Korea (Ohkawa & Rosovsky, 1960).

More examples will be given below, but it should already be apparent that because of variations in both climatic conditions and socio-economic policies as well as in the pace of innovation, no single mathematical expression adequately describes the time course of changes in crop yield.

Yield take-off

Where several slowly diffusing innovations have been adopted gradually over a long period of time, yield may rise slowly at the beginning, without evidence of a sudden take-off, as in the case of wheat in many European countries with a long history of agricultural improvement (Slicher van Bath, 1963, 1977). On the other hand, the yield of rice in Japan underwent a fairly sudden rise following the Meiji Restoration in 1868 (see Figures 1.1 and 2.8), associated with changes in tenure, rent and land tax (Hayami, 1975). Similar 'take-offs', usually associated with the rapid and widespread adoption of a major innovation, have occurred with many crops since then.

With maize in the USA, for example, Brown (1967) ascribed the sharp rise in yield to 'two non-recurring sources of productivity: the replacement of open-pollinated or traditional varieties with hybrids and, to a lesser extent, the use of herbicides'. The time course for these changes is illustrated in Figure 6.4, which shows that the initial rise in yield coincided with the adoption of hybrids, while the ensuing faster rise more or less coincided with the greater use of herbicides on maize crops. However, another probable contributing factor to the initial take-off was the substantial retreat between 1935 and 1950 from areas less favourable for maize. Moreover, the yields of other crops in the USA, such as wheat and cotton, also began to rise in the 1930s (Luttrell & Gilbert, 1976) without the introduction of hybrids, suggesting that a more general improvement in farming practice was also involved, including mechanization and the timeliness of sowing. Likewise, the second phase of more rapid increase in yield coincided not only with the greater use of herbicides but also of fertilizers and other inputs as well (Pimentel *et al.*, 1973). Once again we see the hazards of ascribing yield changes to single factors.

Wheat yields in the USA provide another cautionary example (Figure 6.5). Plant breeders could be, indeed have been, tempted to ascribe the rise in yield to the spread first of the short and then of the semi-dwarf varieties, while agronomists might ascribe it to the wider use of fertilizers. The trends in both factors coincide remarkably closely with that in yield, as may those for several other factors (Dalrymple, 1988), and the advance in yield has come from their joint action.

Examples of more recent take-offs have been presented in Figure 1.8 for paddy rice yields in several countries. In Japan, as we have already seen, there had been an earlier rise in yield followed by a pause, with a renewed rise after

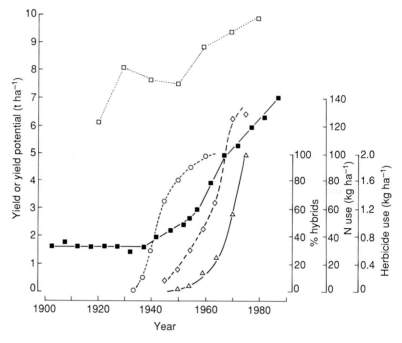

Figure 6.4. Twentieth century maize yields in the USA (by 5-yearly periods) as related to the percentage of area sown to hybrids (○, Frankel, 1983), the use of nitrogenous fertilizers (◇) and herbicides (△) (data from Pimentel *et al.*, 1973; Pimentel, 1979), and the improvement in yield by breeding (□, Russell, 1984).

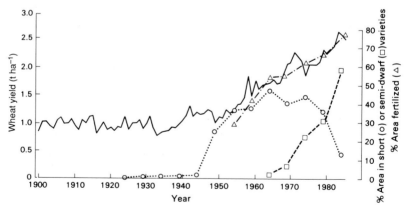

Figure 6.5. Change in US average wheat yield (solid line) as related to the percentage of wheat area fertilized or sown to short or semi-dwarf varieties. Based on data from Dalrymple (1980, 1988) (Evans, 1987*a*).

the second World War until 1976. Australian rice yields were rising even more rapidly at the same time, to reach 7.5 t ha^{-1} in 1971, after which they declined as the crop was extended to less favourable conditions. In the Republic of Korea yields began rising in the late 1950s, rapidly but with severe and serious variation from year to year associated with droughts in the earlier years before irrigation was improved (Chandler, 1979). The take-off for rice yields in Colombia was later still and even faster, associated with a rise in the proportion of the crop irrigated but sustained for only a decade. Indonesian rice yields took off a few years later, and those in Burma (Myanmar) even more recently, while those in Thailand have remained relatively low mainly because of the emphasis on maintaining the traditional quality of Thai rice. Figure 1.8 illustrates the decisive nature of yield take-offs in developing countries in recent years with a crop in which irrigation makes the extensive use of other inputs worth while.

Yield plateaus and declines

Many are the published yield curves in which progress peters out in a dotted extrapolation from a few poor years. In most of these cases there has subsequently been a resumption in the rise of yield following stasis or declines which led pessimists to believe a plateau had been reached. Such apparent plateaus need not imply that yields cannot rise further, for many reasons.

In the first place, a favoured crop may be extended to poorer or less suitable conditions where the lower yields mask a continuing rise under more favourable conditions, as with the shift in US sorghum growing back to dryland conditions since 1973. The reverse can also happen, of course, as with the increasing proportion of the area under wheat, oats and barley in England which is autumn- rather than spring-sown, thereby enhancing the average yield (Silvey, 1986).

Yields per hectare may also be stationary in circumstances where increases in productivity derive more from output per farm worker than from output per hectare (Hayami & Ruttan, 1971). With sugar beet in Britain, for example, the reduction of labour costs assumed greater significance than further increases in yield in the late 1960s, and yields began to decline then (Figure 6.2), although they have since resumed their upwards course. Other constraints to increases in yield may also be operative. For example, where it is important to maintain a particular quality of product, whether for wheat in Canada or rice in Thailand, there may be a penalty in terms of yield improvement. Likewise, the maintenance of environmental quality may limit the amounts of fertilizers or pesticides considered desirable, as in the case of cotton crops in the USA during the 1970s.

Yet another powerful influence on yield trends is the extent to which it is profitable to use inputs, as determined by socio-economic policies. The yield of rice in Asian countries is very responsive to the fertilizer: grain price ratio for example, and many yield plateaus reflect either high input or low output prices. Consider the wheat yields for the United Kingdom and the Yaqui Valley of Mexico in Figure 6.6. Both rose rapidly between 1950 and 1975, when the Mexican yields caught up with the British ones although they started at only half the level. Mexican yields were then stationary for 10 years whereas UK yields continued to rise sharply. Later in this chapter we shall see that the yield potentials of the Mexican and the British wheats have increased to a comparable degree since 1975 (cf. Figure 6.14), so it is likely that the divergent yields have been caused by divergent socio-economic policies. Since the UK entered the European Economic Community in 1974, the increasingly intensive use of inputs has been profitable under the umbrella of the Common Agricultural Policy, which has at least allowed agricultural science to show what can be achieved by way of yield increases.

Early crop yields

Stands of the wild progenitors of several crop plants still exist, but their yields have only rarely been measured. The more or less wild rice in the Jeypore Tract of India may yield about 1 t ha^{-1} of grain (IBPGR/IRRI, 1982), but American wild rice (*Zizania*) stands yield only 0.02–0.14 t ha^{-1} (Hayes *et al.*, 1989). Zohary (1969) mentions yields of 0.5–0.8 t ha^{-1} for the massive stands of wild wheat, barley and oats in Israel. These are similar to estimates of the average yields of wheat crops in England and other northern European countries in the Middle Ages (Figure 1.1), suggesting that there may have been little improvement in yield per hectare over many thousands of years.

Some early records indicate that the genetic potential for higher yields could be expressed with good husbandry. For example, analysis of the extensive Sumerian records on cuneiform tablets indicates that *average* cereal yields at Girsu around 2400 BC were about 2 t ha^{-1} (Jacobsen & Adams, 1958). Yields fell as the land became more saline, to 1.2 t ha^{-1} by 2100 BC and 1.0 t ha^{-1} by 1700 BC. Feliks (1963) records that on good alluvial soils in Galilee in the period of the Mishna and Talmud, about two thousand years ago, wheat yields of up to 45 fold, equivalent to at least 3.6 t ha^{-1}, were obtained with the best of husbandry, possibly after 1–2 years of fallow.

Such yields are higher than the average yields in classical Greece and Rome, commonly only 3–4-fold but rising to 10–15-fold in favourable areas (White, 1963; Fussell, 1967). Although yields of 100–150-fold are occasionally mentioned, they may refer to harvests from single isolated plants (Sperber, 1978). Lewis (1983) points out that although 'Egypt was in the eyes of the Greeks and Romans a place of almost legendary productivity', there is little

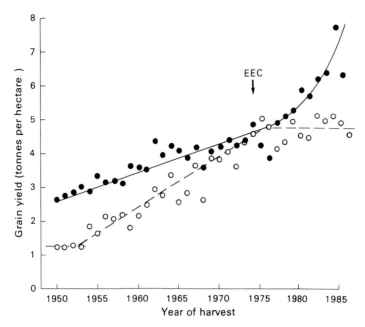

Figure 6.6. Annual wheat yields since 1950 in the United Kingdom (FAO *Production Yearbooks*, solid circles) and Yaqui Valley, Mexico (Secretaria de Agricultura y Ganaderia, open circles) (Evans, 1987a).

evidence on actual yields. He quotes average yields of 9–27 times the planting rate, i.e. up to 2.2 t ha^{-1}, comparable with Egyptian yields in Napoleonic times (Stanhill, 1981a).

The yields of wheat, barley and oats in northern European countries through the Middle Ages have been extensively analysed (e.g. Slicher van Bath, 1963, 1977). Yield levels in still earlier times are not known, but research at the British Museum's ancient farm site at Butser suggests that using cultivation by ard, without manures, yields of up to 2.5 t ha^{-1} of spelt and 3.7 t ha^{-1} of emmer wheat could have been obtained in Britain (Reynolds, 1981). Yields in Roman Britain were probably much lower on average, around 0.6 t ha^{-1} (Bowen, 1961). They may have been even lower in thirteenth century Britain, and there has been considerable debate as to whether they fell between AD 1250 and 1450, owing to diminishing soil fertility (cf. Beveridge, 1927; Bennett, 1935). However, even as early as AD 1211 yields of up to 5.3 t ha^{-1} of wheat may have been attained (Titow, 1972).

Thus, although average yields in earlier times may have been stationary, or even declining when salinization or soil impoverishment occurred, there may have been some improvement in yield potential. Clearer evidence of such

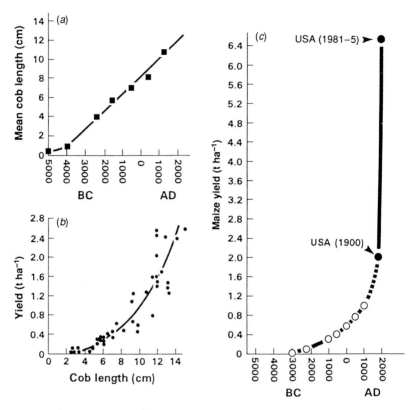

Figure 6.7.   Trend in maize yields over five millennia. (*a*) The relation between archaeological provenance and cob length at Oaxaca, Mexico (data of Mangelsdorf *et al.*, 1967). (*b*) The relation between cob length and yield at Oaxaca (Kirkby, 1973). (*c*) Early yield trends at Oaxaca, estimated from (*a*) and (*b*), and that in the USA from 1900 to 1981–5.

improvement has been found with maize by Kirkby (1973) who combined the excavation data of Mangelsdorf *et al.* (1967) on average corn cob lengths at Tehuacan between 7000 BP and 500 BP, with her field surveys on the relation between cob length and grain yield over four seasons at Oaxaca. The results depicted in Figure 6.7 indicate a progressive improvement in maize yield potential throughout that period, whereas the data of Miksicek *et al.* (1981) for Mayan maize up to AD 250 are not so clear.

In Europe, the ratio of harvested to sown cereal grain seems to have remained fairly stationary until the seventeenth Century, being greatest (*ca.* 7) in England and the Low Countries, whereas in central, northern and eastern European countries the ratio remained close to 4 (Slicher van Bath, 1977), similar to that mentioned by Roman writers on agriculture, such as Columella. In western Europe this ratio then began to rise, to 9.3 in the 1650–99 period, and to 11.1 by 1800–20, with the introduction of legume leys and improved tillage. However, the mid-nineteenth century was the turning point in England, with the introduction of artificial fertilizers, improved drainage, a wider range of implements and the beginnings of more deliberate plant improvement by selection (Craigie, 1883; Bennett, 1935). From then on average wheat yields increased more rapidly (see Figure 1.1), as did record yields. Hallett (1861) mentions a wheat yield of 7.2 t ha^{-1}. Boulaine & Feller (1989) record a yield of 3.2 t ha^{-1} near Paris in 1766, and a questionable yield of 11.5 t ha^{-1} from Tipperary, Ireland, in 1742.

Yield trends: some comparisons

From what has been seen already, there are marked differences between crops, between countries and between environments in their yield levels and in the rate of their improvement. Some of the statistics in the FAO Production Yearbooks are rather arbitrary, but overall there is commonly a 25-fold range among developing countries and up to a 5-fold range among developed countries in the yields of many crops. To some extent this range is associated with the level of GNP per capita, insofar as this affects the resources and inputs available to farmers and the supporting infrastructures for research and extension (Figure 6.8). Environmental and socio-economic factors may over-ride these effects, however, as in the relatively high yields of maize at low GNP per capita for Egypt and China where irrigation is prevalent.

The average yields for all 'developing' countries and for all 'developed' ones, as computed by FAO in recent years, are compared in Table 6.1 for the five staple cereals and major crop groups. With such composite data, many confounding factors are at work, but the influence of environment is apparent in the high yields for rice and the low yields for millet, with the other cereals in between also reflecting the dominant influence of water stress. Among the C_4 cereals the millets tend to be grown in the driest areas, maize in the least stressed, with sorghum in between (cf. Muchow, 1989*a,b*). In developing countries sorghum is mostly grown in areas subject to considerable water stress, too severe for maize, whereas in developed countries it is often grown for feed under more favourable conditions, hence the reversal in the relative yields of wheat and sorghum with development.

Although the average yield level of the five staples clearly reflects the level of environmental stress in which they are grown, the relative increases in average yield over the past two decades do so less clearly. The increase has been greatest for wheat in the developing countries and for millet in the

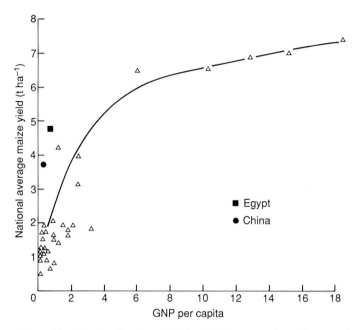

Figure 6.8. National maize yields in 1987 for countries with more than 400 000 hectares of the crop, in relation to gross national product per capita (in thousands of US dollars) (based on data from FAO *Production Yearbook* and the World Bank *World Development Report 1989*).

developed ones. For the cereals overall, the percentage increase has been substantially greater in the developing than in the developed countries, reflecting both the greater need for additional food in the developing countries and the lower initial yield levels. Fluctuations in yield from year to year tend to be greater in the developed countries, owing to their greater responsiveness to global market forces affecting the level of inputs and the environments in which crops are grown. In the developing countries, by contrast, stability of production in predominantly subsistence economies is of over-riding importance.

In the aggregate, the relative increase in C_3 cereal yields has been at least as much as for the C_4 cereals. Average cereal yields have risen over the past 19 years by 3.4% per year in developing countries and by 1.9% per year in developed ones. For the pulses, on the other hand, overall yields have not risen at all in the developing countries but by 2.5% per year in the developed ones. Roots and tubers have undergone a modest increase in yield. However, the FAO statistics for average developed and developing country yields have several traps in them, depending among other things on whether China and

Table 6.1. *Average yields of several crops in developing and developed countries*

Crop	Total area 1989 (Mha)	% total production in developing countries	Yield 1989 (t ha^{-1})		% increase in yield since 1969–71	
			Developing	Developed	Developing	Developed
Wheat	226	42	2.23	2.50	99	39
Rice	146	95	3.38	5.82	52	12
Maize	130	39	2.26	5.93	48	47
Sorghum	44	69	1.03	3.18	29	2
Millet	37	87	0.76	1.45	27	63
Total:						
Cereals	705	53	2.35	3.09	65	36
Roots and tubers	47	66	11.12	16.91	33	13
Pulses	70	63	0.65	1.61	0	47

Source: Data from FAO *Production Yearbook* 43, 1989.

the former USSR are included. The proportion of the global crop grown in developed countries can change substantially over two decades, as it has for barley and the pulses, with the result that the global average yield increased relatively more than either the developed or the developing country yields.

The pattern of yield increases becomes even more complex when the development categories are subdivided into regions. The substantial rise in soybean yields in the developing countries is then seen to be due mainly to the rapid expansion of the crop in South America, especially in Brazil and Argentina. In 1961–5 this region accounted for only about 2.5% of the area under soybean in developing countries when China is included, whereas by 1989 it accounted for 55%. Soybean yields for the South American region have increased only about 15% since the area expanded but because they are much higher than those of Asia and Africa, the apparently rapid increase in soybean yields of developing countries was due largely to a geographic shift in production. Similarly, the apparent increase of 46% in sweet potato yields in 1979 was due to the inclusion of China within the statistics for developing countries. China accounted for 88% of their production of sweet potatoes in 1979 and had a substantially higher yield than most other major producers.

Just as regional statistics conceal a lot of variation between the countries in a region, so do national statistics often conceal large differences between districts. In India, for example, wheat and rice reveal a marked heterogeneity in both yield levels and rates of improvement. The relatively high rice yields in the north of India have doubled since 1960, whereas in the east and west, where much of the rice is still grown, yields have remained low and stationary (David, 1989).

Within a country there can also be considerable differences in yield levels and their progress for a particular crop, depending on its season and type of culture. This is especially true for rice in the tropics, where it is grown at all seasons of the year, in either deep water, irrigated, rainfed bunded or upland culture. In Bangladesh, for example, neither aman (monsoon) nor aus (early summer rainfed) rice crops have, so far, been improved in yield, whereas that of the boro (irrigated winter) crop has doubled. In Indonesia, lowland rice yields have doubled whereas upland yields are stationary (David, 1989).

Experiment stations and good farmers

In earlier times the best yields of crops on experiment stations, with access to advanced varieties and agronomic techniques, were often far greater than average yields in the area and were taken as an indication of the scope for yield improvement, as were the yields obtained by leading farmers. The distribution of yields among farms takes many forms. For wheat in the UK, the distribution about the mean was more or less normal (Tinker &

Widdowson, 1982), with field to field variation being greater than that from year to year. Church & Austin (1983) found the yield of wheat in England to have a standard deviation of about 20% between farms, and of about 15% between fields within farms.

Nguyen & Anderson (1991) suggest that the yield distribution shifts from being skewed towards below-average yields in suboptimal environments where input use is limited to being skewed in favour of above-average yields when input use is high and varieties well adapted. Consequently, agricultural development tends to result in a narrowing of the gap between average and experiment station yields. Examples of such narrowing have been given by Thompson (1975) for maize in Iowa and by Byerlee (1990) for wheat in the Punjab. With rice in Japan there is now little difference between average and experiment station yields but this is not the case in India and many developing countries. However, the highest rice yields obtained each year in both Japan and India are much higher than the experiment station averages (Yoshida, 1981).

As an impetus to self-sufficiency in rice production by Japan, an annual yield contest was sponsored by the Asahi News Co., beginning in 1949, but there was no narrowing of the gap between annual record and average yields up to 1968 (Yoshida, 1981), by which time Japan was self-sufficient in rice and the competition was discontinued. A more recent example of the maintenance of the gap between annual record and average yields is presented in Figure 6.9 for both maize and soybean in Iowa State. Despite the more than two-fold increase in the average state yields of both crops over the past 50 years, the gap between the average and the annual record yields has not narrowed in absolute terms, nor even in relative terms for soybean, for which the state average yields now exceed the annual record yields of 50 years ago.

Yield gaps

The highest yield of paddy rice ever obtained on the IRRI farm at Los Baños in the Philippines is about 11 t ha^{-1}, four times higher than the average national yield in recent years. Such comparisons are often taken to imply that there is ample head-room for further increase in national yield and that technology already available is not being applied by the farmers. The reality is, as usual, more complex and less dramatic.

Scientists from the International Rice Research Institute and their colleagues in the national research systems have analysed the constraints to rice yield through on-farm experiments at many sites in Asia over a period of years (IRRI 1977, 1979b). At each site four or five different packages of inputs (variety, fertilizer, insecticide and herbicide) and their management have been compared with the farmer's own crops, on his land with his water supply and

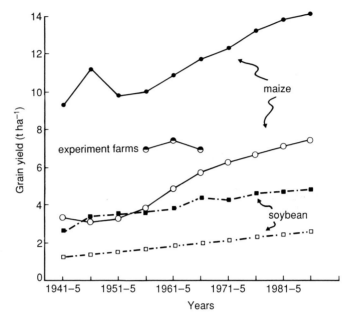

Figure 6.9. Quinquennial state averages for maize and soybean yields in Iowa between 1941 and 1990 (open symbols) compared with those for the highest yields (solid symbols) in the Master Growers' Contests of the Iowa Crop Improvement Association, from data provided by Wayne R. Hansen. Also included are averages for maize yields on seven experiment farms in Iowa, from Thompson (1975).

with due account of both costs and returns. The overall yield gap can then be partitioned into three parts:

(1) the gap between what is currently possible on the experiment farm and what is possible on a substantial sample of actual farms, on which soil, water supply or other site factors may be less favourable;
(2) the gap between what is possible at the farmer's level with the best available technology and what is profitable;
(3) the gap between what is profitable and what is actually done.

For an international agricultural research centre such as IRRI, the first gap is really a composite of how much the yields at the IRRI farm differ from those at the same season in the surrounding province (Laguna) or in the Philippines as a whole or throughout Asia. For 1975 and 1976, the highest yields on the IRRI farm in the wet seasons averaged 5.9 t ha^{-1} whereas those on farms in Laguna averaged 5.4 t ha^{-1}, while in the dry seasons the corresponding figures were 8.4 and 7.5 t ha^{-1}. But Laguna is a very favourable area for rice

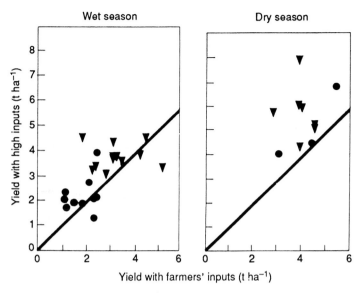

Figure 6.10. Comparison of farm yields with high-input yields of rice in experiments on farmers' fields in Nueva Ecija, Philippines, in both the wet and dry seasons over two years (different symbols) (IRRI, 1977). The sloping lines represent equal yields with the two levels of input.

growing, and the gap between what is possible at IRRI and in the Philippines or Asia at large would be substantially greater (Barker *et al.*, 1977*a*; Herdt, 1979).

Before considering the second yield gap, between what is possible and what is profitable on farmers' fields, let us look first at the gap between farmers' actual yields and those obtainable with high inputs and management, bearing in mind that this gap may have been narrowed somewhat by the farmers observing what the agronomists did on their fields and modifying their practices in later years, and also *vice versa* of course. In some cases no significant yield gap was apparent, as at Subang (Morris *et al.*, 1977) and Yogyakarta, Indonesia and in Taiwan (Herdt, 1979), i.e. not even the most advanced technology package could increase yields. In Subang, for example, fertilizer use was close to the recommended levels, but excess water and insect pests which were not effectively controlled prevented any response to higher input levels during the wet season while water stress kept yields low in the dry season. In many other instances, however, yield was significantly increased at higher management levels as illustrated in Figure 6.10 for Nueva Ecija, but the overall increase from a large number of trials in the wet season was only 25% (Herdt, 1979). The major constraints were fertilizer level and insect control in the wet seasons, and fertilizer level in the dry seasons, often in strong interaction with insecticide levels, fertilizers being most effective when insecticides were also applied and vice versa. Farmers' weed control practices were, in general, highly effective (Barker et al., 1977a).

Returning to the gap between what is possible and what is profitable, the conclusion from the IRRI constraints analyses was that the use of higher levels of input to obtain higher farm yields would generally have reduced profits and increased risk, especially in the wet season crops. With dry season crops the use of additional fertilizer would have increased profits to a significant extent in many cases, whereas additional weed control measures would not. The farmers' practice is economically rational and perceptive, as Schultz (1980) argues, and the gap between actual yields and possible yields depended most of all on the rice:fertilizer price ratio. Losses due to insect pests were high, in general, because the currently available control measures were too costly. The overall conclusion drawn by Herdt (1979) from these constraints studies is that 'it is relatively easy to account for the dramatic gap between what is technically possible and what has been achieved: what is technically possible is more modest than most observers admit; the economics of substantially higher yields is [sic] not attractive; the costs associated with the credit and tenure arrangements that often prevail in developing countries make higher input use totally unattractive for some farmers. Thus the available technology is being used to its potential'. The greatest need, therefore, is not for the diffusion of known technology but for new technology.

Record yields

Like all such records, the highest documented yields have all the fascination of ultimate performance and a few are now included in the Guinness Book of Records. This accolade is merited by the human significance of the subject, but a more comprehensive summary of record crop yields is needed. Table 6.2 is presented as a step in that direction but with acute awareness that recent records, often reported in agricultural news-papers, will have been missed. There is also the problem that yield contests, especially in earlier times, have elicited some ingenious but unhelpful cheating (Boulaine & Feller, 1989).

Record yield contests that run for many years, like those for maize and soybean in Iowa (Figure 6.9), can be valuable for the indication they give of currently attainable maximum yields, and of the extent to which these depend on weather, agronomy and genotype. But only a few of the entries in Table 6.2 came from such contests. Many are quite exceptional yield records of long standing, which reflect a rare combination of conditions and, as such, offer little guide to what is practicable and attainable. In the Japanese rice yield contests, for example, the second best crop each year was, on average, less than 8% below the best one but in 1960, when the Japanese record was set (Table 6.2), the best crop was 22% better than the second (Ishizuka, 1978). In the All-India rice yield competition, in which the winners for three consecutive years all came from the same village in Maharashtra, the record yield of 17.8 t ha^{-1} of paddy was substantially higher than the best crops in preceding years (Suetsugu, 1975). On the other hand, the highest yielding rice crops recorded for Egypt and China were not exceptional (Xu *et al.*, 1984; Tanaka *et al.*, 1987). Many of the record yields publicized in agricultural newspapers do not meet the criteria for long-running yield contests, leading to uncertainty about their status. For example, the record maize yield of 23.2 t ha^{-1} given in Table 6.2 is 15% higher than that recorded in 1985 by the National Corn Growers Association and 6% higher than the highest yield they have ever recorded (D.K. Johnson, personal communication).

Consequently it is misleading to compare these record yields with national averages and to conclude, because they are two to ten times higher, that there is already ample scope for increasing national yields. That is the most common misuse of record yields, but there are others. For example, Boyer (1982, 1983) has taken the gap between the record and the average yields for eight crops in the USA, subtracted amounts estimated for the average losses due to diseases, pests and weeds, and concluded that the substantial residual gap (which varied from 51% of the record yield for sugar beet to 82% for wheat) represented losses caused by environmental stress. The preceding discussion on yield gaps should have made clear just how dubious such an assumption is, particularly as socio-economic constraints frequently depress average yields far below those that are agronomically possible. Similar considerations apply to the use of record yields to define attainable yields for the purpose of crop loss assessment (cf. Loomis & Adams, 1983), to make cases for the support of research on stress, or to raise hopes for quick increases in average yields.

The fact that several record crop yields are of quite long standing has led to suggestions that they too have reached their plateau. However, in at least a few of the better documented cases, such as maize and wheat, some upward movement is apparent, and more systematic monitoring is needed.

Eventually it may also be possible to make valid comparisons between crops, but we are not yet in that position. Record yields are often given in units (e.g. bushels per acre) which require assumptions to be made in converting them to tonnes per hectare. Further assumptions must then be made, e.g. in going from paddy to brown rice, from grain or tuber weight to dry weight, or from cane or root weight to sugar. But even if the edible dry weight conversions were satisfactory, there is still the problem of relating yields when

Table 6.2. *Some record crop yields*

Crop	Country	Year	Yield (publ. units)	t ha^{-1}	t dw ha^{-1}	Reference
Maize	USA	1955	304 bus ac^{-1}	19.1		Walden, 1966
	USA	1973	307	19.3		Matlick, 1974
	USA	1975	338	21.3		Swank et al., 1982
	USA	1977	353	22.2		Pollard, 1977
	USA	1985		23.2		Connell et al., 1987
Rice	India	1970	14.3 t ha^{-1} (paddy)	10.7		Suetsuga, 1975
	India	1971	15.9 (paddy)	11.9		Suetsuga, 1975
	India	1973	16.9 (paddy)	12.7		Suetsuga, 1975
	India	1974	17.8 (paddy)	13.4		Suetsuga, 1975
	Australia		17 (paddy)			McDonald & Woodward, 1977
	Egypt	1982/5	15.3 (paddy)	11.5		Tanaka et al., 1987
	China	1983	2000 jin mu^{-1}	11.3		Xu et al., 1984
	Japan	1960	13.2 t ha^{-1}	10.5		Ishizuka, 1978
	Philippines	1972	(paddy)	11.0		Evans & De Datta, 1979
Wheat	UK	1211		5.3		Stanhill, 1976
	UK	1893		6.3		Stanhill, 1976
	UK	1969		9.5		Stanill, 1976
	UK	1974		10.8		Stanhill, 1976
	UK	1981	111 cwt ac^{-1}	14.0		Blackbeard, 1982
	UK	1982		15.7		Trow-Smith, 1982
	USA	1964	169 bus ac^{-1}	11.4		Reitz, 1967
	USA	1965	209	14.1		Reitz, 1967
	USA	1966	216	14.5		Anon, 1966a
	China	1978	2026 jin mu^{-1}	15.3		Cheng et al., 1979
	phytotron + CO			14.0		Gifford, 1977

Crop	Country	Year	Yield		Reference
Sorghum	USA		320 bus ac^{-1}	20.1	Wittwer, 1975
	USA		347 bus ac^{-1}	21.8	Nelson, 1967
Pearl millet	Kenya	1966		7.3	Doggett & Majisu, 1967
	India			8	D.J. Andrews, pers. comm.
Barley	UK	1980		10.5	McWhirter & McWhirter, 1983
	UK	1984		11.8	McWhirter & McWhirter, 1987
Oats			296 bus ac^{-1}		Wittwer, 1975
Soybean	USA	1966		5.6	Anon, 1966b
	USA	1968		6.7	Cooper, 1985
	USA			7.4	Whigham, 1983
	Italy	1977		6.1	Whigham, 1983
Groundnut	Rhodesia	1973/4		6.4	Smartt, 1978
Cowpea				3.4	Kwapata & Hall, 1990
Potato	UK	1973	102 t ha^{-1}	20.4	Evans & Neild, 1981
	USA		126		Haws et al., 1983
Cassava	Colombia	1977	79.2	28	Kawano, 1978
	Colombia	1978	82.2	29	Veltkamp, 1985
	Indonesia		96		Bruijn & Dharmaputra, 1974
Sweet potato	Papua New Guinea			17	Enyi, 1977
	Japan			14	Kodama et al., 1970
Sugar cane	USA			25.6	Burr et al., 1957
	Swaziland			23.7	Humbert, 1972
	USA (Hawaii)		250 t ha^{-1} y^{-1}		HSPA, 1967
	Australia		220 t ha^{-1} y^{-1}	24.4	Ham, 1970
	Colombia		219 t ha^{-1} y^{-1}		Irvine, 1983
Sugar beet	USA		115 t ha^{-1} (roots)	19.1	Fick et al., 1975
	USA		132 t ha^{-1} (roots)	17.2	Fick et al., 1975
	USA		140 t ha^{-1} (roots)		McWhirter & McWhirter, 1987

ı differs so much, e.g. for cassava, wheat, soybean and oil palm
ields of more than 10 t ha^{-1} of oil have been obtained) (H. von
rsonal communication) and is not known for the record crops.
ᴐ the important factor of crop duration. In Table 6.2, for example,
duration was a full year for the highest sugar cane yields, 240 and
290 days for the two beet crops yielding 19.1 t ha^{-1} of sugar (Loomis &
Gerakis, 1975), about 300 days for the 1982 record wheat crop, 188 days for
the best sweet potato crop, and only 85–90 days for the record pearl millet
yield (D.J. Andrews, personal communication). In terms of kilograms of
edible dry weight per day, therefore, the pearl millet and sweet potato crops
were the highest yielding and wheat the lowest of this élite group.

These record crops not only indicate the ultimate yield potential of existing
varieties; they have also been used to estimate maximum rates of biomass
production (Loomis & Gerakis, 1975; Monteith, 1977). Here again the
assumptions are frail because most record crops are poorly documented, and
there is no evidence of exceptionally high rates of growth and storage by the
record crops. These rates were 18.5 and 10.7 g m^{-2} d^{-1}, respectively, for the
record sugar beet crop (Fick et al., 1975), while for the sweet potato crops the
maximum growth rate was 21–23 g m^{-2} d^{-1} (Hahn & Hozyo, 1983). For the
record cassava crop the storage rate was less than 10 g m^{-2} d^{-1} (Cock, 1983).
Presumably, therefore, both environmental conditions and agronomy
favoured longer than usual storage periods in many of the record crops, as
would clear skies and cool nights in the case of the Maharashtra and
Australian record rice crops. Likewise, the record Chinese wheat crop of 1978
took 62 days from anthesis to maturity (Cheng et al., 1979).

For the record 1982 wheat crop, 260 kg N ha^{-1} was applied over seven
dressings, as well as heavy applications of K, P and Mg, fungicide and
chlormequat (Trow-Smith, 1982). The 1977 record maize crop received 425
kg N ha^{-1} for an estimated internal requirement of 407 kg ha^{-1} (Stanford &
Legg, 1984) plus other major and micro-nutrients, insecticides, herbicides and
almost continual irrigation (Pollard, 1977). Thorough soil preparation, good
establishment, high plant density, ample fertilization, avoidance of water
deficit or excess, and effective protection are common elements in the
management of record crops.

Estimates of potential yield

For a variety of reasons we often want to know what are the likely
limits to the yield of crops: they are central to any estimate of the ultimate
human carrying capacity of this world and also to long-term policies for land
management and for the conservation of biological diversity and other
resources.

As indicators of the limits to yield we may use experiment station or annual record yields, but these are evanescent and, increasingly, attention is being turned to various ways of estimating potential yields. It is here that Arthur C. Clarke's failures of nerve and of imagination (see epigraph) become constricting. Indeed, many early predictions of the limits to yield have long since been surpassed. Potential yield estimators must sail between the Scylla of being realistic in their assumptions and responsible in the sense of not raising false hopes among policy-makers and global planners, and the Charybdis of being perceptive enough not to be trapped by received wisdom and imagined limitations.

Maximum yield estimates may be derived simply by the extrapolation of recent trends. These tend to occur in epidemic proportions whenever there is a pause, albeit temporary, in yield curves. If the assumption is made that yields should be an exponential function of time, as is done by Blaxter (1976), a more or less linear increase in yield implies a deceleration in the exponential rate, from which yield limits can be predicted. For wheat in the UK, Blaxter predicted 5.5 t ha^{-1} which was surpassed, fortunately, within four years.

Simulation models are being used increasingly to estimate crop yields for many purposes but they may be less helpful in relation to potential yields because of our lack of understanding of the limits to many crucial yield-determining processes. By and large, simulation models can now deal fairly satisfactorily with crop photosynthesis and growth, and the limits to these processes are relatively well understood. But when it comes to the processes which control the partitioning of photosynthetic assimilates, the determination of storage capacity, the mobilization of reserves and the duration of storage, the models must still be based on empirical relations. It is probably because of this that several estimates of maximum yields are so close to already-achieved yields, as in the case of wheat (Austin, 1978, 1982) and cassava (Veltkamp, 1985). But the empirical relations on which they are based can be changed by plant breeding and new agronomic opportunities.

For example, earlier sowing – whether made possible by minimum tillage techniques, protected nurseries (Kondo *et al.*, 1975) or breeding for seedling tolerance to cool temperatures – may permit more prolonged development and growth of the storage organs, as may the use of grain drying at the other end of the life cycle. Such prolongation can have a powerful effect on estimates of maximum yield, and there are no clearly defined limits to what can be assumed or effected. It is in these regulatory, as against assimilatory, areas that many potential yield estimates are flawed, my own (Evans, 1972) no less than others, hence the wide range in the estimates that have been made. Those for rice, which were reviewed by Ishizuka (1971), varied from 24 to 42 t ha^{-1}. Gilland (1985) has estimated the maximum theoretical yields of maize, wheat and rice to be 27, 18 and 14 t ha^{-1}, respectively, but I suspect that these will prove to have been rather conservative estimates, especially as the atmospheric CO_2 level rises. Salisbury & Bugbee (1988) have already recorded far higher wheat yields with CO_2 enrichment under artificial light, raising the problem of how potential yield and yield potential should be defined.

Genetic yield potential and its improvement

Introduction

Pronounced increase in the genetic yield potential of the major crops is a relatively recent phenomenon and, at least in the cereals, one with a gathering momentum even though other objectives may have a higher priority in many plant breeding programmes.

Yield potential may be defined as the yield of a cultivar when grown in environments to which it is adapted, with nutrients and water non-limiting and with pests, diseases, weeds, lodging and other stresses effectively controlled. When defined in this way, yield potential may seem to be rather remote from actual yields, but as agriculture becomes more controlled and intensive this definition becomes more immediately relevant, e.g. to much cereal growing in Europe today.

The definition is arbitrary in several ways. For example, varieties selected at low latitudes may display a higher yield potential in cooler climates than in those for which they were selected. Older varieties were selected with low levels of inputs such as fertilizers and may perform better at those levels than at higher ones. Indeed, there may even be crossovers in relative performance with older varieties yielding more than recent ones at low input levels (cf. Chapter 4). Yield potential might therefore be more fairly defined as the yield of a cultivar when grown in the environments and with the agronomy for which it was selected, but the operational problem with such a definition is that, in a world of rapidly changing agronomy and pest and disease biotypes, it would be extremely difficult to make the relevant comparisons.

By comparing the yield of old and new varieties when grown with modern agronomic methods, plant breeders estimate the yield increase due to breeding and allocate the residual increase in average yields to agronomic improvement. Such a procedure has various traps, e.g. in ignoring genotype × environment interactions (Simmonds, 1981) and the influence on yield of socio-economic factors, or of long-term weather trends such as the more favourable conditions in the US Corn Belt from 1930 to 1972 (Thompson, 1986). The progressive rise in atmospheric CO_2 level may already have raised crop yields by 18% or so (Monteith, 1977).

However, the major flaw in such partitioning of credit is the implication that improvements by breeding and by agronomy proceed independently of one another, whereas they are so intimately inter-dependent that we should consider them separately only within the context of recognizing that it is the continuing reciprocation between them that makes progress possible. New agronomic opportunities, such as the availability of cheaper nitrogen fertilizers, create a need for a change in plant type which, after a certain time lag (cf. Jensen, 1967), is progressively accomplished, e.g. in the reduced height of modern cereal varieties. In turn, however, these genetic changes require agronomic developments – such as improved weed control or closer spacing (cf. Cooper, 1985) – in order to be fully exploited, and these in turn may require further genetic tailoring, e.g. in herbicide resistance or delayed senescence of leaves. And so the reciprocation continues.

With maize, cheaper nitrogenous fertilizers and heavier applications eventually led to closer row spacing and higher plant populations than were possible when rows had to be the width of a horse apart in order to control weeds by cultivation. Herbicides permitted higher plant densities and led to the need for quite different characteristics in relation to leaf posture, tillering and inflorescence fertility, which are still being developed; these in turn have required further agronomic changes. In the Philippines there was a dwarf, heavy-tillering, upright-leaved rice variety, Ramai, available in 1930 (Evans *et al.*, 1984), but it was ahead of its time agronomically, and has not even been conserved in gene banks. To partition the credit for increased yields between breeding and agronomy is, therefore, a rather arbitrary procedure. Nevertheless, it is helpful to derive estimates of the progress in raising the yield potential of various crops in various climatic and socio-economic environments, and at various stages of the rise in national yields, *provided* that we don't forget their dependence on agronomic innovations.

There is another important caveat. Would natural selection have raised the yield potential as much as the plant breeders have done, in response to improved agronomy, rising CO_2 level etc? Any impulse to reply 'Of course not' must take account of the results of Suneson (1956) and Allard (1988) showing that the yield of barley composites has been raised substantially by natural selection over the years. Soliman & Allard (1991) found the yield of two composite cross populations to increase by 20–25% in the course of 25–30 generations, and commend such an approach where disease resistance and yield stability are the main objectives of a breeding program, but not for raising yield potential.

Indirect estimates

Before turning to direct comparisons of yield potential, let us consider some indirect ones based on long series of national variety trials, like those conducted by the National Institute of Agricultural Botany (NIAB) in the UK. In these, new varieties were compared at many locations with a standard variety, which was changed every so often. The level of agronomic support also changed, with the result that varieties were compared under the conditions for which they were bred and for which their resistances were appropriate. For example, winter wheat varieties were compared with Heine VII in the years when it was still resistant to yellow rust and with Cappelle-Desprez while it was still resistant to loose smut (Elliott, 1962). The latter variety was the standard against which others were compared for 15 years (cf. Figure 6.11) and it is likely that its yield towards the end of its period as the standard variety may have begun to fail, leading to some over-estimation of the advance in yield potential among the varieties replacing it. On the other hand, if the level of agronomy in the yield trials is conservative, as it tends to be, the advance in yield potential may be under-estimated.

By combining the results of the NIAB yield trials with estimates of the extent to which each variety was grown, and then comparing these with trends in national yields of wheat and barley, Elliott (1962) concluded that improved varieties contributed about 62% and 44%, respectively, to the rise in yield of these crops in England and Wales from 1940 to 1957. These analyses have been extended by Silvey (1978, 1981, 1986), whose figures on varietal use highlight the faster turn-over and shorter effective life of new varieties of both wheat and barley in the UK (Figure 6.11). Over the whole period from 1947 to 1983 Silvey (1986) estimates that varietal improvement accounted for 45, 38 and 23% of the increases in national yield for wheat, barley and oats, respectively, rather striking differences which probably reflect the extent of breeding effort devoted to these crops. In her analysis up to 1975, Silvey (1978) concluded that whereas 'other factors' contributed more to yield advance in wheat than did breeding in the decade from 1947, the position was reversed in the decade from 1967, with their contributions being about equal in the intervening decade. However, when longer periods were compared, the proportion of increase attributable to varieties showed little change (Silvey, 1986). Agronomic trends in recent years, such as the shift to earlier sowing, have restored the advances due to 'other factors', and will no doubt lead to further genetic changes whose effect will be apparent in the next round of estimations. For rice varieties grown in the Kinki district of Japan between 1961 and 1986, the increase in yield potential accounted for less than 20% of the regional yield increase (Hasegawa et al., 1991).

MacKey (1979a) used similar data from Sweden since 1886, when commercial plant breeding began, to construct graphs of the increase in relative yield potential of the various cereals, which illustrate the great differences among them. The yield potential of winter wheat has increased steadily at a *compound* rate of 0.42% per year; that of spring wheat, by

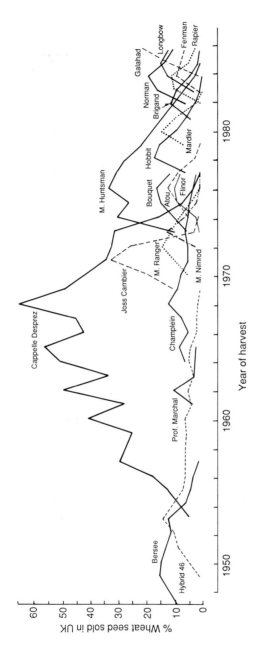

Figure 6.11. The succession of most important winter wheat cultivars in the United Kingdom (adapted from Silvey, 1978, 1986).

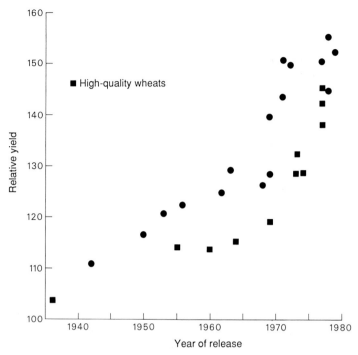

Figure 6.12.   Indirect estimates of the increase in yield potential (relative to Little Joss, 1908 = 100) of varieties of winter wheat released in the UK at various times (Evans, 1981).

contrast, has shown an acceleration in recent years and spring barley a deceleration, whereas spring oats is emerging from a long period of stasis after substantial improvement in the nineteenth century.

A comparable variation is evident among the cereals in Britain when the accumulated NIAB trial data are ordered in the same way for varieties of winter wheat (Figure 6.12), spring wheat (Evans, 1980*b*), and spring barley (L.T. Evans, 1983). Besides the relative stasis through the 1950s, and the rapid rise in yield potential since the late 1960s, Figure 6.12 illustrates the yield penalty for high-quality wheats (cf. Frey, 1971).

Direct estimates

Direct comparisons of the yield potential of old and new varieties have been made for many crops in many countries in recent years. It is extremely important in comparisons of this kind to ensure the full protection from pests, diseases and lodging of the older varieties, which are not adapted to current conditions and biotypes. If they are not fully protected, as sometimes appears to have been the case, the gains in genetic yield potential may be overestimated. Sandfaer & Haahr (1975) present a striking example of this from their comparisons of old and new barley varieties. In their earlier work the older varieties were infected with barley yellow stripe mosaic virus, whereas in later comparisons the older varieties were either free or infected. When free of virus they proved to be just as responsive to nitrogenous fertilizers as the modern varieties, which was not the case when they were infected. Consequently, in the earlier comparison of new varieties with infected older ones, the advance in yield potential was estimated to be 26%, whereas it was only 8% when virus-free plants were compared. Likewise, in their comparisons of hard red winter wheats, Cox et al. (1988) obtained their highest estimate of gain in yield potential when the varieties were compared during an epidemic of stem and leaf rust.

Another possible source of confusion in comparisons of rates of increase in yield potential derives from its not always being clear whether simple or compound rates are being quoted. In the case of the Swedish winter wheats quoted above, division of the overall percentage increase in yield potential by the number of years between the dates of release of the oldest and newest varieties gives a simple rate of increase of 0.54% per year, compared with a compound rate of 0.43% per year, but the two rates may differ by much more than that in other circumstances. Although the increase in yield potential is approximately exponential in some cases, it has often been linear or too irregular – as in sorghum (Krieg & Dalton, 1990) – to justify the estimation of a compound rate, and most investigators appear to have calculated simple rates.

The rate of improvement also varies across environments, being generally faster in terms of tonnes per hectare per year for varieties grown in more favourable conditions (Feyerherm et al., 1984, 1988), and even as percentage gain per year (Austin et al., 1989), suggesting that it is easier to select for yield improvement in environments where the genetic potential can be more fully expressed.

Wheat

Many comparisons of the yield potential of wheat varieties have been made (cf. Austin et al., 1989; CIMMYT, 1989). Austin et al. (1980a) compared the yield potentials of twelve British winter wheat varieties released at various times since 1900. All varieties were effectively protected from pests, diseases and lodging, and were grown both on a heavy soil with high fertilizer application rates and on a lighter soil with lower rates. Their yields under these two conditions are illustrated in Figure 6.13, which shows that, with the

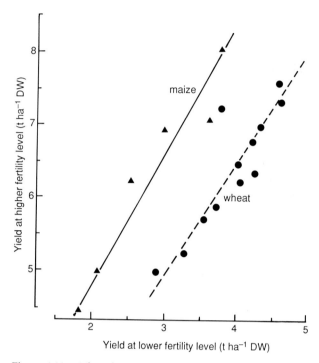

Figure 6.13. The relative effect of high and low soil fertility on the yield of old and new varieties of winter wheat in the UK and of maize hybrids in the USA by decade. Data of Austin *et al.* (1980*a*) for wheat, Castleberry *et al.* (1984) for maize.

exception of one variety, the gains in yield potential were expressed as strongly (in relative terms) at moderately low as at high levels of soil fertility. The indirect estimates of relative yield obtained from the NIAB trials agreed with those from this more direct trial (cf. Figure 6.14). In a further evaluation, which included several nineteenth century varieties, Austin *et al.* (1989) estimated the overall increase in yield potential to be 59%, whereas Ledent & Stoy (1988) found an overall increase of only 6% in the yield potential of Swedish winter wheat varieties released between 1910 and 1976.

Waddington *et al.* (1986, 1987) have conducted two studies of changes in the yield potential of spring wheat varieties bred by CIMMYT in Mexico, since 1950 for the bread wheats and since 1960 for the durum wheats. Their results are compared in Figure 6.14 with those for winter wheat improvement in Britain and in New York state, and with indirect estimates of the improvement of both spring and winter wheat in Britain. In the spring-sown bread wheats the yield potential has increased more than 40% since Yaqui 50 was released in 1950, i.e. at a simple rate of 1.1% per year, comparable with

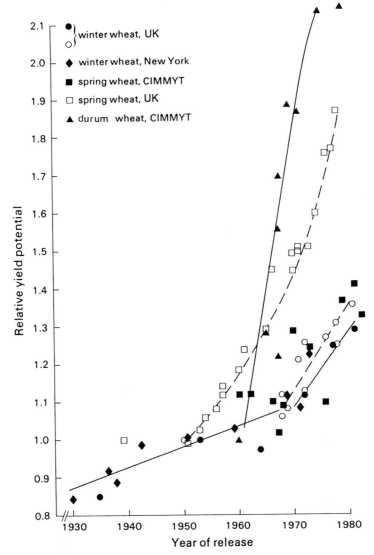

Figure 6.14. Increases in the relative yield potential of wheat cultivars in several breeding programmes with year of release, as assessed by both direct (solid symbols) and indirect (open symbols) comparisons. A value of 1 is assigned to 1950 releases in all cases except the durum wheats, which refer back to 1960 (Evans, 1987*a*). Combined from data of Austin *et al.* (1980*a*) and Evans (1981) for winter wheats in the UK; Jensen (1978) for winter wheat, New York; Waddington *et al.* (1986, 1987) for spring and durum wheats from CIMMYT; Evans (1980*b*) for spring wheats in the UK.

that for winter wheat varieties over the same period in the UK and in New York state (Jensen, 1978). Much faster rates were found with the Mexican durums, and with spring wheat varieties in the UK, in part reflecting the lower yield levels at which they started.

This is not always the case, however, especially when low initial yields reflect adverse growing conditions rather than little plant breeding effort. Reviewing progress in breeding for yield of wheat in the USA, Schmidt (1984) shows that gains in yield potential – which have averaged 0.75% per annum since 1958 across 9 nurseries, equal to about half of the total gain in wheat yields in the USA over the 1958–80 period – have been smallest in the harsher climatic regions and greatest in the more productive regions. This aspect has also been analysed by Feyerherm et al. (1984, 1988, 1989) who conclude that the extent of genetic improvement is greater in the more favourable environments, and that there has been no slowing in the rate of improvement, as found also by Cox et al. (1988) and by Kuhr et al. (1985) among winter wheats over the period from 1970 to 1983. In this latter case, 43% of the rise in yield was estimated to be due to genetic improvement, achieved without any decrease in grain protein content.

Russell (1973) found linear regression of wheat yield by year to account for only 36% of the yield variation in Australia compared with 83% for wheat yields in the USA, and estimated that only 28% of the yield increase in Australia was due to genetic improvement. O'Brien's (1982) estimate of the genetic contribution was also low, accounting for only about one third of the overall improvement in yield, and Donald (1965) found it to account for only 20–24% of the increase in wheat yields in South Australia. However, direct comparisons of the yield potential of varieties grown in Western Australia indicate that their yield potential has increased by 57% over the last century (Perry & D'Antuono, 1989) which is close to the relative increase found in British wheats over the same period (Austin et al., 1989) despite the more adverse and variable conditions. An even greater increase in yield potential is apparent in the direct comparison of varieties grown in Victoria over the last century (Reeves, 1987).

The advances in yield potential of wheat in India are of considerable interest in this context. Kulshrestha & Jain (1982) compared the yield of varieties released over a period of 80 years and found no significant increase in yield potential from the third to the sixth decades of this century (Figure 6.15), which they ascribe to Indian plant breeders having had to work within the framework of traditional low-input agriculture. Since then, however, the emphasis has been on selection for high grain yield in response to fertilizer application and irrigation, and there has been a 74% increase in yield potential over the last two decades, a dramatic illustration of the interaction between agronomic improvement and breeding in the raising of yield potential.

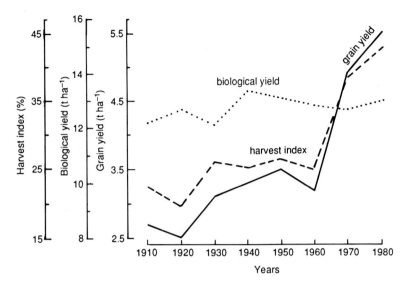

Figure 6.15. Trend in varietal yield potential of wheats in India over 80 years compared with changes in their biological yield and harvest index (Kulshrestha & Jain, 1982).

Barley

Part of the recent increase in the yield of barley in England has been due to the shift from spring to autumn sowing, which rose from 14 to over 50% between 1977 and 1983, during which period the national yield increased by 28% and yield potential by 11% (Silvey, 1986). In a direct comparison of spring barley varieties protected from disease and lodging, Riggs *et al.* (1981) found that over the whole period from 1880 to 1980 yield potential increased by 0.39% per year, but at 0.84% per year from 1953 to 1980. In fact, yield potential has increased more or less exponentially, with no sign of an approaching asymptote (Figure 6.16). Over the period from 1947 to 1985 varietal improvement accounted for 49% of the increase in spring barley yields (Silvey, 1986), but in recent years it has accounted for nearly all of it.

With winter barley varieties in Germany, Hesselbach (1985) found their yield potential to have risen by 60% between 1880 and 1980, at a *compound* rate of 0.47% per year, whereas average yields had increased four-fold. Thus, the agronomic component of the rise in yield exceeded the genetic one. Martiniello *et al.* (1987) compared the yield potential of winter barley varieties grown in Italy over four 'epochs' and found it had increased by 0.75 and 1.1% per year, respectively, in the six- and two-rowed types. Comparisons of malting barley varieties released in the USA since 1920 suggest a more or less linear increase in yield potential totalling 51% (0.9%

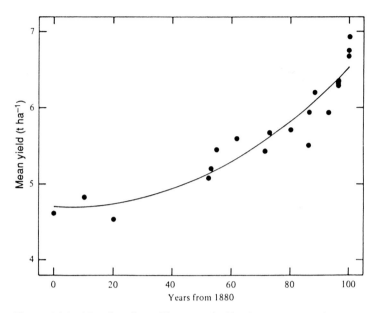

Figure 6.16.   Trend in the yield potential of barley varieties in the
United Kingdom since 1880 (Riggs *et al.*, 1981).

per year), equivalent to 73% of the overall yield increase in spite of the
stringent quality requirements (Wych & Rasmusson, 1983).

Oats

The national yields of barley and oats in the UK increased by the
same proportion (116%), over the period 1947–83, but the genetic component
of this increase was much smaller in oats, only 23% (Silvey, 1986). In direct
comparisons of UK oat varieties of varying vintage, Lawes (1977) found the
yield potential to have increased by 30% since 1908. Wych & Stuthman (1983)
found a 56% increase in Minnesota varieties released since 1923. The increase
in yield potential was particularly rapid over the 1976–80 period (Rodgers *et
al.*, 1983).

Rice

Less comparative work has been done with rice varieties, possibly
because differences in vintage are often confounded with differences in both
growth duration and disease resistance. The comparisons of Hokkaido
varieties by Tanaka *et al.* (1968) and Samoto (1971) focused mainly on

physiological changes. In the tropics it is difficult to protect the older varieties against pests, diseases and lodging sufficiently well for valid estimates to be made of the increase in yield potential. Among the varieties bred at IRRI in more recent years, none surpasses IR8 in yield per crop when pests and diseases can be controlled (e.g. Evans & De Datta, 1979), probably because the emphasis in recent years has been on the breeding of shorter season varieties. A comparison of 50 varieties grown in the Philippines this century indicated that although yield per crop has not increased, grain yield per hectare per day has been substantially increased by shortening the life cycle (Evans *et al.*, 1984; cf. also Calabio & De Datta, 1985).

Maize

In view of the substantial resources devoted to maize breeding in the USA – encouraged by the almost universal adoption of hybrids – and the favourable conditions and innovative agronomy applied to the crop, we should expect a substantial increase in yield potential.

Russell (1974) compared 25 open pollinated or hybrid lines released between 1929 and 1970, and found a fairly linear increase in yield potential which accounted for 63–79% of the increase in the average Iowa yield. However, his subsequent comparisons, included in Figure 6.4, suggest that the yield potential did not increase substantially until the 1960s (Russell, 1984), when the use of inputs such as nitrogenous fertilizers and herbicides became more intensive, yet another indication of the significance of agronomy × breeding interactions in the raising of yield potential.

Duvick (1977) compared a series of hybrids released over the period 1939–71 and estimated the advance in yield potential to account for 58–60% of the increase in yield. However, his estimates of the increase in yield potential (Duvick, 1977, 1984*b*) depended greatly on the density at which the crop was sown (Figure 6.17), the time course approaching a plateau at 30 000, being linear at 47 000 and exponential at 64 000 plants per hectare. The newer hybrids out-performed the older ones not only at high densities with high nitrogen but also at low densities with low nitrogen levels (Duvick, 1984*c*). The yield potential of the inbreds was found to have increased in parallel with that of the single cross hybrids. Lamkey & Smith (1987) found inbreeding depression for yield to have increased steadily over the years, but not on a proportional basis (Russell, 1991). Duvick (1984*c*) comments that whereas the yield potential has risen at an average rate of 1.4% per annum, the human resources required to achieve this increase have risen by about 4% per annum, and their equipment resources even more rapidly.

The greater advantage of the modern hybrids at high densities is also apparent in the comparisons made by Russell (1984), Meghji *et al.* (1984) and

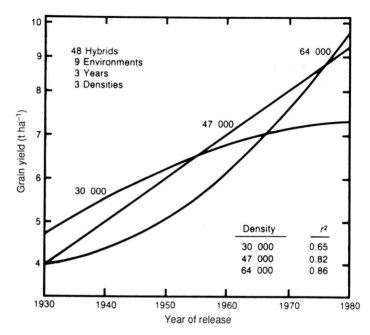

Figure 6.17.   Regressions of grain yield on year of release for 47 maize hybrids and one open-pollinated variety tested at three plant populations over three years (Duvick, 1984*b*).

Carlone & Russell (1987) with other hybrid series. With hybrids grown in Ontario between 1959 and 1988, Tollenaar (1989) showed that the later the year of their introduction the higher was the optimum density for grain yield. He also found that whereas the genetic gain in total grain yield at optimum density had been 1.7% per year, that for machine-harvestable yield had increased by 2.6% per year.

Using hybrids from a commercial source different from those compared by Russell and Duvick, Castleberry *et al.* (1984) obtained similar results although their comparisons were made over a much wider geographical area and at sites ranging from irrigated to water-stressed, and from high to low fertility. The newer hybrids out-performed the older ones in all environments. The proportional relation between yields at the high and low fertility sites, evident in Figure 6.13, indicates that there has been no bias towards relatively better performance at high fertility levels in the more recent hybrids despite the bias towards higher densities.

With a focus on Minnesota, Cardwell (1982) has also estimated the proportion of yield gain due to breeding, and has partitioned the remainder amongst the various agronomic innovations. He concludes that the 'genetic gain in yield potential through plant breeding has been far more significant than the adoption of hybrids *per se*'. For hybrids released in France between 1950 and 1985, Derieux *et al.* (1987) estimated that the increase in yield potential accounted for 55% of the increase in yield and was mainly due to greater resistance to lodging and adaptation to higher planting densities. There was also a trend towards earlier flowering, resulting in more stable performance in adverse environments.

Sorghum

Genetic progress in yield potential from old to recent hybrids and inbreds of sorghum in the USA has been examined by Miller & Kebede (1984), who found it to have increased by an average of 39% since the 1950s. The adoption of hybrids in the USA was so rapid – rising from 15% of the area under sorghum in 1957 to 80–95% by 1960 – that their impact on average yield was pronounced, particularly because they made high-input, irrigated crops worth while. Subsequent improvement of the hybrids has raised yield potential twice as much as did the original shift to hybrids, although this is less obvious in the yield trend (Figure 6.1) because sorghum growing has largely moved from irrigated back to dry land conditions (Krieg & Dalton, 1990).

Soybean

Luedders (1977) compared 21 soybean varieties released over a period of 50 years, grouping them according to two cycles of recurrent selection, the first of which resulted in a yield potential increase of 26% on average, and the second in an increase of 16%, although no variety yielded more than cv. Lee, released in 1954. Overall, this gain represented less than 30% of the increase in average yield over the period. Wilcox *et al.* (1979), Boerma (1979) and Gay *et al.* (1980) also compared old and new soybean varieties and, like Luedders, found the increase in yield potential to vary from 0.5 to 0.9% per year among the various maturity groups. From their more comprehensive survey, Specht & Williams (1984) estimated average rates of increase in yield potential by maturity group and state. Their analysis suggests that much of the overall incease in yield derived from increased genetic potential, but that most of this came from varieties released in the 1940s, as Luedders (1977) had shown.

Peanut

Duncan *et al.* (1978) compared four peanut varieties released between 1943 and 1977, and concluded that the yield potential may have doubled over that interval. However, two more comprehensive comparisons found it to have risen since the 1940s by only 19% (Mozingo *et al.*, 1987) and 27% (Wells *et al.*, 1991), recent progress being slow because of the emphasis on pest resistance and quality.

Cotton

Ramey (1972) found the yields of six old varieties of cotton to be 60% greater than they had been 30 years earlier, reflecting the improvements in agronomy over that interval, but varietal improvement has also been marked. Obsolete and current varieties have been compared by Bridge *et al.* (1971), and by Bridge & Meredith (1983), who estimated the average rise in yield potential to be 9.5 kg lint $ha^{-1} y^{-1}$. This was greater than the actual rise in average yield in Mississippi of 8.6 kg $ha^{-1} y^{-1}$. In further studies Wells & Meredith (1984b) found somewhat different rates of increase in yield potential between two series of varieties, Stoneville and Deltapine, associated with initial differences in the number of bolls per square metre and percentage lint. Meredith & Bridge (1984), in their comprehensive survey, compared the rates of improvement in genetic yield potential from 1910 to 1980 estimated from fifteen series of trials, which averaged 7 kg lint $ha^{-1} y^{-1}$, with those from three direct comparisons of old and new varieties, which averaged 10.4 kg $ha^{-1} y^{-1}$. There was no evidence of any fall in the rate of genetic yield improvement, and Meredith & Wells (1989) found the yield potential of modern varieties to be 32% more than that of the obsolete varieties with which they were compared.

Sugar beet

Published data comparing the yield potential of old and new varieties of sugar beet are sparse and difficult to assess because of the two paths to greater sugar production, via sugar content or root yield. Ulrich (1961) compared two modern Californian varieties with two old German varieties, one of each vintage being a yield type and the other a sugar type. The modern varieties yielded more sugar in the field, but that could have been because the old varieties were extremely susceptible to curly top virus. When grown under disease-free conditions in a phytotron, the results depended on the conditions because the Californian varieties had a greatly reduced ability to grow at cool temperatures.

Rubber

Imle (1977) indicates that the yield potential of the first unselected Wickham trees was only 225 kg ha^{-1} even with the best tapping methods, whereas yields had risen above 400 kg ha^{-1} by the 1930s, to 1600 kg ha^{-1} from the 1950s clone (RRIM 501), 2300 kg ha^{-1} from the 1960s clone (RRIM 600) and over 3000 kg ha^{-1} from RRIM 703 in the 1970s.

Conclusions

Substantial progress in raising the genetic yield potential is apparent in all the crops in which old and new varieties or hybrids have been compared under conditions which protect the earlier varieties while allowing the newer ones to express their potential.

The rise in yield potential can be irregular, more or less linear, accelerating or decelerating depending on plant breeding history and on agronomic conditions. Progress is often slower in adverse conditions and is highly dependent on the level of agronomic improvement. The rate of genetic progress varies from a few per cent (for wheat in Sweden) to more than 100% (for cotton in the USA) of the rate of increase in average yields but is commonly about half of it. This proportion can vary substantially from one decade to another (e.g. for wheat in England), probably owing to the shifting reciprocation between breeding progress and agronomic innovation. Where the level of agronomic support is low, as in many developing countries, or falls, whether for environmental reasons (e.g. cotton in the USA) or because of socio-economic policies (e.g. wheat in Mexico), the rate of increase in yield potential may exceed the rate of increase in yield for a period. Taken overall, there is no indication that the genetic yield potential of any of the major crops is reaching its limit. Indeed, for wheat, barley and maize in some environments, the rate of improvement is accelerating.

Yield variation in relation to yield level

The important issues of vulnerability to fluctuations in yield from year to year and the relation between varietal stability and adaptability have already been discussed in Chapter 4. Here we are concerned with only one aspect of yield stability, namely its relation to yield level as influenced by variety and agronomic support.

Public interest in such fluctuations, and that of research administrators, tends to wax and wane rather rapidly as crop yields and the adequacy of food supplies fall and then rise again, but the impact of such oscillations is so great that they merit cooler and more continuous attention. Over and over again, pauses in the rise of yields have been interpreted as indicating that breeding and agronomy have reached their limit, and the often over-riding influences of climate and socio-economic policies have been neglected.

Stability of yield and production is desirable in both developing and developed countries. To the subsistence farmer it is probably the most important characteristic of all, more so than high yield potential because it assures his family's food supply. Extreme instability of yield and the possibility of crop failure have been associated historically with pest or disease epidemics or with extreme weather conditions (Sen, 1981). The epidemics, whether of wheat rust in Biblical times or of late blight in the Irish potato famine of 1846, arise from a combination of unusual genetic and environmental circumstances. The genetic ones should now be more manageable, yet the crown rust epidemic on Victoria oats in the USA in 1946, the southern corn leaf blight epidemic which devastated maize hybrids based on the Texas cytoplasm in 1970 despite forewarnings from the Philippines (National Academy of Sciences, 1972), the epidemic of soybean leaf rust in Brazil, and many others indicate that genetic vulnerability remains a problem. A high proportion of IRRI rice varieties is still based on cytoplasm derived from one variety (Hargrove et al., 1980), as are hybrid sugar beets (Plucknett et al., 1987). The genetic diversity among American soybean varieties is still limited to a small group of ancestral introductions (Delannay et al., 1983; Specht & Williams, 1984).

The problem was highlighted in 1972 by the US National Academy of Sciences report on *Genetic Vulnerability of Major Crops.* Duvick's (1984a) survey of American plant breeders indicates both awareness of the apparently narrow genetic base in many crops and confidence in their ability to deploy the requisite germplasm resources as new pest and disease problems arise. The fact that the epidemic of southern corn leaf blight had so little impact on maize yield in the USA in subsequent years has increased the confidence of plant breeders in their techniques and resources, yet some lay scepticism is in order.

The vulnerability due to extreme weather conditions remains a greater problem, for two reasons. In the first place, plant breeding has made much slower and less certain progress in selecting for resistance to climatic extremes than for resistance to new pests and pathogens. Secondly, the progress that has been made tends to exact a higher price in terms of yield potential than does resistance to disease. For example, escape from drought or heat stress may be achieved by selection for earlier maturity, but often at the cost of reduced yield potential.

Turning from the more extreme vulnerability associated with rare conditions to the more usual variation in yield from year to year, the issue is whether such variation has risen disproportionately as the yields and yield potentials of crops have been increased. It seems likely that some of the biological forces at work have tended to reduce yield variation while others could have increased it, and the net outcome may in any case be over-ridden by institutional and socio-economic factors.

Both agronomic and genetic changes may have enhanced stability. Higher levels and a wider range of inputs confer greater control over yield, especially if irrigation is available, thereby tending to enhance stability. Among the crops grown in Japan, yields of rice have a lower coefficient of variation (9%) than those of upland crops (which range from 10 to 20%), while among regions those with poor agro-climatic resources are characterized by the greatest variation (Uchijima, 1981). On the other hand, selection for greater yield potential has involved progressively reduced investment by the crop in organs other than those harvested, made possible by the rising level of agronomic support and control. When that fails, however, modern varieties may sometimes be more vulnerable to climatic stress, particularly if selected for responsiveness to high-input levels rather than for stability of yield.

The net result of these opposed effects depends a great deal on socio-economic factors. Barker *et al.* (1981) and Mehra (1981) analysed the stability of cereal production in developing countries before and after the adoption of the new agricultural technology, and concluded that much of the increased instability in food grain production since the 1960s could be attributed to the widespread adoption of modern varieties and more intensive fertilizer use, as did Kulkarni & Pandit (1988) for sorghum in India. Michaels (1981) analysed wheat production in Sonora, Mexico, to see whether its sensitivity to climatic variation was greater as a result of the 'Green Revolution' and found an increased response to inter-annual fluctuations.

In a further analysis of the Indian food grain situation, however, Hazell (1982) showed that the changes in standard deviation of production at the aggregate level were almost entirely attributable to factors other than increases in the variation of yield of individual crops at the state level. Most of the increased variation at the aggregate level was due to increased covariance of production between crops and between states, i.e. to a reduction in the mutually offsetting variations in yields and areas of the different crops and regions. The new varieties and agronomic practices were not the villains of the piece, and the cure for the problem lay in policy formulation rather than in breeding more 'stable' varieties. Such problems are not peculiar to developing countries. Hazell (1984) found the causes of increased variance in aggregate cereal production in the USA to be similar to those in India, except that they focused more narrowly on yield correlations between states, reflecting the greater role played by yield increase in enlarging US cereal production.

These studies, like many others, use the standard deviation of yield or production from trend as the appropriate measure of variation. But for comparisons between crops, varieties, countries, environments or levels of input, it is the size of the variance relative to yield or production – the coefficient of variation – that is most useful, and this has fallen substantially for world production of the three major staples since the 1950s.

At a less aggregated level, let us consider the case of wheat yields in the UK. Stanhill (1976) has analysed the trends and deviations in English wheat yields over a period of 750 years, during which the average yield increased 9-fold, and found that the coefficient of variation has remained about 7% of national yield throughout. Elston & Dennett (1977) examined the 80 year period from 1885, during which yields doubled, and found no increase in the residual standard deviation as a percentage of mean yield, nor did Dennett (1980) in a further analysis in which the average coefficient of variation was 8% of yield. Singh & Byerlee (1990) have analysed the coefficient of variation of wheat yields in 57 countries for the period 1951–86. The relative variability of yields was found to be influenced by country size and by environmental conditions, but not by the level of adoption of modern varieties or fertilizer use. Although the widespread adoption of new technology may have an initial destabilizing effect, there has been a consistent fall in the coefficient of variation of wheat yields in developing countries since 1975.

One problem with such analyses is that long-term changes in variation associated with rising average yield levels may be confounded with those associated with changes in climatic variation (e.g. Schneider & Mesirow, 1976). In order to examine the evidence for changing variation in climate, Waggoner (1979) analysed the variation in wheat yields since 1909 in Argentina, Australia, Canada, India and the USA. He concluded that these gave no evidence of a global increase in yield variation, the annual deviations from the trends of yields being largely independent across countries. Because of the relatively low wheat yields in the countries chosen, variation associated with the rise in national yield *per se* is likely to have been small, if indeed there was any such rise. Newman's (1978) analysis of inter-annual variation over four decades in corn, wheat, sorghum and soybean yields in the USA, and in wheat and barley yields in Canada, indicates that although the variance has increased as yields have increased, the coefficient of variation has decreased in all cases, by 35% overall, which Newman attributes to improved agricultural technology.

Two other possible sources of increased variation during agricultural development should also be remembered. Variation between farms is likely to be greater during the period when a major agronomic innovation or new variety is spreading than before this begins or after it has finished. Similarly, where the proportion of a crop grown under irrigation is changing, or when a crop is undergoing a shift in geographical or ecological location – like soybean in South America or sorghum and cotton in the USA in recent years – both average yield and its variation may change substantially.

We may conclude, however, that there is no strong evidence for any rise in the coefficient of variation of yields as they are raised by either breeding or agronomy. The absolute variance tends to rise in proportion to the rise in average yields, as is to be expected, and these rising variances tend increasingly, at the national level, to covary across regions and crops. The problems caused by reduced off-setting and greater covariance are important ones but they are not to be laid at the door of greater yield *per se*, and they are best dealt with by improved information and socio-economic policies (Anderson *et al.*, 1987).

Yield penalties

In plant breeding programmes in which higher yield potential is not the only (or even the main) objective, there may be both direct and indirect constraints on the rate of progress towards higher yields. The indirect constraints arise because the need to select for several other desirable characteristics at the same time inevitably means that less attention can be given to yield and that selections for higher yield may be compromised by other criteria. Selection for resistance to new pest or disease biotypes may become so urgent that selection for yield potential has to be de-emphasized. For example, the rapid evolution of virulent races of crown rust of oats and the focus on selecting for resistance to them in the USA minimized the improvement of yield potential until the 1960s, after which it became possible and rapid (Rodgers *et al.*, 1983).

Besides such indirect constraints there are more direct ones in which selection for other characteristics may involve a yield penalty.

(1) Crop yield is often positively correlated with crop duration, at least up to a point (Figures 4.3 and 4.4) and the need to select for shorter duration may impose a substantial yield penalty. With rice in the tropics, for example, shorter duration varieties are in demand particularly for the farming systems flexibility which they confer, and there has been a trend towards progressively earlier maturity among the rice varieties bred at IRRI. Selection for yield has therefore focused on minimizing the penalty associated with earlier maturity.

(2) Because proteins and oils have a higher calorific content than carbohydrates, selection for higher protein or oil content involves a penalty when yields are measured in terms of dry weight rather than calories (Penning de Vries, 1975*a*; Sinclair & de Wit, 1975). Yield comparisons between crops or varieties need to bear their differing composition in mind. The record peanut yield of 6.4 t ha^{-1}, for example, is probably equivalent to about 12 t ha^{-1} of wheat.

(3) High-protein crops may exact a further yield penalty, particularly when soil nitrogen levels are limiting towards the end of the life cycle. Up to a certain level of protein in the harvested product, additional storage of protein may be met by additional uptake of nitrogen from the soil. Swiecicki *et al.* (1981) found a negative correlation in peas between seed yield and % protein

content only when the latter exceeded 27%. At high protein levels in the seed, leaf proteins may have to be mobilized sooner, thereby reducing leaf photosynthesis or shortening its duration, with the likelihood of also reducing yield. However, as the level of nitrogenous fertilizer application is raised, so is the negative correlation between yield and % protein likely to be weakened (e.g. Benzian et al., 1983).

Although such negative correlations imply a yield penalty from selection for higher protein content, they do not mean that progress in raising the yield potential of high-protein cereals need be any slower. The yield potential of soft red winter wheats has increased more than that of the hard wheats in the USA (Feyerherm et al., 1984), but the data in Figure 6.12 suggest that the rise in yield potential has been quite as rapid in the higher as in the lower quality British wheats. In both wheat (Cox et al., 1985) and rice (Calabio & De Datta, 1985) there is a negative correlation between time to maturity and percentage protein in the grain. The yield penalty from selection for earlier maturity may, to some extent, be confounded with that for higher protein content.

Although there is a yield penalty for higher protein content in wheat, no direct penalty has been found for other criteria of bread-making quality (O'Brien & Ronalds, 1984). However, the indirect penalties in breeding programmes with a strong focus on maintaining traditional qualities are well known, e.g. for wheat in Canada and rice in Thailand.

(4) There may sometimes be a yield penalty associated with the genes for pest or disease resistance, but whether this is due to the resistance genes themselves or to those brought along with them by linkage drag, even in near-isogenic lines (Zeven et al., 1983), is an open question. Out of 24 near-isogenic lines of barley differing in their resistance genes to powdery mildew, only one had a yield significantly lower (by 9%) than its recurrent parent (Kølster et al., 1986), and that had the ml-o resistance gene, which can cause as much yield loss as the disease itself. However, its adverse effect can be reduced by modifier genes in some genotypes (Schwarzbach, 1976). Comparing eight sources of stem rust resistance in nine wheat cultivars, The et al. (1988) found yields to be reduced with four of the sources, especially with Sr 26 (by 9%). None of the sources increased yield compared with the susceptible isolines. From his population studies with several crops, Allard (1990) concludes that most of the alleles that confer disease resistance have negative effects on yield in the absence of disease.

Plant breeders often sense that the introduction of resistance genes carries a yield penalty, e.g. in the case of greenbug resistance in sorghum (Miller & Kebede, 1984; Krieg & Dalton, 1990). Crossa & Gardner (1987) found the yield of maize to be substantially reduced as the ratio of exotic to adapted germplasm increased from 0.33 to 1, and such penalties are considered by many plant breeders to be likely to accompany wide crosses made for the purpose of securing new resistance genes (Duvick, 1984a). However, Brinkman & Frey (1977) found the yield potential to be increased in a crown rust-resistant isogenic line of oats derived from unadapted *Avena* germplasm. Frey *et al.* (1984), Cox *et al.* (1984) and Bramel-Cox *et al.* (1986) argue that wide crosses may also be beneficial to yield potential.

The procedures involved in selecting for pest or disease resistance may also affect yield potential. Ulrich (1961) found that by selecting sugar beet in warm temperatures at the seedling stage for resistance to curly top virus, Californian plant breeders had unwittingly reduced their ability to grow at cool temperatures, although the effect of this on yield had been masked by the over-riding advantage of resistance to the virus.

(5) Genetically engineered resistances to pests and diseases, and also to herbicides, could involve yield penalties, but these have yet to be shown. Selection for resistance to some herbicides has involved substantial yield penalties, up to 20% in the Canola cv. Triton associated with its resistance to atrazine (Mazur & Falco, 1989). The yield penalty for atrazine resistance can be even higher, up to 65% in dense plant communities of foxtail millet (Reboud & Till-Bottraud, 1991).

(6) Whether selection for resistance to soil or climatic stresses may also reduce yield potential is also a much-debated question, of considerable significance to internationally oriented plant breeding programmes (cf. Chapter 4).

Yield maintenance and declines

So far we have been focusing on the ways in which yield levels have been raised through interactions between plant breeding and agronomy, when socio-economic pressures and incentives permit. But as yields rise and agriculture becomes more intensive, the effort required merely to defend previous gains becomes a more significant component of the overall effort. This has been referred to as maintenance, sustaining or defensive research in order to draw the attention of research administrators to the need for its active support (Ruttan, 1982; Plucknett & Smith, 1986), but resource maintenance has always been an integral component of agricultural research. Evenson (1982) estimates that about half of the funds spent on crop research at experiment stations in the USA has been for maintenance research.

What Plucknett & Smith (1986) refer to as 'the varietal relay race' is more the result of maintenance research than of breeding for greater yield potential. Figure 6.11 illustrates the sequence of winter wheat varieties in the UK, and there are many similar figures for other crops and other countries (e.g. Silvey, 1978, 1981, 1986; Dalrymple, 1980; Plucknett & Smith, 1986). As British agriculture began to intensify in the 1950s, one variety of wheat (Cappelle Desprez) became dominant and remained so for many years. Maris

Huntsman succeeded it in ascendancy, but not to so great an extent nor for so long, and since then there has been an increasingly rapid succession of new varieties.

The traditional practice of many local varieties and land races used over long periods has been replaced by varieties with wider distribution but shorter effective lives. Temporal change is said to be replacing the spatial diversity of varieties (Apple, 1977). However, the number of wheat varieties in use in the USA has tripled over the past 70 years and the proportion of wheat sown to the one, five or ten leading varieties has decreased substantially (Siegenthaler et al., 1986), which hardly suggests any loss of spatial diversity in the US wheat crop. The situation may be at least temporarily different in developing countries, in that IR36 rice and the lines derived from one CIMMYT wheat cross have been grown on more than 11 million hectares. The chance of pest and disease mutation with such widely grown varieties is greatly increased, and their effective life could be quite short. Duvick's (1984a) survey indicates that modern cultivars have a life span of 7 (for maize) to 9 (for soybean and wheat) years, and that breeders expect the effective life of varieties to be even shorter in the future. Wolfe & Barrett (1980) note that barley varieties resistant to powdery mildew last an average of only 3–4 years before resistant biotypes of the pathogen evolve and yields decline, and there have even been cases where varietal resistance to a pathogen has broken down before the variety emerged from varietal testing and recommendation procedures. Nevertheless, the value of resistance breeding, an important component of 'maintenance research', is extremely high both nationally and internationally (Silvey, 1986; Ruttan, 1982).

Indeed, it is particularly significant in the tropics where environmental conditions favour pest and disease build-up, especially when the crop can be grown without a break throughout the year, as in the case of irrigated rice. A major component of IRRI plant breeding programmes has therefore been the continuing incorporation of genes for resistance to more and more pests and pathogens as their biotypes have evolved under the challenge from widely grown resistant varieties.

The losses of rice caused by new strains of blast disease or brown plant hopper are obvious to the eye and quickly generate a sense of urgency of action, but intensive agriculture in areas where there has not been a long tradition of it may generate other problems with a less obvious but longer-term and possibly cumulative impact. As one example of this, consider the trend in the highest yield of high-input rice plots on the IRRI experimental farm (Figure 6.18). In the course of relating yield to irradiance in a long-term 'date of planting' experiment, we observed that the highest yield of IR8 at IRRI began to decline after 1967, soon falling to that of the older Sri Lankan variety H-4. That was to be expected as pest and pathogen pressures built up

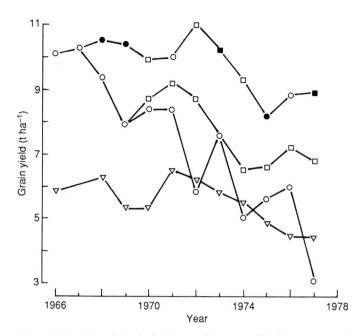

Figure 6.18. Trends in the highest paddy rice yield of any variety (□), IR8 (○) and the older Sri Lankan variety H-4 (▽) for January plantings at IRRI, and in the highest yield obtained with any variety at any site in agronomy experiments in the Philippines. The solid symbols in this uppermost curve indicate crops grown away from the IRRI farm (Evans & De Datta, 1979).

and changed on the IRRI farm, but we also found that the highest yield of any variety also fell substantially on the IRRI farm as IR8 was succeeded by IR24, IR36 and IR42, although less markedly at off-site experimental locations (Evans & De Datta, 1979).

This trend could simply be due to the earlier or greater build-up of pest and disease pressures on the experimental farm, but it could also have been due to other effects of intensive management, such as declining soil fertility or increasing zinc deficiency or boron toxicity associated with the source of the irrigation water. The more rapid decline in yields at the IRRI experiment station was confirmed by Flinn & De Datta (1984), and occurred in spite of an increase in soil organic matter and nitrogen under intensive rice–rice cropping (De Datta *et al.*, 1988). The causes of the yield decline are still being investigated, but it is clear that the long-term impact of intensive agronomy in the tropics requires monitoring and will generate a need for maintenance research as surely as plant breeding does.

Meredith & Bridge (1984) discuss a comparable situation with the yields of cotton in the USA. Their analysis indicates that the decline in yield since 1965 is a widespread phenomenon and is not restricted only to older varieties. Across ten states and many varieties, the average decline in lint yield was 27 kg ha^{-1} year^{-1}. This decline has occurred in spite of continuing increase in varietal yield potential (Bridge & Meredith, 1983) and in atmospheric CO_2 level. As with rice at IRRI, many factors are likely to be involved, including reductions in the use of irrigation water and nitrogenous fertilizers; soil compaction, erosion and loss of fertility; herbicide injury; environmental pollution; and climatic change. Simulation models have been used to assess the causal factors in the decline. Long-term trends in the weather have been ruled out (Wanjura & Barker, 1988; Reddy & Baker, 1990), and increases in atmospheric CO_2 and ultra-violet radiation have had only relatively small effects so far. Reduced irrigation and fertilizer applications seem to account for much of the decline, in fact for 82% of it in the simulation by Wanjura & Barker (1988). The impact of these reductions has been accentuated, especially in the cooler areas, by root injury caused by residual herbicide, according to the simulations by Reddy et al. (1990).

Conclusion

The sustainability of agriculture is currently so fashionable a theme that the efforts by many generations of earlier agriculturists to maintain soil fertility, environmental quality and long-term productivity tend to be overlooked. Long term experiments, so crucial to these concerns (e.g. Jenkinson, 1991), are in many cases being discontinued. Those who currently press for more research on sustainable agriculture should be willing and able to match their demands with an equivalent sustainability of funding and enthusiasm for such research which is, by nature, long-term.

The question of sustainability is not by any means confined to intensive agriculture. We saw in Chapter 2 that even subsistence agriculture is not, in one important sense, sustainable in the long term, and there are many historical examples of declines in low-input agriculture. 'Low-input, high-output' agriculture would also generate sustainability problems were it able to spread from offices to fields.

The question of the sustainability in the long term of high-input, high-output agriculture is an important one, especially in the tropics. It is not a cause for alarm but for calm analysis of the new problems that arise as genetic and agronomic innovation continues.

CHAPTER 7

Inputs and the efficient use of resources

'If these (cultivated) forms are no
longer tended, if the gardeners do not
bring daily sacrifices to their idols, then these playful shadows fall into a
void'   Carl Linnaeus (1737)

Introduction

We have just seen in Chapter 6 that, for a number of crops over the past 50 years or so, about half of the increase in yield is attributable to plant breeding and half to improved agronomy and management. Moreover, the improvement in yield potential by breeding has hinged, to a large extent, on the provision of better agronomic support for the crop. This does not mean that in the absence of inputs the improved varieties 'fall into a void', as Linnaeus puts it in the epigraph, but their relative advantage may be reduced, eliminated or even turned to disadvantage without agronomic support. In this chapter we focus on the range of agronomic inputs, on improvements in their manufacture and use, and on the synergistic interactions between them which have helped to avoid the onset of diminishing returns to input energy.

Ever since Malthus it has been recognized that there is a diminishing response to any one agricultural input as its use increases. Mitscherlich curves, like that illustrated in Figure 1.9 for phosphorus, are very common in agriculture. How are they to be evaded? The way formulated by Rostow (1956) in his 'leading sector' theory of development was through a continual succession of new sources of growth, new sectors of the economy. For particle accelerators, repeated escape from the plateau has come from new approaches to their design (Figure 1.10). A succession of new inputs has been essential for the continuing rise in crop yields, but perhaps even more important than the succession of inputs in agriculture is their synergism and the opportunities they create for further positive interactions.

317

An earlier agricultural revolution

The agricultural revolution which took place in Norfolk in the eighteenth century provides a good example of synergistic interactions. Throughout the Middle Ages there was only slow progress in English wheat yields until the eighteenth century. After that they rose more rapidly, beginning in Norfolk, which by 1794 grew 90% of the wheat 'exported' from the English countryside, thanks to what Arthur Young (1771) called the Norfolk system. He described this as having seven points: first, enclosures without assistance from Parliament; second, use of marl and clay; third, proper rotation of crops; fourth, culture of turnips, hand-hoed; fifth, culture of clover and ray[rye]-grass; sixth, long leases; and seventh, large farms. In fact, enclosure had been progressing for several centuries, marling was an ancient practice, and the benefits of the four-course rotation had been known for at least a century. The really innovative feature of the Norfolk system was the way in which the seven elements were brought together in an interactive way. This is described by Naomi Riches (1937) as follows:

> 'In the first place, the agricultural revolution had little to do with machinery and thus is in sharp contrast to the industrial revolution. The increase sought in productivity was achieved by a more rapid abandonment of the old open-field strip system than had characterized the earlier centuries and by the gradual substitution of elaborate crop rotation by which land was not allowed to lie idle to regain its fertility. Such crops as turnips, lucerne [alfalfa], and clover, the latter especially valuable, were used to restore soil fertility. Productivity was also increased by a more general use of marl, which individual ownership made more practical, and animal manures, made more available by the additional cattle supported on the turnips, lucerne, and clover. The introduction of the so-called artificial grasses also made possible the keeping of large numbers of cattle. Indeed, to some observers early in the century it seemed that meat rather than grain production was characteristic of the new agriculture, but the real goal was a convertible husbandry, in which livestock helped to increase grain production by furnishing more manure and by consuming the crops necessary in the new scheme of rotation. These changes were fundamental; and no mere invention of "drill rollers" or mowing machines, or other ways of getting seed into the ground and crops into the barns, could have increased production as crop rotation did, for it increased the productivity of the soil itself.
>
> The improvement of livestock was incidental to the great discovery that animal and grain production could be successfully combined by using the new crops, clover, turnips, and the artificial grasses as the connecting links. The work of Robert Bakewell and his followers, who did so much to improve livestock by inbreeding, is important but not fundamental to the revolution. The same may be said of Jethro Tull, whose great contribution was to demonstrate that crops, by means of his drill, should and could be planted in rows, making frequent cultivation, horse-hoeing, possible.'

Agricultural development is still too often seen in terms of individual innovations, whereas this quotation from Riches brings out the linkages and synergisms between them. The beneficial effects of innovations on returns to labour in more traditional systems have been discussed by Boserup (1965) and Barker & Cordova (1978), for example, while Loomis (1984) highlights other interactions. We shall see many examples of synergistic interactions as we consider various aspects of the individual inputs in the following pages.

Fertilizers

Animal bones began to be replaced as a source of nutrients in European agriculture in the 1830s by guano, by John Bennett Lawes' phosphate fertilizers after 1843, by sodium nitrate from Chile in the 1860s, by ammonium sulphate from the effluents of gas works in the 1870s, and by increasing use of potash in the 1890s (Thompson, 1968; Goodman *et al.*, 1987). There was a 30-fold increase in the use of fertilizers in Britain over the 50 years up to 1880. In 1898 Sir William Crookes suggested that the fixation of atmospheric nitrogen could be the chemist's major contribution to the maintenance of civilizations based on wheat. The chemist's achievement which he foresaw has denied the doom he had predicted.

The manufacture of nitrogenous fertilizers is energy-intensive, currently requiring about ten times more energy per tonne than the manufacture of phosphatic and potash fertilizers (Mudahar & Hignett, 1985). Considerable progress has been made on the energetic efficiency with which ammonia is produced from methane and air, as indicated in Figure 7.1. The efficiency of production has continued to increase, encouraging more widespread and intensive use. Trends in the world-wide consumption of fertilizers in recent years are indicated in Figure 7.2. Comparable amounts of the major fertilizer elements (in terms of the N, P_2O_5 and K_2O used in FAO statistics) were consumed in the 1950s, but since the mid-1960s the use of nitrogenous fertilizers has increased far more rapidly.

Fertilizer consumption on a global scale has increased linearly, at most, since the mid-1960s (cf. Figure 7.2) whereas total cereal production has increased exponentially (Figure 2.1), so the additional tonnes of cereal produced per additional tonne of fertilizer cannot have fallen. Over the period 1948–86, 9.9 tonnes more grain were produced for each additional tonne of fertilizer used, but over the last 7 years of that period the ratio has been 15.0 t t^{-1}. These figures give a very different picture from that of rapidly diminishing returns to global fertilizer use as presented in several editions of *State of the World* and by Ehrlich & Ehrlich (1990), which simply divide world grain production by world fertilizer use.

Fertilizers are the predominant input into crop production and account for

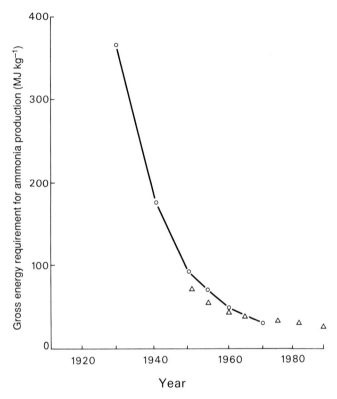

Figure 7.1. The fall in gross energy required for fertilizer ammonia production, based on data of Fleming (1973, circles) and Mudahar & Hignett (1985, triangles).

almost half of the energy used in world agriculture (Mudahar & Hignett, 1985). Until the mid-1970s, fertilizers were consumed mainly in developed countries, but since then much of the growth in their use has been in developing countries. In 1987–8 these consumed 48, 38 and 24% of the total use of N, P and K, respectively. Presumably the faster growth in the use of nitrogenous fertilizers by developing countries will soon have to be balanced by greater use of other fertilizers as well.

There are still very large differences between countries in their consumption of fertilizers per hectare of arable land, as indicated in Figure 7.3. These partly reflect differences in their stage of development but they are also influenced by such factors as how favourable and reliable are the conditions for crop growth, the extent of irrigation, the proportion of arable area

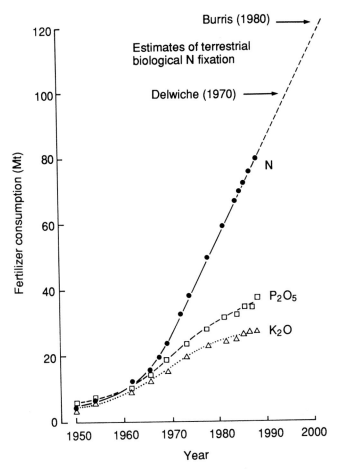

Figure 7.2. Trends in the annual global use of fertilizer N, P₂O₅ and K₂O, together with estimates of terrestrial biological N fixation by Delwiche (1970) and Burris (1980). (Data from FAO *Fertilizer Yearbooks*.)

devoted to cereals, and the use of fertilizers on pastures and non-arable land. For example, two developed countries, Australia and Canada, are among those with low cereal yields and low fertilizer use because of their relatively adverse environmental conditions. On the other hand Egypt, a developing country, has high cereal yields and high fertilizer use associated with its almost total reliance on irrigated cropping. Japan and many West European countries have even higher cereal yields and fertilizer use associated with their

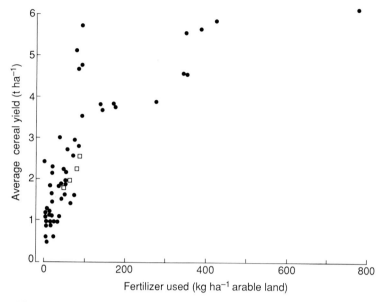

Figure 7.3. The relation between national average cereal yields and fertilizer used per hectare of arable land in various countries in 1985. The open squares indicate the world averages for 1970, 1975, 1980 and 1985. Data from FAO *Production* and *Fertilizer Yearbooks*.

advanced development, favourable environments for cropping, and heavily subsidized agriculture. Finally, with countries such as the Netherlands and New Zealand, the apparently high fertilizer use per hectare of arable land partly reflects the substantial proportion of fertilizer which is applied to pastures rather than crops.

Consequently, Figure 7.3 and others like it should not be interpreted as international Mitscherlich curves illustrating diminishing returns to fertilizer application. There are far too many complications for such figures to support that interpretation, given the differences between the various countries in their climates, crops and use of other inputs. In fact, when the relation between average cereal yield and fertilizer use per hectare of arable land is examined for individual countries over a period of years, as in Figure 7.4, it is often found to be linear even for countries such as the United Kingdom and Japan, which apply large amounts of fertilizer. Clearly, for these and many other countries, diminishing returns to fertilizer applications at the national level have been avoided so far, presumably because of synergistic interactions with the greater use of irrigation and other inputs.

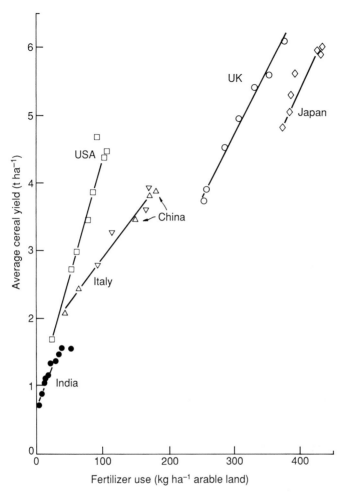

Figure 7.4. Trends in national average cereal yields of Japan, UK, USA, China, Italy and India as related to changes in average fertilizer use betwen 1950 and 1986. Based on data from FAO *Production* and *Fertilizer Yearbooks*.

Nitrogenous fertilizers as a case history

Use

The rapid and sustained growth in the consumption of nitrogenous fertilizer (Figure 7.2) has made it the predominant energy input into crop

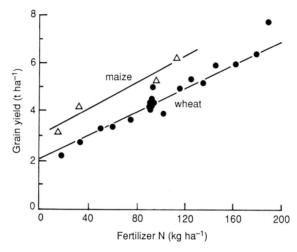

Figure 7.5. Trends in the average grain yield of maize in Minnesota and of wheat in England and Wales as related to changes over the years in the rate of application of fertilizer N. Based on data of Cardwell (1982) for maize and Tinker (1983) and Jenkinson (1986) for wheat.

production despite the impressive reduction in the energy requirement for its manufacture (Figure 7.1). Indeed, this improvement in manufacturing efficiency has been a driving force in its greater use, through the lowering of the fertilizer:grain price ratio (e.g. Welch, 1979).

As a result of these changes, fertilizer nitrogen already exceeds, by at least two-fold, the total amount of nitrogen estimated to be fixed by arable crops, and may surpass the estimated total terrestrial biological nitrogen fixation of 100–120 Mt per year by about the end of this century (Figure 7.2), with some variation depending on which estimate of global N fixation is used (Delwiche, 1970; Burns & Hardy, 1975; Burris, 1980; Rosswall, 1983). The additional input of fixed N into the global economy on this scale has immense consequences not only for agriculture but also for the biosphere as a whole and for many components of our environment, ground water, river and ocean composition and even for stratospheric nitrous oxide and ozone levels (Whitten *et al.*, 1980). By contrast, fertilizer P represents only about 7% of the global flux of that element each year (cf. Richey, 1983). At the national level, fertilizer N already exceeds biological N fixation in many developed countries: in the UK, for example, fertilizer N use is about five times greater than the amount of N fixed biologically (Jenkinson, 1986).

The relation between the increased use of nitrogenous fertilizers and the yield of crops can be illustrated for maize in Minnesota and wheat in England and Wales (Figure 7.5). Thompson (1969) concluded that the rate of fertilizer

N application was the technological factor most closely associated with the rise in Corn Belt maize yields, the overall response being 25.5 kg of maize per kilogram of fertilizer N. Cardwell (1982) reviews other estimates of this ratio, ranging from 19.4 to 25.6 kg grain per kilogram fertilizer N, and concludes from his analysis of the various sources of yield increase that greater use of nitrogenous fertilizers accounted for 47% of the gain in yield from 1930 to 1980 in Minnesota. His analysis disregarded both non-linear responses and the interactions between factors, but his mean values per decade, given in Figure 7.5, indicate a more or less linear yield response to fertilizer N. A linear response is also apparent for British winter wheat yields from 1968 to 1985 in relation to the amount of fertilizer N applied to wheat crops in the preceding year. For both maize and wheat the slope is close to 25 kg grain per kilogram N applied. Without N fertilizer, wheat yields in Britain would be about 2 t ha^{-1}, similar to those in the zero N plots at Rothamsted (Jenkinson, 1982).

Such linear relations between yield and fertilizer application rates contrast sharply with the diminishing returns found in many trials where only fertilizer dose is varied. Diminishing returns may be less pronounced when yield is related to total available N (Schön *et al.*, 1985) or P (Figure 1.9) rather than to the amount applied, and even less pronounced when related to N uptake by the crop (Rhoads & Stanley, 1984). However, linear relations such as those in Figure 7.5 reflect the fact that the rise in fertilizer N applications over the years has been accompanied by varietal improvement and that other inputs have been introduced, augmented, and improved in their formulation and management, multiplying the opportunities for synergistic interactions.

Some interactions

One interaction that has received a lot of attention is that between N application rate and grain yield as influenced by variety. A predominant factor in this case is stem height, the older taller varieties of many cereals being more prone to crop lodging and yield reduction as fertilizer N increases, as may be seen in Figure 7.6 for Peta rice and Tehuacan durum wheat. In these two trials the shorter modern varieties displayed no yield advantage over the older, taller ones when no fertilizer N was applied, but a progressively greater advantage the higher the level of application.

Many experiments of this kind have been reported; the form of the results varies with crop, season, disease and pest incidence and other factors. Sometimes the modern varieties yield more than the older ones even when no N is applied (especially on fertile soils), sometimes there is no difference (as in Figure 7.6), and sometimes the older varieties are superior in the absence of fertilizer, especially under adverse conditions. But in nearly all cases there is a strong interaction between variety and application level such that the

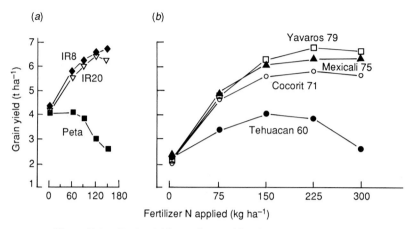

Figure 7.6. Grain yield as influenced by the amount of fertilizer N applied in trials with old (e.g. Peta, Tehuacan) and new varieties of (*a*) rice in the Philippines (Chandler, 1979); (*b*) durum wheat in Mexico (CIMMYT, 1980).

response to N fertilizers is enhanced in the more recent varieties. Moreover, the wider use of the shorter modern varieties has acted as a Trojan horse for greater use of fertilizers, as in the case of rice in the Philippines (Figure 7.7) and wheat in the Punjab (Byerlee, 1987). Higher-order interactions may also be operative in the shifts over time, as with variety × fertilizer N × density of planting interactions in US maize crops (Carlone & Russell, 1987).

More extreme dwarfing genes are known – such as the Rht3 gene in wheat – and may be used in future breeding programmes if still higher levels of N fertilizer application are used. But just as these heavier dressings will depend on further shortening of the stems, so may that in turn depend on more effective control of weeds. The yield of modern dwarf varieties may fall at high levels of N application in the absence of effective weed control (e.g. Moody, 1981) whereas taller varieties more readily suppress competition by shading the weeds.

There are positive interactions on yield between N fertilizer applications and the levels of P and K (Cooke, 1982) and on some soils with other fertilizer nutrients such as sulphur (Randall *et al.*, 1981) and magnesium (Singh & Gill, 1987). These can enhance the effect of N dressings, although not to the dramatic extent that molybdenum enhances symbiotic N fixation by legume crops on some soils.

The supplies of N fertilizer and water also interact positively in the determination of yield in many crops, as Crowther (1934) found with cotton in the Sudan. He concluded that 'In no sense does the Liebig law of limiting

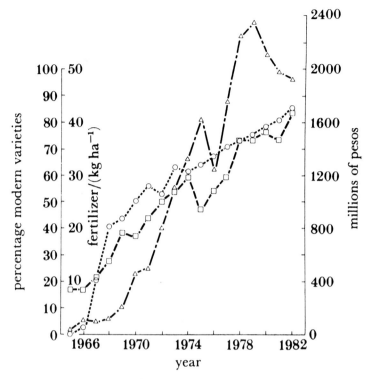

Figure 7.7. Changes in the percentage of rice area in the Philippines planted to modern varieties (○), in fertilizer use (kg N + P + K per hectare of rice, □), and in annual investment in irrigation (10^6 pesos, 1981 price, △) (L.T. Evans, 1986).

factors hold'. A similar response for cotton in California is illustrated in Figure 7.8, and Hexem & Heady (1978), Stutler *et al.* (1981) and Eck (1984) present many other illustrations of fertilizer N × irrigation interactions. Some reach above-optimal N applications, and others reach super-optimality for both fertilizer N and irrigation. Such a situation can arise even in the absence of lodging if pests or diseases become more serious with heavy irrigation and N fertilization, as they often do (e.g. Scriber, 1984; Zadoks, 1985). Effective control of pests and diseases can then enhance the yield response to fertilizer N even more, as shown by Widdowson & Penny (1965) for irrigated wheat crops assailed by take-all disease and cereal cyst nematodes, by Tinker & Widdowson (1982) with the control of leaf diseases, and by Barker *et al.* (1977a) for Asian rice crops. Higher order interactions may also be found with such factors as time and density of sowing, as shown long ago by Gregory *et al.* (1932) with cotton.

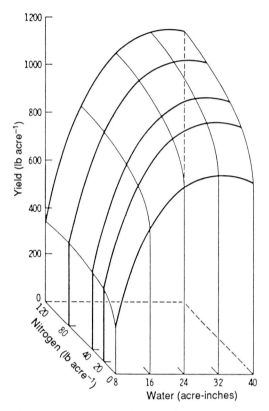

Figure 7.8. Interaction between fertilizer N and irrigation water application rates in determining lint yield of cotton in California (Hexem & Heady, 1978).

The continuing growth in the number and scope of these positive interactions through plant breeding and agronomic innovation helps to keep diminishing returns at bay. Nitrogenous fertilizers accelerate early growth and the closing of the canopy, e.g. in maize and sorghum (Muchow, 1988a). They raise the crop growth rate e.g. in maize (Lemcoff & Loomis, 1986) and wheat (Green & Vaidyanathan, 1986). They increase light interception and the efficiency of radiation use (e.g. Muchow & Davis, 1988). Nitrogenous fertilizers hasten inflorescence initiation in some crops. Even where they do not do so directly, their acceleration of crop establishment allows the plant breeder to select for earlier initiation.

At the other end of the life cycle, an adequate supply of N reduces the need for the crop to mobilize N from its leaves into its storage organs. The plants can grow on current N uptake instead, provided that enough assimilate is partitioned to the roots to maintain nitrate uptake and reduction (Reed *et al.*, 1988). The duration of leaf photosynthetic activity is thereby extended making more prolonged storage possible (e.g. Lemcoff & Loomis, 1986; Muchow, 1988*b*). The plant breeder is then able to select for more prolonged reproductive or storage phases and longer leaf life, provided seasonal conditions make this possible (Cavalieri & Smith, 1985). More scope for optimizing the time of flowering in relation to seasonal sequences is a further consequence.

Nitrogenous fertilizers also permit the plant breeder to break the frequently negative relation between yield and percentage protein in storage organs. Benzian *et al.* (1983) found heavy applications of fertilizer N to result in positive relations between grain yield and percentage protein in wheat. When the crops were sprayed with fungicides, yields frequently exceeded 9 t ha^{-1} without any reduction in percentage protein, yet another example of positive interaction between different kinds of inputs.

The levels of all proteins in the plant–metabolic, photosynthetic, storage or protective – are likely to be raised when additional N is provided, unless selection and breeding are directed to the specific enhancement of only some of these. In wheat the relative proportions of soluble protein, rubisco and other photosynthetic proteins are relatively unaffected by N supply (J.R. Evans, 1983) but this may not be so in other crops such as rice, beans and spinach (see Figure 5.14). Thus, crops differ to some extent in the way in which they partition additional N.

For both wheat and rice, increasing N supply initially enhances rubisco content and photosynthetic rate per unit leaf area until these reach near-maximum values at about 3 mM nitrogen in the nutrient solution (cf. Figure 7.9). Individual leaf area continues increasing up to about 6 mM N, while total leaf area, tiller number, and dry weight per plant increase up to the highest N levels. Photosynthetic rate may, however, fall slightly at high N, as has also been found in the field (Pearman *et al.*, 1979; Morgan, 1988). A similar pattern of response is apparent across a wide array of rice genotypes (Cook & Evans, 1983*a*) and may be fairly general among the cereals. Greater use of fertilizer N can therefore act as a surrogate for selection of cultivars with higher photosynthetic rates in the early stages of agronomic advance. With higher levels of N application, the gains in yield come more from greater leaf growth and tillering and from effects on the *duration* of photosynthetic and metabolic activity.

Another consequence of the greater use of N fertilizers with implications for plant breeding is the reduction in root growth as a proportion of biomass as the level of N availability increases, e.g. in soybean (Rufty *et al.*, 1984; Tolley-Henry & Raper, 1986). Figure 7.10 illustrates this reduction for several wild and cultivated rices (cf. also Brouwer, 1966; Givnish, 1986). Such

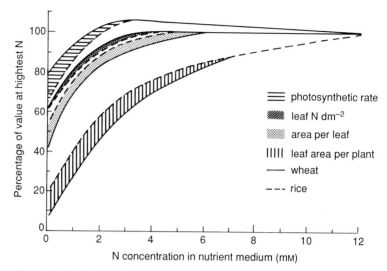

Figure 7.9. Relation between N concentration in the nutrient medium and photosynthetic rate, area per leaf, total leaf area per plant and leaf N dm^{-2} in wheat (data of J.R. Evans, 1983) and rice (data of Cook & Evans, 1983a).

an effect can be expected only when other nutrients and water are in ample and reliable supply, but it can be quite striking. Salisbury & Bugbee (1988) found the roots of wheat to comprise only 3–4% of the biomass when plants were grown under high levels of irradiance, CO_2 and nutrients. Ample fertilizer inputs should therefore allow plant breeders to select for reduced investment in root growth, freeing assimilates to augment yield potential through yet another agronomy × breeding interaction. They should also enhance the response of crops to a rise in atmospheric CO_2 level (Goudriaan & de Ruiter, 1983).

Efficiency of fertilizer nitrogen use

In developed countries such as the United Kingdom there is little evidence of losses of potassium applied in fertilizers, and although more phosphorus may be applied than is needed, it mostly accumulates in the soil for future use. With nitrogenous fertilizers, however, there are substantial losses – sometimes more than 50% (Bock, 1984) – by leaching, denitrification of nitrate and volatilization of ammonia, reducing the efficiency with which they are used as well as exacerbating environmental problems in the hydrosphere, troposphere and stratosphere.

However, along with the increasing efficiency of their manufacture (Figure

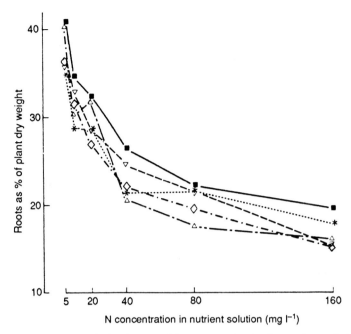

Figure 7.10. Relation between N concentration in the nutrient solution and the proportion of plant dry weight in the roots of several species of *Oryza*: *O. rufipogon* ($\triangle$), *O. nivara* (∇), *O. sativa-spontanea* ($\diamond$), *O. sativa* (■) and African species (*) (Cook & Evans, 1983*a*).

7.1) there has also been a rise in the efficiency with which N fertilizers are used. Nitrogen balance sheets for the Rothamsted Continuous Wheat Experiment show that over the period 1852–1967 only 32% of the 144 kg N ha^{-1} applied each year was recovered in the grain and straw, whereas in 1979–80 the recovery averaged 86% (Jenkinson, 1982), associated with improved varieties, better control of pests and diseases, and earlier sowing in autumn. Liming of acid soils improves N use by crops; Powlson *et al.* (1986) found the % N recovered by wheat crops to be much higher when the phosphorus level did not limit growth, another example of positive interactions between inputs. The uptake of N by high-yielding crops of wheat in the UK and of maize in the USA can equal the amount applied as fertilizer.

Losses of fertilizer N by leaching can be substantial in humid environments, amounting to about one third of the amount applied in the earlier Rothamsted experiments (Jenkinson, 1982). Most of the loss occurred in winter and early spring, hence the large quantities of nitrate-N carried by the Thames river in those seasons (Jenkinson, 1986). While some of this derives from directly leached fertilizer, much of the winter flush comes from the mineralization in autumn of nitrogen in the stubble and roots of the previous crop, derived mostly from applied fertilizers. Earlier sowing of the following crop allows much of this N to be recovered, as shown by Widdowson *et al.* (1987), thereby increasing yield and the return on N fertilizer while at the same time reducing environmental problems. In a multifactorial experiment on combinations of seven factors applied to barley crops, Widdowson *et al.* (1986) found earlier sowing in the autumn to have the greatest effect on yield and to interact strongly with several inputs, especially with pesticide, fungicide and the timing of fertilizer applications.

Substantial amounts of fertilizer nitrate can also be lost by denitrification, especially from waterlogged soils, again leading to both inefficient fertilizer use and environmental problems from the effects of nitrous oxide in the troposphere and stratosphere. Such losses may, however, be substantially reduced by the use of nitrification inhibitors such as nitrapyrin and dicyandiamide as an additional input (e.g. Bremner *et al.*, 1981; Huber *et al.*, 1980; Hoeft, 1984). Such inhibitors may also be used to achieve other agronomic ends, e.g. to control nitrate levels in high-input vegetable growing, to control leaching losses from early fertilizer applications, and to reduce the flush of ammonia after grassland is ploughed up (Tinker, 1985). Once again we see the interactive potential of new inputs. Huber *et al.* (1980), for example, found that nitrapyrin could double the yield response by winter wheat to fertilizer N applications.

Since the early 1970s urea has become increasingly predominant in global N fertilizer use, accounting for over 35% of the world total in 1984. On soils with high urease activity much of the N applied as urea can be lost by ammonia volatilization. Inhibitors of urease greatly reduce N loss under such conditions (Bremner & Douglas, 1971). Such inhibitors, e.g. phenyl phosphorodiamidate, reduce the adverse effects of ammonia on seed germination (Bremner & Krogmeier, 1988). Urea is now being used widely for irrigated crop production in the tropics, with losses of up to 56% of the N applied to rice paddies and irrigated sugar cane fields. However, by appropriate water management, careful timing of fertilizer application in relation to irrigation, deep placement of slow-release fertilizer and the use of urease and nitrification inhibitors, substantial losses by volatilization, denitrification and leaching can now be avoided (De Datta, 1987). Urease inhibitors increased the yield of rice per kilogram of fertilizer N by more than 20% in one instance (Buresh *et al.*, 1988).

Nitrogenous fertilizers are such an important component of modern crop production that many other aspects of their use could be considered, but the purpose of this case history has been to focus on the ways in which they interact with other inputs, and on improvements in the efficiency of their manufacture and use. The outcome of these features has been the avoidance, so far, of diminishing returns to their use in crop production (e.g. Figure 7.5).

Other aspects of fertilizer use

The greater the uptake of N by crops the greater is the requirement for other nutrients, so that as the yield level is raised more elements have to be supplied exogenously, and at greater rates. As the yield of maize is raised by fertilizer N, for example, so is the requirement for P (Fox, 1979).

The other plant nutrients also interact with one another and with water supply, pest and disease incidence and environmental conditions. For example, mass flow of nutrient ions in the soil solution is important for the transport of nitrate, Ca and Mg to the roots of maize – but less so for K and much less so for P (Barber, 1964) – giving positive interactions with irrigation. On the other hand, dry soil conditions may so restrict the uptake of P that its deficiency becomes a major factor in drought stress (Cooke, 1982). Phosphorus level affects Mg uptake. It interacts with disease severity, e.g. in the case of take-all disease of wheat in which high rates of P application reduce the effects of the disease (Mattingley *et al.*, 1980; Hornby, 1985). High K levels also restrict the incidence of many diseases.

Higher levels of P encourage root growth but may reduce the proportion of plant biomass invested in root growth. With barley, for example, the root:shoot ratio declined progressively with age when the P supply was adequate, but increased progressively when it was inadequate (Chapin *et al.*, 1989). Genetic differences in the ability of crops to gain nutrients and grow under various deficiencies, excesses and other adverse soil conditions are well known, e.g. for deficiencies of P, Cu, Mn and Fe in wheat and rye (Graham, 1984), and for toxicities of Al in wheat and barley (Foy *et al.*, 1965), soybean (Armiger *et al.*, 1968) and cotton (Foy *et al.*, 1967), and of Mn in cotton (Foy *et al.*, 1969). It is quite likely that some of the genes for such tolerances have been accumulated during the process of empirical selection for yield potential, with the result that modern cultivars may be less adversely affected than their wild progenitors by low nutrient status, as in rice (Cook & Evans, 1983*a*). On the other hand, more recent selection for yield at high nutrient levels could have reduced the relative investment in roots so that performance under low soil fertility might be adversely affected. Depending on the net effect of these opposed trends, there could be a crossover in relative yields of old and new cultivars as the soil fertility or fertilizer application level changes.

Comparison of the yields of British winter wheat varieties of different vintage when grown under high and low fertility conditions (Figure 6.13) gave no evidence that selection under favourable conditions has prejudiced performance at lower fertility levels. Likewise, comparison of the yields of US maize hybrids of differing vintage in high and low fertility conditions did not reveal any bias towards relatively better performance with high soil fertility in the more recent hybrids (Castleberry *et al.*, 1984).

The examples given in Figure 7.6 give no evidence of crossovers in performance at low fertilizer levels. There are cases where the modern cultivars are superior even in the absence of fertilizers (e.g. De Datta, 1987) and others where they are inferior (e.g. Evans & De Datta, 1979). In situations where the soil is fertile and has received fertilizer dressings in previous years, as in many of the rice varietal comparisons at IRRI, the absence of a crossover in performance with zero fertilizer is not a sufficient test. However, the durum wheat trial in Figure 7.6, which followed a crop of unfertilized sorghum grown to reduce residual fertility, was a better test. The problems of such trials, and several examples of crossovers in the relative yield of wheat and barley varieties with rising N applications, are considered by Anderson (1985).

Whether cultivars with superior yield under high soil fertility are also superior in low-fertility soils or with little or no fertilizer is an important question on which there are more strong opinions than strong data. The question has a significant bearing on plant breeding strategies for low input and adverse conditions, and has been answered in two quite different ways. Many plant breeders would agree with Frey (1964) that the most effective environment for selection is that in which differential performance is most manifest, and that a non-stress, high-fertility environment meets this condition best. However, the results of Gotoh & Osanai (1959) suggest that selection of wheat at low fertility levels more frequently resulted in superior and adaptable lines. Maize synthetics based on selections at low soil N levels were found by Muruli & Paulsen (1981) to yield more under such conditions than synthetics based on selections at high N levels. Atlin & Frey (1989) conclude that direct selection in the presence of stress is superior to selection in a non-stress environment for yield in low P conditions.

This question clearly deserves more attention. Weeds adapt to different fertility levels (e.g. Snaydon, 1970) and wild plants to a great variety of deficiencies and excesses. Plant breeding specifically for such conditions is likely to be more prominent in the future, as will selection for performance with heavy fertilizer applications, and we should know whether and how such selection is likely to prejudice performance under low fertility conditions.

Irrigation

Advantages and use

The practice of irrigation may be almost as old as agriculture itself, the earliest evidence being the mace-head of the Egyptian Scorpion King, which shows one of the last pre-Dynastic Kings ceremonially cutting an irrigation ditch about 5000 years ago. A millennium later the tomb of the Assyrian Queen Semiramis was inscribed: 'I constrained the mighty river to flow according to my will and led its waters to fertilize lands that before had been barren and without inhabitants' (Shanan, 1987). Initially based on techniques for the diversion of water from rivers, impounding by dams associated with other uses of water and with complex reticulation systems then became dominant. In the latter part of this century, energy-intensive pumping systems – with much lower capital costs and greater efficiency of water use – have also become prominent, leading to the greater use of sprinkler systems and to a draw-down in ground-water sources (Ambroggi, 1980; Pierce, 1990).

The major benefits from irrigation are as follows:

(1) Crop yields are substantially increased. FAO estimates suggest that food production per unit land area is on average 2.5 times greater with irrigation than on land without it (Stanhill, 1986). The 13% of cropland which is irrigated in the USA is estimated to yield crops worth 30% of the total production (National Research Council, 1989).

(2) Food production is more stable and less risky with irrigation. Many crops and environments could be used to illustrate this important feature, but Stanhill's (1986) comparison of wheat yields in Egypt and Israel makes the point well (Figure 7.11). Yields in the two countries, with not dissimilar environments, have increased at about the same rate, but have been about 1 t ha^{-1} greater in Egypt, where the crop is wholly irrigated, than in Israel, where it is predominantly rainfed. The year-to-year variation is far less with irrigation, the coefficient of variation for yield being only 7% in Egypt compared with 26% in Israel, associated with a coefficient of 22% in the annual total volume of rainfall in Israel. Colville (1967) presents a similar comparison of maize yields with and without irrigation.

(3) Associated with this greater reliability and level of yield, irrigation makes the use of other inputs worthwhile. The rates of application of fertilizer to crops such as wheat, rice, cotton and sugar cane in India are 4–8 times higher when irrigated than when rainfed (Desai, 1982). In the Philippines, there has been a parallel increase in irrigation, fertilizer application and the use of modern varieties of rice (Figure 7.7), as also in the Punjab for wheat. Irrigation encourages the use of other inputs: it often makes them easier to apply, and it enhances their effect.

(4) Irrigation makes possible the growth and development of crops outside their usual season and environment, thereby enhancing the scope for multiple cropping and year-round food production, especially in the tropics where temperature and irradiance make this feasible. In such areas, greater cropping intensity may contribute as much as greater yields per crop to increased food production.

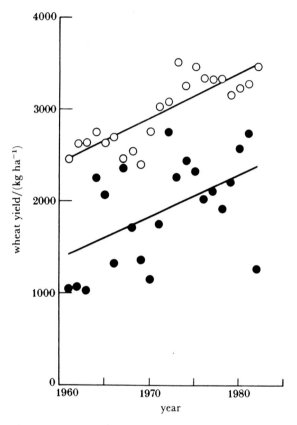

Figure 7.11. Trends in the level and variation of wheat yields under irrigation in Egypt (open circles) compared with dryland conditions in Israel (solid circles). From Stanhill (1986).

For these and other reasons the extension of irrigation looms large in agricultural development, despite the environmental and health problems often associated with it and the low efficiency of water use in many schemes. Over 40% of World Bank loans in the agricultural sector are for irrigation and drainage (Robertson, 1986), but this is not surprising when 50–80% of the total food supply of countries such as Indonesia, India, China and Pakistan is dependent on irrigation (Rangeley, 1986).

For the world as a whole the area irrigated has increased about 6–fold in this century, and by about 20% in each of the past two decades. It is currently about 229 Mha, equivalent to one sixth of the total arable area. The whole of the arable area in Egypt is irrigated, compared with 76% in Pakistan, 63% in Israel, 46% in China, 36% in Indonesia, 25% in India, 10% in the USA and

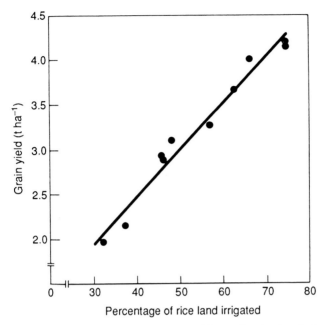

Figure 7.12. Average national rice yield in Colombia in relation to percentage of rice land irrigated between 1966 and 1975 (after Chandler, 1979).

4% in Australia. For Asia as a whole this proportion is 31%, far higher than in other regions.

Effects on yield

The close relation between national rice yields in South and East Asian countries and the proportion of rice area irrigated in each country indicates a strong effect of irrigation on yield with the qualification that, as illustrated in Figure 7.7, other inputs also vary with the extent of irrigation. The same associations are likely to confound the relation between irrigation and yield for any one country across years, such as that for rice yields in Colombia (Figure 7.12).

At the individual crop level, more or less linear relations between accumulated growth and total evapotranspiration have been found with many crops (Hanks, 1983), the slope being influenced by environmental conditions, species and crop duration. Differences in early growth and time to maturity contribute to apparent varietal differences in water use efficiency (e.g. Rasmussen & Hanks, 1978), but for closed stands under comparable

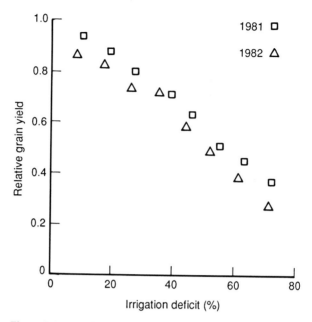

Figure 7.13. Relative grain yield of pearl millet crops at Hyderabad as influenced by irrigation deficit during a stress period beginning at flowering in two years (Mahalakshmi *et al.*, 1988). (American Society of Agronomy, Inc. Reprinted with permission.)

conditions varietal differences in water use efficiency for biomass production are generally small (e.g. Retta & Hanks, 1980).

The relation between evapotranspiration as influenced by the amount of irrigation and grain yield is often less close than that with biomass (Hanks, 1983), but is reasonably linear nevertheless. This is apparent for pearl millet in Figure 7.13 (cf. also Mahalakshmi *et al.*, 1990) and in the survey by Doorenbos & Kassam (1979) covering more than 500 experiments conducted throughout the world on many crops over many years. The C_4 crops maize, sorghum and sugar cane emerge from the survey as the ones with the greatest yield response to additional irrigation, as is to be expected from their lower transpiration ratio (cf. Stanhill, 1985).

Interactions

The strong interactions on yield between water supply and N fertilizer applications have already been considered (Figure 7.8). These vary between crops, as well as for any one crop (Bielorai *et al.*, 1983), but they probably constitute one of the most powerful components in the escape from diminishing returns. Water supply also interacts with the levels of P and several other nutrients. Energy requirements and costs can be reduced by incorporating fertilizers in irrigation water ('fertigation'), a practice that may also increase fertilizer efficiency, especially for N applications to sandy soils (Dasberg *et al.*, 1983; Feigin *et al.*, 1982; Rehm & Wiese, 1975). As Stanhill (1986) points out, fertigation allows the nitrogen content around the root system to be repeatedly optimized, thereby reducing volatilization, leaching and environmental pollution. Losses of ammonia by volatilization following application of urea can be reduced by the use of additives retarding its hydrolysis (Rappaport & Axzley, 1984).

Yet another beneficial interaction between water and N supply is that faster leaf area growth at high N levels allows the ground to be covered and the canopy closed more rapidly, thereby minimizing ammonia volatilization (Denmead *et al.*, 1982) and evaporative water loss from the soil surface.

The interactions between irrigation and pest and disease injury are both positive and negative in their effects on yield. With denser crops and higher humidity within the canopy some problems are exacerbated, such as powdery mildew of wheat and tungro virus in rice, whereas other diseases such as brown spot and rice blast become less serious with irrigation. Irrigation also influences pest and disease incidence by extending the duration of crop growth, which may exacerbate pest build-up by the end of the season, as in the case of the cotton boll worm. On the other hand, irrigation can be withheld at particular times as one component of strategies for the control of both the bollworm and the pink bollworm in cotton crops (Bariola *et al.*, 1981). Irrigation in the tropics makes possible the continuous cultivation of crops such as rice. This could allow the natural enemies of pests to build-up, optimizing population regulation by parasites and predators, but the regular sowing of a sequence of crops may also exacerbate problems with some pests, such as the brown planthopper (Perfect, 1986).

Many insecticides can be effectively applied by using overhead sprinkler systems (Chalfant & Young, 1982); such 'chemigation' may reduce both costs and energy use. Nematicides applied by drip irrigation may be more effective than those applied by conventional methods (Overman, 1976). The flooding of rice paddies suppresses non-aquatic weeds, especially many potentially troublesome C_4 species (Tanaka, 1976). Irrigation may exacerbate other weed problems, but it may also be used as the medium for effective herbicide applications (Dowler *et al.*, 1982).

These and other interactions allow irrigation to enhance the effects of other inputs as well as having a substantial direct effect on crop yields.

Efficiency

In 1965, 81% of all water use in the world was for irrigation, but by the year 2000 this proportion may be only 33% in spite of a doubling in the amount of water used for irrigation (Stanhill, 1985), reflecting the increasing competition for water and a likely rise in its cost. More efficient use of water in irrigation will be essential.

Stanhill (1986) has illustrated the overall increase in the efficiency of water use in Israeli agriculture (Figure 7.14). In 1950 almost 9000 m^3 of water was applied to each hectare irrigated – about half as much as the average application in Egypt – but this amount has been progressively reduced by about 40%, thereby reducing the cost of water from 12% to 6% of total input costs. On top of this, the productivity of Israel's irrigated agriculture per cubic metre of water has increased two-and-a-half-fold, associated with an increase in average water use efficiency from about 30% to 80% through a comprehensive research programme leading to a legally enforced water allocation policy.

Surface irrigation was the norm at the beginning of this period, whereas about 80% of the area is now irrigated by sprinklers and the remainder by drip systems. However, sprinkler systems require considerable energy, and more efficient surface irrigation systems, such as surge-flow and cable-irrigation, are being developed (Stanhill, 1986).

Irrigated agriculture in arid lands is highly productive and, as we have seen, characterized by stability, at least in the short term. Yet, despite the striking improvements in its efficiency of water use, it is still expensive in terms of both energy and cost. Stanhill (1981b) and Tanner & Sinclair (1983) have pointed out that irrigation water would be more efficiently used for crop production in areas where the potential transpiration rate is less, i.e. in those seasons and regions where demand is smaller, and where irrigation is supplementary or conjunctive rather than essential. It is in these humid and subhumid areas that irrigation has been extended most rapidly in the USA in recent years (Stewart & Musick, 1982).

Plant breeding implications

Despite the similar rate of improvement of wheat yields in Egypt and Israel, the one irrigated and the other not so (Figure 7.11), progress towards higher yields is usually faster the more favourable the water regime. This is illustrated in Figure 7.15 for rice in the Philippines; similar figures have been presented for rice in Japan, and in India, Bangladesh and Indonesia (David, 1989), as well as for maize in the USA (Colville, 1967).

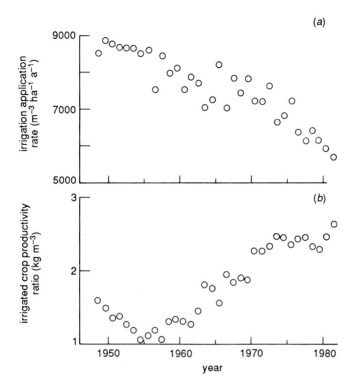

Figure 7.14. Changes in (*a*) the rate of water application and (*b*) the efficiency of water use in irrigated crop production in Israel, 1948–82 (Stanhill, 1986).

The nature of improved crop performance with irrigation is an important question, with relevance to selection procedures in breeding programmes. By assuring that the crop will not run out of water during grain filling, irrigation could allow the plant breeder to select for longer grain growth duration or later maturity, which could increase yield potential although possibly at the expense of yield under dryland, shorter season conditions. Likewise, irrigation, combined with fertilizer application, might make it possible for plant breeders to select plants with reduced root growth, thereby freeing assimilates for investment in the grains. O'Toole (1982) found that relatively little root growth took place after anthesis in an irrigated rice crop, whereas in crops that were water-stressed after anthesis there was more root growth throughout the profile the more the stress. With some crops, such as cotton, it is the distribution of roots in the profile rather than their total weight that

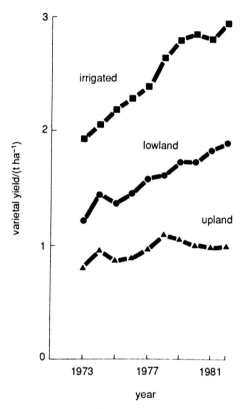

Figure 7.15. Yield trends for irrigated, lowland and rainfed upland rice crops in the Philippines (L.T. Evans, 1986).

changes with irrigation (Klepper *et al.*, 1973), but for many plants the root:shoot ratio is reduced by irrigation (Brouwer, 1966).

Evidence that the root:shoot ratio has been reduced by selection under irrigated conditions is sparse. Zuño-Altoveros *et al.* (1990) found improved lowland varieties of rice to have only about half the root volume of upland varieties at the same age, associated with their reduced resistance to drought stress, but whether the root:shoot ratio is also less in the lowland varieties is unclear. The ratio adjusts so quickly to changes in the water and nutrient regime that genetic change may not be needed.

If a reduced root system has been unconsciously selected for during the improvement of yield potential with irrigation, a crossover in the relative performance of varieties could take place in the more water-stressed environments. As with the possible crossover in performance with fertilizer levels and soil fertility, relevant data are as slender as opinions are strong. Mederski & Jeffers (1973) support Frey's (1964) assertion that because differences in yield between varieties are greater in non-stressed conditions, selection is more effectively done in them. However, their yield comparisons between eight varieties within each of four soybean maturity groups revealed many crossovers in performance between low and high water stress. Cooper (1985) also records crossovers in the relative performance of soybean varieties with increasing water stress. Byth *et al.* (1969) found that although heritability and predicted genetic advance were greatest in the non-stress environment, actual yield advance was greater in the stressed environments.

Similar results have been obtained by Fischer *et al.* (1989) with maize lines grown under either mild, medium or severe soil moisture deficits. Pronounced crossovers in performance were found even after only three cycles of selection. When grown under severe deficits, the selections made under that condition out-yielded those made under mild deficits by 26%, whereas these latter were only 13% better than the former when grown under mild deficits, in which yields were four times higher. Evidence of crossovers in the relative yield of wheat varieties as site mean yield decreased has been presented by Laing & Fischer (1977), Fischer & Maurer (1978), Fischer (1981), and Bruckner & Frohberg (1987). Since site mean yield in these studies was correlated with growing season rainfall, such crossovers suggest a trade-off between yield potential and performance under water stress.

Irrigation has many other implications for plant breeding programmes. In more tropical environments it permits multiple cropping and leads to greater emphasis on shorter-duration varieties, rapid establishment, daylength-insensitivity and greater resistance to those pests and diseases exacerbated by irrigation. It permits change in the balance between the various phases of the life cycle. By extending the growing season and allowing crops to be grown in arid environments, irrigation may expose them to extremes of heat or cold not previously encountered. Thus, not only does irrigation have a profound direct effect on crop yields, and large indirect effects through its interactions with other inputs, but it also influences plant breeding strategies.

Crop protection

Introduction

The formidable global survey of yield losses due to pests, diseases and weeds made by Cramer (1967) will no doubt continue to be cited for many years to come, until another academic masochist is willing to undertake as comprehensive an assessment. Some of Cramer's global estimates of percentage yield loss are indicated in the following table:

Crop	Losses due to			
	Insects	Diseases	Weeds	Total
All	13.8	11.6	9.5	34.9%
Wheat	5.0	9.1	9.8	23.9
Rice	26.7	8.9	10.8	46.4
Maize	12.4	9.4	13.0	34.8
Millet and sorghum	9.6	10.6	17.8	38.0
Sugar cane	20.1	19.2	15.7	55.0
Potatoes	6.5	21.8	4.0	32.3
Groundnuts	17.1	11.5	11.8	40.4
Cotton	16.1	12.0	5.8	33.9

Total yield losses ranged from 55% for sugar cane to less than 15% for rye. Whereas the overall losses due to pests, diseases and weeds were of comparable magnitude, their relative impact varied greatly between crops. Insects took heaviest toll of rice, sugar cane and groundnuts, whereas diseases were worst for grapes (23.4%), potatoes and cocoa (20.8%). Weeds caused greatest yield loss in millet, sorghum and sugar cane. Even with a particular crop, however, there may be great variation in the relative constraints to yield by pests, diseases and weeds between locations, years and seasons, as shown for rice by the IRRI project on farm level constraints to high yields in Asia (IRRI, 1977, 1979b). According to Mochida (1974), potential yield losses due to pest and disease attacks on rice crops at the Kyushu National Experiment Station in Japan averaged 52.7% over a 10 year period, while spraying with insecticides increased rice yields by 54% in India, and by an average of 61% in the Philippines.

The relative magnitude of yield losses is likely to have changed since Cramer made his survey. Parker & Fryer (1975) estimated the overall yield loss due to weeds at 11%, somewhat higher than Cramer's figure, but the rapid rise in the use of herbicides since their survey is likely to have reduced overall weed losses. By contrast, losses due to pests and diseases may have increased somewhat with intensification and the frequent development of resistance. Indeed, Pimentel's (1976) estimates indicate a rise in the percentage loss due to pests and diseases combined with a fall in that due to weeds in the USA (cf. also Wheatley, 1987).

Changes in the overall use of agrichemicals have been described by Green et al. (1977), Braunholtz (1981), Hutson & Roberts (1985) and Jutsum (1988), who show that much of the overall growth in recent years has been due to the greater use of herbicides, which increased by 500% between 1960 and the mid-1970s, overtaking insecticides in value in 1970. Whether the use of other agrichemicals, particularly growth regulants, will expand as much as has been forecast (Hardy, 1978) remains to be seen.

Another trend in the agrichemical industry is a move away from the production of compounds for specific use on minor crops or pests. There is a sharper focus on a small number of major crop needs, such as herbicides for maize, soybeans and cotton, insecticides for cotton, rice, maize, and fruit and vegetable crops, and fungicides for the latter (Braunholtz, 1981). Thus, even though more compounds than ever before are still being screened, perhaps 150 000–200 000 per year, this activity is increasingly focused on a few major problems (Morrod, 1981). While this is regrettable, particularly in limiting the opportunities for less important crops, it is an inevitable consequence of the more comprehensive testing requirements which mean that it now requires more than US$20 million and an increasing proportion of the patent rights time to launch a new product, hence the fall in the number of new compounds coming on the market.

As general indications of the impact of agrichemicals on yield, it has been estimated that the elimination of all pesticide use in the USA would reduce crop and livestock production by about 30%, while similar action in the UK would reduce yields of potatoes by 42%, cereals by 45% and sugar beet by 67% (Braunholtz, 1981). Several figures indicating a close relation between national crop yields and the use of pesticides have been published (e.g. Marstrand & Pavitt, 1973; Green, 1978), but given the problems with national data on pesticide use, the fact that most pesticides are used on relatively few crops, and the parallel changes in the use of other inputs, such figures are unlikely to be valid indicators of the impact of pesticides on crop yields.

Insecticides and fungicides

Efficiency and selectivity

The quite remarkable increase in the activity of crop protection chemicals during this century is shown in Figure 7.16, which is salutary in also reminding us of the undesirable, unspecific and dangerous compounds used to control pests and weeds at the turn of the century. The more or less parallel reduction in application rates for insecticides, fungicides and herbicides through three or more orders of magnitude since then reflects attack on increasingly specific and sensitive sites within the organisms. There is no reason to suppose that the limits of activity have yet been reached, although the lethal dose for deltamethrin on *Anopheles* mosquitoes is now only 0.04 ng per insect (Graham-Bryce, 1981).

As the required rate of application has diminished, it has been possible to improve the methods of application, especially by shifting to low and ultra-low volume sprays. This has greatly increased efficiency (Geissbuhler, 1981), while at the same time eliminating the requirement for large volumes of

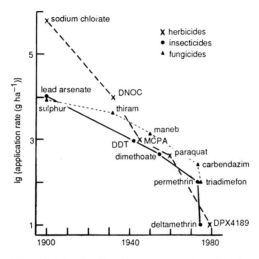

Figure 7.16. Application rates, on a logarithmic scale, for representative crop protection chemicals plotted against the year of their introduction (Graham-Bryce, 1981).

water, permitting more flexible and timely application procedures, and reducing environmental contamination. There is still considerable scope for improvements in application, and new low energy systems to deliver ultra-low volumes of electrostatically charged sprays are being developed (Matthews, 1981). Such systems, made possible by the advances in pesticide activity, have in turn amplified their effectiveness in the field, and are making pest and disease control feasible in situations where it was previously impractical, e.g. on cotton in parts of Africa and in many pest management systems. They constitute yet another illustration of the synergism between innovations.

With smaller doses, the risk of environmental contamination has been reduced, as it has been also by modifying the sorption, accumulation and degradation rates of more recent pesticides. Figure 7.17 illustrates the evolution of some of these properties for a number of major insecticides, such as the reduced potential for accumulation and the increase in degradability.

Integrated pest and disease management

Integrated pest management is an example, *par excellence*, of the value of synergistic interactions between inputs, of which management skills and research insights are as important as any. One cornerstone of integrated pest management is the concept of a damage threshold, originally 'the lowest

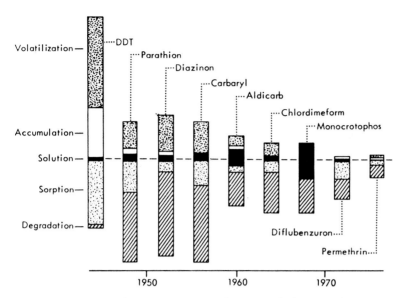

Figure 7.17. Changes over time in the potential of insecticides to be distributed and accumulated in the environment. This potential is increased by volatility and the likelihood of accumulation in non-target organisms, and reduced by a lower rate of application (bar length), adsorption by soil particles, and greater degradability (Geissbühler, 1981).

pest density that will cause economic damage' (Stern *et al.*, 1959). This definition was later refined to 'the level of pest attack at which the benefit of control just exceeds its cost' (Mumford & Norton, 1984). It has, therefore, become more an economic than a physiological threshold. This concept of a threshold was an important innovation but, as Zadoks (1985) points out, it implies a repeatability which is unrealistic, because the nature of the relation between damage and pest/disease populations varies greatly with environmental and agronomic conditions, stage of crop development and the presence of other pests and/or diseases.

Many different ways of assessing the incidence–severity relation have been devised (cf. James & Teng, 1979). Tammes (1961) presented a generalized curve relating yield to increase in the injurious agent, such that there was an initial plateau with no change in yield as pest/disease incidence increased, followed by a more or less linear fall in yield as incidence intensified, and then by a lower plateau where yield was unaffected by further pest/disease build-up. Bardner & Fletcher (1974) suggest that no single case exhibits all these features. They cite two cases where the upper threshold is supposedly clear, one with sugar beet (Jones *et al.*, 1955) and the other with cucumbers (Hussey & Parr, 1963), but neither set of data is convincing on this score. With sugar beet the upper threshold was apparent only when whole plants, not leaves, were removed from the crop, while there were no data for the assumed upper plateau in the cucumber crop. For this, the relation between leaf damage and yield loss was approximately linear, as it often is, e.g. for powdery mildew in oats (I.T. Jones, 1977), stem borer in rice (Soejitno, 1977) and late blight in potatoes (Waggoner & Berger, 1987).

In cases where the assimilate supply is limiting to yield – a situation that should become increasingly common as plant breeding for higher yield potential proceeds – no threshold yield plateau should be evident with leaf injury except in crops with a greater leaf area index than is needed for full light interception late in their life cycle. At lower yield levels, where assimilate supply may be less limiting, an upper threshold may be found in some circumstances. In fact, low pest and disease infections may even stimulate photosynthetic activity and growth, as in peas with aphids (Barlow *et al.*, 1977), soybeans (Turnipseed, 1972), tobacco with mosaic virus (Owen, 1957) and peaches infected with several other viruses (Smith & Neales, 1977), leading to 'overcompensation' in yield with light infections (Pedigo *et al.*, 1986).

In crops that produce an excess of flowers or fruits, like cotton and many legumes, a yield threshold – or 'damage boundary' as Pedigo *et al.* (1986) prefer to call it – may be apparent with pests that preferentially attack young flowers and fruits. Indeed, the assumption that such thresholds are characteristic and common probably owes too much to crops like cotton, as Stern (1973) admits. Even with cotton, however, Stern records that thresholds are known for only five of 19 pests which attack the crop.

Far more common and characteristic are examples of yield loss increasing more or less linearly with increase in pest or disease incidence and offering no evidence of a damage threshold or boundary (Bardner & Fletcher, 1974; Brown, 1976; Pedigo *et al.*, 1986). Existing data are not sufficient to decide whether yield thresholds are more characteristic of lower-yielding crops, but it seems likely that any such thresholds will become less common in high-yielding crops where sink and source are closely matched.

Relation to plant breeding for yield

Breeding for genetic resistance to many pests and diseases almost inevitably reduces the emphasis that can be given to the improvement of yield potential. Moreover, the incorporation of resistance genes sometimes exacts a yield penalty. Quite apart from their direct effects on yield, therefore, the availability of effective pesticides and fungicides may benefit yield by allowing the plant breeder to give more attention to yield potential. Pests and pathogens can, however, develop resistance to agrichemicals as quickly as they can overcome genetic resistances (Conway, 1982; Metcalf, 1984; Dekker, 1985). The most effective long term strategy is probably to combine the use of agrichemicals with genetic resistance in order to optimize the use and extend the effective life of both.

Herbicides

In recent years herbicides have constituted the fastest growing class of inputs to crop production. They have also transformed agriculture, allowing many time-honoured and previously essential husbandry practices to be abandoned (Fryer, 1981). Before the introduction of herbicides, weeds often dictated what crops were grown, and in what rotations because, as Shakespeare put it: 'sweet flow'rs are slow and weeds make haste'.

Effective herbicides, along with other inputs, have reduced the need for crop rotation. They have also reduced the need for tillage, even allowing crops to be grown without any cultivation. In many developing countries, the labour required for hand weeding constitutes one of the most serious bottlenecks to crop production (e.g. Trenbath, 1985), especially if it must be done in the 'hungry season' or when malaria and other debilitating conditions are at their peak. Herbicides have much to offer in such circumstances (Clark & Haswell, 1970), although their use may have some adverse environmental and social impacts, e.g. by displacing women and landless labourers from one of their traditional tasks (Barker & Cordova, 1978).

Reliance on herbicides for weed control has tended to become absolute in developed countries (Auld *et al.*, 1987) and new problems have appeared. Whereas some traditional weeds such as charlock and poppy have become of little consequence with herbicides, others have become more important, such as the wild oat and weedy relatives of several other crops (Fryer, 1981). These may pose even greater problems in the future because of introgression of herbicide resistance and other genes from the crop to its weedy relative (cf. Chapter 8), but modern herbicides have transformed crop production and have interacted with other inputs to create new opportunities not only for enhancing yield per crop but also for minimizing the turn-around time between crops as well as soil loss by erosion. What they do not appear to have done, so far, is to reduce the population of weed seeds in the soil, of which only a small fraction (e.g. 6–9%) germinates each year for many species (Chancellor, 1981), but this may be achieved in future with chemicals which break seed dormancy or regulate vegetative regeneration.

Effects on yield

Cramer (1967) and Parker & Fryer (1975) estimated an overall yield loss due to weeds of 10–11%, but the extent of loss varies greatly between crops, environments and years. Quite heavy infestations of weeds may cause little loss in yield provided they are removed before the crop enters its rapid growth phase. Likewise, late infestations may also have little effect on yield. Crops vary in the duration of the sensitive period between these early and late phases (e.g. Scott & Wilcockson, 1974; Roberts, 1976). There is usually a well-defined critical period for weed control, during which the effectiveness of herbicide applications is maximal.

The relation between weed density and yield also varies with crop variety, weed, fertilizer level and crop density. Figure 7.18 illustrates the differing impact of the density of four weeds on the grain yield of Newbonnet rice. The adverse effects of even small infestations by red rice and barnyard grass are apparent, and the yield losses from these weeds were even greater with a semi-dwarf cultivar (Smith, 1988). According to Auld et al. (1987) the hyperbolic form of the yield × weed density relation, evident with red rice in Figure 7.18, is the most common one, and they regard the sigmoid form with a threshold (e.g. Harper, 1977) as hypothetical.

The form of the relation changes when yield is plotted against weed biomass rather than density. Although three weeds of sugar beet crops displayed quite different relations when beet yield was plotted against the number of weed plants per square metre, all the data fell on one curve when yield was related to weed dry weight per square metre (Scott & Wilcockson, 1976). The taller weeds increased the proportion of shoot to root weight in the crop, and reduced the optimum level of N fertilizer application, illustrating the potential for synergistic interactions between herbicide and fertilizer application.

Some interactions

The remarkable reduction in the application rate of herbicides, illustrated in Figure 7.16, has many implications for crop production: in relation to energy use in agriculture as herbicides replace cultivation and management for weed control; by making ultra-low volume spraying possible and thereby reducing the need to transport and use relatively large amounts of water, so difficult in many developing countries, particularly in semi-arid regions; by making the direct seeding of rice feasible, thereby reducing planting costs and creating a need to modify plant type; and by making zero-tillage practices more effective, thereby allowing more timely planting and faster turn-around of crops, reducing soil erosion and allowing steeper land to

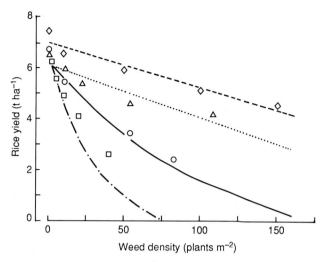

Figure 7.18. The reduction of rice yields caused by different densities of weeds: red rice (*Oryza sativa*, □), barnyard grass (*Echinochloa crusgalli*, ○), bearded sprangletop (*Leptochloa fascicularis*, △) and broadleaf signal grass (*Brachiaria platyphylla* ◇). Adapted from Smith (1988).

be cropped (cf. Unger & McCalla, 1980; Cannell, 1981; Smil *et al.*, 1983; Phillips *et al.*, 1980; Phillips & Phillips, 1984; Sprague & Triplett, 1986). In fact, modern herbicides provide one of the best examples of the synergisms between inputs, often unforeseen, as well as of the ultimate impact of curiosity-motivated research (Evans, 1982).

Herbicides have one great advantage over insecticides and systemic fungicides, namely their much longer useful life. This is associated with the longer generation times of weeds compared with those of pests and pathogens, with their irregular germination over a period of years, with the use of different herbicides in the course of crop rotation, and with the reduced 'fitness' of some resistant biotypes (Gressel, 1984). Resistance to them has developed so slowly and so infrequently that the shift from older to newer herbicides reflects more the greater efficiency of the new than the declining effectiveness of the older compounds.

Genetic resistance to insecticides, fungicides and antibiotics developed so quickly that ecologists warned there would be selection for herbicide resistance in weeds if farmers were not careful (e.g. Harper, 1957). But although the chlorinated phenoxy herbicides such as 2,4-D have been continuously deployed since the mid-1940s, resistance to them has not become a problem. With the S-triazine herbicides introduced 15 years later, which replaced 2,4-D for maize cultivation because they gave better control of the same weeds over a longer period, resistance has appeared in over 50 weed species (Le Baron & Gressel, 1982; Darmency & Gasquez, 1990). However, triazine resistance in weeds is associated with reduced photosynthesis, greater sensitivity to high temperatures (Ducruet & Lemoine, 1985) and other disadvantages.

Besides the appearance of resistant biotypes, it was suggested that a greater problem would be posed by changes in the weed flora, with tolerant or resistant species becoming more prominent. Changes in the weed flora have certainly occurred with intensification of agriculture, but the investigation by Haas & Streibig (1982) suggests that these changes were due more to changes in fertilizer use, drainage and varieties than to the use of phenoxy herbicides.

Weed control has traditionally been the province of agronomy, except insofar as tall or lax-leaved cultivars may have been selected for their greater competitiveness with weeds. Now, however, the possibility of breeding crop varieties resistant to specific herbicides is being actively explored. Gressel (1978) has tabulated many cases of differential tolerance or resistance to various herbicides among crop plants, and others have been found since then, e.g. of cotton varieties to glyphosate (Jordan & Bridge, 1979). More recently, however, deliberate attempts have been made to select or to engineer resistant varieties, using a variety of approaches (cf. Mazur & Falco, 1989).

Mutant forms of the target enzyme which are active but less sensitive to the herbicide have been selected, e.g. of acetolactate synthase to the sulfonylurea and imidazolinone herbicides by Chaleff & Ray (1984). An alternative strategy is the overproduction of the target enzyme, as in the case of 5-enol-pyruvylshikimate-3-phosphate synthase, conferring tolerance of glyphosate on transgenic petunia plants (Shah et al., 1986). A third approach has been to transform crops so that they can detoxify the herbicide, as has been achieved by De Block et al. (1987) with tobacco, potato and tomato by transferring the bar gene in Streptomyces, which confers resistance to phosphinothricin and bialaphos. While such developments may be of great commercial value in presaging specific variety × herbicide packages, they may also alleviate the problem of the after-effects of herbicide treatments used in zero-tillage systems.

Energy and crop production

Since energy analysis of agriculture was launched by Odum in 1967, it has spawned a sophisticated and extensive literature. Of many aspects related to crop production, we can focus here on only a few, which bear on the efficiency and interaction of inputs in relation to yield. Issues relating to crop quality, environmental health, agricultural development and social justice are not considered, but the use of agriculture for fuel or energy production is discussed in the next chapter.

Some conventions of energy analysis

Energy analysis of agriculture commonly follows conventions decided on at the International Federation of Institutes for Advanced Study workshop (IFIAS, 1975). Gross energy requirements are estimated without taking into account the solar energy either received or absorbed by the crop for photosynthesis. This exclusion of solar energy as an input may outrage some ecologists, as Leach (1976) suggests, and it certainly changes the energy return (ER, output/input energy) ratios. For British agriculture as a whole, for example, this ratio is 0.34 without the solar input and 0.0002 when incident radiation is included (Leach, 1976). For Corn Belt maize crops in 1970, the industrial inputs would constitute only 11% of the total energy input if the energy used in photosynthesis was included (Pimentel *et al.*, 1973). The solar energy term would so dominate the inputs that the analyses would be of little policy value, but its inclusion would change the ecological perspective. Its exclusion shifts the balance towards Odum's (1971) dictum that 'man no longer eats potatoes made from solar energy; now he eats potatoes partly made of oil'.

Energy analysis of agriculture may exclude solar energy but it does move upstream far enough to include the energy 'embodied' in the manufacture and transport of machinery, fertilizers, etc. Just how far upstream the analysis is taken depends on judgement of how significant the components are likely to be, e.g. for capital investment in irrigation schemes or phosphate mining. Such items are excluded from the 'universe' accounted by Smil *et al.* (1983).

Another convention is that energy analysis of agriculture stops at the farm gate for outputs. Beyond that we enter the realm of the larger or at least more energy-consuming 'food system'. In the USA, for example, the food system accounted for 12.8% of total energy use in 1970, but the agricultural sector consumed only a quarter of that proportion, processing and transport accounting for 39% while the commercial and home sectors used 37% (Steinhart & Steinhart, 1974). In Australia these proportions are 11% to the farm gate, 38% from there to the retail store and 57% from the store to the table (Gifford & Millington, 1975). Although the Australian food production system has an ER ratio of 2.4 at the farm gate, this falls to 0.6 at the retail store and to 0.36 at the dining table. Those who castigate agriculture for its energy use should recognize that profligacy is much greater beyond the farm gate.

Energy costs for transport become particularly significant in comparisons between high input cropping close at hand and lower input cropping at a distance, especially in view of the unrealistic relative prices that often characterize such situations. Slesser (1984) has examined this aspect in relation to local production *vis-à-vis* imports by the European Community, and concludes that although there would be advantages, in terms of total energy use, from importing mutton and beef from Australia or Argentina, this is not so clear cut for cereals in spite of the energy-intensiveness of their production in Europe.

The conventions of energy analysis with regard to human labour also assume considerable significance in comparisons of ER ratios across the full range of agricultural systems. The usual convention is to include only the metabolic energy expended by farm labourers during their work, not the accessory energy utilized by the worker himself, nor the total energy associated with the worker as a member of society, or with his family as such. By the usual convention, the energy intake for human labour is 5–8 MJ per man per day, whereas the full charge for lifestyle support energy in the USA would be about 600 MJ per farm worker per day in the USA in 1974 (Fluck, 1981).

By the usual convention, human labour has become a negligible proportion of the total energy input into modern farming systems, for example less than 0.1% of that for Corn Belt maize crops in the USA, but the lifestyle support figure would make it a significant component. Most energy analysts follow the usual convention on the grounds that rural lifestyle is separate from the business of producing food, and because the lifestyle estimate introduces some double accounting (e.g. Leach, 1975; Pimentel, 1980; Stanhill, 1984). However, Wit (1975) and Fluck (1981) argue that full lifestyle accounting is more appropriate, as do Wit & van Heemst (1976) and Fluck & Baird (1980) in relation to the high marginal substitution ratio of 344–543 MJ per man day for the replacement of labour by machines. When lifestyle energy use is high, total rural energy use may be reduced by replacing labour with machines. The way in which human energy is accounted is also of significance in comparisons between organic and conventional farming. Although the ER ratio may be higher for organic farming – at least of wheat and maize – by the usual convention (Pimentel *et al.*, 1984*a*), it would be lower than that of conventional farming in many cases if life-style energy costs were used.

The mechanization and intensification of cropping has led to a huge increase in energy returns per joule expended by humans in food gathering or production. The ratio of energy gained to energy expended by the gatherer generally ranges from 4 to 20 among birds, bees and fish (Lawton, 1972), and commonly falls within that range for hunters and gatherers such as !Kung bushmen, Australian aborigines or atoll fishermen. So do the returns on labour for shifting cultivation in some environments, e.g. for the Tsembaga of Papua New Guinea, in a heavily forested area of poor soils and steep slopes (ER = 16–20; Rappaport, 1968, 1971), or for rice growing by the Dayak or

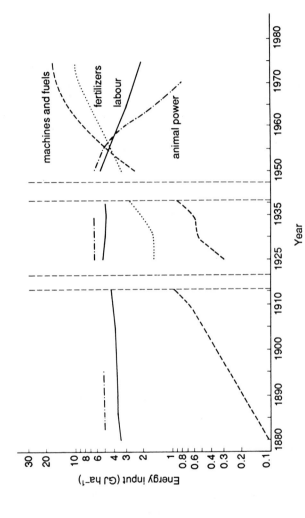

Figure 7.19. Changes in the energy input profile of German agriculture since 1880 (Weber, 1979, as cited by Slesser, 1984).

Iban (ER = 14–18, Leach, 1975). However some bush-fallow systems have an energy return on human labour of 60–65, e.g. with cassava crops in Africa.

The several examples analysed by Leach (1975) of subsistence systems based wholly on human labour showed energy returns of 11–31, whereas those using animal power showed returns on human labour ranging from 23 to 60. With high input crops the energy returns on human labour ranged from 50 to more than 3400. For US maize in 1945 the return was 641, and this increased to more than 4000 in 1970. In fact, the return on human labour in cropping is now so high that the human labour input is commonly neglected. Its progressive fall along with that in animal power accompanying the rise in other inputs in German agriculture over the past century is illustrated in Figure 7.19.

Energy return ratios

We have just compared agricultural systems in terms of their ratio of harvested energy to human metabolic energy input, a ratio which has increased almost a thousand fold with agricultural intensification. However, this trend has been accompanied by a fall in the ratio of output to total input energy, as may be seen from Figure 7.20. For one of the sugar beet systems in Heichel's (1973) study the ER ratio fell below one if the energy cost of processing the sugar was included. A ratio of one or more is viewed in some quarters as a moral imperative for agriculture, below which it is seen to be failing to capture a net energy intake from the sun.

Apart from its symbolic value, there is no real reason why field crops should exceed an ER ratio of one. The ratio for fresh peas in England is 0.9, for Brussels sprouts 0.2, for winter lettuce 0.002, and for various forms of fishing 0.006–0.06 (Leach, 1975). For tomato production it may vary from 0.64 in a Californian field to less than 0.01 in an English glasshouse (Stanhill, 1980). Why should we not expend net energy in producing foods we want as we do on other wants? Why should the farmer spend less energy in producing the wheat for a loaf of bread than we might spend in buying the loaf?

In fact, many field crops grown in environments to which they are adapted return 3–4-fold as much energy as is expended on them. Figure 7.21 relates the output and input energies per hectare of most of the crop examples analysed with a more or less uniform basis of estimation in Pimentel's (1980) handbook. The solid line to the left hand side fits the most energy-efficient C_3 crops of barley, wheat, oats, rice and soybean and of the C_4 crops maize and sorghum over a ten-fold range of harvested energy per hectare. Its slope represents an ER of 4.0, and there is no evidence of diminishing returns in the higher-yielding crops. With three exceptions these crops were non-irrigated. Most irrigated crops have substantially lower ER ratios, those of rice and

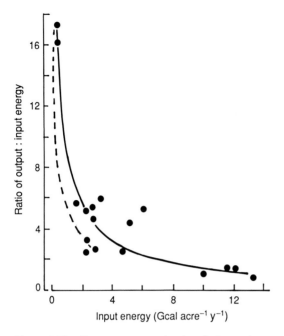

Figure 7.20. Energy return ratios in relation to input energy for 15 different agricultural systems analysed by Heichel (1973). The broken curve indicates the relation found by Flinn & Duff (1985) for 21 rice production systems with a 22-fold range in input energy (cf. Figure 7.23).

sorghum falling close to one and those of sugar beet even lower (Austin *et al.*, 1978).

Sugar crops are, in one sense, the ideal candidates for energy analysis because their primary output is a pure source of metabolic energy, although beet tops and pulp and cane bagasse are also useful outputs. Energy return ratios become less appropriate as other criteria such as protein content and quality become more significant. This is why ER ratios of substantially less than one are characteristic of glasshouse crops such as tomatoes and lettuce, as well as of intensive meat and milk production (e.g. Leach, 1975; Lewis & Tatchell, 1979) or fishing, as well as of national agricultural systems in which such activities are prominent. The overall ER of the agricultural systems of the United Kingdom, Israel, the Netherlands and the USA range from 0.5 to 0.7, whereas that of Australia is 2.8 (Gifford, 1976). This positive energy balance for Australia is a reflection of relatively low input cropping, free-range animal rearing, and the extensive use of legume leys rather than N fertilizer. Such a low-input system is no longer a realistic option for many countries.

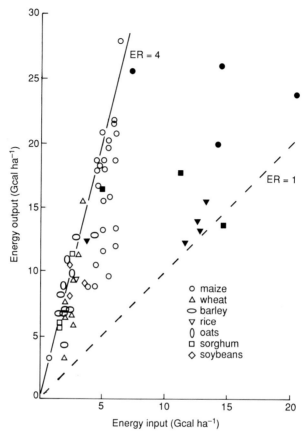

Figure 7.21. The relation between output and input energies for a wide range of farming systems reviewed by Pimentel (1980). Dryland crops, open symbols; irrigated crops, solid symbols.

Nor, increasingly, is subsistence agriculture in Third World countries. Often seen as frugal in its use of energy, subsistence farming is in fact not so when its total energy use is taken into account, as Makhijani (1975) has shown for villages in India, China, Tanzania, Nigeria, Bolivia and Mexico. Their main problem is of access to more energy at critical times. Makhijani adds, 'The idea that subsistence agriculture is harmless to the environment is another misapprehension', witness the long history of deforestation, over-grazing, erosion and flooding. Revelle (1976) estimated an energy return ratio of 0.37 for India as a whole before agricultural intensification became prominent.

Amish farmers in the USA offer another window on low-input farming because, for religious reasons, they have largely retained the labour intensive methods and lifestyles of their sixteenth-century forebears, although in some regions they now use stationary motors. Comparisons with modern farms nearby indicated somewhat higher ER ratios on Amish farms. They used less fossil fuel, and their crop yields were 38% lower in spite of their much higher labour input. Their frugal lifestyle was more important than their agricultural systems for energy conservation (Johnson *et al.*, 1977).

The ER ratio does not provide a satisfactory basis for the comparison of agricultural systems because it takes no account of the solar energy input and, in particular of the way in which other inputs may enhance the capture of solar energy by crops. Simply because it ignores the solar input, the ER ratio displays high values for low input systems and a progressive fall as inputs increase, as in Figure 7.20. Such figures are often misinterpreted as indicating a diminishing return on inputs in modern agriculture. Examination of the marginal rates of return on additional inputs is what is needed.

Yield response to higher energy inputs

There are many examples of the effects on yield when one factor is varied, and almost invariably they exhibit diminishing returns (e.g. Figure 1.9). What energy analysis allows us is – by expressing all the inputs on a common scale – to see to what extent their integration and interactions have kept diminishing returns at bay. Unfortunately, there are as yet too few comparative studies with uniform energy accounting either across years or across systems to sustain a strong conclusion.

The early energy analysis of US maize production by Pimentel *et al.* (1973), at five-yearly intervals over three decades, provides the best series of profiles through time of energy use in a rapidly intensifying but geographically stable and important crop production system. The original study examined the years 1945, 1950, 1954, 1959, 1964 and 1970 in analysing inputs, but related these to three yearly means of yield centred on those years. Comparable data for 1975 have since been presented by Pimentel & Pimentel (1979) and Pimentel & Terhune (1977*a*), and for 1980 and 1985 by Pimentel *et al.* (1984*a*, 1990). This seminal piece of work has been qualified and criticized for various shortcomings (e.g. by Loomis, 1984) and the whole analysis has been reworked by Smil *et al.* (1983). It constitutes the best case available.

One weakness of the original study was that it used a constant value of 77.5 MJ per kilogram N for the energy cost of nitrogenous fertilizer throughout the whole period whereas, as indicated in Figure 7.1, there was a substantial fall in the average energy cost per kilogram N in the USA between 1945 and 1974. There was also a significant shift in use towards higher-analysis fertilizers. The outcome of these advances was that the energy cost of N fertilizers applied to the US corn crop fell progressively from 99 MJ kg^{-1} in 1945 to 61 MJ kg^{-1} in 1974, as estimated by Smil *et al.* (1983). Since N fertilizers represented more than one third of the total energy inputs by 1970, this revision of the original analysis has a substantial impact on its outcome.

Many other components have also been modified in the revised analysis by Smil *et al.* (1983), such as family fuel use and some double counting of the electricity used in irrigation. The original estimates of energy equivalents for farm machinery have been almost halved. The shift towards reduced tillage may also have been under-estimated, as also the shift to diesel fuel. As Smil *et al.* (1983) point out, the use of liquid fuels in US maize production is unique in comparison with all other major non-renewable energy inputs in that the energy consumed in 1974 was no greater than that in 1945.

The outcome of these and other changes in the estimations is illustrated in Figure 7.22. For each level of output energy (i.e. yield), the analysis by Smil *et al.* (1983) suggests a substantially lower energy input and a higher energy return ratio than those estimated in 1973 by Pimentel *et al.*: 5.5 (cf. 3.7) for 1945, and 4.1 (cf. 2.4) for 1974. However, the important feature of this form of presentation is the absence of any evidence of diminishing returns. This is not what Pimentel *et al.* (1973) had concluded. They emphasized that the output per unit energy input had decreased from 3.7 in 1945 to 2.8 in 1970, but their conclusion that the energy efficiency of US corn production is declining was considered by Smil *et al.* (1983) to be premature and unsupportable. These latter authors conclude that 'the energy efficiency of modern, fossil fuel-subsidized grain corn production in the United States has basically remained the same during the past two decades'. When I first depicted the data of Pimentel *et al.* (1973) in a form comparable with Figure 7.22 (Evans, 1975c), it was suggested that the 1970 data indicated that diminishing returns were setting in, but the data of Pimentel *et al.* (1990) for 1980 and 1985 hint at rising rather than falling returns.

Comparable analyses of other major crop systems during a period of intensification are needed. The nearest approach is that of Udagawa (1976) for rice production in Japan, but it is not fully documented. Dazhong's (1988) energy accounting of the crop systems of China indicates a fairly linear relation between input and output energies up to the highest yield levels.

Rather than comparing one production system over a period of years, the uniform analysis of energy inputs into one crop grown at various stages of intensification may be revealing, although differences in environmental conditions and yield potential could confound the analysis. Figure 7.23 presents the results of such a comparison of 28 rice crops by Flinn & Duff (1985; IRRI, 1988b), ranging from upland rice in Latin America with a yield of only 1.2 t ha^{-1} to irrigated, intensively managed crops with yields of 8 t ha^{-1} in China using organic manures and in Egypt using fertilizers. The relation

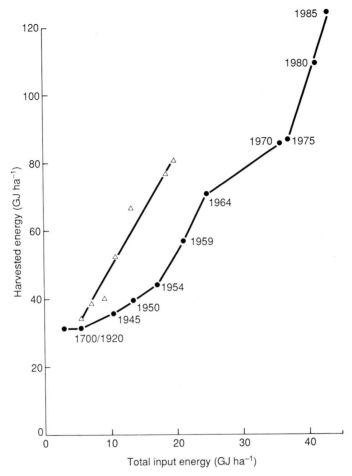

Figure 7.22. The relation between total input energy and harvested energy for the US maize crop at about 5-yearly intervals between 1945 and 1985, plus estimates for the years 1700 and 1920. Derived from data of Pimentel *et al.* (1990, ●), and of Smil *et al.* (1983, △) for the period 1945–75.

between the ER ratio and input energy for these crops is indicated in Figure 7.20. Despite the sharp fall in the ER ratio there is no indication of diminishing marginal returns to input energy in the intensively managed crops (Figure 7.23). The rate of return on input energy – i.e. the slope of the linear regression, about 3:1 – is similar to that for the US maize crop in Figure 7.22, but somewhat less than that for the wide range of crops in favourable environments illustrated in Figure 7.21.

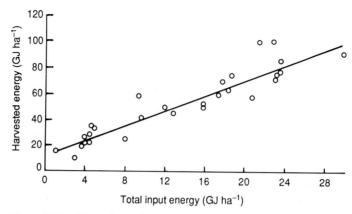

Figure 7.23.   The relation between total energy input and harvested energy for 28 rice production systems (adapted from IRRI, 1988*b*).

Thus, in spite of Slesser's (1984) conclusion that the intensification of European agriculture is associated with diminishing returns on energy input, these examples – the best we have – suggest that high input cropping need not be so.

Sustaining the marginal return to inputs

How has the avoidance of diminishing returns apparent in Figures 7.21, 7.22, and 7.23 been achieved? Many factors are at work, but three major ones are probably: (1) the continuing succession of innovations, both agronomic and genetic; (2) the progressive improvements in the efficiency of input manufacture, formulation and use; (3) the synergistic interactions between inputs.

The key role of a succession of innovations has been illustrated in Figure 1.10 for particle accelerators, and is apparent in the sequence of inputs into US maize production indicated in Figure 6.4 and highlighted by the energy analyses of Pimentel *et al.* (1973), Green (1978) and Smil *et al.* (1983). The major rise in tractor power was succeeded by those in N and then, K fertilizers, insecticides, herbicides, grain drying, and irrigation. Another successive adoption of innovations is illustrated for sugar beet in the UK (Figure 7.24).

Progressive reductions in the energy cost of several inputs have already been referred to, e.g. in the energy required for the manufacture of nitrogenous fertilizers (Figure 7.1) and by the introduction of more highly active compounds, as in the case of insecticides, fungicides and herbicides (Figure 7.16). The newer pesticides may require more energy per kilogram for

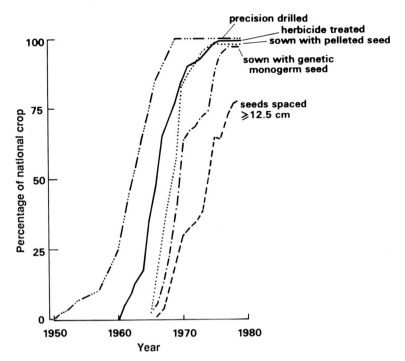

Figure 7.24. Successive adoption of agronomic innovations for sugar beet crops in the United Kingdom (adapted from Milford *et al.*, 1980).

their manufacture – among the herbicides, for example, 2,4-D, atrazine and glyphosate require 85, 190 and 454 GJ per tonne respectively (Green *et al.*, 1977) – but such increases are more than compensated by the reductions in dose rate.

As for the synergistic interactions between various inputs, and the opportunities opened up by these, many examples have been given in the preceding pages. They have often been unforeseen and serendipitous. A good example of such interactions is provided by the development of minimum tillage techniques. Quite apart from its other advantages, e.g. in improving timeliness, increasing flexibility, reducing erosion and expanding the arable area, minimum tillage reduces the energy input. Only about one sixth to one tenth as much energy is required for spraying as for mechanical tillage (e.g. Green & McCulloch, 1976; Frye *et al.*, 1982), depending on soil type and topography (Richey *et al.*, 1977). The spraying operation can be combined with others to reduce its cost still further. The extent of the overall energy saving depends on the energy required for herbicide manufacture. In the case examined by Green & McCulloch (1976), one mechanical weeding required

0.56 GJ ha^{-1}, the spraying operation only 0.056 GJ ha^{-1}, while the herbicides (except diuron) ranged from 0.10 to 0.21 GJ ha^{-1}. Minimum tillage techniques could therefore save about 1 GJ ha^{-1} of energy, although this may be reduced in situations where extra fertilizer is then required, as found by Frye *et al.* (1982). Other trends, besides the rising real cost of fuel, are likely to favour the adoption of reduced tillage practices, including the rising real cost of farm machinery and the decline in the rural labour force. Minimum tillage was applied to 33 million hectares of US cropland in 1980, and to more than a third of the winter cereal area in Britain (Cannell, 1981). Although its adoption has been less rapid than early enthusiasts predicted, minimum tillage is likely to have a major impact not only on the energetics of agriculture but also on many other aspects, such as soil loss, water infiltration and storage, fertilizer practices, and pest, disease and weed problems in the future (Phillips & Phillips, 1984; Sprague & Triplett, 1986).

Conclusions

Agronomic inputs have had a powerful impact on the enhancement of yield, which shows no sign of abating. Their undiminishing returns owe much to the continuing innovations of chemists, engineers and biologists of many kinds, and to interactions between agronomy and plant breeding.

The continuing response to input energy evident in Figures 7.21–7.23 is a remarkable achievement, even if diminishing returns of yield eventually set in. Over and over again the designers of particle accelerators (Figure 1.10) have thought the asymptote was at hand, and so have agronomists, but continuing innovation and the synergistic interactions between inputs have rescued the advance in yield. In this respect the evolution of yield has been like the evolution of evolution, with ever-increasing opportunities for synergisms as organisms become more complex (cf.. Vermeij, 1987).

Interactions may, of course, be negative as well as positive, so the art of agronomy lies in the effective integration of an ever-widening range of inputs and techniques. We have seen this trend at work in the integration of pest management. Agronomic management is also bound to become increasingly informed, integrated, and quantitative as a result of continuous advances in understanding, information flow and management techniques.

CHAPTER 8

The future of yield

*'They are ill discoverers that think
there is no land, when they can see
nothing but sea.'* Francis Bacon (1605)

Introduction

Scientific research cannot avoid inventing the future, but to predict the future is another matter. The only thing of which we can be certain about the future of agriculture is that there will be elements in it which we could not even envisage now, so why discuss the future of yield? As Gabor (1963) puts it: 'The future cannot be predicted, but futures can be invented'. It is the alternative futures, the choices and the opportunities, that we should consider. Also, by peering ahead we may foresee the costs and consequences of these choices a little more clearly, and improve our husbandry of the world's resources.

'Which future?' also depends on how far ahead we try to look. Here we shall peer beyond the many confident projections of present trends but not so far as a wholly transformed agriculture, say to about AD 2040 when the population of the world may be approximately double its present level. Even with that modest time frame, however, such prophecies tend to be coloured by the circumstances when they are made. In the mid 1970s, for example, both grain and oil were in short supply, grain prices were high and predictions of famine abounded. Ten years later grain stocks were high, prices low, oil ample and fears of a chronic glut widespread. Since then there has been another faltering in the growth of world food supplies, tempering the optimism of the early 1980s.

Windows on the future of agriculture

There have been several comprehensive projections of recent trends in agriculture over a short period, such as FAO's *Agriculture: Toward 2000*

365

(1981) and *The Global 2000 Report to the President of the US* (Barney, 1980/
1). The FAO projection is, as Hrabovszky (1984) put it, 'a normative,
demand-driven analysis', which deals more with what needs to be done than
with how it could be done. Its central concern with energy in and from
agriculture in the future reflects the aftermath of the oil crisis. By contrast, the
Global 2000 report is more concerned about the serious and accelerating loss
of resources essential for agriculture through erosion, salinization, desertification, extinction and pollution. Even so, it assumes a linear increase in
average crop yields, continuing past trends. The summary of its major
findings and conclusions says:

> 'If present trends continue, the world in 2000 will be more crowded, more
> polluted, less stable ecologically, and more vulnerable to disruption than the
> world we live in now. Serious stresses involving population, resources, and
> environment are clearly visible ahead. Despite greater material output, the
> world's people will be poorer in many ways than they are today. For
> hundreds of millions of the desperately poor, the outlook for food and other
> necessities of life will be no better. For many it will be worse.'

However, in another comprehensive projection called 'The Resourceful
Earth', Simon & Kahn (1984) reach very different conclusions:

> 'If present trends continue, the world in 2000 will be *less crowded* (though
> more populated), *less polluted*, *more stable ecologically* and *less vulnerable
> to resource-supply disruption* than the world we live in now. Stresses
> involving population, resources, and environment *will be less in the future
> than now* ... The world's people will be *richer* in most ways than they are
> today ... The outlook for food and other necessities of life will be *better* ...
> life for most people on earth will be *less precarious* economically than it is
> now.'

Simon & Kahn may take too little cognizance of the differences of
opportunity between the highly developed and the developing countries, but
it is clear that the conclusions from such projections are highly dependent on
the assumptions and innate optimism of those who make them.

The plausible art of simulation modelling on a global scale was introduced
by Forrester (1971) in his *World Dynamics*, in which he concluded that the
expansion of food production would be one of the last limits to growth to be
encountered. Forrester's methods, if not his conclusions, have been prominent ever since, particularly following the publication of *The Limits to
Growth* (Meadows *et al.*, 1972) and its successor *Mankind at the Turning
Point* (Mesarovic & Pestel, 1974). These early models were based on rather
gloomy assumptions about resources, erosion, exponentially increasing
pollution and the costs of expanding the arable area, so it was not surprising
that they predicted disaster. Gloom in, doom out. The spurious authority lent
to this outcome by such heroic simulations stimulated public discussion. It

also led in time to a wider appreciation of just how sensitive such conclusions are to the assumptions in simulation models, and of the inadequate basis for many of the assumptions that had to be made (e.g. Cole *et al.*, 1973; Encel *et al.*, 1975; Freeman & Jahoda, 1978).

Some of the comments by Marstrand & Pavitt (1973) on the agricultural sub-system of the World 3 model highlight these problems:

> 'The assumptions about the physical limits of the critical variables in the agricultural sub-system of World 3 are pessimistic. By making more optimistic but, on the basis of available information, equally plausible assumptions about them, any physical limits to agricultural production recede beyond the time horizon of the model. The major problems of feeding the less developed world are seen to lie in political rather than in physical limits.'
>
> 'The agricultural sub-system has a very important effect on the overall behaviour of World 3. If the resource depletion and the pollution modes of "collapse" are avoided, the combination of diminishing returns in agriculture and a growing population leads to the draining of all investible resources into agriculture and to yet another "collapse". The only conclusion that one can draw with any certainty is that in its present form the agricultural sub-system is an unsatisfactory tool for forecasting the future. . .'.

Although such global simulation models are being continually refined, they remain in thrall to many uncertain assumptions, such as that of diminishing returns to inputs (cf. Chapter 7).

For more comprehensive and detailed – albeit less quantitative – assessments of the shape of agriculture over the next 30–50 years, the views of experienced agriculturists such as Duckham (1966) and Yates (1968) are helpful. Some of the changes they envisaged have yet to be realized, large-scale desalination being one of these. This was confidently expected soon after World War II, and Yates reaffirmed that expectation 20 years later. In a rather different category are the many changes foreseen by Duckham and Yates to flow from the introduction of minimum tillage. These may yet be achieved within their time frame of 50 years, but adoption has been less extensive than they envisaged. However, other potential benefits of minimum tillage, such as the speed and flexibility of operations, may eventually give these techniques even greater impact than Duckham and Yates foresaw.

This example highlights a common feature of agricultural research, that its eventual benefits turn out to be quite different from those foreseen or intended. Another case in recent years has been the development of pesticides and herbicides which are effective at very low dose rates (Figure 7.16). These have not only reduced energy input and pollution but have made possible low-volume spraying, which has transformed the practicability of pest and weed control in many parts of the developing world.

One other cautionary example should also be mentioned. Yates (1968), quite rightly in the context of received wisdom in the 1960s, considered that the major problem in relation to world food needs would be the supply of protein. He therefore emphasized the possibilities of proteins from leaves, algae, yeast, bacteria and even chemical synthesis. However, our perspective on the world food problem was substantially changed when the human nutrition standards for protein *vis-à-vis* calories were radically revised (FAO–WHO, 1973), leading to a shift of emphasis from protein to calorie limitation in many situations, and from pulse crops and animal products back to cereals.

These are the hazards of predicting the future of agriculture for even the most judicious. That is why the visionaries like Francis Bacon, H.G. Wells, Aldous Huxley or Herman Kahn have often got the best of the exercise by looking further than they could see. However, we need elements of all these approaches if we are to glimpse the future shape of agriculture.

Overall perspective

Many features of today's agriculture will remain, such as the reliance on a few well-tried staple crops, and some of the trends apparent in recent years will develop further. Present inputs will be used more intensively, but also more effectively, and new kinds of inputs will be introduced, such as chemical regulators of growth and development.

Such inputs may supplement or even replace some of the functions of plant breeding, e.g. in relation to crop height, quality characteristics or response to certain stresses. Likewise, plant breeding will supplement agronomy in some of its traditional tasks, e.g. by engineering crop varieties with genetic resistance to specific herbicides or by breeding beneficial predatory or parasitic insects resistant to pesticides for use in integrated pest management (Pickett, 1988).

Consequently, although the use of chemical inputs will be enhanced and diversified, it will also become more efficient, specific and sparing, lessening the adverse effects on the environment. Having passed through a chemical phase agriculture will, with the help of genetic engineering techniques, become more biological. Varietal characteristics are likely to assume greater significance, and the seed industry will extend the industrial appropriation of agriculture, and of agricultural research, that began with mechanization, reliance on fertilizers, and the use of hybrids (Goodman *et al.*, 1987).

Associated with input intensification, the scale of operations in and around agriculture is bound to increase. Farms, fields and equipment will continue to get bigger in spite of our sentimental attachment to the concept of small independent farmers. At every stage of agricultural development, even at its dawn 10 000 years ago, there has probably been nostalgia for the preceding

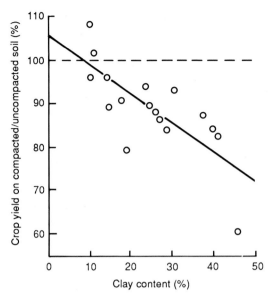

Figure 8.1. Reductions in crop yields associated with vehicle compaction of the soil during the previous season (adapted from Hakansson, 1980).

stage. Many agricultural innovations are inherently scale-neutral, but the socio-economic forces at work tend to favour the large farmer no less than the large agrichemical corporation.

With increase in scale will come the need for more expert and informed farm management. Farm work will become progressively less manual, more cerebral. With many interacting inputs to be combined with variable weather and market demands, farmers will require greater education and skills. To achieve high crop yields, every decision along the way must be right (Tinker & Widdowson, 1982). Programmed and automated equipment, controlled and monitored from the farm's information and computer centre, is likely to be used increasingly. As farm equipment has got larger and heavier, and crop spraying more frequent, soil compaction and the yield loss it causes have become a more serious problem, especially on clay soils (Figure 8.1). One solution to the problem, especially in flatter areas, may be the use of mobile gantries with wheel bases 10–20 m wide and travelling on a fixed pattern of reinforced tracks, useable for all operations and in all weathers. Alternatively, deep tillage may be used to restore the compacted tracks (Spoor *et al.*, 1987).

Longer term weather and market forecasting, the use of remote sensing and simulation models to determine irrigation and pest control schedules, and the need for more current awareness of agricultural research, will challenge even the best trained and most able of farm managers. In addition, there are likely to be greater demands for their cooperation beyond the farm, at the catchment or regional level, in order to control pests and diseases, water use, ground water quality, pollution, erosion or farm sales more effectively.

Environmental changes that may be associated with rising CO_2 levels could force further shifts in the pattern of farming over the next 50 years. So too will the almost inevitable revisitation of the liquid fuels crisis, which may cause energy production to become a more significant component of agriculture until alternative sources of energy are developed.

The staple crops

The relatively few staple crop plants – those that loom large in this book – are likely to remain our main sources of food, feed and fibre with a higher probability than was envisaged 20 years ago. Wholly new food crops and the revival of old ones such as *Coix*, *Amaranthus* or *Chenopodium* are unlikely to displace the current staples. Indeed, our dependence on wheat, rice and maize will probably increase. They are so productive and adaptable, and their further development is backed by such a breadth of knowledge and of genetic resources, that they are likely to remain dominant. The search for new agrichemicals is heavily focused on the major staples.

They are likely to spread to new environments, wheat to more tropical, rice to more temperate, and maize to cooler or drier areas. Their uses may also change, as illustrated by the extent to which maize is now competing with sugar cane and beet as a source of sweeteners and with other crops for other products (Pomeranz, 1987). Cassava is becoming a significant feed crop. Tobacco, the vignette at the head of this chapter, may yet be used as a source of crystalline rubisco, a highly nutritious protein, in a form like table sugar, as Wildman (1983) hopes. Given the ease with which tobacco can be genetically engineered to produce a great variety of new compounds in significant amounts – e.g. up to 1.3% of its total leaf protein as a functional antibody (Hiatt *et al.*, 1989) – it could, as a ratoon crop, become the preferred vehicle for the production of a great variety of pharmaceutical, dietary and industrial compounds.

Other staple crops are also likely to be selected or genetically engineered as sources of specific products other than those for which we now harvest them, such as special carbohydrate molecules or intensely sweet proteins like monellin and thaumatin (Kinghorn & Soejarto, 1986; Ogata *et al.*, 1987). There are inherent compromises in producing both starch and protein from one crop, as we have seen, and selection for production of the one or the other, or for oil (cf. Figure 3.14) may become more common, especially if the staple cereals are also used as energy crops. Similarly, rather than attempt to adapt wheat to the humid, lowland tropics, it may prove more feasible to insert the genes for the polypeptides that confer breadmaking quality into maize or rice, should the price relativities for these staples change sufficiently.

The future of the most productive oilseed crops also seems assured, but that of the pulses is less clear. Even in countries like India where they are particularly esteemed and valuable components of the diet, pulses are being displaced by cereals whose yield is more responsive to intensive agriculture. In developed countries the greatest demand for them is for animal feed. Twenty years ago it seemed that the world would not be able to grow enough pulse crops (Yates, 1968), but this now seems an unlikely bottleneck in the future. Of the root crops, sweet potato and cassava may become more significant. The potato is likely to remain a staple crop, and if it could be selected for adaptation to the lowland humid tropics it would, as with wheat, become even more important in the future.

So significant to the future of mankind are these few staple crops that they deserve at least as much research attention as each of the major human diseases (cf. Mangelsdorf, 1966). That they have not received it is possibly because we have taken their productivity for granted. They may also have suffered from what Medawar (1967) refers to as 'an intellectual class distinction in science' which led, in effect, to two biologies, of *Neurospora*, *Drosophila* and *Arabidopsis* on the one hand and of late blight, brown planthopper and wheat on the other. Fortunately, the distinction between them is now weakening and the future will probably see more profound research attention given to the staple crops.

New crops

New crops have great appeal, like new continents or frontiers in the past, but evangelists for the new often overlook the virtues of the old. Hinman (1986) claims that new crops are needed because the genetic pool of existing crops is too narrow. With more than 80 000 accessions of varieties, land races and wild progenitors of rice and large collections of the genetic resources of other staple crops in long term storage (Plucknett *et al.*, 1987), such an argument is nonsense. Another of Hinman's arguments is that our present staples are environmentally demanding or too narrowly adapted. This too is unconvincing because one of the outstanding features of staple crops, and a key to their success, is their adaptability (cf. Chapters 3 and 4). Existing crops are also claimed to 'need' high inputs, but we have seen that these are unavoidable if we want sustained high yields (Chapters 5 and 7).

The hazards of excessive monoculture are sometimes given as a reason why new crops are needed, but we already have a sufficient variety of staple crops to deal with that problem. Our reliance on non-renewable resources is given as another reason for research on new crops, but again our existing staple crops are likely to out-perform new crops in any attack on that problem. The variety of their climatic requirements and responses should also ensure sufficient adaptation to changing climatic conditions. Better nutrition is also used as a justification for 'new' crops, but without adequate analysis of their likely acceptance and contribution to dietary deficiencies (Tripp, 1990).

One valid case for new crops is that they could be useful sources of new products, industrial compounds, pharmaceutical precursors, flavours, sweeteners, lubricants, polymers, cosmetics, etc. (e.g. Seigler, 1977; Ritchie, 1979). However, it is quite likely that genetic engineering techniques will allow many of these compounds to be produced more effectively by the staple crops.

Perhaps the most likely niche for new crops or, rather, for rehabilitated old ones such as many of the food plants of the Incas (Vietmeyer, 1989), will be in meeting the demands for a greater variety of food as living standards rise. In the case of wild rice, this demand has led to greater efforts to domesticate it (Hayes et al., 1989). The further improvement of many tropical fruits might well create much-needed export opportunities for developing countries.

Supposedly promising new crops may turn out to be no better than their long-domesticated relatives, as in the case of the *Echinochloa* millets (Evans & Bush, 1985). The amount of time and research needed to turn promise into reality should not be under-estimated, as Stewart & Lucas (1986) show for guayule. Whatever the hypothetical attractions of a genetically engineered *Pseudocerealis* (O'Type et al., 1984), they are unlikely to surpass those of a long-domesticated staple cereal.

Crops for energy

In the 1960s, when Duckham (1966) and Yates (1968) were looking 50 years ahead, they foresaw the use of fossil fuels as a supplement to agriculture for the production of single cell protein, synthetic fibres and dietary components. Now the perspective is reversed, and agriculture is being looked to not only as a source of feedstocks and industrial compounds but also as a supplementary source of liquid fuels and energy. The implied compliment to agriculture is welcome, but the expectation that 'energy crops' and greater utilization of crop residues could match even present levels of petroleum consumption is quite unrealistic. Gifford (1984) has shown that this would require a supply of biomass in excess of the total terrestrial primary productivity.

Pimentel et al. (1984b) point out that it would require 8 times more arable land to fuel one car than to feed one person in the USA. They estimate that biomass resources could provide up to 11% of US energy needs by AD 2000, but at the cost of greater soil erosion, nutrient loss, air pollution and land use conflicts. However, the more modest aim of replacing a significant proportion of the liquid fuels and energy needed in agriculture by products from crop residues is feasible (Gifford, 1984). Given the greater political clout of the urban areas in any competition for liquid fuels, it could also be desirable for agriculture to become more self-reliant in this respect.

Many different approaches to the conversion of crop and animal residues into more convenient energy sources are being tried (e.g. Slesser & Lewis, 1979). The extent of their use in developed countries is still rather limited, and seems likely to remain so compared with that of other sources such as wind power, solar panels and liquid fuels from coal. In developing countries, on the other hand, biomass items such as wood, cereal straw and animal dung still constitute a major source of energy, and are already extensively used for fuel, at some cost to soil fertility and crop yields. Their more efficient use, e.g. via biogas or ethanol production, will be important in the years ahead.

Although crop and animal residues may be used more completely and efficiently in the agriculture of the next 50 years, the growing of crops specifically for fuel is unlikely to be prominent except in particular cases like sugar cane in Brazil or when grain or sugar prices fall to low levels. Vegetable oils can be extracted on-farm from crops such as sunflower, peanut and rapeseed for blending or direct use as liquid fuels at reasonable cost (Stewart *et al.*, 1981). However, the demand for these oilseed crops for food, cooking and industrial uses is growing so rapidly that their extensive use for liquid fuels seems unlikely. The direct production of hydrocarbon resins by new crops such as *Euphorbia lathyris* for use as fuels, as proposed by Calvin (1979), is also likely to be rather limited in scale, given their greater value for industrial uses and the low proportion of resin in the crop (Stewart *et al.*, 1982).

Many staple crops are also being evaluated as potential energy sources, the most notable being sugar cane, cereals such as maize and sorghum, root crops such as cassava, oilseeds, coconut and palm oil, and even the pineapple (Marzola & Bartholomew, 1979). Future oil prices will influence the extent to which crops are grown for fuel, but so too will the extent of further increases in the yield of the staple crops. Only if these can get ahead of population growth will it be possible to release significant amounts of arable land from the demands of food production.

Environmental change

Today's press is replete with predictions of the adverse effects of environmental change, especially of rising CO_2 levels, whereas the possible beneficial effects on agriculture are rarely mentioned.

The accelerating rise in atmospheric CO_2 content since the industrial

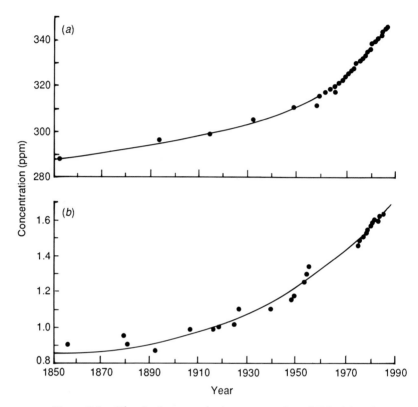

Figure 8.2.   The rise in atmospheric concentration of CO_2 (*a*) and methane (*b*) since 1850. (From Houghton & Woodwell (1989)). Copyright © 1989 by Scientific American, Inc. All rights reserved.)

revolution is indicated in Figure 8.2, along with the rise in atmospheric methane content. Most of the additional CO_2 has come from the combustion of fossil fuels and to a lesser extent (currently about one fifth as much) from changes in land use, especially the clearing of forests (Houghton & Woodwell, 1989). The atmospheric CO_2 content bears a close relation to the world population (Figure 8.3) and is likely to continue rising. It could reach twice the pre-industrial level of about 280 ppm before the middle of the twenty-first century.

In assessments of the effects of rising atmospheric CO_2 levels on crop photosynthesis and yield, we need to remember that the longer-term effects may be substantially less than the immediate response to additional CO_2. Besides the feed-back inhibition of photosynthesis that may occur over hours or days, there may also be slower adaptive changes in other plant

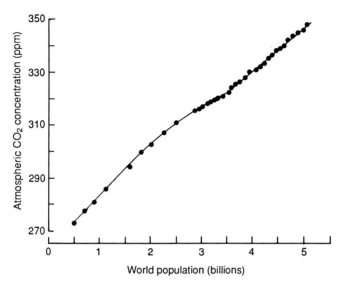

Figure 8.3. The relation between world population and atmospheric CO_2 concentration (Idso, 1989).

characteristics, such as reduced stomatal density and leaf nitrogen content. Woodward (1987) and Penuelas & Matamala (1990) have found a 40% reduction in stomatal density and a 20% reduction in leaf N content in herbarium specimens of leaves of a variety of plants collected over the past 200 years or so. Since similar changes are found when plants develop under high CO_2 levels, they cannot be taken as evidence of natural or unconscious selection for adaptation to additional CO_2 but this may, nevertheless, be occurring.

Direct effects of the rise in CO_2 on yield

Past increases in CO_2 level could account for about 20% of the rise in UK wheat yields (Monteith, 1977). Although this is a relatively small proportion of the total increase in the yield of crops like wheat in England, rice in Japan (Figure 1.1) or maize in the USA (Figure 6.4) since 1900, all crops in all environments should have benefited from the rise in CO_2, and for some of them it may have been the major source of increase in yield so far.

The higher the CO_2 level rises over the next 50 years, at least, the faster the rates of photosynthesis and growth are likely to be. However, down-regulation of photosynthesis already occurs in field crops (e.g. Sage & Sharkey, 1987), and may become even more pronounced with the rise in CO_2 level.

Reductions in stomatal density and leaf N contents, increased partitioning to reserves (e.g. Rowland-Bamford *et al.*, 1990) or a rise in respiration rate (cf. Amthor, 1991) can also mute the photosynthetic and growth responses to additional CO_2. However, enhancing responses have also been found, such as a fall in crop respiration rates (Gifford *et al.*, 1985; Reuveni & Gale, 1985; Bunce & Caulfield, 1991). Musgrave *et al.* (1986*b*) have suggested that only those genotypes lacking the cyanide-resistant respiratory overflow pathway could respond to a rise in CO_2 (but cf. Obenland *et al.*, 1988).

The net result of these and other effects can be that the relative growth rate may be unaffected by CO_2 level (Bunce, 1990*b*), implying that there is scope for the selection of cultivars more responsive to higher levels of atmospheric CO_2. Indeed, such selection may well be occurring empirically already.

Greater photosynthesis and yield at higher CO_2 levels have been observed with many crop plants (Kimball, 1983; Cure & Acock, 1986; Enoch & Kimball, 1986). From his extensive survey, Kimball (1983) concluded that crop yields would increase about 33% overall with a doubling of atmospheric CO_2 content. Cure & Acock (1986) suggested that doubling of the CO_2 content would cause an initial increase in photosynthetic rate of 52% across ten major crops, a 29% increase in their acclimated rate, and a 41% increase in their yield. Adams *et al.* (1990) anticipate that doubling of the CO_2 level will increase the photosynthesis of soybean, wheat and maize crops by 35, 25 and 10%, respectively. Cotton appears to be particularly responsive, a doubling of the CO_2 level doubling growth – cf. only a 20% increase for maize (Wong, 1979) – and seed yield (Radin *et al.*, 1987). The rise in CO_2 level *per se* is likely to benefit C_3 crops more than those with the C_4 photosynthetic pathway, but C_4 crops may adapt more readily to the accompanying increases in temperature or aridity, making the net effect hard to predict.

Synergistic interactions with agronomic inputs may also enhance the beneficial effects of additional CO_2. High nutrient levels increase the yield response to elevated CO_2 (Sionit *et al.*, 1981; Sionit, 1983; Goudriaan & de Ruiter, 1983) and high CO_2 increases the response to N (Wong, 1979; Hocking & Meyer, 1991) and P (Norby & O'Neill, 1991). Crops should also be able to take greater advantage of high irradiance when the atmospheric CO_2 level is higher, particularly after selection for these conditions.

Higher CO_2 levels may enable crops to perform better at high temperatures (Idso *et al.*, 1987*b*; Gifford, 1992), as a consequence of the more favourable balance between carboxylation and oxygenation by rubisco, and between photosynthesis and respiration. This will be important not only because higher temperatures are likely to be associated with higher atmospheric CO_2 levels through the 'greenhouse effect' but also because partial stomatal closure at high CO_2 will further raise leaf temperatures (Idso *et al.*, 1987*a*). Such stomatal closure will increase water use efficiency (e.g. Morison &

Gifford, 1983; Morison, 1985; Jones *et al.*, 1985; King & Greer, 1986; Baker *et al.*, 1990*b*), but also the danger of injury by high temperatures.

Although the absolute increase in yield following a rise in CO_2 level may be greater in well-watered crops than in water-stressed ones, the reverse may be true for the relative increase in growth and yield, as in the case of wheat (Gifford, 1979; Sionit *et al.*, 1980). The earlier flowering of crops at higher CO_2 levels (Marc & Gifford, 1984) should also be advantageous in areas subject to drought stress, and perhaps also under irrigation when fewer leaves are formed, as in rice (Baker *et al.*, 1990*a*). Thus, the rise in atmospheric CO_2 content is likely to have substantially beneficial *direct* effects on crop yields in both favoured and adverse environments (Gifford, 1992). These beneficial effects should be enhanced by varietal selection and agronomic intensification but may be diminished by the *indirect* effects of the rise in CO_2 and other components of the atmosphere, such as methane.

Indirect effects of the rise in CO_2

Despite the many predictions of environmental disaster from the so-called 'greenhouse effect', considerable uncertainty still characterizes the outputs from even the most heroic general circulation models of the relations between atmospheric CO_2 and climate (Crane, 1985; Idso, 1989). Uncertainties about the magnitudes of both the sources and sinks for atmospheric CO_2 remain, as well as of the many and complex processes involved, such as the effects of rising temperature via atmospheric water vapour content on cloud formation and the trapping of heat on the one hand and on the reflection from cloud cover on the other (cf. Mitchell *et al.*, 1989). Whether the higher CO_2 level compensates for the rise in temperature, in terms of crop growth and yield, will depend to a considerable degree on the associated changes in irradiance.

Analyses of prehistoric changes in temperature and CO_2 content, using ice cores, suggest that the two parameters have been linked in the past, but without indicating which was cause and which was effect. Time series analyses for the last century also reveal that temperature and atmospheric CO_2 have been significantly correlated over the past 30 years, but that the changes in CO_2 content seem to follow rather than precede those in temperature (Kuo *et al.*, 1990).

Adams *et al.* (1990) have combined two general circulation models (GCMs) with three crop simulation models and an economic model to assess the effects of CO_2-induced climate change on US agriculture. Their results highlight the substantial differences between the two GCMs in relation to crop yields. With one GCM the beneficial effects of the rise in CO_2 outweighed the adverse effects of changes in temperature and rainfall for

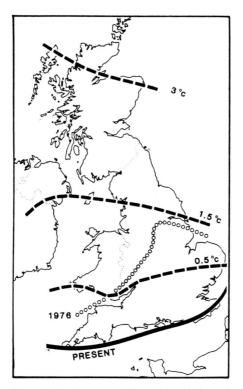

Figure 8.4.   Present and hypothetical future limits to the ripening of maize grain crops in the UK with various degrees of warming. The limit for 1976, a particularly warm summer, is also shown (Parry *et al.*, 1990).

dryland crops of wheat, maize and soybean to result in increased yields, whereas yields decreased with the other GCM. Both models predicted reduced supplies but greater demands for irrigation water. In areas with low yields, little change should be expected.

Looking 50 or so years ahead, it seems likely that:

1. The doubling of the atmospheric CO_2 content will have led to a rise in the global average temperature of 2 °C or more, with larger increases at higher latitudes. This will have caused a progressive shift in cropping patterns, with a likely extension of cropping at high latitudes. An indication of the possible northwards shift in maize growing in the UK with higher temperatures is illustrated in Figure 8.4.

2. Monsoon rains may penetrate to higher latitudes, resulting in greater rainfall in drought-prone areas such as the Sahel and Northern India, but not necessarily in longer growing seasons (Parry, 1990).

3. The middle and high latitudes may experience reduced soil moisture

levels during summer (Crane, 1985), which could have a major impact on grain production in the current granary areas. The general circulation models are still too crude to predict regional changes in greater detail, but these are likely to have complex implications, both biological and political.

4. Similarly, no confident predictions can yet be made about the changes in year-to-year variability, which could be more important for the world food supply than the changes in average climatic conditions (Reifsnyder, 1989). The occurrence of injurious high temperatures is bound to increase (Mearns *et al.*, 1984).

5. The extent to which mean sea level will rise is also uncertain, and likewise therefore the extent to which arable land will be lost. Much will depend on the rate at which the sea level rises and coastlines adjust.

6. Increases in several other gases besides CO_2 will contribute to the greenhouse effect, such as water vapour, methane and nitrous oxide, which are all influenced by agriculture as well as influencing it. Methane, for example, is rising much faster, relatively, than CO_2 (Figure 8.2*b*) and could become the major greenhouse gas within 50 years. Cattle, especially lot-fed animals, and their farmyard manure piles and rice paddies are significant sources of methane, and management techniques to reduce their contributions to the atmospheric rise may become important in the future, as also for nitrous oxide, which comes largely from agricultural activity.

Whether the indirect effects of the rise in CO_2 level will outweigh, counterbalance or enhance the direct effects on crop productivity is quite uncertain at this stage (cf. Idso, 1989; Reifsnyder, 1989). The simulation modelling of rice yields in Asia by Jansen (1990) suggests that the indirect effects of changes in temperature and radiation after AD 2020 could outweigh the beneficial effects of the rise in atmospheric CO_2. The rise in CO_2 will create opportunities for synergistic interactions in crop production, but its net effect on yields will remain Delphic until our understanding of the many processes involved is greatly improved.

Inputs

In most temperate developed countries there is little scope for increase in either arable area or the number of crops per year, so greater yields will be the key to greater production, as also in those developing countries where only limited increases in crop area are possible. About 90% of the potentially arable land in Asia is already cultivated, as we saw in Chapter 2, compared with only 15–20% in Latin America and Africa. Admittedly these estimates depend on arbitrary definitions of what is potentially arable, and we can be sure that in the future, at least in Asia, land thought not to be potentially arable now will become so when the accumulation of needs and resources makes it worthwhile. Large-scale desalination of sea-water, for example, and abundant power for pumping water, could transform the agricultural opportunities in many arid areas.

The drive for higher yields will come from population pressure, rising expectations and standards of living, rising labour costs and land values and greater capitalization. The loss of arable land to cities, roads, airports, conservation and recreation, as well as from erosion, waterlogging, salinization and toxification will have to be balanced by higher yields, as will any requirement for arable land for crops to supply fuels, feedstocks or pharmaceuticals. Higher yields will be an ecological as well as an economic necessity.

Further increases in yield will require further intensification and diversification of input use. Many developing countries see a shift to modern, high-input agriculture as the preferred solution to their food supply problems, whereas in developed countries there are many who urge a return to a lower-input, less polluting, more conserving and renewable form of agriculture.

The charms of traditional agriculture are too readily romanticized. All too often it is characterized by poverty, uncertainty, hard work, monotony, poor health, lack of education and lack of opportunity. Also, in one very important sense subsistence agriculture is not truly self-sufficient because its high requirement for labour at certain times is associated with large families, which can be accommodated in the following generations only by a progressive increase in the area of arable land. It is this lack of equilibrium which has caused so much deforestation in the past and so much erosion, desertification and unstable agriculture. Makhijani (1975) comments that 'The idea that subsistence agriculture is harmless to the environment is a ... misapprehension', and Leach (1975) that 'most pre-industrialists have not achieved the environmental bliss and harmony that some claim for them'. Only by raising yield levels beyond those of subsistence agriculture, with the help of inputs from beyond the farm gate, can the need for ever more arable land be avoided, and the present level of world population provided for (Buringh & van Heemst, 1977).

However, although reversion to self-sufficient agriculture is no longer possible, concern for the conservation of resources on the one hand and of environmental quality on the other, as well as the increasing cost-price squeeze on farmers, has led to widespread interest in less intensive farming systems. These go by many names, such as low-input, natural, biological, sustainable, organic, alternative or ecological agriculture.

Emphases vary but, in general, these alternative agricultures seek to:

> reduce the use of those off-farm inputs most harmful to the environment or health;
> integrate and take advantage of naturally occurring beneficial interactions;

better adapt agricultural systems to the total environment of the
region;
optimize the use of biological and physical resources (National
Research Council, 1989; Pimentel *et al.*, 1989).

These are all valid objectives, but modern agriculture is already moving in
the same directions, to a more sparing, timely and effective use of inputs; to a
more integrated approach as in pest management; to closer local adaptation;
and towards the optimum use of the resources available through more
informed and expert management.

The differences between the prevailing and alternative agricultures are of
degree rather than kind. Comparisons between them depend heavily on value
systems, management ability and on whether farmers, society as a whole, or
future generations are the focus (Lockeretz, 1989). The National Research
Council Report (1989) emphasizes that many federal policies in the USA
discourage the adoption of alternative approaches by penalizing various
conservation practices and encouraging 'unrealistically high' yield goals,
inefficient fertilizer and pesticide consumption, and unsustainable use of land
and water. Its strictures probably apply also to Europe, Japan and other areas
of high-input agriculture, and they need to be addressed. Nevertheless, the
intensification and diversification of inputs will no doubt continue in the
future, generating new problems along the way.

Irrigation

Given its power to enhance the response to other inputs and the
stability of production, irrigation is bound to increase in extent despite its
high cost, energy use and impact on environmental quality, human health and
national debt. If the desalination of sea water becomes feasible, far more land
could be irrigated. Yates (1968) foresaw an additional 1.5 billion hectares of
irrigated land and Marstrand & Rush (1978) a total of 1.9 billion hectares,
about eight times more than is currently irrigated. If this were achieved it
would increase the proportion of arable land which is irrigated from about
16% now to over 50%, which would have a huge impact on average yields.

However, until that radical advance is made, the potentially irrigable area
is about 0.47 billion hectares (Moen & Beek, 1974; Buringh *et al.*, 1975), only
twice the area currently irrigated. Of this potentially irrigable land, 60% is
accounted for by only three countries, China, the former USSR and India, but
these estimates may have to be substantially modified in the light of the
changes in rainfall distribution induced by rising CO_2 and methane levels. In
the USA, for example, it is predicted that the water supplies available for
irrigation will be substantially reduced (Adams *et al.*, 1990).

Because of the growing competition from urban and industrial water use, there will be an escalating concern for water use efficiency in agriculture. Irrigation techniques that distribute and apply water more uniformly and reduce evaporative and other losses will be used more, increasing the overall efficiency of water use from its present level of 20–40% to more than 50% (Moen & Beek, 1974). Where the competition for water is severe, the use of supplementary irrigation may increase relative to other forms, because water is used most efficiently for crop production in climates and seasons where the demand is least (Stanhill, 1985). Crop growth simulation models will be used to improve irrigation management, which will be more tightly integrated across catchments or irrigation districts, with the help of better weather forecasting, remote sensing and data management. More cognizance will be taken of the stages when crops are particularly susceptible to water stress.

Irrigation water will be used increasingly for the application of fertilizers (fertigation), herbicides and pesticides (chemigation). More attention will, preferably from the outset, be paid to the adequacy of drainage from irrigation schemes, and to their long-term maintenance. Syr Darya, in Uzbekistan, and the Nile Valley are still being irrigated after more than 5000 years, so irrigation schemes are not inherently unsustainable.

Soil fertility and nutrient supply

Fertilizers and soil amendments will be applied in greater amount and variety, in forms with more controlled release. Placement will be improved and the timing of applications will be based, like those of irrigation water and pesticides, on crop growth and nutrient cycle models. Losses from volatilization and leaching should be reduced by such management models, and also by the greater use of nitrification and urease inhibitors, thereby ameliorating the adverse environmental consequences of heavier fertilizer applications.

Whereas there is likely to be greater emphasis on the conservation and recycling of nutrients such as P, the use of N fertilizers is unlikely to be limited by the supply side, assuming ample energy for their manufacture. Groundwater contamination by nitrates and the magnitude of nitrous oxide losses to the atmosphere will be more likely constraints on their use. If these are not mastered, there may be greater emphasis on biological N fixation and legume crops, especially if they can be genetically engineered not to repress their N-fixation when fertilizer N is applied to them. However, fertilizer N seems bound to exceed the total biological fixation of N on land within the next 50 years (Figure 7.2).

Greater reliance on biological N fixation figures prominently in most alternative agricultures. So too does reliance on farmyard manure, which currently provides 10–15% of global fertilizer requirements. Several estimates of the amount of N in animal manures in the USA suggest it is equivalent to about three quarters of the current use of N fertilizers there (about 10^7 tonnes N), but half of the total N in the manure is lost by NH_3 volatilization (Bouldin *et al.*, 1984). Since higher rates of N fertilizer application are likely in the future, even the complete use of animal manures would not replace more than about a quarter of the fertilizer. Currently, only about 0.2% of all food marketed in the USA is 'organic' (Oelhaf, 1978) and although that proportion may rise, it seems unlikely to become a major fraction.

The long-term Broadbalk experiment at Rothamsted indicates that the continued use of farmyard manure has more or less doubled the amount of organic N in the soil, but soil N and crop yields have not declined when only inorganic fertilizer has been applied (Jenkinson, 1982). Cereal yields with chemical fertilizers match those with farmyard manure when equal quantities of N, P, K and Mg are provided (Benzian & Lane, 1982). With rice crops, both in Japan and in the tropics, yields with chemical fertilizers are at least as high as those with long-term applications of farm yard manure (Kumazawa, 1984; De Datta *et al.*, 1988). In Stanhill's (1990) survey of many crops, 'organic' yields were, on average, 91% as high as those with fertilizers.

However, the relative outcome depends on the crop, on growing conditions and on yield levels. Wheat apparently responds to organic farming less well than do soybeans and maize. Fertilizers display a greater advantage in good years and, for several crops, yields may be higher with farmyard manure in low-yield conditions but higher with fertilizers under favourable conditions, as illustrated in Figure 8.5 (cf. Lockeretz *et al.*, 1981; Kumazawa, 1984; Stanhill, 1990).

Higher yield levels in the future are therefore likely to favour continuing reliance on fertilizers. The operative issues may not be those of yield, however, but of balancing differences in crop quality, pest and weed control, soil fertility maintenance, salinization, waste disposal, environmental quality, energy use, labour and equipment requirements and overall costs (Oelhaf, 1978; Lockeretz *et al.*, 1981; Bouldin *et al.*, 1984; Olson, 1987). The estimates by Pimentel & Terhune (1977*b*), for example, suggest that the energy cost of merely hauling and spreading the manure from the feed lot is about half of the energy cost of using N fertilizers in the USA. Depending on the crop, either conventional or organic farming may be the more energy-efficient (Pimentel *et al.*, 1983, 1984*a*). Both have their environmental costs.

Weed control

The widespread use of herbicides has reduced yield losses due to weeds, a trend which Pimentel (1976) expects to develop further. Future herbicides will be effective at still lower dose rates but may or may not be more

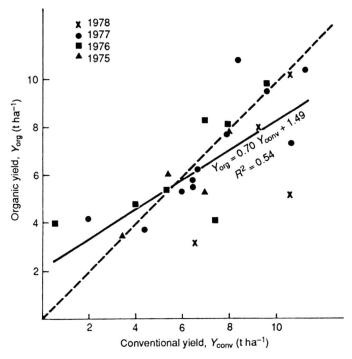

Figure 8.5.   Maize yields on 26 matched pairs of organically and conventionally managed fields. The broken line represents equal yields on the two kinds of fields, the solid line the best fit (Lockeretz *et al.*, 1981). (Copyright © 1981 by the AAAS.)

specific in their action. In cases where a particular weed is predominant over a large area, the introduction of a specific herbicide may be commercially viable, but the already high cost of developing new herbicides will become even higher, favouring compounds with a broader spectrum of action. A herbicide eliminating all plants with the C_4 pathway of photosynthesis from C_3 pathway crops, or vice versa, would have a far larger market than one specifically eliminating a particular weed. Moreover, the advantages of specificity may be conferred on a broad-spectrum herbicide by selecting or engineering genetic resistance to it in the crop plant. This has been done by Chaleff & Ray (1984) with tobacco for resistance to sulfonylurea herbicides, by Comai *et al.* (1985) for resistance to glyphosate, by Bayley *et al.* (1992) with cotton for resistance to 2,4-D in order to protect the crop from wind drift injury, and by others (Mazur & Falco, 1989).

With a herbicide as broadly effective and potentially as cheap as glyphosate, this is an attractive option for the future. For new herbicides under patent cover the strategy would be effective only if resistance to them could be found quickly and incorporated into widely useable varieties. However, weeds are only one of many constraints faced by a crop. Adaptability, quality and resistance to pests, diseases and climatic stresses may be far more important in the farmer's choice of variety than its resistance to a particular herbicide.

There is, however, one situation in which a combination of new herbicide plus variety resistant to it may be demanded, namely when close relatives of a staple crop are among its worst weeds. This is a growing problem with crops such as rice (Oka, 1988; Smith, 1988; cf. Figure 7.18), oats, barley, sorghum (Wet & Harlan, 1975), proso millet (Cavers & Bough, 1985), rapeseed, sugar cane, sunflower (Ellstrand, 1988) and sugar beet (Hornsey & Arnold, 1979; Longden, 1980) and could trouble others in the future. The problem has probably been accentuated by the effective control of other weeds by herbicides. Even highly specific compounds are unlikely to differentiate close relatives, and the combination of broad-spectrum herbicides with genetically engineered resistant varieties may be the most practicable solution. However, although herbicide resistance has been relatively slow to develop in weeds (Gressel, 1978; Le Baron & Gressel, 1982), possibly because of the severe penalty on seed production at high population densities (Reboud & Till-Bottraud, 1991), the transfer of genetic resistance from a crop to its out-breeding wild relatives is likely to be more rapid than transfers in the other direction, and could soon neutralize this strategy (Colwell *et al.*, 1985; Ellstrand, 1988; Tiedje *et al.*, 1989).

Nevertheless, the control of weeds will no longer be entirely agronomic, and the future will see increasingly powerful and discriminating interactions between genetic engineering, herbicide use and biological control systems.

Plant growth regulators

Just as there will be a growing genetic contribution to the previously agronomic function of weed control, so will plant growth regulators offer agronomic control of many crop characteristics previously managed by the plant breeders. The use of plant growth regulators in crop production has been expected to grow rapidly, matching or even exceeding that of herbicides (Hardy, 1978). Except in horticulture, where a variety of regulators is already in widespread use, dwarfing compounds such as chlormequat have so far been the only ones used widely in agriculture (Batch, 1981; Nickell, 1982) but that is likely to change over the next 50 years.

Regulators of crop growth and development will be diversified to control germination, growth habit, tillering, flowering time, fruit and seed set, partitioning, senescence, maturation and quality (Nickell, 1982). Most of these features have, to date, been modified mainly by plant breeding. Compounds breaking the dormancy of seeds in the soil or regulating vegetative regeneration will be used to control the behaviour of the whole potential weed population in the soil, not merely that fraction which germinates or regenerates in any one year (Chancellor, 1981). Similarly, adjuvants such as calcium peroxide will be used to ensure more uniform germination and stand structure of the crop (Batch, 1981).

Another likely class of new inputs will help crops to survive or recover from periods of environmental stress. Here, too, reliance has been mainly on plant breeding, which has achieved a great deal with the more predictable seasonal stresses, for which selection is feasible. But with unseasonal, irregular stresses, improved weather forecasting will make the use of regulators to alleviate stress more practicable. Frost injury at critical stages may be reduced (Batch, 1981), and chilling injury ameliorated by the application of abscisic acid analogues (Flores *et al.*, 1988), possibly even after the chilling period. With further understanding of heat shock and injury, it is likely that these too could be alleviated, perhaps by genetic engineering. If so, it could open the way to the more effective use of anti-transpirants, and they in turn to the control of pests, diseases and parasites such as *Striga* on sorghum (Press *et al.*, 1989). Plant growth regulators may also prove of value in the control of some pests and diseases (e.g. Hedin, 1990).

Other plant growth regulators may be used to modify the plant breeding process itself, as chemical hybridizing agents such as azetidine-3-carboxylic acid and o-methyl threonine are beginning to do for the production of hybrid seed (McRae, 1985). Indeed, this new class of regulators may eventually have a great impact on the yield of crop plants such as wheat by making the large-scale use of hybrids feasible, as well as the use of genes for extreme dwarfness.

Control of pests and diseases

Pest and disease management seems likely to become more biological, taking more account of the need: (i) to preserve or enhance the competitors, predators and parasites of pest organisms; (ii) to tolerate sub-threshold pest populations; (iii) to maximize the impact of pesticides; (iv) to conserve genes for resistance; and (v) to minimize environmental side effects. The integration of these aims will be aided by computerized management, better weather forecasting and more coordinated control across farms.

Resistance genes and/or new pesticides have been the cornerstones of previous control measures. However, the number of effective resistance genes still available for use is often quite small. More will be found, of course, and others may be derived from the genome of the parasites themselves (Sanford & Johnston, 1985). Some have already been transferred from other organisms by the techniques of genetic engineering, such as those for the *Bacillus thuringiensis* toxin (Vaeck *et al.*, 1987; Barton *et al.*, 1987), the coat protein of tobacco mosaic virus (Abel *et al.*, 1986), and satellite RNA of other viruses (Harrison *et al.*, 1987; Gerlach *et al.*, 1987) into tobacco. Greater use of horizontal resistance (Simmonds, 1991), combinations of several different kinds of resistance, or the use of multiline cultivars (Wolfe, 1985) may provide more stable, long term solutions to pest and disease problems and reduce dependence on pesticides.

However, continuing evolution of pest and disease organisms is likely to overwhelm such genetically engineered solutions in the long term (Gould, 1988). New pests and pathogens may fill the vacant niches. The cultivation of individual varieties over greater areas in future will increase the chances of resistances breaking down, so that the varietal relay race may be run even faster.

Integrated pest and disease management systems are already in widespread use on some crops, especially horticultural ones but also on cotton, peanuts and rice, of which more than 40% of the US acreage was under integrated pest management in 1986 (National Research Council, 1989). For high-value processed food crops and for several of the major field crops such as maize, wheat and soybeans, reliance on integrated pest management is low and may remain so for a variety of legislative and market reasons. Moreover, pesticide use is not always greatly reduced by integrated pest management programmes (Wheatley, 1987).

Pesticides and fungicides are, therefore, likely to remain a crucial input, albeit used more sparingly, more specifically and with more skilful timing. Variants of old pesticides, such as the pyrethroids and other botanicals, of newer ones such as the hormone mimics, pheromones and semiochemicals (Pickett, 1988), and whole new classes of insecticides and synergists are bound to be introduced and will, in the process, open up new opportunities for both plant breeders and farm managers.

Management of resources

The agriculture of the future will, in a sense, be more horticultural, i.e. more intensive and with a greater degree of control over the environment. The more this happens, the more it will pay to introduce additional inputs and controls and to manage these more tightly with the help of simulation modelling.

Simulation models will be valuable adjuncts to research and management at all levels, global, regional, farm, crop or process (Rabbinge *et al.*, 1990). At the farm level they will be used increasingly to schedule irrigation, fertilizer applications and pest/disease control measures; to predict yields; to schedule harvests and to coordinate activities.

Climate is unlikely to be manipulated in the next 50 years, but adverse weather will be predicted more surely and its effects ameliorated to a greater extent. Pollution will be reduced by better management and by turning wastes into resources whenever possible. New problems, both environmental and social, are bound to arise as new inputs and practices are introduced, but these should be seen as opportunities for solution rather than for complaint (Wharton, 1969).

Whether the management of agricultural resources will become more oriented to their long term conservation is less clear than might be hoped. The genetic resources of crops have been secured by international action. National and regional governments have protected some vulnerable lands either by alienating them from agriculture or by legislating constraints on their use, and the scale and diversity of these are bound to increase in the future. Yet desertification of agricultural land continues on far too great a scale. Soil erosion afflicts much good crop land, and other forms of deterioration also occur with intensive cropping (Wolman & Fournier, 1987; Pierce, 1990). Better measures to counter the loss or degradation of soil must surely become more prominent in the future.

Rural landscapes will change in ways we may bemoan, of which there is no more telling example than the loss of hedgerows, one of the glories of the English landscape in Victorian times and a haven for wildlife and predators of crop pests. However, the new landscape patterns also have their attractions, not least the large expanses of exuberant crops. Hancock (1972) writes: 'Man has too often spoiled the land on which he lives; but quite often he has fostered and fortified its life-sustaining powers. Sometimes he has been successively the spoiler, the restorer, the improver. The terraced hillsides of Tuscany . . . are the living embodiment of this rhythm'.

Such perceptions could encourage the hope that agriculture will become more concerned with the cultivation of landscapes as well as of crops in the future, especially when it gets ahead of population growth. However, Davidson & Lloyd (1977) argue that: 'Farming can no longer be considered the natural custodian of rural areas, as it was two decades ago . . . its practices are becoming less and less influenced by the constraints of natural systems and more and more capable of inducing major and permanent change in them'. Moore (1977) concludes: 'The only effective compromise will be provided by a mixture of intensive agriculture and intensive conservation'. The pursuit of productivity rather than of beauty will still dominate and change the rural landscape.

Research

The evolution of agriculture has depended successively on man-power, horse power, mechanical power and chemical power. It is entering the era of computer power and a future dominated by the power of scientific research (Lu *et al.*, 1979). What was once a wholly resource-based industry becomes more and more a science-based one. Just as other major inputs into agriculture, such as machinery, fertilizers, new varieties and hybrids, have been increasingly appropriated by the industrial sector (Goodman *et al.*, 1987) so also in all likelihood will agricultural research.

There are sharply differing views on how research can best serve agriculture. Should it be run like a national gallery, buying only what we want to see, or like a national park, allowing it to run wild? The gallery approach is currently in the ascendancy, with much talk of relevance, priorities and imperatives; of harnessing and directing; and of customer–contractor principles. This tidy front garden of science is so often a clichéd, trite, sterile creature of fashion, we are lucky that the chaotic back garden of science survives, even in neglect, to produce the understanding from which springs more radical change (Evans, 1982).

Fashion has a powerful influence, more than it should, not only on how research is done but also on what is done. Currently two paradigms are dominant, at opposite ends of the spectrum of biological organization. Molecular biology has already demonstrated its potential in relation to crop yields, by transferring into tobacco new kinds of resistance to virus diseases and herbicides, new controls on the breeding system, and new functions (Uchimiya *et al.*, 1989), including the ability to produce functional antibodies (Hiatt *et al.*, 1989). However, a greater challenge remains, of improving the efficiency of complex processes like photosynthesis and respiration, which have been honed by prolonged and intense natural selection.

At the other end of the spectrum of biological organization, simulation modelling offers the power to integrate our understanding of complex systems, like crops, and to attempt more holistic approaches to their management. Barker (1985) sees this expansion of 'our capacity to manage increasingly complex systems, to develop conceptual models of such systems and to fit into them the myriad basic discoveries that derive from our total research effort' as crucial to the future of agriculture.

So too, however, will be 'the myriad basic discoveries' from research within the paradigms lost and, I hope, ultimately regained between these two dominant approaches. As I have emphasized throughout this book, it is the interactions and unforeseen synergisms not only between breeding and agronomy, but between all the disciplines, that have kept Malthus at bay.

Agricultural research has, until recently, been largely in the public domain and internationally accessible. Developing countries have therefore been able to benefit from it in spite of their often limited national research capacities. By the year AD 2000, four fifths of the world's population will live in developing countries, and that proportion will rise still higher. The future of their crop yields is crucial to the future of mankind. Most developing countries are in the tropics and sub-tropics where agricultural problems and opportunities, as well as crops, can be quite different from those of the developed temperate-zone countries. Thus, raising crop yields in them may not be simply a matter of transferring the agronomic techniques already in use in the developed world.

Research for agro-ecosystems better adapted to more tropical conditions has been boosted in recent years by the establishment of a system of international agricultural research centres. Through their plant breeding and other networks they have made research on crops a truly collaborative international enterprise. While focused on the problems of developing countries, these international centres have involved research workers from developed countries in their efforts and, in doing so, are forging an increasingly global approach to agricultural research. Although intensification of cropping will follow somewhat different pathways in different environments – e.g. with more emphasis on multiple, relay and intercropping in the tropics than in temperate regions – the major innovations will have wide applicability. These innovations will come, as they have in the past, from all corners of the world and of research, and their synergistic interactions will ensure that yields continue to rise long into the future.

At the dawn of the scientific era, Francis Bacon foresaw in 'New Atlantis' that the only commodity of external trade would be light, the light of understanding. In agriculture it is unlikely to be the only one but it will, increasingly, be the crucial one.

'And deeper than did ever plummet sound
I'll drown my book' (Solemn music).
                    Shakespeare (1612)

References

Abel, P.P., Nelson, R.S., De, B., Hoffmann, N., Rogers, S.G., Fraley, R.T. & Beachy, R.N. (1986). Delay of disease development in transgenic plants that express the tobacco mosaic virus coat protein gene. *Science* **232**, 738–43.

Acevedo, E., Craufurd, P.Q., Austin, R.B. & Perez-Marco, P. (1991). Traits associated with high yield in barley in low rainfall environments. *J. Agric. Sci. (Camb.)* **116**, 23–36.

Ackerson, R.C., Havelka, V.D. & Boyle, M.G. (1984). CO_2 enrichment effects on soybean physiology. II. Effects of stage specific CO_2 exposure. *Crop Sci.* **24**, 1150–54.

Adams, M.W. (1967). Basis of yield component compensation in crop plants with special reference to the field bean, *Phaseolus vulgaris. Crop Sci.* **7**, 505–10.

Adams, M.W. (1975). Plant architecture and yield in the grain legumes. In *Report of the TAC Working Group on the Biology of Yield in Grain Legumes.* DDDR:IAR/75/2. FAO, Rome.

Adams, R.M. (1966). *The Evolution of Urban Society.* Aldine, Chicago.

Adams, R.M., Rosenzweig, C., Peart, R.M., Ritchie, J.T., McCarl, B.A., Glyer, J.D., Curry, R.B., Jones, J.W., Boote, K.J. & Allen, L.H. (1990). Global climate change and US agriculture. *Nature (Lond.)* **345**, 219–24.

Aggarwal, P.K. & Penning de Vries, F.W.T. (1989). Potential and water-limited wheat yields in South East Asia. *Agric. Systems* **30**, 49–69.

Aggarwal, P.K. & Sinha, S.K. (1987). Performance of wheat and Triticale varieties in a variable soil water environment. IV Yield components and their association with grain yield. *Field Crops Res.* **17**, 45–53.

Aggarwal, P.K., Liboon, S.P. & Morris, R.A. (1987). A review of wheat research at the International Rice Research Institute. *IRRI Res. Pap.* No. 124, pp. 1–12.

Aguilar, M.J. & Fischer, R.A. (1975). Analisis de crecimiento y rendimiento de 30 genotipos de trigo bajo condiciones ambientales optimas de cultivo. *Agrociencia* (Mexico) **21**, 185–98.

Aiba, S., Humphrey, A.E. & Millis, N.F. (1973). *Biochemical Engineering.* 2nd edn. University of Tokyo Press, Tokyo.

Aitken, Y. (1977). Evaluation of maturity genotype – climate interactions in maize

(*Zea mays* L.). *Zeitschr. f. Pflanzenzücht.* **78**, 216–37.

Akazawa, T. (1982). Cultural change in prehistoric Japan: receptivity to rice agriculture in the Japanese archipelago. *Adv. World Archaeol.* **1**, 151–210.

Akita, S. (1980). Studies on the differences in photosynthesis and photorespiration among crops. I. The differential responses of photosynthesis and dry matter production to oxygen concentration among species. *Bull. Natl. Inst. Agric. Sci., Ser D.* No. 31, pp. 1–58.

Akita, S. (1989). Improving yield potential in tropical rice. In *Progress in Irrigated Rice Research*, pp. 41–73. IRRI, Los Baños.

Akita, S., Mochizuki, N., Yamada, M. & Tanaka, I. (1986). Variations in heterosis in leaf photosynthetic activity of maize (*Zea mays* L.) with growth stages. *Japan J. Crop Sci.* **55**, 404–7.

Akita, S., Blanco, L. & Katayama, K. (1990). Physiological mechanism of heterosis in seedling growth of *indica* F_1 rice hybrids. *Jap. J. Crop Sci.* **59**, 548–56.

Alberda, T. (1971). Potential production of grassland. In *Potential Crop Production*. (ed. P.F. Wareing & J.P. Cooper), pp. 159–171. Heinemann, London.

Ali, M.A.M., Okiror, S.D. & Rasmusson, D.C. (1978). Performance of semi-dwarf barley. *Crop Sci.* **18**, 418–22.

Allan, R.E. (1983). Harvest indexes of backcross-derived wheat lines differing in culm height. *Crop Sci.* **23**, 1029–32.

Allan, R.E. (1986). Agronomic comparisons among wheat lines nearly isogenic for three reduced-height genes. *Crop Sci.* **26**, 707–10.

Allard, R.W. (1988). Genetic changes associated with the evolution of adaptedness in cultivated plants and their wild progenitors. *J. Hered.* **79**, 225–38.

Allard, R.W. (1990). Future directions in plant population genetics, evolution and breeding. In *Plant Population Genetics, Breeding and Genetic Resources*, (ed. A.H.D. Brown, M.T. Clegg, A.L. Kahler & B.S. Weir), pp. 1–19. Sinauer Associates, Sunderland, Massachusetts.

Allen, T.F.H. (1977). Neolithic urban primacy: the case against the invention of agriculture. *J. Theoret. Biol.* **66**, 169–80.

Ambroggi, R.P. (1980). Water. *Sci. Amer.* **243**(3), 100–17.

Amir, H.A. & Khalifa, F.M. (1991). Performance and yield of sunflower (*Helianthus annuus*) cultivars under rainfed and irrigated conditions in Sudan. *J. Agric. Sci. (Camb.)* **116**, 245–51.

Ammerman, A. & Cavalli-Sforza, L.L. (1971). Measuring the rate of spread of early farming in Europe. *Man* **6**, 674–88.

Amthor, J.S. (1984). The role of maintenance respiration in plant growth. *Plant, Cell & Environ.* **7**, 561–9.

Amthor, J.S. (1989). *Respiration and Crop Productivity*. Springer-Verlag, Berlin.

Amthor, J.S. (1991). Respiration in a future, higher-CO_2 world. *Plant, Cell & Environ.* **14**, 13–20.

An, Z. (1989). Prehistoric agriculture in China. In *Foraging and Farming: The Evolution of Plant Exploitation* (ed. D.R. Harris & G.C. Hillman), pp. 643–9. Unwin Hyman, London.

Anderson, E. (1960). The evolution of domestication. In *Evolution after Darwin*, vol. 2 (ed. Sol Tax), pp. 67–84. University of Chicago Press.

Anderson, J.R., Hazell, P.B.R. & Evans, L.T. (1987). Variability of cereal yields: sources of change and implications for agricultural research and policy. *Food Policy* **12**, 199–212.

Anderson, R.S., Brass, P.R., Levy, E. & Morrison, B.M. (eds) (1982). *Science, Politics, and the Agricultural Revolution in Asia.* Westview Press, Boulder, Colorado.

Anderson, W.K. (1985). Differences in response of winter cereal varieties to applied nitrogen in the field. I. Some factors affecting the variability of responses between sites and seasons. *Field Crops Res.* 11, 353–67.

Anderson, W.K. (1986). Some relationships between plant population, yield components and grain yield of wheat in a Mediterranean environment. *Aust. J. Agric. Res.* 37, 219–33.

Angus, J.F. & Lewin, L.G. (1991). Forecasting Australian rice yields. In *Climatic Variation and Change: Implications for Agriculture in the Pacific Rim* (ed. Shu Geng & C.W. Cady), pp. 1–8. University of California, Davis.

Angus, J.F., Jones, R. & Wilson, J.H. (1972). A comparison of barley cultivars with different leaf inclinations. *Aust. J. Agric. Res.* 23, 945–57.

Anon (1966a). *Farm Journal* 90, 42.

Anon. (1966b). Reiser's second staggering yield: a new record. *Soybean Digest* 26(4), 10–11.

Apel, P. & Lehmann, C.O. (1969). Variabilität und Sortenspezifität der Photosyntheserate bei Sommergerste. *Photosynthetica* 3, 255–62.

Apel, P., Tschape, M., Schalldach, I. & Aurich, O. (1973). Die Bedeutung der Karyopsen für die Photosynthese und Trockensubstanzproduktion bei Weizen. *Photosynthetica* 7, 132–9.

Apple, J.L. (1977). The theory of disease management. In *Plant Disease. An Advanced Treatise* (ed. J.G. Horsfall), pp. 79–101. Academic Press, New York.

Archbold, H.K. (1945). Some factors concerned in the process of starch storage in the barley grain. *Nature (Lond.)* 156, 70–3.

Arjunan, A., Natarajaratnam, N., Nagarajan, M., Sadasiram, R. & Balakrishnan, K. (1990). Photosynthesis and productivity in rice cultivars. *Photosynthetica* 24, 273–5.

Armiger, W.H., Foy, C.D., Fleming, A.L. & Caldwell, B.E. (1968). Differential tolerance of soybean varieties to an acid soil high in exchangeable aluminum. *Agron. J.* 60, 67–70.

Arora, V.K. & Prihar, S.S. (1983). Regression models of dryland wheat yields from water supplies in Ustifluvent in Punjab, India. *Field Crops Res.* 6, 41–50.

Asada, K., Konishi, S., Kawashima, Y. & Kasai, Z. (1960). Translocation of photosynthetic products assimilated by the top leaf to the ear of rice and wheat plants. *Mem. Res. Inst. Food Sci. Kyoto Univ.* No. 22, pp. 1–11.

Asay, K.H., Nelson, C.J. & Horst, G.L. (1974). Genetic variability for net photosynthesis in tall fescue. *Crop Sci.* 14, 571–4.

Ashby, E. (1932). Studies in the inheritance of physiological characters. II. Further experiments upon the basis of hybrid vigour and upon the inheritance of efficiency index and respiration rate in maize. *Ann. Bot.* 46, 1007–32.

Athwal, D.S. (1971). Semidwarf rice and wheat in global food needs. *Quart. Rev. Biol.* 46, 1–34.

Atlin, G.N. & Frey, K.J. (1989). Predicting the relative effectiveness of direct versus indirect selection for oat yield in three types of stress environment. *Euphytica* 44, 137–42.

Atlin, G.N. & Frey, K.J. (1990). Selecting oat lines for yield in low-productivity environments. *Crop Sci.* 30, 556–61.

Atsmon, D., Bush, M.G. & Evans L.T. (1986). Effects of environmental conditions on expression of the 'Gigas' characters in wheat. *Aust. J. Plant Physiol.* 13, 365–79.

Aufhammer, G. & Fishbeck, G. (1964). Ergebnisse von Gefäss- und Foldversuchen mit dem Nachsbau Keinfähiger Gersten- und HaferKörner aus dem Grundstein des 1932 errichteten Nürnberger Stadttheaters. *Zeitschr f. Pflanzenzücht.* **51** 354–73.

Aufhammer, W. & Zinsmaier, P. (1982). Das Speichervermögen von Weizenähren in Abhängigkeit von Beziehung zwischen den einzelnen Kornanlagen. *Zeitschr. für Acker-und Pflanzenbau* **151**, 249–66.

Auld, B.A., Menz, K.M. & Tisdell, C.A. (1987). *Weed Control Economics.* Academic Press, New York.

Austin, R.B. (1972). The relationship between dry matter increment and sugar concentrations in beetroot leaves. *Photosynthetica* **6**, 123–32.

Austin, R.B. (1978). Actual and potential yields of wheat and barley in the United Kingdom. *ADAS Quart. Rev.* **29**, 76–87.

Austin, R.B. (1982). Crop characteristics and the potential yield of wheat. *J. Agric. Sci. (Camb.)* **98**, 447–53.

Austin, R.B. (1989). Genetic variation in photosynthesis. *J. Agric. Sci. (Camb.)* **112**, 287–94.

Austin, R.B. (1990). Is improved agronomy needed for realising the benefits from improved varieties? In *Proc. Intl. Congr. Plant Physiol., New Delhi, India.* (ed. S.K. Sinha, P.V. Sane, S.C. Bhargava & P.K. Agrawal), vol 1, pp. 93–100. Indian Agricultural Research Institute, New Delhi.

Austin, R.B. & Edrich, J. (1975). Effects of ear removal on photosynthesis, carbohydrate accumulation, and on the distribution of assimilated ^{14}C in wheat. *Ann. Bot.* **39**, 141–52.

Austin, R.B., Ford, M.A., Edrich, J.A. & Hooper, B.E. (1976). Some effects of leaf posture on photosynthesis and yield in wheat. *Ann. Appl. Biol.* **83**, 425–46.

Austin, R.B., Ford, M.A., Edrich, J.A. & Blackwell, R.D. (1977). The nitrogen economy of winter wheat. *J. Agric. Sci. (Camb.)* **88**, 159–67.

Austin, R.B., Kingston, G., Longden, P.C. & Donovan, P.A. (1978). Gross energy yields and the support energy requirements for the production of sugar from beet and cane; a study of four production areas. *J. Agric. Sci. (Camb.)* **91**, 667–75.

Austin, R.B. Bingham, J., Blackwell, R.D., Evans, L.T., Ford, M.A., Morgan, C.L. & Taylor M. (1980*a*). Genetic improvements in winter wheat yields since 1900 and associated physiological changes. *J. Agric. Sci. (Camb.)* **94**, 675–89.

Austin, R.B., Morgan, C.L., Ford, M.A. & Blackwell, R.D. (1980*b*). Contributions to grain yield from pre-anthesis assimilation in tall and dwarf barley phenotypes in two contrasting seasons. *Ann. Bot.* **45**, 309–19.

Austin, R.B., Morgan, C.L., Ford, M.A. & Bhagwat, S.G. (1982). Flag leaf photosynthesis of *Triticum aestivum* and related diploid and tetraploid species. *Ann. Bot.* **49**, 177–89.

Austin, R.B., Morgan, C.L. & Ford, M.A. (1986). Dry matter yields and photosynthetic rates of diploid and hexaploid *Triticum* species. *Ann. Bot.* **57**, 847–57.

Austin, R.B., Ford, M.A. & Morgan, C.L. (1989). Genetic improvement in the yield of winter wheat: a further evaluation. *J. Agric. Sci. (Camb.)* **112**, 295–301.

Austin, R.B., Craufurd, P.Q., Hall, M.A., Acevedo, E., Silveira Pinheiro, B. & Ngugi, E.C.K. (1990). Carbon isotope discrimination as a means of evaluating drought resistance in barley, rice and cowpeas. *Bull. Soc. Bot. France* **137** (*Act. Bot.* (1)), 21–30.

Avratovscukova, N. (1968). Differences in photosynthetic rate of leaf discs in five tobacco varieties. *Photosynthetica* **2**, 149–60.

Azcon-Bieto, J. (1983). Inhibition of photosynthesis by carbohydrates in wheat leaves. *Plant Physiol.* **73**, 681–6.

Azcon-Bieto, J. (1986). Effect of oxygen on the contribution of respiration to the CO_2 compensation point in wheat and bean leaves. *Plant Physiol.* **81**, 379–82.

Babu, R.C., Srinivasan, P.S., Natarajaratnam, N. & Rangasamy, S.R.S. (1985). Relationship between leaf photosynthetic rate and yield in blackgram [*Vigna mungo* (L.) Hepper] genotypes. *Photosynthetica* **19**, 159–63.

Bacon, F. (1605). *The Advancement of Learning.*

Baenziger, P.S., Wesenberg, D.M., Sicher, R.C. (1983). The effect of genes controlling barley leaf and sheath waxes on agronomic performance in irrigated and dryland environments. *Crop Sci.* **23**, 116–20.

Bagga, A.K. & Rawson, H.M. (1977). Contrasting responses of morphologically similar wheat cultivars to temperatures appropriate to warm temperate climates with hot summers: a study in controlled environment. *Aust. J. Plant Physiol.* **4**, 877–87.

Bagnall, D.J. & King, R.W. (1991). Response of peanut (*Arachis hypogaea*) to temperature, photoperiod and irradiance. 2. Effect on peg and pod development. *Field Crops Res.* **26**, 279–93.

Bagnall, D.J., King, R.W. & Farquhar, G.D. (1988). Temperature-dependent feedback inhibition of photosynthesis in peanut. *Planta* **175**, 348–54.

Bailey, G. (1983). Hunter-gatherer behaviour in pre-history: problems and perspectives. In *Hunter-gatherer Economy in Prehistory. A European Perspective* (ed. G. Bailey), pp. 1–6. Cambridge University Press.

Baker, J.T., Allen, L.H., Boote, K.J., Jones, P. & Jones, J.W. (1990*a*). Developmental responses of rice to photoperiod and carbon dioxide concentration. *Agric. For. Meteorol.* **50**, 201–10.

Baker, J.T., Allen, L.H., Boote, K.J., Jones, P. & Jones, J.W. (1990*b*). Rice photosynthesis and evapotranspiration in subambient, ambient and superambient carbon dioxide concentrations. *Agron. J.* **82**, 834–40.

Baldwin, J.E. & Krebs, H. (1981). The evolution of metabolic cycles. *Nature (Lond.)* **291**, 381–2.

Bamakhramah, H.S., Halloran, G.M. & Wilson, J.H. (1984). Components of yield in diploid, tetraploid and hexaploid wheats (*Triticum* spp.). *Ann. Bot.* **54**, 51–60.

Barah, B.C., Binswanger, H.P., Rana, B.S. & Rao, N.G.P. (1981). The use of risk aversion in plant breeding: concept and application. *Euphytica* **30**, 451–8.

Barber, S.A. (1964). Water essential to nutrient uptake. *Plant Fd. Rev.* **10**, 5–7.

Bardner, R. & Fletcher, K.E. (1974). Insect infestations and their effects on the growth and yield of field crops: a review. *Bull. Ent. Res.* **64**, 141–60.

Bariola, L.A., Heneberry, T.J. & Kittock, D.L. (1981). Chemical termination and irrigation cut off to reduce over-wintering populations of pink boll worms. *J. Econ. Entomol.* **74**, 106–9.

Barker, R. (1985). The changing world of research opportunities. In *Crop Productivity – Research Imperatives Revisited.* (ed. M. Gibbs & C. Carlson), pp. 103–10.

Barker, R. & Cordova, V.G. (1978). Labor utilization in rice production. In *Economic Consequences of the New Rice Technology*, pp. 113–36. IRRI, Los Baños.

Barker, R., De Datta, S.K., Gomez, K.A. & Herdt, R.W. (1977*a*). Philippines 1974, 1975, 1976. In *Constraints to High Yields on Asian Rice Farms: An Interim Report*, pp. 121–56. IRRI, Los Baños.

Barker, R., Apraksirikul, S. & Antiporta, D. (1977*b*). Source of output growth in Asian food grains. IRRI Agric. Econ. Dept. Paper No. 77–2.

Barker, R., Gabler, E.C. & Winkelmann, D. (1981). Long term consequences of technological change on crop yield stability. In *Food Security for Developing*

Countries (ed. A. Valdes), pp. 53–78. Westview, Boulder, Colorado.

Barlow, C.A., Randolph, P.A. & Randolph, J.C. (1977). Effects of pea aphids, *Acyrthosiphon pisum* (Homoptera, Aphididae) on growth and productivity of pea plants, *Pisum sativum. Canad. Entomol.* **109**, 1491–502.

Barneix, A.J., Cooper, H.D., Stulen, I. & Lambers, H. (1988). Metabolism and translocation of nitrogen in two *Lolium perenne* populations with contrasting rates of mature leaf respiration and yield. *Physiol. Plantar.* **72**, 631–6.

Barnett, K.H. & Pearce, R.B. (1983). Source-sink alteration and its effect on physiological parameters in maize. *Crop Sci.* **23**, 294–9.

Barney, G.O. (1980–1). *The Global 2000 Report to the President of the US: Entering the 21st Century.* A report prepared by the Council on Environmental Quality and the Department of State. Pergamon, New York. (3 vols.)

Barrau, J. (1965*a*). L'Humide et le Sec. *J. Polynesian Soc.* **74**, 329–46.

Barrau, J. (1965*b*). Histoire et préhistoire horticoles de l'Océanie tropicale. *J. Soc. des Océanistes* **21**, 55–78.

Barrau, J. (1979). Coping with exotic plants in folk taxonomies. In *Classifications in their Social Context.* (ed. R.F. Ellen & D. Reason), pp. 139–44. Academic Press, New York.

Barton, C.M., Gomis, F.R., Mikcicek, C.A. & Donahue, D.J. (1990). Domestic olive. *Nature (Lond.)* **346**, 518–19.

Barton, K.A., Whiteley, H.R. & Yang, N.-S. (1987). *Bacillus thuringiensis* δ-endotoxin expressed in transgenic *Nicotiana tabacum* provides resistance to Lepidopteran insects. *Plant Physiol.* **85**, 1103–9.

Batch, J.J. (1981). Recent developments in growth regulators for cereal crops. *Outlook on Agric.* **10**, 371–8.

Bayley, C., Trolinder, N., Ray, C., Morgan, M., Quisenberry, J.E. & Ow, D.W. (1992) Engineering 2,4-D resistance into cotton. *Theor. Appl. Genet.* **83**, 645–9.

Beadle, G.W. (1977). The origin of *Zea mays.* In *Origins of Agriculture* (ed. C.A. Reed), pp. 615–35. Mouton, The Hague.

Beaven, E.S. (1920). Breeding cereals for increased production. *J. Farmers Club., Whitehall Court, London,* part 6, pp. 107–31.

Beevers, H. (1970). Respiration in plants and its regulation. In *Prediction and Measurement of Photosynthetic Productivity,* pp. 209–14. Pudoc, Wageningen.

Begg, J.E. & Burton, G.W. (1971). Comparative study of five genotypes of pearl millet under a range of photoperiods and temperatures. *Crop Sci.* **11**, 803–5.

Begg, J.E. & Torssell, B.W.R. (1974). Diaphotonastic and parahelionastic leaf movements in *Stylosanthes humilis* H.B.K. (Townsville stylo). In *Mechanisms of Regulation of Plant Growth* (ed. R.L. Bieleski, A.R. Ferguson & M.M. Cresswell), pp. 277–83. Bulletin 12, Roy. Soc. N.Z, Wellington.

Begonia, G.B., Aldrich, R.J. & Nelson, C.J. (1988). Effects of simulated weed shade on soybean photosynthesis, biomass partitioning and axillary bud development. *Photosynthetica* **22**, 309–19.

Bellwood, P. (1985). *Prehistory of the Indo-Malaysian Archipelago.* Academic Press, New York.

Bennett, M.D. (1972). Nuclear DNA content and minimum generation time in herbaceous plants. *Proc. R. Soc. Lond.* B**181**, 109–35.

Bennett, M.D. (1976). DNA amount, latitude, and crop plant distribution. *Environ. Exp. Bot.* **16**, 93–108.

Bennett, M.D. (1987). Variation in genomic form in plants and its ecological implications. *New Phytol.* **106** (suppl.), 177–200.

Bennett, M.D. & Smith, J.B. (1976). Nuclear DNA amounts in angiosperms. *Phil. Trans. R. Soc. London* B274, 227–74.

Bennett, M.D., Smith, J.B. & Heslop-Harrison, J.S. (1982). Nuclear DNA amounts in angiosperms. *Proc. R. Soc. Lond.* B216, 179–99.

Bennett, M.K. (1935). British wheat yield per acre for seven centuries. *Econ. Hist.* 3(10), 12–29.

Ben Ze'ev, N. & Zohary, D. (1973). Species relationships in the genus *Pisum* L. *Israel J. Bot.* 22, 73–91.

Benzian, B. & Lane, P. (1982). Effects of husbandry treatments on nitrogen concentration of grain and related yields in winter-wheat experiments made in South-East England. *J. Sci. Food Agric.* 33, 1063–71.

Benzian, B. Darby, R.J., Lane, P., Widdowson, F.V. & Verstraeten, L.M.J. (1983). Relationship between N concentration of grain and grain yield in recent winter wheat experiments in England and Belgium, some with large yields. *J. Sci. Food Agric.* 34, 685–95.

Benzioni, A. & Dunstone, R.L. (1986). Jojoba: adaptation to environmental stress and the implications for domestication. *Quart. Rev. Biol.* 61, 177–99.

Besford, R.T., Withers, A.C. & Ludwig, L.J. (1985). Ribulose bisphosphate carboxylase activity and photosynthesis during leaf development in the tomato. *J. Exp. Bot.* 36, 1530–41.

Beveridge, W. (1927). The yield and price of corn in the Middle Ages. *Econ. Hist.* 1, 155.

Bhagsari, A.S. (1981). Relation of photosynthetic rates to yield in sweet potato genotypes. *Hort. Sci.* 16, 779–80.

Bhagsari, A.S. & Brown, R.H. (1976a). Photosynthesis in peanut (*Arachis*) genotypes *Peanut Science* 3 (1), 1–5.

Bhagsari, A.S. & Brown, R.H. (1976b). Translocation of photosynthetically assimilated ^{14}C in peanut (*Arachis*) genotypes. *Peanut Sci.* 3 (1), 5–9.

Bhagsari, A.S. & Brown, R.H. (1986). Leaf photosynthesis and its correlation with leaf area. *Crop Sci.* 26, 127–32.

Bhagsari, A.S. & Harmon, S.A. (1982). Photosynthesis and photosynthate partitioning in sweet potato genotypes. *J. Amer. Soc. Hort. Sci.* 107, 506–70.

Bhagsari, A.S., Ashley, D.A., Brown, R.H. & Boerma, H.R. (1977). Leaf photosynthetic characteristics of determinate soybean cultivars. *Crop Sci.* 17, 929–32.

Bhan, S., Singh, H.G. & Singh, A. (1973). Note on root development as an index of drought resistance in sorghum (*Sorghum bicolor* (L.) Moench). *Indian J. Agric. Sci.* 43, 828–30.

Bhardwaj, S.N., Singh, M. & Ruwali, K.N. (1975). Physiological parameters for breeding for higher productivity in upland cotton. *Indian J. Agric. Res.* 45, 124–7.

Bhattacharya, S., Bhattacharya, N.C., Biswas, P.K. & Strain, B.R. (1985). Response of cowpea (*Vigna unguiculata* L.) to CO_2 enrichment environment on growth, dry matter production and yield components at different stages of vegetative and reproductive growth. *J. Agric. Sci. (Camb.)* 105, 527–34.

Bidinger, F., Musgrave, R.B. & Fischer, R.A. (1977). Contribution of stored preanthesis assimilate to grain yield in wheat and barley. *Nature (Lond.)* 270, 431–3.

Bidinger, F.R., Mahalakshmi, V. & Rao, G.D.P. (1987). Assessment of drought resistance in pearl millet (*Pennisetum americanum* (L.) Leeke). I. Factors affecting yields under stress. *Aust. J. Agric. Res.* 38, 37–48.

Bielorai, H., Mantell, A., Moreshet, S. (1983). Water relations of cotton. In *Water Deficits and Plant Growth*, vol. 7 (ed. T.T. Kozlowski), pp. 49–87. Academic

Press, New York.

Binford, L.R. (1968). Post-Pleistocene adaptations. In *New Perspectives in Archaeology* (ed. S.R. Binford & L.R. Binford), pp. 313–41. Aldine, Chicago.

Bingham, J. (1967). Investigations on the physiology of yield in winter wheat, by comparisons of varieties and by artificial variation in grain number per ear. *J. Agric. Sci. (Camb.)* **68**, 411–22.

Birecka, H. & Dakic-Wlodkowska, L. (1963). Photosynthesis, translocation and accumulation of assimilates in cereals during grain development. III. Spring wheat photosynthesis and the daily accumulation of photosynthates in the grain. *Acta Soc. Bot. Polon.* **32**, 631–50.

Birecka, H. & Dakic-Wlodkowska, L. (1966). Photosynthetic activity and productivity before and after ear emergence in spring wheat. *Acta Soc. Bot. Polon.* **35**, 637–62.

Biscoe, P.V., Scott, R.K. & Monteith, J.L. (1975). Barley and its environment. III. Carbon budget of the stand. *J. Appl. Ecol.* **12**, 269–91.

Bjorkman, O. (1981). Responses to different quantum flux densities. In *Physiological Plant Ecology*, vol. 1 (*Responses to the Physical Environment*) (ed. O.L. Lange, P.S. Nobel & C.B. Osmond), pp. 57–107. Springer, Berlin.

Black, J.N. (1957). Seed size as a factor in the growth of subterranean clover (*Trifolium subterraneum* L.) under spaced and sward conditions. *Aust. J. Agric. Res.* **8**, 335–51.

Blackbeard, J. (1982). *Arable Farming* Feb. 1982, pp. 45–9.

Blacklow, W.M., Darbyshire, B. & Pheloung, P. (1984). Fructan polymerized and depolymerized in the internodes of winter wheat as grain filling progressed. *Plant Science Letters* **36**, 213–18.

Blanco, L.C., Casal, C., Akita, S. & Virmani, S.S. (1990). Biomass, grain yield, and harvest index of F_1 rice hybrids and inbreds. *Intl. Rice Res. Newsl.* **15**(2), 9–10.

Blaxter, K.L. (1976). The use of resources. *Animal Production* **23**, 267–79.

Blechschmidt-Schneider, S., Ferrar, P. & Osmond, C.B. (1989). Control of photosynthesis by the carbohydrate level in leaves of the C_4 plant *Amaranthus edulis* L. *Planta* **177**, 515–25.

Bloom, A.J., Chapin, F.S. & Mooney, H.A. (1985). Resource limitation in plants – an economic analogy. *Ann. Rev. Ecol. Sys.* **16**, 363–92.

Blum, A. (1977). Basis of heterosis in the differentiating sorghum panicle. *Crop Sci.* **17**, 880–2.

Blum, A. (1979). Genetic improvement of drought resistance in crop plants: a case for sorghum. In *Stress Physiology in Crop Plants* (ed. H. Mussell & R.C. Staples), pp. 429–45. Wiley, New York.

Blum, A. (1982). Evidence for genetic variability in drought resistance and its implications for plant breeding. In *Drought Resistance in Crops with Emphasis on Rice*, pp. 53–68. IRRI, Los Baños.

Blum, A. (1985). Breeding crop varieties for stress environments. *CRC Crit. Rev. Plant Sci.* **2**, 199–238.

Blum, A. (1989). The temperature response of gas exchange in sorghum leaves and the effect of heterosis. *J. Exp. Bot.* **40**, 453–60.

Blum, A., Arkin, G.F. & Jordan, W.R. (1977a). Sorghum root morphogenesis and growth. I. Effect of maturity genes. *Crop Sci.* **17**, 149–53.

Blum, A., Jordan, W.R. & Arkin, G.F. (1977b). Sorghum root morphogenesis and growth. II. Manifestation of heterosis. *Crop Sci.* **17**, 153–7.

Blum, A., Shpiler, L., Golan, G. & Mayer, J. (1989). Yield stability and canopy temperature of wheat genotypes under drought stress. *Field Crops Res.* 22, 289–96.

Bock, B.R. (1984). Efficient use of nitrogen in cropping systems. In *Nitrogen in Crop Production* (ed. R.D. Hauck), pp. 273–94. Amer. Soc. Agron., Madison, Wisconsin.

Boerboom, B.W.J. (1978). A model of dry matter production in cassava (*Manihot esculenta* Crantz). *Neth. J. Agric. Sci.* 26, 267–77.

Boerma, H.R. (1979). Comparison of past and recently developed soybean cultivars in maturity groups VI, VII, and VIII. *Crop Sci.* 19, 611–13.

Boerma, H.R. & Ashley, D.A. (1988). Canopy photosynthesis and seed fill duration in recently developed soybean cultivars and selected plant introductions. *Crop Sci.* 28, 137–40.

Boonstra, A.E.H.R. (1936). Der Einfluss der verschiedenen assimilierenden Teile auf den Samenertrag von Weizen. *Zeitschr. f. Pflanzenzücht.* A 21, 115–47.

Borchers-Zampini, C., Glamm, A.B., Hoddinott, J. & Swanson, C.A. (1980). Alterations in source-sink patterns by modifications of source strength. *Plant Physiol.* 65, 1116–20.

Borger, H., Huhnke, W., Köhler, D., Schwanitz, F. & von Sengbusch, R. (1956). Untersuchungen über die Ursachen der Leistung von Kulturpflanzen. I. *Der Züchter* 26, 363–70.

Borlaug, N.E. (1971). *The Green Revolution, Peace and Humanity.* Nobel Peace Prize Speech. The Nobel Foundation, Stockholm.

Borrie, W.D. (1970). *The Growth and Control of World Population.* Weidenfield & Nicolson, London.

Boserup, E. (1965). *The Conditions of Agricultural Growth.* Aldine, Chicago.

Boserup, E. (1981). *Population and Technology.* Oxford University Press.

Boukerrou, L. & Rasmusson, D.D. (1990). Breeding for high biomass yield in spring barley. *Crop Sci.* 30, 31–5.

Boulaine, J. & Feller, C. (1989). Établissement du premier record de France de productivité du blé par le sieur Charlemagne-Bobigny, 8 Août 1766. *Comptes Rendus de l'Acad. d'Agric. de France* 75 (4), 31–8.

Bouldin, D.R., Klausner, S.D. & Reid, W.S. (1984). Use of nitrogen from manure. In *Nitrogen in Crop Production* (ed. R.D. Hauck), pp. 221–245. Amer. Soc. Agron., Madison, Wisconsin.

Bowen, H.C. (1961). *Ancient Fields. A Tentative Analysis of Vanishing Earthworks and Landscapes.* Brit. Ass. Adv. Science., London.

Boyer, J.S. (1982). Plant productivity and environment. *Science* 218, 443–8.

Boyer, J.S. (1983). Environment stress and crop yields. In *Crop Reactions to Water and Temperature Stresses in Humid Temperate Climates* (ed. C.D. Raper & P.J. Kramer), pp. 3–7. Westview, Boulder, Colorado.

Boyer, J.S. & McPherson, H.G. (1975). Physiology of water deficits in cereal crops. *Adv. Agron.* 27, 1–24.

Boyer, J.S., Johnson, R.R. & Saupe, S.G. (1980). Afternoon water deficits and grain yields in old and new soybean cultivars. *Agron. J.* 72, 981–6.

Bramel-Cox, P.J., Andrews, D.J., Bidinger, F.R. & Frey, K.J. (1984). A rapid method of evaluating growth rate in pearl millet and its weedy and wild relatives. *Crop Sci.* 24, 1187–91.

Bramel-Cox, P.J., Andrews, D.J. & Frey, K.J. (1986). Exotic germplasm for improving grain yield and growth rate in pearl millet. *Crop Sci.* 26, 687–90.

Braunholtz, J.T. (1981). Crop protection: the role of the chemical industry in an uncertain future. *Phil. Trans. R. Soc. Lond.* B**295**, 19–34.

Bravdo, B. & Pallas, J.E. (1982). Photosynthesis, photorespiration and RuBP carboxylase/oxygenase activity in selected peanut genotypes. *Photosynthetica* **16**, 36–42.

Breaux, R.D. (1984). Breeding to enhance sucrose content of sugar cane varieties in Louisiana. *Field Crops Res.* **9**, 59–67.

Bremner, J.M. & Douglas, L.A. (1971). Inhibition of urease activity in soils. *Soil Biol. Biochem.* **3**, 297–307.

Bremner, J.M. & Krogmeier, M.J. (1988). Elimination of the adverse effects of urea fertilizer on seed germination, seedling growth, and early plant growth in soil. *Proc. Natl. Acad Sci.* USA **85**, 4601–4.

Bremner, J.M., Breitenbeck, G.A. & Blackner, A.M. (1981). Effect of nitrapyrin on emission of nitrous oxide from soil fertilized with anhydrous ammonia. *Geophys. Res. Lett.* **8**, 353–6.

Brenchley, W.E. & Hall, AD (1909). The development of the grain of wheat. *J. Agric. Sci. (Camb.)* **3**, 195–217.

Bretting, P.K. (1990). New perspectives on the origin and evolution of New World domesticated plants: Introduction. *Econ. Bot.* **44** (suppl. 3), 1–5.

Brewbaker, J.L. (1979). Diseases of maize in the wet lowland tropics and the collapse of the classic Maya civilization. *Econ. Bot.* **33**, 101–18.

Bridge, R.R. & Meredith, W.R. (1983). Comparative performance of obsolete and current cotton cultivars. *Crop Sci.* **23**, 949–52.

Bridge, R.R., Meredith, W.R. & Chism, J.R. (1971). Comparative performance of obsolete varieties and current varieties of upland cotton. *Crop Sci.* **11**, 29–32.

Briggs, L.J. & Shantz, H.L. (1913). The water requirements of plants. II. A review of the literature. *US Dept. Agric. Bur. Plant Ind. Bull.* No. 285, pp. 1–96.

Brinkman, M.A. & Frey, K.J. (1977). Growth analysis of isoline-recurrent parent grain yield differences in oats. *Crop Sci.* **17**, 426–30.

Brinkman, M.A. & Frey, K.J. (1978). Flag leaf physiological analysis of oat isolines that differ in grain yield from their recurrent parents. *Crop Sci.* **18**, 69–73.

Brocklehurst, P.A. (1977). Factors controlling grain weight in wheat. *Nature (Lond.)* **266**, 348–9.

Brockway, L.H. (1979). *Science and Colonial Expansion. The Role of the British Royal Botanic Gardens.* Academic Press, New York.

Bronson, B. (1977). The earliest farming: demography as cause and consequence. In *Origins of Agriculture* (ed. C.A. Reed), pp. 23–48. Mouton, The Hague.

Brooking, I.R. & Kirby, E.J.M. (1981). Interrelationships between stem and ear development in winter wheat: the effects of a Norin 10 dwarfing gene, Gai/Rht2. *J. Agric. Sci. (Camb.)* **97**, 373–81.

Brooks, A. & Farquhar, G.D. (1985). Effect of temperature on the CO_2/O_2 specificity of ribulose-1,5-bisphosphate carboxylase/oxygenase and the rate of respiration in the light. Estimates from gas exchange measurements on spinach. *Planta* **165**, 397–406.

Brouwer, R. (1962*a*). Distribution of dry matter in the plant. *Neth. J. Agric. Sci.* **10**, 361–76.

Brouwer, R. (1962*b*). Nutritive influences on the distribution of dry matter in the plant. *Neth. J. Agric. Sci.* **10**, 399–408.

Brouwer, R. (1966). Root growth of cereals and grasses. In *The Growth of Cereals and Grasses* (ed. F.L. Milthorpe & J.D. Ivins), pp. 153–66. Butterworth, London.

Brown, A.H. (1953). The effects of light on respiration using isotopically enriched oxygen. *Amer. J. Bot.* **40**, 719–29.

Brown, A.H.D., Daniels, J. & Latter, B.D.H. (1968). Quantitative genetics of sugar cane. *Theoret. Appl. Genet.* **38**, 361–9.

Brown, E.B. (1976). Assessment of damage by nematodes with economic considerations. *Ann. Appl. Biol.* **84**, 448–51.

Brown, L.R. (1967). The world outlook for conventional agriculture. *Science* **158**, 604–11.

Brown, L.R. (1974). *By Bread Alone.* Praeger, New York.

Brown, L.R. (1975). The world food prospect. *Science* **190**, 1053–9.

Brown, L.R. (1985). Reducing hunger. In *State of the World 1985* (ed. L.R. Starke), pp. 23–41; 249–51. Norton, New York.

Brown, L.R. (1989). Re-examining the world food prospect. In *State of the World 1989* (ed. L. Starke), pp. 41–58. Norton, New York.

Brücher, H. (1969). Die Evolution der Gartenbohne *Phaseolus vulgaris* L. aus der südamerikanischen Wildbohne *Ph. aborigineus* Burk. *Angew. Bot.* **42**, 119–28.

Bruckner, P.L. & Frohberg, R.C. (1987). Stress tolerance and adaptation in spring wheat. *Crop Sci.* **27**, 31–6.

Bruijn, G.H. de & Dharmaputra, T.S. (1974). The Mukibat system, a high-yielding method of cassava production in Indonesia. *Neth. J. Agric. Sci.* **22**, 89–100.

Brunken, J., de Wet, J.M.J. & Harlan, J.R. (1977). The morphology and domestication of pearl millet. *Econ. Bot.* **31**, 163–74.

Bruyn, L.P. de & de Jager, J.M. (1978). A meteorological approach to the identification of drought sensitive periods in field crops. *Agric. Meteorol.* **19**, 35–40.

Bugbee, B.G. & Salisbury, F.B. (1988). Exploring the limits of crop productivity. I. Photosynthetic efficiency of wheat in high irradiance environments. *Plant Physiol.* **88**, 869–78.

Bull, T.A. (1965). *Taxonomic and physiological relationships in the Saccharum complex.* Ph.D. thesis, University of Queensland, Brisbane.

Bull, T.A. (1971). The C_4 pathway related to growth rates in sugar cane. In *Photosynthesis and Photorespiration* (ed. M.D. Hatch, C.B. Osmond & R.O. Slatyer), pp. 68–75. Wiley, New York.

Bull, T.A. & Glasziou, R.T. (1963). The evolutionary significance of sugar accumulation in *Saccharum*. *Aust. J. Biol. Sci.* **16**, 737–42.

Bunce, J.A. (1981). Relationships between maximum photosynthetic rates and photosynthetic tolerance of low leaf water potentials. *Can. J. Bot.* **59**, 769–74.

Bunce, J.A. (1983). Photosynthetic characteristics of leaves developed at different irradiances and temperatures: an extension of the current hypothesis. *Photosynth. Res.* **4**, 87–97.

Bunce, J.A. (1985). Effects of weather during leaf development on photosynthetic characteristics of soybean leaves. *Photosynth. Res.* **6**, 215–20.

Bunce, J.A. (1986). Measurements and modelling of photosynthesis in field crops. *CRC Crit. Rev. Plant Sci.* **4**, 47–77.

Bunce, J.A. (1989). Growth rate, photosynthesis and respiration in relation to leaf area index. *Ann. Bot.* **63**, 459–63.

Bunce, J.A. (1990a). The effect of leaf size on mutual shading and cultivar differences in soybean leaf photosynthetic capacity. *Photosynth. Res.* **23**, 67–72.

Bunce, J.A. (1990b). Short- and long term inhibition of respiratory carbon dioxide efflux by elevated carbon dioxide. *Ann. Bot.* **65**, 637–42.

Bunce, J.A. & Caulfield, F. (1991). Reduced respiratory carbon dioxide efflux during growth at elevated carbon dioxide in three herbaceous perennial species. *Ann. Bot.* **67**, 325–30.

Bunemann, G. & Grassia, A. (1973). Growth and mineral distribution in grafted tomato/potato plants according to sink number. *Sci. Hort.* **1**, 13–24.

Bunting, A.H. (1975). Time, phenology and the yield of crops. *Weather* **30**, 312–25.

Buresh, R.J., De Datta, S.K., Padilla, J.L. & Samson, M.I. (1988). Field evaluation of two urease inhibitors with transplanted lowland rice. *Agron. J.* **80**, 763–8.

Buringh, P. & Dudal, R. (1987). Agricultural land use in space and time. In *Land Transformation in Agriculture* (ed. M.G. Wolman & F.G.A. Fournier), pp. 9–43. Wiley, Chichester.

Buringh, P. & van Heemst, H.D.J. (1977). *An Estimation of World Food Production based on Labour-Oriented Agriculture.* (45 pp.) Centre for World Food Market Research, Wageningen.

Buringh, P., van Heemst, H.D.J. & Staring, G.J. (1975). *Computation of the Absolute Maximum Food Production of the World.* (59 pp.) Dept. of Tropical Soil Sci., Agric. Univ., Wageningen.

Burkill, I.H. (1951). The rise and decline of the greater yam in the service of man. *Adv. Sci. (London)* **7**, 443–8.

Burkill, I.H. (1952). Habitats of man and the origins of the cultivated plants of the Old World. *Proc. Linn. Soc. (Lond.)* **164**, 12–42.

Burkill, I.H. (1960). The organography and the evolution of Dioscoraceae, the family of yams. *J. Linn. Soc. (Bot.)* **56**, 319–412.

Burns, R.C. & Hardy, R.W.F. (1975). *Nitrogen Fixation in Bacteria and Higher Plants.* Springer, Berlin.

Burr, G.O., Hartt, C.E., Brodie, H.W., Tanimoto, T., Kortschak, H.P., Takahashi, D., Ashton, F.M. & Coleman, R.E. (1957). The sugar cane plant. *Ann. Rev. Plant Physiol.* **8**, 275–308.

Burris, J.S., Edje, O.T. & Wahab, A.H. (1973). Effects of seed size on seedling performance in soybeans. II. Seedling growth and photosynthesis and field performance. *Crop Sci.* **13**, 207–10.

Burris, R.H. (1980). The global nitrogen budget – science or seance? In *Nitrogen Fixation*, vol 1 (ed. W.E. Newton & W.H. Orme-Johnson), pp. 7–16. University Park Press, Baltimore.

Burt, R.L. (1964). Carbohydrate utilization as a factor in plant growth. *Aust. J. Biol. Sci.* **17**, 867–77.

Burt, R.L. (1966). Some effects of temperature on carbohydrate utilization and plant growth. *Aust. J. Biol. Sci.* **19**, 711–14.

Burton, G.W. & Powell, J.B. (1968). Pearl millet breeding and cytogenetics. *Adv. Agron.* **20**, 49–89.

Busch, R.H. & Chamberlain, D.D. (1981). Effects of daylength response and semi-dwarfism on agronomic performance of spring wheat. *Crop Sci.* **21**, 57–60.

Busch, R.H. & Kofoid, K. (1982). Recurrent selection for kernel weight in spring wheat. *Crop Sci.* **22**, 568–72.

Busch, R.H., Elsayed, F.A. & Heiner, R.E. (1984). Effect of daylength insensitivity on agronomic traits and grain protein in hard red spring wheat. *Crop Sci.* **24**, 1106–9.

Bush, M.G. & Evans, L.T. (1988). Growth and development in tall and dwarf isogenic lines of spring wheat. *Field Crops Res.* **18**, 243–70.

Butterfass, T. (1964). Die Korrelation zwischen der Chloroplastenzahl und der Zellgrosse bei diploiden, triploiden und tetraploiden Zuckerrüben (*Beta vulgaris* L.) *Naturwiss.* **51**, 70–1.

Buttery, B.R. & Buzzell, R.I. (1977). The relationship between chlorophyll content and rate of photosynthesis in soybeans *Can. J. Plant Sci.* **57**, 1–5.

Buttery, B.R., Buzzell, R.I. & Findlay, W.I. (1981). Relationships among photosynthetic rate, bean yield and other characters in field grown cultivars of soybean. *Can. J. Plant Sci.* **61**, 191–8.

Buttrose, M.S. (1962*a*). Physiology of cereal grain. III. Photosynthesis in the wheat ear during grain development. *Aust. J. Biol. Sci.* **15**, 611–18.

Buttrose, M.S. (1962*b*). The influence of environment on the shell structure of starch granules. *J. Cell Biol.* **14**, 159–67.

Buzzell, R.I. & Buttery, B.R. (1977). Soybean harvest index in hill plots. *Crop Sci.* **17**, 968–70.

Byerlee, D. (1987). *Maintaining the momentum in post-green Revolution agriculture: A micro-level perspective from Asia.* Michigan State Univ., East Lansing, Intl. Development Paper No. 10.

Byerlee, D. (1990). *Technical change, productivity and sustainability in irrigated cropping systems of South Asia: emerging issues in the post-Green Revolution era.* CIMMYT Economics Working Paper 90/06. CIMMYT, Mexico.

Byrne, M.C., Nelson, C.J. & Randall, D.D. (1981). Ploidy effects on anatomy and gas exchange of tall fescue leaves. *Plant Physiol.* **68**, 891–3.

Byth, D.E., Caldwell, B.E. & Weber, C.R. (1969). Specific and non-specific index selection in soybeans, *Glycine max* L. (Merrill). *Crop Sci.* **9**, 702–705.

Cabauatan, P.Q., Hibino, H., Lapis, D.B., Omura, T. & Tsuchizaki, T. (1985). Rice grassy stunt virus 2: a new strain of rice grassy stunt in the Philippines. *IRRI Res. Paper Ser.* No. 106, pp. 1–8.

Caemmerer, S. von & Farquhar, G.D. (1981). Some relationships between the biochemistry of photosynthesis and the gas exchange of leaves. *Planta* **153**, 376–87.

Calabio, J.C. & De Datta, S.K. (1985). Increasing productivity and protein content using early-maturing rices and efficient nitrogen management. *Fertilizer Res.* **6**, 73–84.

Calder, J.W. (1953). Trend of wheat yields in New Zealand. *N.Z. Wheat Review* **1950–52**, 17–18.

Calder, N. (1967). *The Environment Game.* Secker & Warburg, London.

Callen, E.O. (1967). The first New World cereal. *Amer. Antiquity* **32**, 535–8.

Callen, E.O. (1973). Dietary patterns in Mexico between 6500 BC and 1580 AD. In *Man and his Foods* (ed. C.E. Smith), pp. 29–49. University of Alabama, Alabama.

Calvin, M. (1979). Petroleum plantations for fuel and materials. *Bioscience* **29**, 533–8.

Candolle, A. de (1882). *Origine des Plantes Cultivées.* (transl. 1886). Kegan Paul, London.

Cannell, R.Q. (1981). Potentials and problems of simplified cultivation and conservation tillage. *Outlook on Agric.* **10**, 379–84.

Cannell, R.Q., Brun, W.A. & Moss, D.N. (1969). A search for high net photosynthetic rate among soybean genotypes. *Crop Sci.* **9**, 840–1.

Canny, M.J. (1973). *Phloem Translocation.* Cambridge University Press.

Cardwell, V.B. (1982). Fifty years of Minnesota corn production: sources of yield increase. *Agron. J.* **74**, 984–90.

Carlone, M.R. & Russell, W.A. (1987). Response to plant densities and nitrogen levels for maize cultivars under varying fertility and climatic conditions. *Crop Sci.* **27**, 465–70.

Carter, G.F. (1977). A hypothesis suggesting a single origin of agriculture. In *Origins of Agriculture* (ed. C.A. Reed), pp. 89–133. Mouton, The Hague.

Carver, B.F. & Nevo, E. (1990). Genetic diversity of photosynthetic characters in native populations of *Triticum dicoccoides*. *Photosynth. Res.* **25**, 119–28.

Carver, B.F., Smith, E.L. & England, H.O. (1987). Regression and cluster analysis of environmental responses of hybrid and pure line winter wheat cultivars. *Crop Sci.* **27**, 659–64.

Castleberry, R.M., Crum, C.W. & Krull, C.F. (1984). Genetic yield improvement of US maize cultivars under varying fertility and climatic environments. *Crop Sci.* **24**, 33–6.

Cavalieri, A.J. & Smith, O.S. (1985). Grain filling and field drying of a set of maize hybrids released from 1930 to 1982. *Crop Sci.* **25**, 856–60.

Cavalli-Sforza, L.L. & Feldman, M.W. (1981). *Cultural Transmission and Evolution.* Princeton University Press, Princeton, New Jersey.

Cavers, P. & Bough, M.A. (1985). Proso millet (*Panicum miliaceum* L.). – a crop and a weed. In *Studies in Plant Demography: a Festschrift for John L. Harper* (ed. J. White), pp. 143–55. Academic Press, New York.

Ceccarelli, S. (1989). Wide adaptation: How wide? *Euphytica* **40**, 197–205.

Ceccarelli, S. & Grando, S. (1991). Selection environment and environmental sensitivity in barley. *Euphytica* **57**, 157–67.

Ceppi, D., Sala, M., Gentinetta, E., Verderio, A. & Motto, M. (1987). Genotype-dependent leaf senescence in maize. *Plant Physiol.* **85**, 720–5.

Chaleff, R.S. & Ray, T.B. (1984). Herbicide-resistant mutants from tobacco cell cultures. *Science* **223**, 1148–51.

Chalfant, R.B. & Young, J.R. (1982). Chemigation, or application of insecticide through overhead sprinkler irrigation systems, to manage insect pests infesting vegetable and agronomic crops. *J. Econ. Entomol.* **75**, 237–41.

Chalmers, D.J. & van den Ende, B. (1975). Productivity of peach trees. Factors affecting dry weight distribution during tree growth. *Ann. Bot.* **39**, 423–32.

Chalmers, D.J., Canterfore, R.L., Jerie, P.H., Jones, T.R. & Ugalde, T.D. (1975). Photosynthesis in relation to growth and distribution of fruit in peach trees. *Aust. J. Plant Physiol.* **2**, 635–45.

Chalmers, D.J., Olsson, K.A. & Jones, T.R. (1983). Water relations of peach trees and orchards. In *Water Deficits and Plant Growth* (ed. T.T. Kozlowski), vol. 7, pp. 197–232. Academic Press, New York.

Chaloupka, G. (1983). Kakadu rock art: its cultural, historic and prehistoric significance. In *The Rock Art Sites of Kakadu National Park – Some Preliminary Research Findings for Park Conservation and Management* (ed. D. Gillespie), pp. 1–34. Aust. Natl. Parks and Wild Life Service, Special Publication No. 10.

Chamberlin, R.J. & Wilson, G.L. (1982). Development of yield in two grain sorghum hybrids. I. Dry weight and carbon-14 studies. *Aust. J. Agric. Res.* **33**, 1009–18.

Chambers, R. (1984). Beyond the Green Revolution: a selective essay. In *Understanding Green Revolutions* (ed. T. Bayliss-Smith & S. Wanmali), pp. 362–79. Cambridge, University Press.

Chancellor, R.J. (1981). The manipulation of weed behaviour for control purposes. *Phil. Trans. R. Soc. Lond.* **B295**, 103–10.

Chandler, R.F. (1979). *Rice in the Tropics: A Guide to the Development of National Programs.* Westview, Boulder, Colorado.

Chang, J.-H. (1981). Corn yield in relation to photoperiod, night temperature and solar radiation. *Agric. Meteorol.* **24**, 253–62.

Chang, T.T. (1976). The rice cultures. *Phil. Trans. R. Soc. Lond.* **B275**, 143–55.

Chang, T.T. (1983). The origins and early cultures of the cereal grains and food legumes. In *The Origins of Chinese Civilization* (ed. D.N. Keightley), pp. 65–94. University of California, Berkeley.

Chang, T.T. (1989). Domestication and spread of the cultivated rices. In *Foraging and Farming – The Evolution of Plant Exploitation*, (ed. D.R. Harris & G.C. Hillman), pp. 408–17, Unwin Hyman, London.

Chapin, F.S., Groves, R.H. & Evans, L.T. (1989). Physiological determinants of growth rate in response to phosphorus supply in wild and cultivated *Hordeum* species. *Oecologia* 79, 96–105.

Chatterton, N.J. (1973). Product inhibition of photosynthesis in alfalfa leaves as related to specific leaf weight. *Crop Sci.* 13, 284–5.

Chatterton, N.J. & Silvius, J.E. (1979). Photosynthate partitioning into starch in soybean leaves I. Effects of photoperiod versus photosynthetic period duration. *Plant Physiol.* 64, 749–53.

Chaudhuri, U.N. & Kanemasu, E.T. (1985). Growth and water use of sorghum (*Sorghum bicolor* (L.) Moench) and pearl millet (*Pennisetum americanum* (L.) Leeke). *Field Crops Res.* 10, 113–24.

Cheng, D.-Z., Bao, X.K. & Cheng, Z. (1979). A preliminary study on morphological and physiological indices of high yielding spring wheat in Chaidamu Basin. *Aelea Agriculture* 2, 29–39. (In Chinese with English abstract.)

Chevalier, A. (1932). *Resources Végétales du Sahara et de ses Confins nord et sud.* Musée d'Histoire Naturelle, Paris. (Cited by Harlan, 1975.)

Childe, V.G. (1934). *New Light on the Most Ancient East. The Oriental Prelude to European Prehistory.* Kegan Paul, London.

Cholick, F.A., Welsh, J.R. & Cole, C.V. (1977). Rooting patterns of semi-dwarf and tall winter wheat cultivars under dryland field conditions. *Crop Sci.* 17, 637–9.

Chowdhury, S.I. & Wardlaw, I.F. (1978). The effect of temperature on kernel development in cereals. *Aust. J. Agric. Res.* 29, 205–23.

Christensen, L.E., Below, F.E. & Hageman, R.H. (1981). The effects of ear removal on senescence and metabolism of maize. *Plant Physiol.* 68, 1180–5.

Christy, A.L. & Porter, C.A. (1982). Canopy photosynthesis and yield in soybean. In *Photosynthesis*, vol. 2 (*Development, Carbon Metabolism and Plant Productivity*) (ed. Govindjee), pp. 499–511. Academic Press, New York.

Church, B.M. & Austin, R.B. (1983). Variability of wheat yields in England and Wales. *J. Agric. Sci. (Camb.)* 100, 201–4.

CIAT (1979). *Annual Report 1978*, p. C-34. CIAT, Cali, Colombia.

CIMMYT (1980). *Report on Wheat Improvement.* CIMMYT, Mexico D.F.

CIMMYT (1981). *World Maize Facts and Trends.* CIMMYT, Mexico D.F.

CIMMYT (1989). *1987–88 CIMMYT World Wheat Facts and Trends. The wheat revolution revisited: Recent trends and future challenges.* CIMMYT, Mexico D.F.

Clark, C. (1967). *Population Growth and Land Use.* MacMillan, London.

Clark, C. & Haswell, M. (1970). *The Economics of Subsistence Agriculture.* MacMillan, London.

Clark, J.D. & Brandt, S.A. (eds) (1984) *From Hunters to Farmers. The Causes and Consequences of Food Production in Africa.* University of California Press, Berkeley.

Clark, J.G.D. (1965). Radiocarbon dating and the spread of farming economy. *Antiquity* 39, 45–8.

Clarke, A.C. (1962). *Profiles of the Future.* Gollancz, London.

Claussen, W. (1977). Einfluss der Frucht auf Netto-assimilationsleistung und Netto-photosyntheseraten der Aubergine (*Solanum melongena* L). *Gartenbauwissenschaft.* 42 (2), 61–5.

Claussen, W. & Biller, E. (1977). Die Bedeutung der Saccharose- und Starkegehälte der

Blätter für die Regulierung der Netto-photosyntheseraten. *Z. Pflanzenphysiol.* **81**, 189–98.

Cock, J.H. (1983). Cassava. In *Potential Productivity of Field Crops Under Different Environments*, pp. 341–59. IRRI, Los Baños.

Cock, J.H. (1984). Cassava. In *The Physiology of Tropical Crops* (ed. P.R. Goldsworthy & N.M. Fisher), pp. 529–49. Wiley, New York.

Cock, J.H. & Yoshida, S. (1972). Accumulation of ^{14}C-labelled carbohydrate before flowering and its subsequent redistribution and respiration in the rice plant. *Proc. Crop Sci. Soc. Japan* **41**, 226–34.

Cock, J.H. & Yoshida, S. (1973). Changing sink and source relations in rice (*Oryza sativa* L.) using carbon dioxide enrichment in the field. *Soil Sci. Plant Nutr.* **19**, 229–34.

Cohen, C.J., Chilcote, D.O. & Frakes, R.V. (1982). Gas exchange and leaf area characteristics of four tall fescue selections differing in forage yield. *Crop Sci.* **22**, 709–11.

Cohen, M.N. (1977*a*). *The Food Crisis in Prehistory*. Yale University Press, New Haven.

Cohen, M.N. (1977*b*). Population pressure and the origins of agriculture: an archaeological example from the coast of Peru. In *Origin of Agriculture* (ed. C.A. Reed), pp. 135–77. Mouton, The Hague.

Colbert, T.R., Darrah, L.L. & Zuber, M.S. (1984). Effect of recurrent selection for stalk crushing strength on agronomic charactersitics and soluble stalk solids in maize. *Crop Sci.* **24**, 473–8.

Cole, H.S.D., Freeman, C., Jahoda, M. & Pavitt, K.L.R. (1973). *Thinking about the Future*. Sussex University Press/Chatto & Windus, London.

Colville, W.L. (1967). Environment and maximum yield of corn. In *Maximum Crop Yields – The Challenge*, pp. 21–36. Amer. Soc. Agron., Madison, Wisconsin. (Special Publication No. 9.)

Colwell, R.K., Norse, E.A., Pimentel, D., Sharples, F.E. & Simberloff, D. (1985). Genetic engineering in agriculture. *Science* **229**, 111–12.

Comai, L., Facciotti, D., Hiatt, W.R., Thompson, G., Rose, R.E. & Stalker, D.M. (1985). Expression in plants of a mutant *aro A* gene from *Salmonella typhimurium* confers tolerance to glyphosate. *Nature (Lond.)* **317**, 741–4.

Condon, A.G., Richards, R.A. & Farquhar, G.D. (1987). Carbon isotope discrimination is positively correlated with grain yield and dry matter production in field-grown wheat. *Crop Sci.* **27**, 996–1001.

Conklin, H. (1957). *Hanunoo Agriculture in the Philippines*. FAO, Rome.

Connell, T.R., Below, F.E., Hageman, R.H. & Willman, M.R. (1987). Photosynthetic components associated with differential senescence of maize hybrids following ear removal. *Field Crops Res.* **17**, 55–62.

Conway, G. (ed.) (1982). *Pesticide Resistance and World Food Production*. Imperial College Centre for Environmental Technology, London.

Cook, M.G. & Evans, L.T. (1978). Effect of relative size and distance of competing sinks on the distribution of photosynthetic assimilates in wheat. *Aust. J. Plant Physiol.* **5**, 495–509.

Cook, M.G. & Evans, L.T. (1983*a*). Nutrient responses of seedlings of wild and cultivated *Oryza* species. *Field Crops Res.* **6**, 205–18.

Cook, M.G. & Evans, L.T. (1983*b*). Some physiological aspects of the domestication and improvement of rice (*Oryza* spp.). *Field Crops Res.* **6**, 219–38.

Cook, M.G. & Evans, L.T. (1983*c*). The roles of sink size and location in the partitioning of assimilates in wheat ears. *Aust. J. Plant Physiol.* **10**, 313–27.

Cooke, G.W. (1982). *Fertilizing for Maximum Yield*. Granada, London.

Cooper, R.L. (1985). Breeding semi-dwarf soybeans. *Plant Breeding Rev.* 3, 289–311.

Corke, H. & Kannenberg, L.W. (1989). Selection for vegetative phase and actual filling period duration in short season maize. *Crop Sci.* 29, 607–12.

Corley, R.H.V. (1983). Oil palm and other tropical tree crops. In *Potential Productivity of Field Crops Under Different Environments*, pp. 383–99. IRRI, Los Baños.

Coursey, D.G. (1967). *Yams*. Longmans, London.

Coursey, D.G. (1976). The origin and domestication of yams in Africa. In *Origins of African Plant Domestication* (ed. J.R. Harlan, J.M.J. de Wet & A.B.L. Stemler), pp. 383–408. Mouton, The Hague.

Coursey, D.G. & Coursey, C.K. (1971). The New Yam festivals of West Africa. *Anthropos* 66, 444–84.

Cowan, I. (1984). Optimization of productivity: carbon and water economy in higher plants. In *Control of Crop Productivity* (ed. C.J. Pearson), pp. 13–32. Academic Press, Sydney.

Cox, M.C. Qualset, C.O. & Rains, D.W. (1985). Genetic variation for nitrogen assimilation and translocation in wheat. I. Dry matter and nitrogen accumulation. *Crop Sci.* 25, 430–5.

Cox, T.S., House, L.R. & Frey, K.J. (1984). Potential of wild germplasm for increasing yield of grain sorghum. *Euphytica* 33, 673–84.

Cox, T.S., Murphy, J.P. & Rodgers, D.M. (1986). Changes in genetic diversity in the red winter wheat regions of the United States. *Proc. Natl. Acad. Sci. USA* 83, 5583–6.

Cox, T.S., Shroyer, J.P., Liu, B.-H., Sears, R.G. & Martin, T.J. (1988). Genetic improvement in agronomic traits of hard red winter wheat cultivars from 1919 to 1987. *Crop Sci.* 28, 756–60.

Coyne, D.P. (1967). Photoperiodism: inheritance and linkage studies in *Phaseolus vulgaris*. *J. Hered.* 58, 313–14.

Crafts-Brandner, S.J. & Egli, D.B. (1987). Sink removal and leaf senescence in soybean. Cultivar effects. *Plant Physiol.* 85, 662–6.

Crafts-Brandner, S.J. & Poneleit, C.G. (1987a). Carbon dioxide exchange rates, ribulose bisphosphate carboxylase/oxygenase and phosphoenolpyruvate carboxylase activities, and kernel growth characteristics of maize. *Plant Physiol.* 84, 255–60.

Crafts-Brandner, S.J. & Poneleit, C.G. (1987b). Effect of ear removal on CO_2 exchange and activities of ribulose bisphosphate carboxylase/oxygenase and phosphoenolpyruvate carboxylase of maize hybrids and inbred lines. *Plant Physiol.* 84, 261–5.

Crafts-Brandner, S.J., Salvucci, M.E. & Egli, D.B. (1990). Changes in ribulose bisphosphate carboxylase/oxygenase and ribulose 5-phosphate kinase abundances and photosynthetic capacity during leaf senescence. *Photosynthesis Res.* 23, 223–30.

Craigie, P.G. (1883). Statistics of agricultural production. *Statistical Journal* 46, 1–47.

Cramer, H.H. (1967). *Plant Protection and World Crop Production*. Bayer, Leverkusen.

Crane, A.J. (1985). Possible effects of rising CO_2 on climate. *Plant Cell Environ.* 8, 371–9.

Craufurd, P.Q. & Bidinger, F.R. (1988). Effect of the duration of the vegetative phase on crop growth, development and yield in two contrasting pearl millet hybrids. *J. Agric. Sci. (Camb.)* 110, 71–9.

Craufurd, P.Q. & Bidinger, F.R. (1989). Potential and realized yield in pearl millet (*Pennisetum americanum*) as influenced by plant population density and life cycle duration. *Field Crops Res.* 22, 211–25.

Craufurd, P.Q., Austin, R.B., Acevedo, E. & Hall, M.A. (1991) Carbon isotope discrimination and grain yield in barley. *Field Crops Res.* 27, 301–13.

Criswell, J.G. & Hume, D.J. (1972). Variation in sensitivity to photoperiod among early maturing soybean strains. *Crop Sci.* 12, 657–60.

Criswell, J.G. & Shibles, R.M. (1971). Physiological basis for genotypic variation in net photosynthesis of oat leaves. *Crop Sci.* 11, 550–3.

Criswell, J.G. & Shibles, R.M. (1972). Influence of sink-source on flag-leaf net photosynthesis in oats. *Iowa State J. Sci.* 46, 405–15.

Crookes, W. (1898/1917). *The Wheat Problem.* 3rd edn, 1917. Longmans Green, London.

Crosbie, T.M. & Mock, J.J. (1981). Changes in physiological traits associated with grain yield improvement in three maize breeding programmes. *Crop Sci.* 21, 255–9.

Crosbie, T.M. & Pearce, R.B. (1982). Effects of recurrent phenotypic selection for high and low photosynthesis on agronomic traits in two maize populations. *Crop Sci.* 22, 809–13.

Crosbie, T.M., Mock, J.J. & Pearce, R.B. (1977). Variability and selection advance for photosynthesis in Iowa stiff stalk synthetic maize population. *Crop Sci.* 17, 511–14.

Crosbie, T.M., Mock, J.J. & Pearce, R.B. (1978a). Inheritance of photosynthesis in a diallel among eight maize inbred lines from Iowa Stiff Stalk Synthetic. *Euphytica* 27, 657–64.

Crosbie, T.M., Pearce, R.B. & Mock, J.J. (1978b). Relationships among CO_2 exchange rate and plant traits in Iowa stiff stalk synthetic maize population. *Crop Sci.* 18, 87–90.

Crosbie, T.M., Pearce, R.B. & Mock, J.J. (1981a). Selection for high CO_2 exchange rate among inbred lines of maize. *Crop Sci.* 21, 629–31.

Crosbie, T.M., Pearce, R.B. & Mock, J.J. (1981b). Recurrent phenotypic selection for high and low photosynthesis in two maize populations. *Crop Sci.* 21, 736–40.

Cross, H.Z. (1975). Diallel analysis of duration and rate of grain filling of seven inbred lines of corn. *Crop Sci.* 15, 532–5.

Crossa, J. (1988). A comparison of results obtained with two methods for assessing yield stability. *Theoret. Appl. Genet.* 75, 460–7.

Crossa, J. & Gardner, C.O. (1987). Introgression of an exotic germplasm for improving an adapted maize population. *Crop Sci.* 27, 187–90.

Crossa, J., Westcott, B. & Gonzalez, C. (1988). Analyzing yield stability of maize genotypes using a spatial model. *Theoret. Appl. Genet.* 75, 863–8.

Crowther, F. (1934). Studies in growth analysis of the cotton plant under irrigation in the Sudan. I. The effects of different combinations of nitrogen applications and water supply. *Ann. Bot.* 48, 877–913.

Cure, J.D. & Acock, B. (1986). Crop responses to carbon dioxide doubling: A literature survey. *Agric. Forest Meteorol.* 38, 127–45.

Curtis, D.L. (1968). The relation between the date of heading of Nigerian sorghums and the duration of the growing season. *J. Appl. Ecol.* 5, 218–26.

Curtis, P.E., Ogren, W.L. & Hageman, R.M. (1969). Varietal effects in soybean photosynthesis and photorespiration. *Crop Sci.* 9, 323–7.

Dalrymple, D.G. (1980). *Development and spread of semi-dwarf varieties of wheat* and rice in the United States. An international perspective. USDA Agric. Econ. Report, No. 455.

Dalrymple, D.G. (1986a). *Development and spread of high-yielding wheat varieties in developing countries.* USAID, Washington, DC.

Dalrymple, D.G. (1986b). *Development and spread of high-yielding rice varieties in developing countries.* USAID, Washington, DC.

Dalrymple, D.G. (1988). Changes in wheat varieties and yields in the United States, 1919–1984. *Agric. Hist.* 62 (4), 20–36.

Danckwerts, J.E. & Gordon, A.J. (1987). Long-term partitioning, storage and remobilization of ^{14}C-assimilated by *Lolium perenne* (cv Melle). *Ann. Bot.* 59, 55–66.

Dantuma, G., von Kittlitz, E., Frauen, M. & Bond, D.A. (1983). Yield, yield stability and measurement of morphological and phenological characters of Faba bean (*Vicia faba* L.) varieties grown in a wide range of environments in Western Europe. *Zeitschr. f. Pflanzenzücht.* 90, 85–105.

Darlington, C.D. (1963). *Chromosome Botany and the Origins of Cultivated Plants.* Allen & Unwin, London.

Darmency, H. & Gasquez, J. (1990). Résistances aux herbicides chez les mauvaises herbes. *Agronomie* 6, 457–72.

Darwin, C.R. (1859). *On the Origin of Species by means of Natural Selection.* John Murray, London.

Darwin, C.R. (1868). *The Variation of Animals and Plants under Domestication.* 2 vols. John Murray, London.

Darwinkel, A. (1978). Patterns of tillering and grain production of winter wheat at a wide range of plant densities. *Neth. J. Agric. Sci.* 26, 383–98.

Dasberg, S., Bielorai, H. & Erner, J. (1983). Nitrogen fertilization of Shamouti oranges. *Plant Soil* 75, 41–9.

Dash, C.R. & Rao, C.N. (1990). Effect of varying light intensities on yield and yield components of rice plant types. *Oryza* 27, 90–3.

David, C.C. (1989). The global rice situation. In *Progress in Irrigated Rice Research*, pp. 9–24. IRRI, Los Baños.

Davidson, J. & Lloyd, R. (1977). *Conservation and Agriculture.* Wiley, New York.

Davidson, J.L. & Philip, J.R. (1958). Light and pasture growth. (Proc. UNESCO Sympos. Climatology and Microclimatology, Canberra, 1956). *Arid Zone Research* 11, 181–7.

Davies, D.R. (1975). Studies of seed development in *Pisum sativum*. I. Seed size in reciprocal crosses. *Planta* 124, 297–302.

Day, D.A., Vos, O.C., Wilson, D. & Lambers, H. (1985). Regulation of respiration in the leaves and roots of two *Lolium perenne* populations with contrasting mature leaf respiration rates and crop yields. *Plant Physiol.* 78, 678–83.

Daynard, T.B. & Kannenberg, L.W. (1976). Relationships between length of the actual and effective grain filling periods and the grain yield of corn. *Can. J. Plant Sci.* 56, 237–42.

Daynard, T.B., Tanner, J.W. & Duncan, W.G. (1971). Duration of the grain filling period and its relation to grain yield in corn, *Zea mays* L. *Crop Sci.* 11, 45–8.

Dazhong, W. (1988). Energy use in the crop systems of China. *Crit. Rev. Plant Sci.* 7, 25–53.

Dean, C. & Leech, R.M. (1982). Genome expression during normal leaf development. 2. Direct correlation between ribulose biphosphate carboxylase content and nuclear ploidy in a polyploid series of wheat. *Plant Physiol.* 70, 1605–8.

De Block, M., Botterman, J., Vandewiele, M., Dockx, J., Thoen, C., Gossele, V., Movva, N.R., Thompson, C., Van Montagu, M. & Zeemans, J. (1987). Engineering herbicide resistance in plants by expression of a detoxifying enzyme. *EMBO J.* 6, 2513–18.

De Datta, S.K. (1970). *Nature (Lond.)* 227, 230.

De Datta, S.K. (1987). Advances in soil fertility research and nitrogen fertilizer management for lowland rice. In *Efficiency of Nitrogen Fertilizers for Rice*, pp. 27–41. IRRI, Los Baños.

De Datta, S.K. & Zarate, P.M. (1970). Environmental conditions affecting the growth characteristics, nitrogen response and grain yield of tropical rice. *Biometeorology* 4, 71–89.

De Datta, S.K., Gomez, K.A. & Descolsota, J.P. (1988). Changes in yield response to major nutrients and in soil fertility under intensive rice cropping. *Soil Sci.* 146, 350–8.

Deinum, B. (1976). Photosynthesis and sink size: an explanation for the low productivity of grass swards in autumn. *Neth. J. Agric. Sci.* 24, 238–46.

Dekker, J. (1985). The development of resistance to fungicides. *Progress in Pesticide Biochem. Toxicol.* 4, 165–218.

Delaney, R.H. & Dobrenz, A.K. (1974a). Morphological and anatomical features of alfalfa leaves as related to CO_2 exchange. *Crop Sci.* 14, 444–7.

Delaney, R.H. & Dobrenz, A.K. (1974b). Yield of alfalfa as related to carbon exchange. *Agron. J.* 66, 498–501.

Delannay, X., Rodgers, D.M. & Palmer, R.G. (1983). Relative genetic contributions among ancestral lines to North American soybean cultivars. *Crop Sci.* 23, 944–9.

DeLoughery, R.L. & Crookston, R.K. (1979). Harvest index of corn affected by population density, maturity rating, and environment. *Agron. J.* 71, 577–80.

Delwiche, C.O. (1970). The nitrogen cycle. *Sci. Amer.* 223, 136–46.

Denbow, J.R. & Wilmsen, E.N. (1986). Advent and course of pastoralism in the Kalahari. *Science* 234, 1509–15.

Denmead, O.T., Freney, J.R. & Simpson, J.R. (1982). Dynamics of ammonia volatilization during furrow irrigation of maize. *Soil Sci. Soc. Amer. J.* 46, 149–55.

Dennett, M.D. (1980). Variability of annual wheat yields in England and Wales. *Agric. Meteorol.* 22, 109–11.

Dennett, M.D., Elston, J. & Speed, C.B. (1981a). Climate and cropping systems in West Africa. *Geoforum* 12, 193–202.

Dennett, M.D., Elston, J. & Speed, C.B. (1981b). Rainfall and crop yields in seasonally arid West Africa. *Geoforum* 12, 203–9.

Derieux, M., Darrigrand, M., Gallais, A., Barriere, Y., Bloc, D. & Montalant, Y. (1987). Estimation du progres genetique realisé chez les mais grain en France entre 1950 et 1985. *Agronomie* 7, 1–11.

Desai, G. (1982). *Sustaining rapid growth in India's fertilizer consumption: a perspective based on composition of use.* IFPRI Res. Report No. 31, pp. 1–71.

Detwiler, R.P. & Hall, C.A.S. (1988). The global carbon cycle. *Science* 241, 1738–9.

Dev, G., Saggar, S., Singh, R. & Sidhu, B.S. (1980). Root distribution patterns of some wheat varieties in arid brown soil under rainfed conditions. *J. Nuclear Agric. Biol.* 9 (3), 81–90.

Dey, S.K., Singh, V.P. & Sahu, G. (1990). Photosynthetic rate and associated leaf characters of rice varieties at seedling stage. *Oryza* 27, 355–7.

Dickinson, W. (1847). On a variety of Italian ryegrass. *J. Roy. Agric. Soc. England* 8,

572–82.

Diethelm, R. & Shibles, R. (1989). Relationship of enhanced sink demand with photosynthesis and amount and activity of ribulose-1,5-bisphosphate carboxylase in soybean leaves. *J. Plant Physiol.* **134**, 70–4.

Dingkuhn, M., Schnier, H.F., De Datta, S.K., Wijanco, E. & Dorffling, K. (1990). Diurnal and developmental changes in canopy gas exchange in relation to growth in transplanted and direct-seeded flooded rice. *Aust. J. Plant Physiol.* **17**, 119–34.

Dingkuhn, M., Schnier, H.F., De Datta, S.K., Dorffling, K. & Javellana, C. (1991). Relationships between ripening-phase productivity and crop duration, canopy photosynthesis and senescence in transplanted and direct-seeded lowland rice. *Field Crops Res.* **26**, 327–45.

Djisbar, A. & Gardner, F.P. (1989). Heterosis for embryo size and source and sink components of maize. *Crop Sci.* **29**, 985–92.

Dobben, W.H. van (1962). Influence of temperature and light conditions on dry-matter distribution, development rate and yield of arable crops. *Neth. J. Agric. Sci.* **10**, 377–89.

Dobben, W.H. van (1965). The photoperiodic reaction of wheat in Mediterranean region and in North-western Europe. *Mediterranea* **5**, 54–9.

Dodd, J.L. (1980). Grain sink size and predisposition of *Zea mays* to stalk rot. *Phytopathol.* **70**, 534–5.

Doebley, J. (1990). Molecular evidence and the evolution of maize. *Econ. Bot.* **44** (suppl. 3), 6–27.

Doggett, H. (1965). The development of the cultivated sorghums. In *Crop Plant Evolution* (ed. J.B. Hutchinson), pp. 50–69. Cambridge University Press.

Doggett, H. & Majisu, B.N. (1967). Bulrush millet breeding. *Rec. Res. East Agric. Agric. For. Res. Org.*, **1966**, 74–7.

Donald, C.M. (1962). In search of yield. *J. Aust. Inst. Agric. Sci.* **28**, 171–8.

Donald, C.M. (1965). The progress of Australian agriculture and the role of pastures in environmental change. *Aust. J. Sci.* **27**, 187–98.

Donald, C.M. (1968). The breeding of crop ideotypes. *Euphytica* **17**, 385–403.

Donald, C.M. (1981). Competitive plants, communal plants, and yield in wheat crops. In *Wheat Science – Today and Tomorrow* (ed. L.T. Evans & W.J. Peacock), pp. 223–247. Cambridge University Press.

Donald, C.M. & Hamblin, J. (1976). The biological yield and harvest index of cereals as agronomic and plant breeding criteria. *Adv. Agron.* **28**, 361–405.

Done, A.A. & Muchow, R.C. (1988). Grain yield and assimilate supply in grain sorghum. *Ninth Australian Plant Breeding Conference Proc*, pp. 271–2.

Doney, D.L. & Theurer, J.C. (1976). Hypocotyl diameter as a predictive selection criterion in sugar beet. *Crop Sci.* **16**, 513–15.

Doney, D.L., Theurer, J.C. & Wyse, R.E. (1985). Respiration efficiency and heterosis in sugar beet. *Crop Sci.* **25**, 448–50.

Doorenbos, J. & Kassam, A.H. (1979). *Yield response to water.* FAO, Rome.

Dornhoff, G.M. & Shibles, R.M. (1970). Varietal differences in net photosynthesis of soybean leaves. *Crop Sci.* **10**, 42–5.

Dornhoff, G.M. & Shibles, R. (1976). Leaf morphology and anatomy in relation to CO_2 exchange rate of soybean leaves. *Crop Sci.* **16**, 377–81.

Dow el-madina, I.M. & Hall, A.E. (1986). Flowering of contrasting cowpea (*Vigna unguiculata* (L.) Walp.) genotypes under different temperatures and photoperiods. *Field Crops Res.* **14**, 87–104.

Dowler, C.C., Rohde, W.A., Fetzer, L.E., Scott, D.E., Sklany, T.E. & Swann, C.W.

(1982). The effect of sprinkler irrigation on herbicide efficacy, distribution, and penetration in some Coastal Plain soils. *Univ. Georgia Coll. Agric. Res. Bull.* No. 281.

Downes, R.W. (1971). Relationship between evolutionary adaptation and gas exchange characteristics of diverse *Sorghum* taxa. *Aust. J. Biol. Sci.* 24, 843–52.

Downton, W.J.S. & Tregunna, E.B. (1968). Photorespiration and glycolate metabolism: A reexamination and correlation of some previous studies. *Plant Physiol.* 43, 923–9.

Dreger, R.H., Brun, W.A. & Cooper, R.L. (1969). Effect of genotype on the photosynthetic rate of soybean. (*Glycine max*) (L.) Merr.). *Crop Sci.* 9, 429–31.

Dreyer, J., Duncan, W.G. & McCloud, D.E. (1981). Fruit temperature, growth rates, and yield of peanuts. *Crop Sci.* 21, 686–8.

Duarte, R. & Adams, M.W. (1963). Component interaction in relation to expression of a complex trait in a field bean cross. *Crop Sci.* 3, 185–6.

Duckham, A.N. (1966). Forty years on: agriculture in retrospect and prospect. *J. Roy. Agric. Soc. England* 127, 7–16.

Ducruet, J.M. & Lemoine, Y. (1985). Increased heat sensitivity of the photosynthetic apparatus in triazine-resistant biotypes from different plant species. *Plant Cell Physiol.* 26, 419–29.

Dudley, J.W., Lambert, R.J. & Alexander, D.E. (1974). Seventy generations of selection for oil and protein concentration in the maize kernel. In *Seventy Generations of Selection for Oil and Protein in Maize*, (ed. J.W. Dudley) pp. 181–211. Crop Sci. Soc. Amer, Madison, Wisconsin.

Duncan, W.G. & Hesketh, J.D. (1968). Net photosynthetic rates, relative leaf growth rates, and leaf numbers of 22 races of maize grown at eight temperatures. *Crop Sci.* 8, 670–4.

Duncan, W.G., McCloud, D.E., McGraw, R.L. & Boote, K.J. (1978). Physiological aspects of peanut yield improvement. *Crop Sci.* 18, 1015–20.

Dunstone, R.L. & Evans, L.T. (1974). Role of changes in cell size in the evolution of wheat. *Aust. J. Plant Physiol.* 1, 157–65.

Dunstone, R.L., Gifford, R.M. & Evans, L.T. (1973). Photosynthetic characteristics of modern and primitive wheat species in relation to ontogeny and adaptation to light. *Aust. J. Biol. Sci.* 26, 295–307.

Duvick, D.N. (1977). Genetic rates of gain in hybrid maize yields during the past 40 years. *Maydica* 22, 187–96.

Duvick, D.N. (1984*a*). Genetic diversity in major farm crops on the farm and in reserve. *Econ. Bot.* 38, 161–78.

Duvick, D.N. (1984*b*). Progress in conventional plant breeding. In *Gene Manipulation in Plant Improvement*, (ed. J.P. Gustafson), pp. 17–31. Plenum, New York.

Duvick, D.N. (1984*c*). Genetic contributions to yield gains of US hybrid maize, 1930–1980. In *Genetic Contributions to Yield Gains of Five Major Crop Plants*, (ed. W.R. Fehr), pp. 15–47. Crop Sci. Soc. Amer., Madison, Wisconsin. (Special Publication. No. 7.)

Dwelle, R.B. Kleinkopf, G.E., Steinhorst, R.K., Pavek, J.J. & Hurley, P.J. (1981). The influence of physiological processes on tuber yield of potato clones (*Solanum tuberosum* L.). *Potato Res.* 24, 33–47.

Dwelle, R.B., Hurley, P.J. & Pavek, J.J. (1983). Photosynthesis and stomatal conductance of potato clones (*Solanum tuberosum* L.). Comparative differences in diurnal patterns, response to light levels and assimilation through upper and lower leaf surfaces. *Plant Physiol.* 72, 172–6.

Dwyer, L.M. & Tollenaar, M. (1989). Genetic improvement in photosynthetic response of hybrid maize cultivars, 1959 to 1988. *Can. J. Plant Sci.* **69**, 81–91.

Dyson, P.W. (1977). An investigation into the relations between some growth parameters and yield of barley. *Ann. Appl. Biol.* **87**, 471–83.

Eagles, C.F. & Wilson, D. (1982). Photosynthetic efficiency and plant productivity. In *Handbook of Agricultural Productivity* (ed. M. Rechcigl), vol. 1. pp. 213–47. CRC Press, Boca Raton, Florida.

Earle, F.R. & Jones, Q. (1962). Analyses of seed samples from 113 plant families. *Econ. Bot.* **16**, 221–50.

Eastin, J.D. (1983). Sorghum. In *Potential Productivity of Field Crops under Different Environments*, pp. 181–204. IRRI, Los Baños.

Eastin, J.D. & Sullivan, C.Y. (1984). Environmental stress influences on plant persistence, physiology and production. In *Physiological Basis of Crop Growth and Development* (ed. M.B. Tesar), pp. 201–36. Amer. Soc. Agron., Madison, Wisconsin.

Eberhart, S.A. & Russell, W.A. (1966). Stability parameters for comparing varieties. *Crop Sci.* **6**, 36–40.

Eck, H.V. (1984). Irrigated corn yield response to nitrogen and water. *Agron. J.* **76**, 421–8.

Ecochard, R., Cavalie, G., Nicco, C., Piquemal, M. & Sarrafi, A. (1991). Rubisco content and specific activity in barley (*Hordeum vulgare* L.). *J. Exp. Bot.* **42**, 39–43.

Edmeades, G. & Tollenaar, M. (1990). Genetic and cultural improvements in maize production. In *Proc. Intl. Congr. Plant Physiol., New Delhi, India* (ed. S.K. Sinha, P.V. Sane, S.C. Bhargava & P.K. Agrawal), vol. 1, pp. 164–80 Indian Agricultural Research Institute, New Delhi.

Edwards, K.J.R. & Cooper, J.P. (1961). The genetic control of leaf development in ryegrass. *Heredity* **16**, 236–7.

Egli, D.B. (1981). Species differences in seed growth characteristics. *Field Crops Res.* **4**, 1–12.

Egli, D.B. (1988). Plant density and soybean yield. *Crop Sci.* **28**, 977–81.

Egli, D.B., Pendleton, J.W. & Peters, D.B. (1970). Photosynthetic rate of three soybean communities as related to carbon dioxide levels and solar radiation. *Agron. J.* **62**, 411–14.

Egli, D.B., Fraser, J., Leggett, J.E. & Poneleit, C.G. (1981). Control of seed growth in soya beans (*Glycine max* (L.) Merrill). *Ann. Bot.* **48**, 171–6.

Egli, D.B., Swank, J.C. & Pfeiffer, T.W. (1987). Mobilization of leaf N in soybean genotypes with varying durations of seedfill. *Field Crops Res.* **15**, 251–8.

Ehdaie, B., Hall, A.E., Farquhar, G.D., Nguyen, H.T. & Waines, J.G. (1991). Water-use efficiency and carbon isotope discrimination in wheat. *Crop Sci.* **31**, 1282–8.

Ehleringer, J. & Pearcy, R.W. (1983). Variation in quantum yield for CO_2 uptake among C_3 and C_4 plants. *Plant Physiol.* **73**, 555–9.

Ehrlich, P.R. & Ehrlich, A.H. (1990). *The Population Explosion.* Simon & Schuster, New York.

Ekanayake, I.J., O'Toole, J.C., Garrity, D.P. & Masajo, T.M. (1989). Inheritance of root characters and their relations to drought resistance in rice. *Crop Sci.* **25**, 927–33.

Elliott, C.S. (1962). The importance of variety testing in relation to crop production. *J. Natl. Inst. Agric. Bot.* **9**, 199–206.

Elliott, W.A. & Perlinger, G.J. (1977). Inheritance of shattering in wild rice. *Crop Sci.* **17**, 851–3.

Ellis, J.R. & Leech, R.M. (1985). Cell size and chloroplast size in relation to chloroplast replication in light-grown wheat leaves. *Planta* **165**, 120–5.

Ellstrand, N.C. (1988). Pollen as a vehicle for the escape of engineered genes? *Trends in Biotech.* **6**, 530–2.

Elmore, C.D. (1980). The paradox of no correlation between leaf photosynthetic rates and crop yields. In *Predicting Photosynthesis for Ecosystem Models*, vol. 2, (ed. J.D. Hesketh & J.W. Jones), pp. 155–67. CRC Press, Boca Raton, Florida.

Elmore, C.D., Hesketh, J.D. & Muramoto, H. (1967). A survey of rates of leaf growth, leaf aging and leaf photosynthetic rates among and within species. *J. Arizona Acad. Sci.* **4**, 215–19.

El-Sharkawy, M.A. & Hesketh, J.D. (1964). Effect of stomatal differences among species on leaf photosynthesis. *Crop Sci.* **4**, 619–21.

El-Sharkawy, M. & Hesketh, J. (1965). Photosynthesis among species in relation to characteristics of leaf anatomy and CO_2 diffusion resistances. *Crop Sci.* **5**, 517–21.

El-Sharkawy, M., Hesketh, J.D. & Muramoto, H. (1965). Leaf photosynthetic rates and other growth characteristics among 26 species of *Gossypium*. *Crop Sci.* **5**, 173–5.

El-Sharkawy, M.A., Loomis, R.S. & Williams, W.A. (1968). Photosynthetic and respiratory exchanges of carbon dioxide by leaves of the grain amaranth. *J. Appl. Ecol.* **5**, 243–51.

El-Sharkawy, M.A., Cock, J.H. & Held, A.A. (1984). Photosynthetic responses of cassava cultivars (*Manihot esculenta* Crantz) from different habitats to temperature. *Photosynthesis Res.* **5**, 243–50.

El-Sharkawy, M.A., Cock, J.H., Lynam, J.K., Hernandez, A.P. & Cadavid, L.F. (1990). Relationships between biomass root-yield and single leaf photosynthesis in field-grown cassava. *Field Crops Res.* **25**, 183–201.

Elston, J. & Dennett, M.D. (1977). A weather watch for semi-arid lands within the tropics. *Phil. Trans. R. Soc. Lond.* B**278**, 593–609.

Elston, J.F., Greenland, D.J. & Dennett, M.D. (1980). Long term trends in the aerial and edaphic environment. In *Opportunities for Increasing Crop Yields* (ed. R.G. Hurd, P.V. Biscoe & C. Dennis), pp. 87–96. Pitman, Boston.

Encel, S., Marstrand, P.K. & Page, W. (1975). *The Art of Anticipation*. Martin Robertson, London.

Engels, F. (1844). *Outlines of a Critique of Political Economy*. (Cited by Flew (1970).)

Engelbrecht, T.H. (1917). Uber die Entstehung des Kulturroggens In *Studien und Forschungen zur Menschen- und Völkerkunde* (ed. G. Buschan), vol. 14, pp. 17–21.

Engledow, F.L. & Wadham, S.M. (1923). Investigations on yield in the cereals. Part I. *J. Agric. Sci. (Camb.)* **13**, 390–439.

Enoch, H.Z. & Kimball, B.A. (eds). (1986). *Carbon Dioxide Enrichment of Greenhouse Crops*, vol. 1. *Status and CO_2 Sources*. CRC Press, Boca Raton, Florida.

Enyi, B.A.C. (1977). Analysis of growth and tuber yield in sweet potato (*Ipomea batatas*) cultivars. *J. Agric. Sci. (Camb.)* **89**, 421–30.

Erskine, W., Ellis, R.H., Summerfield, R.J., Roberts, E.H. & Hussain, A. (1990). Characterization of responses to temperature and photoperiod for time to flowering in a world lentil collection. *Theoret. Appl. Genet.* **80**, 193–9.

Evans, A. (1991). Whole-plant responses of *Brassica campestris* (Cruciferae) to altered sink-source relations. *Amer. J. Bot.* **78**, 394–400.

Evans, J.R. (1983). Nitrogen and photosynthesis in the flag leaf of wheat (*Triticum aestivum* L.). *Plant Physiol.* **72**, 297–302.

Evans, J.R. (1986). The relationship between CO_2-limited photosynthetic rate and ribulose-1,5-bisphosphate carboxylase content in two nuclear-cytoplasm substitution lines of wheat, and the coordination of ribulose-bisphosphate-carboxylation and electron-transport capacities. *Planta* **167**, 351–8.

Evans, J.R. (1989). Photosynthesis and nitrogen relationships in leaves of C_3 plants. *Oecologia* **78**, 9–19.

Evans, J.R. & Austin, R.B. (1986). The specific activity of ribulose-1,5-bisphosphate carboxylase in relation to genotype in wheat. *Planta* **167**, 344–50.

Evans, J.R. & Seemann, J.R. (1984). Differences between wheat genotypes in specific activity of ribulose-1,5-bisphosphate carboxylase and the relationship to photosynthesis. *Plant Physiol.* **74**, 759–65.

Evans, J.R., von Caemmerer, S. & Adams, W.W. (eds). (1988). *Ecology of Photosynthesis in Sun and Shade.* (*Aust. J. Plant Physiol.* **15**, 1–358.)

Evans, L.T. (1972). Storage capacity as a limitation on grain yield. In *Rice Breeding*, pp. 499–514. IRRI, Los Baños.

Evans, L.T. (1975a). Crops and world food supply, crop evolution, and the origins of crop physiology. In *Crop Physiology. Some Case Histories* (ed. L.T. Evans), pp. 1–22. Cambridge University Press.

Evans, L.T. (1975b). The physiological basis of crop yield. In *Crop Physiology: Some Case Histories* (ed. L.T. Evans), pp. 327–55. Cambridge University Press.

Evans, L.T. (1975c). Crop plants, an international heritage and opportunity. *Search* **6**, 272–80.

Evans, L.T. (1976). Physiological adaptation to performance as crop plants. *Phil. Trans. R. Soc. Lond.* B**275**, 71–83.

Evans, L.T. (1978a). The influence of irradiance before and after anthesis on grain yield and its components in microcrops of wheat grown in a constant daylength and temperature regime. *Field Crops Res.* **1**, 5–19.

Evans, L.T. (1978b). Learning from Tachai: agricultural research in China. *Search* **9**, 24–33.

Evans, L.T. (1978c). The yield of crops – trends and limits. In *Proc. Lincoln College Centennial Seminar: Agriculture – Resources and Potential*, pp. 18–30. Caxton, Christchurch.

Evans, L.T. (1979). An agriculturist in the Botanic Gardens. *Search* **10**, 341–6.

Evans, L.T. (1980a). The natural history of crop yield. *Amer. Sci.* **68**, 388–97.

Evans, L.T. (1980b). Response to challenge: William Farrer and the making of wheats. *J. Aust. Inst. Agric. Sci.* **46**, 3–13.

Evans, L.T. (1981). Yield improvement in wheat: empirical or analytical? In *Wheat Science – Today and Tomorrow* (ed. L.T. Evans & W.J. Peacock), pp. 203–22. Cambridge University Press.

Evans, L.T. (1982). Science and the suburban spirit. *Search* **13**, 307–11.

Evans, L.T. (1983). Raising the yield potential: by selection or design? In *Genetic Engineering of Plants* (ed. T. Kosuge, C.P. Meredith & A. Hollaender), pp. 371–89. Plenum, New York.

Evans, L.T. (1984). Physiological aspects of varietal improvement. In *Gene Manipulation in Plant Improvement* (ed. J.P. Gustafson), pp. 121–46. Plenum, New York.

Evans, L.T. (1986). Irrigation and crop improvement in temperate and tropical environments. *Phil. Trans. R. Soc. Lond.* A**316**, 319–30.

Evans, L.T. (1987a). Opportunities for increasing the yield potential of wheat. In *The Future Development of Maize and Wheat in the Third World*, pp. 79–93. CIMMYT, Mexico.

Evans, L.T. (1987b). Short day induction of inflorescence initiation in some winter wheat varieties. *Aust. J. Plant Physiol.* **14**, 277–86.

Evans, L.T. & Bush, M.G. (1985) Growth and development of channel millet (*Echinochloa turneriana*) in relation to its potential as a crop plant and compared with other *Echinochloa* millets, rice and wheat. *Field Crops Res.* **12**, 295–317.

Evans, L.T. & De Datta, S.K. (1979). The relation between irradiance and grain yield of irrigated rice in the tropics, as influenced by cultivar, nitrogen fertilizer application and month of planting. *Field Crops Res.* **2**, 1–17.

Evans, L.T. & Dunstone, R.L. (1970). Some physiological aspects of evolution in wheat. *Aust. J. Biol. Sci.* **23**, 725–41.

Evans, L.T. & King, R.W. (1975). Factors affecting flowering and reproduction in grain legumes. In *Report of TAC Working Group on the Biology of Yield of Grain Legumes*. DDDR:IAR/75/2. FAO, Rome.

Evans, L.T. & Knox, R.B. (1969). Environmental control of reproduction in *Themeda australis*. *Aust. J. Bot.* **17**, 375–89.

Evans, L.T. & Rawson, H.M. (1970). Photosynthesis and respiration by the flag leaf and components of the ear during grain development in wheat. *Aust. J. Biol. Sci.* **23**, 245–54.

Evans, L.T. & Wardlaw, I.F. (1976). Aspects of the comparative physiology of grain yield in cereals. *Adv. Agron.* **28**, 301–59.

Evans, L.T., Dunstone, R.L., Rawson, H.M. & Williams, R.F. (1970). The phloem of the wheat stem in relation to requirements for assimilate by the ear. *Aust. J. Biol. Sci.* **23**, 743–52.

Evans, L.T., Bingham, J., Jackson, P. & Sutherland, J. (1972). Effect of awns and drought on the supply of photosynthate and its distribution within wheat ears. *Ann. Appl. Biol.* **70**, 67–76.

Evans, L.T., Wardlaw & Fischer, R.A. (1975). Wheat. In *Crop Physiology: Some Case Histories* (ed. L.T. Evans), pp. 101–49. Cambridge University Press.

Evans, L.T., Visperas, R.M. & Vergara, B.S. (1984). Morphological and physiological changes among rice varieties used in the Philippines over the last seventy years. *Field Crops Res.* **8**, 105–24.

Evans, S.A. & Neild, J.R.A. (1981). The achievement of very high yields of potatoes in the UK. *J. Agric. Sci. (Camb.)* **97**, 391–6.

Evenson, R.E. (1982). Agriculture. In *Government and Technical Progress* (ed. R. Nelson), pp. 233–82. Pergamon Press, New York.

Evenson, R.E., O'Toole, J.C., Herdt, R.W., Coffman, W.R. & Kauffman, H.E. (1978). Risk and uncertainty as factors in crop improvement research. IRRI Res. Paper. No. 15, pp. 1–19.

Fakorede, M.A.B. & Mock, J.J. (1978). Changes in morphological and physiological traits associated with recurrent selection for grain yield in maize. *Euphytica* **27**, 397–409.

FAO (annual) *Fertilizer Yearbook*. FAO, Rome.

FAO (annual) *Production Yearbook*. FAO, Rome.

FAO (1979). Utilization of grains in the livestock sector: trends, factors and development issues. FAO Committee on Commodity Problems. Intergovernmental Group on Grains. Rome.

FAO (1981). *Agriculture: Toward 2000*. FAO, Rome.

FAO (1983). *Changing Patterns and Trends in Feed Utilization*. FAO Economic and Social Development Paper No. 37. FAO, Rome.

FAO (1986). *The State of Food and Agriculture, 1986*. FAO, Rome.

FAO (1987). *The Fifth World Food Survey*. FAO, Rome.

FAO/WHO (1973). *Ad Hoc Expert Committee on Energy and Protein Requirements*. Tech. Rep. Ser. No. 522. W.H.O., Geneva.

FAO/WHO/UNU (1985). Energy and protein requirements. WHO Technical Report, Series 724. WHO, Geneva.

Farmer, B.H. (1969). Available food supplies In *Population and Food Supply* (ed. J. Hutchinson), pp. 75–95. Cambridge University Press.

Farmer, B.H, (ed.) (1977). *Green Revolution? Technology and Change in Rice-growing Areas of Tamil Nadu and Sri Lanka*. Westview Press, Boulder, Colorado.

Farquhar, G.D. & Richards, R.A. (1984). Isotopic composition of plant carbon correlates with water-use efficiency of wheat genotypes. *Aust. J. Plant Physiol.* 11, 539–52.

Farquhar, G.D. & Sharkey, T.D. (1982). Stomatal conductance and photosynthesis. *Ann. Rev. Plant Physiol.* 33, 317–45.

Farquhar, G.D. & von Caemmerer, S. (1982). Modelling of photosynthetic response to environmental conditions. In *Encyclopedia of Plant Physiology*, (New Series 12 B) (ed. O.L. Lange, P.S. Nobel, C.B. Osmond & H. Ziegler), pp. 549–87. Springer, Berlin.

Farquhar, G.D., von Caemmerer, S. & Berry, J.A. (1980). A biochemical model of photosynthetic CO_2 assimilation in leaves of C_3 species. *Planta* 149, 78–90.

Feigin, A., Letey, J. & Jarrell, W.M. (1982). Celery response to type, amount, and method of N-fertilizer application under drip irrigation. *Agron. J.* 74, 971–7.

Felger, R.S. (1979). Ancient crops for the twenty-first century. In *New Agricultural Crops* (ed. G.A. Ritchie), pp. 5–20. Westview Press, Boulder, Colorado.

Feliks, J. (1963). *Agriculture in Palestine in the Period of the Mishna and Talmud*. Magnes Press, Jerusalem.

Ferguson, H. (1974). Use of variety isogenes in plant water use-efficiency studies. *Agric. Meteorol.* 14, 25–9.

Ferraris, R., Norman, M.J.T. & Andrews, A.C. (1973). Adaptation of pearl millet (*Pennisetum typhoides*) to coastal New South Wales. 1. Preliminary evaluation. *Aust. J. Expt. Agric. An. Hus.* 13, 685–91.

Feyerherm, A.M. & Paulsen, G.M. (1981). An analysis of temporal and regional variation in wheat yields. *Agron. J.* 73, 863–7.

Feyerherm, A.M. Paulsen, G.M. & Sebaugh, J.L. (1984). Contribution of genetic improvement to recent wheat yield increases in the USA. *Agron. J.* 76, 985–90.

Feyerherm, A.M., Kemp, K.E. & Paulsen, G.M. (1988). Wheat yield analysis in relation to advancing technology in the mid-west United States. *Agron. J.* 80, 998–1001.

Feyerherm, A.M., Kemp, K.E. & Paulsen, G.M. (1989). Genetic contribution to increased wheat yields in the USA between 1979 and 1984. *Agron. J.* 81, 242–5.

Fick, G.W., Loomis, R.S. & Williams, W.A. (1975). Sugar beet. In *Crop Physiology: Some Case Histories* (ed. L.T. Evans), pp. 259–95. Cambridge University Press.

Field, C. & Mooney, H.A. (1986). The photosynthesis–nitrogen relationship in wild plants. In *On the Economy of Plant Form and Function* (ed. T.J. Givnish), pp. 25–55. Cambridge University Press.

Finlay, K.W. & Wilkinson, G.N. (1963). The analysis of adaptation in a plant breeding programme. *Aust. J. Agric. Res.* **14**, 742–54.

Fischer, K.S. & Palmer, A.F.E. (1983). Maize. In *Potential Productivity of Field Crops Under Different Environments*, pp. 155–80. IRRI, Los Baños.

Fischer, K.S. & Palmer, A.F.E. (1984). Tropical maize. In *The Physiology of Tropical Field Crops* (ed. P.R. Goldsworthy & N.M. Fisher), pp. 213–48. John Wiley, New York.

Fischer, K.S. & Wilson, G.L. (1971). Studies of grain production in *Sorghum vulgare*. I. The contribution of pre-flowering photosynthesis to grain yield. *Aust. J. Agric. Res.* **22**, 33–7.

Fischer, K.S. & Wilson, G.L. (1975a). Studies of grain production in *Sorghum bicolor* (L. Moench.). III. The relative importance of assimilate supply, grain growth capacity and transport system. *Aust. J. Agric. Res.* **26**, 11–23.

Fischer, K.S. & Wilson, G.L. (1975b). Ibid. IV. Some effects of increasing and decreasing photosynthesis at different stages of the plant's development on the storage capacity of the inflorescence. *Aust. J. Agric. Res.* **26**, 25–30.

Fischer, K.S., Johnson, E.C. & Edmeades, G.O. (1982). Breeding and selection for drought resistance in tropical maize. In *Drought Resistance in Crops with Special Emphasis on Rice*, pp. 377–99. IRRI, Los Baños.

Fischer, K.S., Edmeades, G.O. & Johnson, E.C. (1989). Selection for the improvement of maize yield under moisture-deficits. *Field Crops Res.* **22**, 227–43.

Fischer, R.A. (1973). The effect of water stress at various stages of development on yield processes in wheat. In *Plant Responses to Climatic Factors* (ed. R.O. Slatyer), pp. 233–41. UNESCO, Paris.

Fischer, R.A. (1975). Yield potential in a dwarf spring wheat and the effect of shading. *Crop Sci.* **15**, 607–13.

Fischer, R.A. (1981). Optimizing the use of water and nitrogen through breeding of crops. *Plant and Soil* **58**, 249–78.

Fischer, R.A. (1983). Wheat. In *Potential Productivity of Field Crops under Different Environments*, pp. 129–54. IRRI, Los Baños.

Fischer, R.A. (1985a). Number of kernels in wheat crops and the influence of solar radiation and temperature. *J. Agric. Sci. (Camb.)* **105**, 447–61.

Fischer, R.A. (1985b). Physiological limitations to producing wheat in semi-tropical and tropical environments and possible selection criteria. In *Wheats for More Tropical Environments*, pp. 209–30. CIMMYT, Mexico.

Fischer, R.A. & I. Aguilar, M. (1976). Yield potential in a dwarf spring wheat and the effect of carbon dioxide fertilization. *Agron. J.* **68**, 749–52.

Fischer, R.A. & Hille Ris Lambers, D. (1978). Effect of environment and cultivar on source limitation to grain weight in wheat. *Aust. J. Agric. Res.* **29**, 443–58.

Fischer, R.A. & Kertesz, Z. (1976). Harvest index in spaced populations and grain weight in microplots as indicators of yielding ability in spring wheat. *Crop Sci.* **16**, 55–9.

Fischer, R.A. & Kohn, G.D. (1966). The relationship of grain yield to vegetative growth and post flowering leaf area in the wheat crop under conditions of limited soil moisture. *Aust. J. Agric. Res.* **17**, 281–95.

Fischer, R.A. & Laing, D.R. (1976). Yield potential in a dwarf spring wheat and response to crop thinning. *J. Agric. Sci. (Camb.)* **87**, 113–22.

Fischer, R.A. & Maurer, O.R. (1976). Crop temperature modification and yield potential in a dwarf spring wheat. *Crop Sci.* **16**, 855–9.

Fischer, R.A. & Maurer, O.R. (1978). Drought resistance in spring wheat cultivars. I. Grain yield responses. *Aust. J. Agric. Res.* **29**, 897–912.

Fischer, R.A. & Quail, K.J. (1989). The effect of major dwarfing genes on yield potential in spring wheats. *Euphytica* **46**, 51–6.

Fischer, R.A. & Stockman, Y.M. (1986). Increased kernel number in Norin 10-derived dwarf wheat. Evaluation of the cause. *Aust. J. Plant Physiol.* **13**, 767–84.

Fischer, R.A. & Wood, J.T. (1979). Drought resistance in spring wheat cultivars. III Yield associations with morpho-physiological traits. *Aust. J. Agric. Res.* **30**, 1001–20.

Fischer, R.A., Aguilar, I. & Laing, D.R. (1977). Post-anthesis sink size in a high-yielding dwarf wheat: yield responses to grain number. *Aust. J. Agric. Res.* **28**, 165–75.

Fischer, R.A., Bidinger, F., Syme, J.R. & Wall, P.C. (1981). Leaf photosynthesis, leaf permeability, crop growth, and yield of short spring wheat genotypes under irrigation. *Crop Sci.* **21**, 367–73.

Fisher, R.A. (1924). The influence of rainfall on the yield of wheat at Rothamsted. *Phil. Trans. R. Soc. Lond.* B**213**, 89–142.

Flannery, K.V. (1968). Archaeological systems theory and early MesoAmerica. In *Anthropological Archaeology in the Americas* (ed. B.J. Meggers), pp. 67–87. Anthropol. Soc., Washington, DC.

Flannery, K.V. (1972). The cultural evolution of civilizations. *Ann. Rev. Ecol. Systematics* **3**, 399–426.

Fleming, J.C. (1973). *The energy requirements for ammonia production.* B.Sc. Thesis, Univ. Strathclyde. (Cited by Slesser (1984).)

Flew, A.G.N. (1970). Introduction to *Malthus, An Essay on the Principle of Population.* Penguin Books, Harmondsworth.

Flight, C. (1976). The Kintampo culture and its place in the economic prehistory of West Africa. In *Origins of African Plant Domestication* (ed. J.R. Harlan, J.M.J. de Wet & A.B.L. Stemler), pp. 211–21. Mouton, The Hague.

Flinn, A.M. (1974). Regulation of leaflet photosynthesis by developing fruit in the pea. *Physiol. Plantar.* **31**, 275–8.

Flinn, J.C., De Datta, S.K. (1984). Trends in irrigated rice yields under intensive cropping at Philippine research stations. *Field Crops Res.* **9**, 1–15.

Flinn, J.C. & Duff, B. (1985). Energy analysis, rice production systems, and rice research. IRRI Res. Paper Ser. No. 114, pp. 1–11.

Flinn, J.C., Garrity, D.P. (1986). Yield stability and modern rice technology. In *Variability in Grain Yields.* (ed. J.R. Anderson & P.B.R. Hazell), pp. 251–64. Johns Hopkins, Baltimore.

Flores, A., Grau, A., Laurich, F. & Dörffling, K. (1988). Effect of new terpenoid analogues of abscisic acid on chilling and freezing resistance. *J. Plant Physiol.* **132**, 362–9.

Flower, D.J., Rani, U.A. & Peacock, J.M. (1990). Influence of osmotic adjustment on the growth, stomatal conductance and light interception of contrasting sorghum lines in a harsh environment. *Aust. J. Plant Physiol.* **17**, 91–105.

Fluck, R.C. (1981). Net energy sequestered in agricultural labour. *Trans. Amer. Soc. Agric. Eng.* **24**, 1449–55.

Fluck, R.C. & Baird, C.D. (1980). *Agricultural Energetics.* AVI Publications, Westport, Connecticut.

Ford, D.M. & Shibles, R. (1988) Photosynthesis and other traits in relation to chloroplast number during soybean leaf senescence. *Plant Physiol.* **86**, 108–11.

Ford, D.M., Shibles, R. & Green, D.E. (1983). Growth and yield of soybean lines selected for divergent leaf photosynthetic ability. *Crop Sci.* **23**, 517–20.

Ford, M.A., Austin, R.B., Angus, W.J. & Sage, G.C.M. (1981). Relationships between the responses of spring wheat genotypes to temperature and photoperiodic treatments and their performance in the field. *J. Agric. Sci. (Camb.)* **96**, 623–34.

Forrester, J.W. (1971). *World Dynamics*. Wright-Allen Press, Cambridge, Massachusetts.

Fousova, S. & Avratovscukova, N. (1967). Hybrid vigour and photosynthetic rate of leaf disks in *Zea mays* L. *Photosynthetica* **1**, 3–12.

Fox, R.L. (1979). Comparative responses of field grown crops to phosphate concentrations in soil solutions. In *Stress Physiology in Crop Plants* (ed. H. Mussell, R.C. Staples), pp. 81–106. Wiley, New York.

Foy, C.D., Armiger, W.H., Briggle, L.W. & Reid, D.A. (1965). Differential aluminum tolerance of wheat and barley varieties in acid soils. *Agron. J.* **57**, 413–17.

Foy, C.D., Armiger, W.H., Fleming, A.L. & Lewis, C.F. (1967). Differential tolerance of cotton varieties to an acid soil high in exchangeable aluminum. *Agron. J.* **59**, 415–18.

Foy, C.D., Fleming A.L. & Armiger, W.H. (1969). Differential tolerance of cotton varieties to excess manganese. *Agron. J.* **61**, 690–4.

Foyer, C.H. (1987). The basis for source-sink interaction in leaves. *Plant Physiol. Biochem.* **25**, 649–57.

Francis, C.A. (1972). Photoperiod sensitivity and adaptation in maize. *Proc. 27th Ann. Corn and Sorghum Res. Conf.* **27**, 119–31.

Francis, C.A., Saeed, M., Nelson, L.A. & Moomaw, R. (1984). Yield stability of sorghum hybrids and random-mating populations in early and late planting dates. *Crop Sci.* **24**, 1109–12.

Frankel, F.R. (1971). *India's Green Revolution. Economic Gains and Political Costs*. Princeton University Press.

Frankel, O.H. (1947). The theory of plant breeding for yield. *Heredity* **1**, 109–20.

Frankel, R, (ed.) (1983). *Heterosis: Reappraisal of Theory and Practice*. Springer, Berlin.

Freeman, C. & Jahoda, M. (eds.) (1978). *World Futures: The Great Debate*. Martin Robertson, London.

Freeman, J.D. (1955). *Iban Agriculture*. HMSO, London.

French, R.J. & Schultz, J.E. (1984). Water use efficiency of wheat in a Mediterranean-type environment. II. Some limitations to efficiency. *Aust. J. Agric. Res.* **35**, 765–75.

Frey, K.J. (1964). Adaptation reaction of oat strains selected under stress and non-stress environmental conditions. *Crop Sci.* **4**, 55–8.

Frey, K.J. (1971). Improving crop yields through plant breeding. In *Moving off the Yield Plateau* (ed. J.D. Eastin & R.D. Munson), pp. 15–18. Amer. Soc. Agron., Madison, Wisconsin. Special Publication 20.

Frey, K.J., Cox, T.S., Rodgers, D.M. & Bramel-Cox, P. (1984). Increasing cereal yields with genes from wild and weedy species. In *Proc. 15th Intl. Genetics Congress, New Delhi, India*, pp. 51–68. Oxford/IBH, New Delhi.

Fritsch, R. (1977). Über morphologische Wurzelmerkmale bei *Triticum* L. und *Aegilops* L. (Gramineae). *Kulturpflanze* **25**, 45–70.

Frye, W.W., Walker, J.N. & Duncan, G.A. (1982). Comparison of energy requirements of no-tillage and conventional tillage. In *No-Tillage Research: Research Reports and Reviews* (ed. R.E. Phillips, G.W. Thomas & R.L. Blevins), pp. 76–83. University of Kentucky, Lexington.

Fryer, J.D. (1981). Weed management: fact or fable? *Phil. Trans. R. Soc. Lond.* **B295**, 185–97.

Fujii, J.A. & Kennedy, R.A. (1985). Seasonal changes in the photosynthetic rate in apple trees. *Plant Physiol.* **78**, 519–24.

Fussell, G.E. (1967). Farming systems of the Classical Era. *Technology and Culture* **8**, 16–44.

Fussell, L.K. & Pearson, C.J. (1978). The course of grain development and its relationship to black region appearance in *Pennisetum americanum*. *Field Crops Res.* **1**, 21–31.

Fussell, L.K., Pearson, C.J. & Norman, M.J.T. (1980). Effects of grain development and thermal history on grain maturation and seed vigour of *Pennisetum americanum*. *J. Exp. Bot.* **31**, 621–33.

Gabel, M. (1979). *Ho-Ping: Food for Everyone*. Anchor Books, New York.

Gabor, D. (1963). *Inventing the Future*. Secker and Warburg, London.

Gale, M.D. & Youssefian, S. (1983). Pleiotropic effect of the Norin 10 dwarfing genes, Rht1 and Rht2, and interactions in response to chlormequat. *Proc. 6th Intl. Wheat Genetics Sympos, Kyoto, Japan*, pp. 271–7.

Gale, M.D. & Youssefian, S. (1985). Dwarfing genes in wheat. In *Progress in Plant Breeding*, vol. 1 (ed. G.E. Russell), pp. 1–35. Butterworths, London.

Gale, M.D., Edrich, J. & Lupton, F.G.H. (1974). Photosynthetic rates and the effects of applied gibberellin in some dwarf, semi-dwarf and tall wheat varieties (*Triticum aestivum*). *J. Agric. Sci. (Camb.)* **83**, 43–6.

Gales, K. (1983). Yield variation of wheat and barley in Britain in relation to crop growth and soil conditions – a review. *J. Sci. Food Agric.* **34**, 1085–104.

Galinat, W.C. (1974). The domestication and genetic erosion of maize. *Econ. Bot.* **28**, 21–7.

Galinat, W.C. (1988). The origin of corn. In *Corn and Corn Improvement*, 3rd edn (ed. G.F. Sprague & J.W. Dudley), pp. 1–31. Amer. Soc. Agron., Madison, Wisconsin.

Gallagher, J.N. & Biscoe, P.V. (1978). Radiation absorption, growth and yield of cereals. *J. Agric. Sci. (Camb.)* **91**, 47–60.

Gallagher, J.N., Biscoe, P.V. & Scott, R.K. (1975). Barley and its environment. V. Stability of grain weight. *J. Appl. Ecol.* **12**, 319–36.

Gallagher, J.N., Biscoe, P.V. & Scott, R.K. (1976a). Barley and its environment. VI. Growth and development in relation to yield. *J. Appl. Ecol.* **13**, 563–83.

Gallagher, J.N., Biscoe, P.V. & Hunter, B. (1976b). Effects of drought on grain growth. *Nature (Lond.)* **264**, 541–2.

Gallaher, R.N., Ashley, D.A. & Brown, R.H. (1975). ^{14}C photosynthate translocation in C_3 and C_4 plants as related to leaf anatomy. *Crop Sci.* **15**, 55–9.

Gangopadhyaya, M. & Sarker, R.P. (1965). Influence of rainfall distribution on the yield of wheat crop. *Agric. Meteorol.* **2**, 331–50.

Gardner, F.P., Valle, R. & McCloud, D.E. (1990). Yield characteristics of ancient races of maize compared to a modern hybrid. *Agron. J.* **82**, 864–8.

Garner, W.W. & Allard, H.A. (1920). Effect of the relative length of day and night and other factors of the environment on growth and reproduction in plants. *J. Agric. Res.* **18**, 553–606.

Garner, W.W. & Allard, H.A. (1930). Photoperiod responses of soybeans in relation to temperature and other environmental factors. *J. Agric. Res.* **41**, 719–35.

Gaskel, M.L. & Pearce, R.B. (1981). Growth analysis of maize hybrids differing in photosynthetic capability. *Agron. J.* **73**, 817–21.

Gaskel, M.L. & Pearce, R.B. (1983). Stomatal frequency and stomatal resistance of maize hybrids differing in photosynthetic capability. *Crop Sci.* **23**, 176–7.

Gavin, W. (1951). The way to higher crop yields. *J. Min. Agric.* **58**, 105–11.

Gawronska, H., Dwelle, R.B., Pavek, J.J. & Rowe, P. (1984). Partitioning of photoassimilates by four potato clones. *Crop Sci.* **24**, 1031–6.

Gay, S., Egli, D.B. & Reicosky, D.A. (1980). Physiological aspects of yield improvement in soybeans. *Agron. J.* **72**, 387–91.

Gebeyehou, G., Knott, D.R. & Baker, R.J. (1982). Relationships among durations of vegetative and grain filling phases, yield components, and grain yield in durum wheat cultivars. *Crop Sci.* **22**, 287–90.

Geertz, C. (1963). *Agricultural Involution. The Process of Ecological Change in Indonesia.* University of California Press, Berkeley.

Geiger, D.R. (1966). Effect of sink region cooling on translocation of photosynthate. *Plant Physiol.* **41**, 1667–72.

Geiger, D.R. (1976). Effects of translocation and assimilate demand on photosynthesis. *Can. J. Bot.* **54**, 2337–45.

Geiger, D.R. (1987). Understanding interactions of source and sink regions of plants. *Plant Physiol. Biochem.* **25**, 659–66.

Geissbuhler, H. (1981). The agrochemical industry's approach to integrated pest control. *Phil. Trans. R. Soc. Lond.* B**295**, 111–23.

Gent, M.P.N. & Kiyomoto, R.K. (1985). Comparison of canopy and flag leaf net carbon dioxide exchange of 1920 and 1977 New York winter wheats. *Crop Sci.* **25**, 81–6.

Gent, M.P.N. & Kiyomoto, R.K. (1989). Assimilation and distribution of photosynthate in winter wheat cultivars differing in harvest index. *Crop Sci.* **29**, 120–5.

Gentry, H.S. (1942). *Rio Mayo Plants.* Carnegie Inst. Publicn. No. 527.

Gentry, H.S. (1969). Origin of the common bean, *Phaseolus vulgaris. Econ. Bot.* **23**, 55–69.

Gepts, P. (1990). Biochemical evidence bearing on the domestication of *Phaseolus* (Fabaceae) beans. *Econ. Bot.* **44** (supp. 3), 28–38.

Gerbaud, A., André, M., Gaudillère, J.-P. & Daguenet, A. (1988). The influence of reduced photorespiration on long term growth and development in wheat. *Physiol. Plantar.* **73**, 479–85.

Gerik, T.J. & Eastin, J.D. (1985). Temperature effects on dark respiration among diverse sorghum genotypes. *Crop Sci.* **25**, 957–61.

Gerlach, W.L., Llewellyn, D. & Haseloff, J. (1987). Construction of a plant disease resistance gene from the satellite RNA of tobacco ringspot virus. *Nature (Lond.)* **328**, 802–5.

Gifford, R.M. (1974*a*). A comparison of potential photosynthesis, productivity and yield of plant species with differing photosynthetic metabolism. *Aust. J. Plant Physiol.* **1**, 107–17.

Gifford, R.M. (1974*b*). Photosynthetic limitations to cereal yield. In *Mechanisms of Regulation of Plant Growth* (ed. R.L. Bieleski, A.R. Ferguson & M.M. Cresswell), pp. 887–93. Bull. 12, Roy. Soc. New Zealand.

Gifford, R.M. (1976). An overview of fuel used for crops and national agricultural systems. *Search* **7**, 412–17.

Gifford, R.M. (1977). Growth pattern, carbon dioxide exchange and dry weight distribution in wheat growing under different photosynthetic environments. *Aust. J. Plant Physiol.* **4**, 99–110.

Gifford, R.M. (1979). Growth and yield of CO_2-enriched wheat under water-limited conditions. *Aust. J. Plant Physiol.* **6**, 367–78.

Gifford, R.M. (1984). Energy in different agricultural systems: renewable and non-renewable sources. In *Energy and Agriculture* (ed. G. Stanhill), pp. 84–112. Springer, Berlin.

Gifford, R.M. (1986). Partitioning of photoassimilate in the development of crop yield. In *Plant Biology*, vol. 1. (*Phloem Transport*) (ed. J. Cronshaw, W.L. Lucas & R.T. Giaquinta), pp. 535–549. Allan R. Liss, New York.

Gifford, R.M. (1992). Interaction of carbon dioxide with growth limiting environmental factors in vegetation productivity: implications for the global carbon cycle. *Adv. Bioclimatol.* **1**, 25–58.

Gifford, R.M. & Bremner, P.M. (1981). Accumulation and conversion of sugars by developing wheat grains. I. Liquid culture of kernels over several days. *Aust. J. Plant Physiol.* **8**, 619–29.

Gifford, R.M. & Marshall, C. (1973). Photosynthesis and assimilate distribution in *Lolium multiflorum* Lam. following differential tiller defoliation. *Aust. J. Biol. Sci.* **26**, 517–26.

Gifford, R.M. & Millington, R.J. (1975). Energetics of agriculture and food production. CSIRO Bull. No. 288.

Gifford, R.M., Bremner, P.M. & Jones, D.B. (1973). Assessing photosynthetic limitation to grain yield in a field crop. *Aust. J. Agric. Res.* **24**, 297–307.

Gifford, R.M., Thorne, J.H., Hitz, W.D., Giaquinta, R.T. (1984). Crop productivity and photoassimilate partitioning. *Science* **225**, 801–8.

Gifford, R.M., Lambers, H. & Morison, J.I.L. (1985). Respiration of crop species under CO_2 enrichment. *Physiol. Plantar.* **63**, 351–6.

Gilland, B. (1985). Cereal yields in theory and practice. *Outlook on Agric.* **14**, (2), 56–60.

Givnish, T.J. (ed.) (1986). *On the Economy of Plant Form and Function.* Cambridge University Press.

Gladstones, J.S. (1967). Selection for economic characters in *Lupinus angustifolius* and *L. digitatus*. 1. Non-shattering pods. *Aust. J. Exp. Agric. An. Husb.* **7**, 360–6.

Gladstones, J.S. (1970). Lupins as crop plants. *Field Crop Abs.* **23**, 128–48.

Godwin, W. (1793). *An Enquiry concerning Political Justice.* Book 8, Chapter 7.

Goedmakers, A. (1989). Ecological perspectives of changing agricultural land use in the European community. *Agriculture, Ecosys. & Env.* **27**, 99–106.

Goldsworthy, P.R. (1970). The growth and yield of tall and short sorghums in Nigeria. *J. Agric. Sci. (Camb.)* **75**, 109–22.

Golson, J. (1977). No room at the top: agricultural intensification in the New Guinea highlands. In *Sunda and Sahul* (ed. J. Allen, J. Golson & R. Jones), pp. 601–38. Academic Press, London.

Golson, J. (1982). The Ipomoean revolution revisited: society and the sweet potato in the upper Wahgi valley. In *Inequality in New Guinea Highlands Societies* (ed. A. Strathern), pp. 109–85. Cambridge University Press.

Golson, J. (1985). Agricultural origins in South east Asia. In *Recent Advances in Indo-Pacific Prehistory* (ed. V.N. Misra & P. Bellwood), pp. 307–14. Oxford University Press, New Delhi.

Goodman, D., Sorj, B. & Wilkinson, J. (1987). *From Farming to Biotechnology. A Theory of Agro-Industrial Development.* Blackwell, Oxford.

Goodman, M.M. (1988). The history and evolution of maize. *CRC Crit. Rev. Plant Sci.* **7**, 197–220.

Gordon, A.J., Hesketh, J.D. & Peters, D.B. (1982). Soybean leaf photosynthesis in relation to maturity classification and stage of growth. *Photosynthesis Res.* **3**, 81–93.

Gorman, C.F. (1969). Hoabinhian: a pebble tool-complex with early plant associations in southeast Asia. *Science* **163**, 671–3.

Gorman, C. (1977). *A priori* models and Thai prehistory: a reconsideration of the beginnings of agriculture in Southeastern Asia. In *Origins of Agriculture* (ed. C.A. Reed), pp. 321–55. Mouton, The Hague.

Gotoh, K. & Osanai, S.-I. (1959). Efficiency of selection for yield under different fertilizer levels in a wheat cross. *Jap. J. Breeding* **9**, 173–8.

Gotoh, T. (1976). Studies on varietal differences in vernalization requirement in wheat. *Jap. J. Breeding* **26**, 307–27.

Gott, M.B. (1961). Flowering of Australian wheats and its relation to frost injury. *Aust. J. Agric. Res.* **12**, 547–65.

Goudriaan, J., de Ruiter, H.E. (1983). Plant growth in response to CO_2 enrichment, at two levels of nitrogen and phosphorus supply. 1. Dry matter, leaf area and development. *Neth. J. Agric. Sci.* **31**, 157–69.

Gould, F. (1988). Evolutionary biology and genetically engineered crops. *Bioscience* **38** (1), 26–33.

Grafius, J.E. (1965). A geometry of plant breeding. *Mich. State Univ. Agric. Exp. Sta. Res. Bull.* No. 7.

Grafius, J.E. (1978). Multiple characters and correlated response. *Crop Sci.* **18**, 931–4.

Grafius, J.E., Thomas, R.L. & Barnard, J. (1976). Effect of parental component complementation on yield and components of yield in barley. *Crop Sci.* **16**, 673–7.

Graham, D. (1980). Effects of light on 'dark' respiration. In *The Biochemistry of Plants* (ed. P.K. Stumpf & E.E. Conn), vol. 2, pp. 525–79. Academic Press, New York.

Graham, R.D. (1984). Breeding for nutritional characteristics in cereals. *Adv. Plant Nutrition* **1**, 57–102.

Graham-Bryce, I.J. (1981). The current status and future potential of chemical approaches to crop protection. *Phil. Trans. R. Soc. Lond.* B**295**, 5–16.

Grava, J. & Raisanen, K.A. (1978). Growth and nutrient accumulation and distribution in wild rice. *Agron. J.* **70**, 1077–81.

Greeley, M. (1982*a*). Pinpointing post-harvest losses. *Ceres* **15** (1), 30–37.

Greeley, M. (1982*b*). Farm-level post-harvest food losses: the myth of the soft third option. *Inst. Dev. Studies, Sussex Bull.* **13** (3), 51–60.

Green, C.F. (1989). Genotypic differences in the growth of *Triticum aestivum* in relation to absorbed solar radiation. *Field Crops Res.* **19**, 285–95.

Green, C.F. & Vaidyanathan, L.V. (1986). A reappraisal of biomass accumulation by temperate cereal crops. *Speculations in Sci. Technol.* **9**, 193–212.

Green, M.B. (1978). *Eating Oil. Energy Use in Food Production.* Westview Press, Boulder, Colorado.

Green, M.B. & McCulloch, A. (1976). Energy considerations in the use of herbicides. *J. Sci. Food Agric.* **27**, 95–100.

Green, M.B., Hartley, G.S. & West, T.F. (1977). *Chemicals for Crop Protection and Pest Control.* Pergamon, Oxford.

Gregory, A.C. (1886). Memoranda on the Aborigines of Australia. *J. Anthropol. Inst.* **16**, 131–3.

Gregory, F.G., Crowther, F. & Lambert, A.R. (1932). The interrelation of factors controlling the production of cotton under irrigation in the Sudan. *J. Agric. Sci. (Camb.)* **22**, 617–38.

Gressel, J. (1978). Genetic herbicide resistance: projections on appearance in weeds and breeding for it in crops. In *Plant Regulation and World Agriculture* (ed. T.K. Scott), pp. 85–109. Plenum, New York.

Gressel, J. (1984). Evolution of herbicide-resistant weeds. In *Origins and Development of Adaptation*, pp. 73–93. Pitman Books, London.

Grey, G. (1841). *Journals of Two Expeditions of Discovery in North-West and*

Western Australia During the Years 1837, 1838 and 1839, vol. 2. T. & W. Boone, London.

Gribbin, J. (1974). Climate and the world's food. *New Scientist* **64**, 643–5.

Griffin, K. (1975). *The Political Economy of Agrarian Change*. Macmillan, London.

Grigg, D. (1982). *The Dynamics of Agricultural Change. The Historical Experience*. St Martins Press, New York.

Grun, P. (1990). The evolution of cultivated potatoes. *Econ. Bot.* **44** (suppl. 3), 39–55.

Guinn, G. & Mauney, J.R. (1980). Analysis of CO_2 exchange assumptions: feedback control. In *Predicting Photosynthesis for Ecosystem Models* (ed. J.D. Hesketh & J.W. Jones), vol. 2, pp. 1–16. CRC Press, Boca Raton, Florida.

Gupta, S.C., Frey, K.J. & Cox, D.J. (1986a). Changes in several traits of oats caused by selection for vegetative growth rate. *Plant Breeding* **97**, 222–6.

Gupta, S.C., Cox, D.J. & Frey, K.J. (1986b). Association of two measures of vegetative growth rate with other traits in inter and intra-specific matings of oats. *Theoret. and Appl. Genet.* **72**, 756–60.

Haas, H. & Streibig, J.C. (1982). Changing patterns of weed distribution as a result of herbicide use and other agronomic factors. In *Herbicide Resistance in Plants* (ed. H.M. Le Baron & J. Gressel), pp. 57–80. Wiley Interscience, New York.

Habeshaw, D. (1973). Translocation and the control of photosynthesis in sugar beet. *Planta* **110**, 213–26.

Hadley, P., Roberts, E.H. Summerfield, R.J. & Minchin, F.R. (1983). A quantitative model of reproductive development in cowpea (*Vigna unguiculata* (L.) Walp.). in relation to photoperiod and temperature, and implications for screening germplasm. *Ann. Bot.* **51**, 531–43.

Hadley, P., Roberts, E.H., Summerfield, R.J. & Minchin, F.R. (1984). Effects of temperature and photoperiod on flowering in soyabean (*Glycine max* (L.) Merrill): a quantitative model. *Ann. Bot.* **53**, 669–81.

Hahn, S.K. (1977). A quantitative approach to source potentials and sink capacities among reciprocal grafts of sweet potato varieties. *Crop Sci.* **17**, 559–62.

Hahn, S.K. (1982). Screening sweet potato for source potentials. *Euphytica* **31**, 13–18.

Hahn, S.K. & Hozyo, Y. (1983). Sweet potato and yam. In *Potential Productivity of Field Crops Under Different Environments*, pp. 319–40. IRRI, Los Baños.

Hakansson, I. (1980). Swedish experiments with crop response to wheel-induced soil compaction. In *Proc. Working Group on Soil Compaction by Vehicles of High Axle Load*, pp. 38–40. Uppsala.

Hall, A.D. & Unwin, M.J. (1974–6). Another look at wheat yield trends. *New Zealand Wheat Review* No. 13, pp. 60–4.

Hall, A.E. & Loomis, R.S. (1972). An explanation for the difference in photosynthetic capabilities of healthy and beet yellows virus-infected sugar beets (*Beta vulgaris* L.) *Plant Physiol.* **50**, 576–80.

Hall, A.J. & Milthorpe, F.L. (1978). Assimilation source-sink relationships in *Capsicum annuum* L. III. The effects of fruit excision on photosynthesis and leaf and stem carbohydrates. *Aust. J. Plant Physiol.* **5**, 1–13.

Hall, A.J., Whitfield, D.M. & Connor, D.J. (1990). Contribution of preanthesis assimilates to grain-filling in irrigated and water-stressed sunflower crops. II. Estimates from a carbon budget. *Field Crops Res.* **24**, 273–94.

Hallett, F.F. (1861). On "Pedigree" in wheat as a means of increasing the crop. *J. Roy. Agric. Soc.* **22**, 371–81.

Ham, G.J. (1970). Water requirements of sugar cane. In *Water Problems in Tropical Queensland*, Water Res. Foundation of Australia Report No. 32, paper 4, pp. 1–7.

Hamilton, R.I., Subramanian, B., Reddy, M.N. & Rao, C.H. (1982). Compensation in grain yield components in a panicle of rainfed sorghum. *Ann. Appl. Biol.* **101**, 119–25.

Hancock, W.K. (1972). *Discovering Monaro: a Study of Man's Impact on his Environment.* Cambridge University Press.

Hang, A.N., McCloud, D.E., Boote, K.J. & Duncan, W.G. (1984). Shade effects on growth, partitioning and yield components of peanuts. *Crop Sci.* **24**, 109–15.

Hanks, R.J. (1983). Yield and water use relationships. In *Limitations to Efficient Water Use in Crop Production* (ed. H. Taylor, W.R. Jordan & T.R. Sinclair), pp. 393–412. Amer. Soc. Agron., Madison, Wisconsin.

Hansen, P. (1970). ^{14}C studies on apple trees. VI. The influence of the fruit on the photosynthesis of the leaves, and the relative photosynthetic yields of fruits and leaves. *Physiol. Plantar.* **23**, 805–10.

Hanson, H., Borlaug, N.E. & Anderson, R.G. (1982). *Wheat in the Third World.* Westview Press, Boulder, Colorado.

Hanson, P.R., Jenkins, G. & Westcott, B. (1981). Early generation selection in a cross of spring barley. *Zeitschr. f. Pflanzenzücht.* **83**, 64–80.

Hanson, P.R., Riggs, T.J., Klose, S.I. & Austin, R.B. (1985). High biomass genotypes in spring barley. *J. Agric. Sci. (Camb.).* **105**, 73–8.

Hanson, W.D. (1971). Selection for differential productivity among juvenile maize plants: Associated net photosynthetic rate and leaf area changes. *Crop Sci.* **11**, 334–9.

Hanson, W.D. (1985). Association of seed yield with partitioned lengths of the reproductive period in soybean genotypes. *Crop Sci.* **25**, 525–9.

Hanway, J.J. & Weber, C.R. (1971). Dry matter accumulation in eight soybean (*Glycine max* (L.) Merrill) varieties. *Agron. J.* **63**, 227–30.

Hardacre, A.K. & Eagles, H.A. (1986). Comparative temperature response of Corn Belt Dent × Pool 5 maize hybrids. *Crop Sci.* **26**, 1009–12.

Hardman, L.L. & Brun, W.A. (1971). Effect of atmospheric carbon dioxide enrichment at different developmental stages on growth and yield components of soybeans. *Crop Sci.* **11**, 886–8.

Hardy, R.W.F. (1978). Chemical plant growth regulation in world agriculture. In *Plant Growth Regulation and World Agriculture* (ed. T.K. Scott), pp. 165–206. Plenum, New York.

Hargrove, T.R. (1977). Genetic and sociologic aspects of rice breeding in India. IRRI Res. Paper Ser. No. 10, pp. 1–31.

Hargrove, T.R., Coffman, W.R. & Cabanilla, V.L. (1980). Ancestry of improved cultivars of Asian rice. *Crop Sci.* **20**, 721–7.

Harlan, J.R. (1967). A wild wheat harvest in Turkey. *Archaeol.* **20**, 197–201.

Harlan, J.R. (1969). Evolutionary dynamics of plant domestication. *Japan J. Genetics* **44**, (suppl. 1), 337–43.

Harlan, J.R. (1971). Agricultural origins: centers and noncenters. *Science* **174**, 468–74.

Harlan, J.R. (1975). *Crops and Man.* Amer. Soc. Agron., Madison, Wisconsin.

Harlan, J.R. (1976). Genetic resources in wild relatives of crops. *Crop Sci.* **16**, 329–33.

Harlan, J.R. (1977). The origins of cereal agriculture in the Old World. In *Origins of Agriculture* (ed. C.A. Reed), pp. 357–83. Mouton, The Hague.

Harlan, J.R. (1981). The early history of wheat: earliest traces to the sack of Rome. In *Wheat Science – Today and Tomorrow* (ed. L.T. Evans & W.J. Peacock), pp. 1–19. Cambridge University Press.

Harlan, J.R. (1986). Plant domestication: diffuse origins and diffusions. In *The Origin and Domestication of Cultivated Plants* (ed. G. Barigozzi), pp. 21–34. Elsevier, Amsterdam.

Harlan, J.R. & de Wet, J.M.J. (1965). Some thoughts about weeds. *Econ. Bot.* **19**, 16–24.

Harlan, J.R. & de Wet, J.M.J. (1971). Towards a rational classification of cultivated plants. *Taxon* **20**, 509–17.

Harlan, J.R. & de Wet, J.M.J. (1973). On the quality of evidence for origin and dispersal of cultivated plants. *Current Anthropol.* **14**, 51–62.

Harlan, J.R. & Pasquereau, J. (1969). Décrue agriculture in Mali. *Econ. Bot.* **23**, 70–4.

Harlan, J.R. & Stemler, A. (1976). The races of Sorghum in Africa. In *Origins of African Plant Domestication.* (ed. J.R. Harlan, J.M.J. de Wet & A.B.L. Stemler), pp. 465–78. Mouton, The Hague.

Harlan, J.R. & Zohary, D. (1966). Distribution of wild wheats and barley. *Science* **153**, 1074–80.

Harlan, J.R., de Wet, J.M.J. & Price, E.G. (1973). Comparative evolution of cereals. *Evolution* **27**, 311–25.

Harlan, J.R., de Wet, J.M.J. & Stemler, A. (1976). Plant domestication and indigenous African agriculture. In *Origins of African Plant Domestication* (ed. J.R. Harlan, J.M.J. de Wet & A.B.L. Stemler), pp. 3–19. Mouton, The Hague.

Harland, S.C. (1951). Genetics and the world's food. In *Four Thousand Million Mouths* (ed. F. Le Gros Clark & N.W. Pirie), pp. 93–108. Oxford University Press, London.

Harper, J.L. (1957). Ecological aspects of weed control. *Outlook on Agric.* **1**, 197–205.

Harper, J.L. (1977). *Population Biology of Plants.* Academic Press, New York.

Harris, D.R. (1972). The origins of agriculture in the Tropics. *Amer. Sci.* **60**(2), 180–93.

Harris, D.R. (1976). Traditional systems of plant food production and the origins of agriculture in West Africa. In *Origins of African Plant Domestication.* (ed. J.R. Harlan, J.M.J. de Wet & A.B.L. Stemler), pp. 311–56. Mouton, The Hague.

Harris, D.R. (1977*a*). Subsistence strategies across Torres Strait. In *Sunda and Sahul. Prehistoric Studies in Southeast Asia, Melanesia and Australia.* (ed. J. Allen, J. Golson & R. Jones), pp. 421–63. Academic Press, New York.

Harris, D.R. (1977*b*). Alternative pathways toward agriculture. In *Origins of Agriculture* (ed. C.A. Reed), pp. 179–243. Mouton, The Hague.

Harris, R.E., Moll, R.H. & Stuber, C.W. (1976). Control and inheritance of prolificacy in maize. *Crop Sci.* **16**, 843–50.

Harrison, B.D., Mayo, M.A. & Baulcombe, D.C. (1987). Virus resistance in transgenic plants that express cucumber mosaic virus satellite RNA. *Nature (Lond.).* **328**, 799–802.

Harrison, S.A., Boerma, H.R. & Ashley, D.A. (1981). Heritability of canopy apparent photosynthesis and its relationship to seed yield in soybeans. *Crop Sci.* **21**, 222–6.

Hart, R.H., Pearce, R.B., Chatterton, N.J., Carlson, G.E., Barnes, D.K. & Hanson, C.H. (1978). Alfalfa yield, specific leaf weight, CO_2 exchange rate, and morphology. *Crop Sci.* **18**, 649–53.

Hartt, C.E. (1963). Translocation as a factor in photosynthesis. *Naturwiss.* **50**, 666–7.

Hartung, R.C., Poneleit, C.G. & Cornelius, P.L. (1989). Direct and correlated responses to selection for rate and duration of grain fill in maize. *Crop Sci.* **29**, 740–5.

Hartwig, E.E. (1973). Varietal development. In *Soybeans: Improvement, Production and Uses* (ed. B.E. Caldwell), pp. 187–210. Amer. Soc. Agron., Madison, Wisconsin.

Hasegawa, T., Horie, T. & Yandell, B.S. (1991). Improvement of yielding ability in Japonica rice cultivars and its impact on regional yield increase in Kinki District, Japan. *Agric. Systems* **35**, 173–87.

Haudricourt, A.G. & Hedin, L. (1943). *L'homme et les Plantes Cultivées*. Gallimard, Paris.

Hawaian Sugar Planters Association (1967). Transplanting spacing trials. *Ann. Rep.* 1967,p. 14.

Hawkes, J. (1967). The history of the potato. *J. Royal Hortic. Soc.* **92**, 207–24; 249–62; 288–302; 364–5.

Hawkes, J.G. (1985). *Plant Genetic Resources. The Impact of the International Agricultural Research Centers.* CGIAR Study Paper, The World Bank, Washington, DC.

Hawkes, J.G. (1989). The domestication of roots and tubers in the American tropics. In *Foraging and Farming: The Evolution of Plant Exploitation* (ed. D.R. Harris & G.C. Hillman), pp. 481–503. Unwin Hyman, London.

Hawkes, J.G. (1990). *The Potato. Evolution, Biodiversity and Genetic Resources.* Belhaven Press, London.

Haws, L.D., Inoue, H., Tanaka, A. & Yoshida, S. (1983). Comparison of crop productivity in the tropics and temperate zone. In *Potential Productivity of Field Crops Under Different Environments*, pp. 403–13. IRRI, Los Baños.

Hay, R.K.M. & Kirby, E.J.M. (1991). Convergence and synchrony – a review of the coordination of development in wheat. *Aust. J. Agric. Res.* **42**, 661–700.

Hayami, Y. (with Akino, M., Shintani, M. & Yamada, S.) (1975). *A Century of Agricultural Growth in Japan.* University of Tokyo Press, Tokyo.

Hayami, Y. & Ruttan, V.W. (1971). *Agricultural Development: an International Perspective.* Johns Hopkins, Baltimore.

Hayashi, K. & Ito, H. (1962). Studies on the form of rice plant varieties with particular reference to the efficiency of utilizing sunlight. I. The significance of extinction coefficient in rice plant communities. *Proc. Crop Sci. Soc. Japan* **30**, 329–34.

Hayes, J.D. (1970). The architecture of yield in spring barley. *Ann. Rep. Welsh Plant Breeding Station*, 1970, pp. 32–3.

Hayes, P.M. & Stucker, R.E. (1987). Selection for heading date synchrony in wild rice. *Crop Sci.* **27**, 653–8.

Hayes, P.M., Stucker, R.E. & Wandrey, G.G. (1989). The domestication of American wild rice (*Zizania palustris*, Poaceae). *Econ. Bot.* **43**, 203–14.

Hazell, P.B.R. (1982). *Instability in Indian Food Grain Production.* IFPRI Res. Rep. No. 30, pp. 1–60.

Hazell, P.B.R. (1984). Sources of increased instability in Indian and US cereal production. *Amer. J. Agric. Econ.* **66**, 302–11.

Healy, M.J.R. & Jones, E.L. (1962). Wheat yields in England, 1815–1859. *J. Roy. Stat. Soc.* **125**, 574–9.

Hearn, A.B. & Constable, G.A. (1984). Cotton. In *The Physiology of Tropical Food Crops* (ed. P.R. Goldsworthy & N.M. Fisher), pp. 495–527. Wiley, New York.

Heath, O.V.S. & Gregory, F.G. (1938). The constancy of the mean net assimilation rate and its ecological importance. *Ann. Bot. N.S.* **2**, 811–18.

Hedin, P.A. (1990). Bioregulator-induced changes in allelochemicals and their effects on plant resistance to pests. *CRC Crit. Rev. Plant Sci.* **9**, 371–9.

Heichel, G.H. (1971). Confirming measurements of respiration and photosynthesis with dry matter accumulation. *Photosynthetica* 5, 93–8.

Heichel, G.H. (1973). Comparative efficiency of energy use in crop production. *Conn. Agric. Expt. Stn. Bull.* No. 739.

Heichel, G.H. & Musgrave, R.B. (1969). Varietal differences in net photosynthesis of *Zea mays* L. *Crop Sci.* 9, 483–6.

Heide, O.M., Bush, M.G. & Evans, L.T. (1985). Interaction of photoperiod and gibberellin on growth and photosynthesis of high latitude *Poa pratensis*. *Physiol. Plantar.* 65, 135–45.

Heitholt, J.J. & Egli, D.B. (1985). Influence of deflowering on dry matter production of soybeans. *Field Crops Res.* 12, 163–73.

Helbaek, H. (1954). Prehistoric food plants and weeds in Denmark. *Danm. Geologiske Undersogelse*, 2R, 80, 250–61.

Helbaek, M. (1966). Commentary on the phylogenesis of *Triticum* and *Hordeum*. *Econ. Bot.* 20, 350–60.

Helbaek, H. (1969). Plant collecting, dry farming and irrigation agriculture in prehistoric Deh Luran. In *Prehistoric and Human Ecology of the Deh Luran Plain* (ed. F. Hole, K.V. Flannery & J.A. Neely) (Mem. Museum of Anthropol. No. 1.), pp. 383–426. University of Michigan, Ann Arbor.

Hemmer, H. (1990). *Domestication: The Decline of Environmental Appreciation*. Cambridge University Press.

Henle, H.V. (1974). *Report on China's Agriculture*. FAO, Rome.

Herdt, R.W. (1979). An overview of the constraints project results. In *Farm-level Constraints to High Rice Yields in Asia 1974–1977*, pp. 395–411. IRRI, Los Baños.

Herold, A. (1980). Regulation of photosynthesis by sink activity – the missing link. *New Phytol.* 86, 131–44.

Hesketh, J.D. (1963). Limitations to photosynthesis responsible for differences among species. *Crop Sci.* 3, 493–6.

Hesketh, J.D. (1968). Effect of light and temperature during plant growth on subsequent leaf CO_2 assimilation rates under standard conditions. *Aust. J. Biol. Sci.* 21, 235–41.

Hesketh, J.D. & Moss, D.N. (1963). Variation in the response of photosynthesis to light. *Crop Sci.* 3, 107–10.

Hesketh, J.D., Ogren, W.L., Hageman, M.E. & Peters, D.B. (1981). Correlations among leaf CO_2-exchange rates, areas and enzyme activities among soybean cultivars. *Photosynth. Res.* 2, 21–30.

Hesselbach, J. (1985). Breeding progress with winter barley (*Hordeum vulgare* L.) *Zeitschr. f. Pflanzenzücht.* 94, 101–10.

Hewitt de Alcantara, C. (1976). *Modernizing Mexican Agriculture: Socioeconomic Implications of Technological Change 1940–1970*. UNRISD, Geneva.

Hexem, R.W. & Heady, E.O. (1978). *Water Production Functions for Irrigated Agriculture*. Iowa State University, Ames.

Hiatt, A., Cafferkey, R. & Bowdish, K. (1989). Production of antibodies in transgenic plants. *Nature (Lond.)* 342, 76–8.

Hicks, D.R. & Stucker, R.E. (1972). Plant density effect on grain yield of corn hybrids diverse in leaf orientation. *Agron. J.* 64, 484–7.

Higgins, G.M., Kassam, A.H., Naiken, L., Fischer, G. & Shah, M.M. (1982). Potential population supporting capacities of lands in the developing world. Tech. Rep. FPA/INT/513 FAO/UNFPA/IIASA, Rome.

Hill, R.R., Shenk, J.S. & Barnes, R.F. (1988). Breeding for yield and quality. In *Alfalfa and Alfalfa Improvement* (ed. A. Hanson), pp. 809–25. Amer. Soc. Agron., Madison, Wisconsin.

Hillman, G.C., Colledge, S.M. & Harris, D.R. (1989). Plant food economy during the Epipalaeolithic period at Tell Abu Hureyra, Syria: dietary diversity, seasonality and modes of exploitation. In *Foraging and Farming. The Evolution of Plant Exploitation* (ed. D.R. Harris & G.C. Hillman), pp. 240–68. Unwin Hyman, London.

Hinman, C.W. (1986). Potential new crops. *Sci. Amer.* **255** (1), 25–29.

Ho, P.-T. (1969). The loess and the origin of Chinese agriculture. *Amer. Hist. Rev.* **75**, 1–36.

Ho, P.-T. (1977). The indigenous origins of Chinese agriculture. In *Origins of Agriculture* (ed. C.A. Reed), pp. 413–84. Mouton, The Hague.

Hobbs, S.L.A. (1988). Genetic variability in photosynthesis and other leaf characters in *Brassica. Photosynthetica* **22**, 388–93.

Hocking, P.J. & Meyer, C.P. (1991). Carbon dioxide enrichment decreases critical nitrate and nitrogen concentrations in wheat. *J. Plant Nutrition* **14**, 571–84.

Hodanova, D. (1981). Photosynthetic capacity, irradiance and sequential senescence of sugar beet leaves. *Biol. Plantar.* **23**, 58–67.

Hoeft, R.G. (1984). Current status of nitrification inhibitor use in US agriculture. In *Nitrogen in Crop Production* (ed. R.D. Hauck), pp. 561–70. Amer. Soc. Agron., Madison, Wisconsin.

Hofstra, G. & Hesketh J.D. (1975). The effects of temperature and CO_2 enrichment on photosynthesis in soybean. In *Environment and Biological Control of Photosynthesis* (ed. R. Marcelle), pp. 71–80. W. Junk, The Hague.

Hofstra, G. & Nelson, D. (1969). A comparative study of translocation of assimilated ^{14}C from leaves of different species. *Planta* **88**, 103–12.

Holley, R.N. & Goodman, M.M. (1988). Yield potential of tropical hybrid maize derivatives. *Crop Science* **28**, 213–18.

Hoogendoorn, J. (1985*a*). A reciprocal F_1 monosomic analysis of the genetic control of time of ear emergence, number of leaves and number of spikelets in wheat (*Triticum aestivum* L.). *Euphytica* **34**, 545–58.

Hoogendoorn, J. (1985*b*). The physiology of variation in the time of ear emergence among wheat varieties from different regions of the world. *Euphytica* **34**, 559–71.

Hornby, D. (1985). Soil nutrients and take-all. *Outlook on Agric.* **14**, 122–8.

Horner, T.W. & Frey, K.J. (1957). Methods for determining natural areas for oat varietal recommendations. *Agron. J.* **49**, 313–15.

Hornsey, K.G. & Arnold, M.H. (1979). The origins of weed beet. *Ann. Appl. Biol.* **92**, 279–85.

Horst, G.L., Nelson, C.J. & Asay, K.H. (1978). Relationship of leaf elongation to forage yield of tall fescue genotypes. *Crop Sci.* **18**, 715–19.

Horton, D. (1987). *Potatoes. Production, Marketing, and Programs for Developing Countries.* Westview Press, Boulder, Colorado.

Hosaka, K. & Hanneman, R.E. (1988). The origin of the cultivated tetraploid potato based on chloroplast DNA. *Theoret. Appl. Genet.* **76**, 172–6.

Houghton, R.A. & Woodwell, G.M. (1989). Global climatic change. *Sci. Amer.* **260** (4), 18–26.

Housley, T.L. & Peterson, D.M. (1982). Oat stem vascular size in relation to kernel number and weight. I. Controlled environment. *Crop Sci.* **22**, 259–63.

Howell, J.M. (1987). Early farming in Northwestern Europe. *Sci. Amer.* **257** (5), 98–105.

Hozyo, Y. (1977). The influence of source and sink on plant production of *Ipomoea* grafts. *Jap. Agric. Res. Quart.* **11**, 77–83.

Hozyo, Y. & Kato, S. (1976). The interrelationship between source and sink of the grafts of wild type and improved varieties of *Ipomoea*. *Proc. Crop Sci. Soc. Japan* **45**, 117–23.

Hozyo, Y. & Park, C.Y. (1971). Plant production in grafted plants between wild type and improved varieties of *Ipomoea*. *Bull. Natl. Inst. Agric. Sci.* **22**, 145–64.

Hrabovszky, J.P. (1984). Energy, agriculture and rural development. In *Energy and Agriculture: Their Interacting Futures* (ed. M. Lévy & J.L. Robinson), pp. 93–130. Harwood/UNU, Chur, Switzerland.

Huang, L.-K., Wong, S.C., Terashima, I., Zhang, X., Lin, D.-X. & Osmond, C.B. (1989). Chilling injury in mature leaves of rice. I. Varietal differences in the effects of chilling on canopy photosynthesis under simulated dry cold dew wind conditions experienced in South-East China. *Aust. J. Plant Physiol.* **16**, 321–37.

Hubbard, K.R. & Ross, B.L. (1975). Winter wheat – analysis of components of yield of high yielding crops 1975. In *ADAS Experiments and Development in the Eastern Region*, pp. 69–71. MAFF, London.

Hubbard, R.N.L.B. (1980). Development of agriculture in Europe and the Near East: Evidence from quantitative studies. *Econ. Bot.* **34**, 51–67.

Huber, D.M., Warren, H.L., Nelson, D.W., Tsai, C.Y. & Shaner, G.E. (1980). Response of winter wheat to inhibiting nitrification of fall applied nitrogen. *Agron. J.* **72**, 632–7.

Huber, S.C. (1981). Inter- and intra-specific variation in photosynthetic formation of starch and sucrose. *Z. Pflanzenphysiol.* **101**, 49–54.

Huber, S.C. & Israel, D.W. (1982). Biochemical basis for partitioning of photosynthetically fixed carbon between starch and sucrose in soybean (*Glycine max.* Merr.) leaves. *Plant Physiol.* **69**, 691–6.

Hubick, K.T. & Farquhar, G.D. (1989). Carbon isotope discrimination and the ratio of carbon gained to water lost in barley cultivars. *Plant, Cell and Environ.* **12**, 795–804.

Hubick, K.T., Farquhar, G.D. & Shorter, R. (1986). Correlation between water-use efficiency and carbon isotope discrimination in diverse peanut (*Arachis*) germplasm. *Aust. J. Plant Physiol.* **13**, 803–16.

Hubick, K.T., Shorter, R. & Farquhar, G.D. (1988). Heritability and genotype environment interactions of carbon isotope discrimination and transpiration efficiency in peanut (*Arachis hypogaea* L.). *Aust. J. Plant Physiol.* **15**, 799–813.

Hubick, K.T., Hammer, G.L. Farquhar, G.D., Wade, L.J., von Caemmerer, S. & Henderson, S.A. (1990). Carbon isotope discrimination varies genetically in C_4 species. *Plant Physiol.* **92**, 534–7.

Hucl, P. & Baker, R.J. (1987). A study of ancestral and modern Canadian spring wheats. *Can. J. Plant Sci.* **67**, 87–97.

Huda, A.K.S., Ghildyal, B.P. & Tomar, V.S. (1976). Contribution of climatic variables in predicting maize yield under monsoon conditions. *Agric. Meteorol.* **17**, 33–47.

Hughes, W.G., Westcott, B. & Sharp, P.L. (1987). Joint selection for high yield and low sensitivity in winter wheat (*Triticum aestivum* L.). *Plant Breeding* **99**, 107–17.

Humbert, R.P. (1972). Mechanization of sugar cane harvesting. *Outlook on Agric.* **7**, 10–13.

Humphries, E.C. & French, S.A.W. (1969). Photosynthesis in sugar beet depends on root growth. *Planta* **88**, 87–90.

Humphries, E.C. & Thorne, G.N. (1964). The effect of root formation on photosynthesis of detached leaves. *Ann. Bot. N.S.* **28**, 391–400.

Hunt, L.A. (1979). Photoperiodic response of winter wheats from different climatic regions. *Zeitschr. f. Pflanzenzücht.* **82**, 70–80.

Hunt, L.A., Wholey, D.W. & Cock, J.H. (1977). Growth physiology of cassava (*Manihot esculenta* Crantz). *Field Crop Abs.* **30**, 77–91.

Hunter, R.B., Hunt, L.A. & Kannenberg, L.W. (1974). Photoperiod and temperature effects on corn. *Can. J. Plant Sci.* **54**, 71–8.

Hurd, E.A. (1974). Phenotype and drought tolerance in wheat. *Agric. Meteorol.* **14**, 39–55.

Hussey, N.W. & Parr, W.J. (1963). The effect of glasshouse red spider mite (*Tetranychus urticae* Koch) on the yield of cucumber. *J. Hort. Sci.* **38**, 255–63.

Hutchinson, J.B. (ed.) (1965). *Crop Plant Evolution*. Cambridge University Press.

Hutchinson, J.B. (1976). India: local and introduced crops. *Phil. Trans. R. Soc. Lond.* B**275**, 129–41.

Hutchinson, J.B., Silow, R.A. & Stephens, S.G. (1947). *The Evolution of Gossypium*. Oxford University Press, London.

Hutson, D.H. & Roberts, T.R. (1985). Insecticides. In *Insecticides*, vol. 5 (ed. D.H. Hutson & T.R. Roberts), pp. 1–34. John Wiley, Chichester.

Hymowitz, T. (1970). On the domestication of the soybean. *Econ. Bot.* **24**, 408–21.

IBPGR–IRRI (1982). Rice Advisory Committee. Conservation of the wild rices of tropical Asia. *Plant Genetic Resource Newsletter.*, No. 49, pp. 13–18.

Ibrahim, M.E. & Buxton, D.R. (1981). Early vegetative growth of cotton as influenced by leaf type. *Crop Sci.* **21**, 639–43.

Idso, S.B. (1989). *Carbon Dioxide and Global Change: Earth in Transition*. Inst. Biospheric Res., Tempe, Arizona.

Idso, S.B., Kimball, B.A. & Mauney, J.R. (1987*a*). Atmospheric carbon dioxide enrichment effects on cotton mid-day foliage temperature: implications for plant water use and crop yield. *Agron. J.* **79**, 667–72.

Idso, S.B., Kimball, B.A., Anderson, M.G. & Mauney, J.R. (1987*b*). Effects of atmospheric CO_2 enrichment on plant growth: the interactive role of air temperature. *Agriculture, Ecosys. Environ* **20**, 1–10.

IFIAS (1975). *Workshop on Energy Analysis and Economics. Lidingo, Sweden 1975*. Workshop Report No. 9. IFIAS, Stockholm.

Ikehashi, H. (1973). (Studies on the environmental and varietal differences of germination habits in rice seeds with special reference to plant breeding.) *J. Central Agric. Expt. Sta.* **19**, 1–60. (In Japanese.)

Imle, E.P. (1977). Hevea rubber: past and future. In *Crop Resources* (ed. D.S. Seigler), pp. 119–36. Academic Press, New York.

Incoll, L.D. & Neales, T.F. (1970). The stem as a temporary sink before tuberization in *Helianthus tuberosus* L. *J. Exp. Bot.* **21**, 469–76.

Innes, P. & Blackwell, R.D. (1983*a*). Some effects of leaf posture on the yield and water economy of winter wheat. *J. Agric. Sci. (Camb.)* **101**, 367–76.

Innes, P. & Blackwell, R.D. (1983*b*). Effects of differences in date of ear emergence and height on the yield and water use of winter wheat. *Plant Breeding Inst. (Cambridge). Ann. Rep.* 1983, pp. 103–6.

Inosaka, M. (1957). (On the connection between the branches of the first order and upper leaves of rice plant.). *Proc. Crop Sci. Soc. Japan* **26**, 197–8. (In Japanese.)

IRRI (1977). *Constraints to high yields on Asian rice farms: an interim report*. IRRI, Los Baños.

IRRI (1979a). *Annual Report for 1978.* IRRI, Los Baños.
IRRI (1979b). *Farm-level Constraints to High Rice Yields in Asia: 1974–77.* IRRI, Los Baños.
IRRI (1988a). *IRRI Highlights 1987*, pp. 28–29. IRRI, Los Baños.
IRRI (1988b). Rice farming systems. *IRRI Highlights 1987*, pp. 59–60. IRRI, Los Baños.
Irvine, J.E. (1967). Photosynthesis in sugar cane varieties under field conditions. *Crop Sci.* **7**, 297–300.
Irvine, J.E. (1975). Relations of photosynthetic rates and leaf and canopy characters to sugar cane yield. *Crop Sci.* **15**, 671–6.
Irvine, J.E. (1983). Sugarcane. In *Potential Productivity of Field Crops under Different Environments*, pp. 361–81. IRRI, Los Baños.
Isaac, E. (1970). *The Geography of Domestication.* Prentice Hall, Englewood Cliffs.
Ishag, H.M. (1973). Physiology of seed yield in field beans (*Vicia faba* L.). I. Yield and yield components. *J. Agric. Sci. (Camb.)* **80**, 181–9.
Ishizuka, Y. (1971). Physiology of the rice plant. *Adv. Agron.* **23**, 241–315.
Ishizuka, Y. (1978). *The Rice Yield Competition in Japan.* ASPAC Food and Fertilizer Technology Center (Taiwan). Extension Bull. No. 109.
Islam, M.S. & Morison, J.I.L. (1992). Influence of solar radiation and temperature on irrigated rice grain yield in Bangladesh. *Field Crops Res.* **30**, 13–28.
Ito, R. (1975). Significance of grain/straw ratio in rice breeding. *Jap. Agric. Res. Quart.* **9**, 189–90.
Iwama, K. & Nishibe, S. (1989). Comparison of root characters among cultivated potatoes (*Solanum tuberosum*) and their wild relatives. *Jap. J. Crop Sci.* **58**, 126–32.
Izhar, S. & Wallace, D.M. (1967). Studies on the physiological basis for yield differences. III. Genetic variation in photosynthetic efficiency of *Phaseolus vulgaris* L. *Crop Sci.* **7**, 457–60.
Izquierdo, J.A. & Hosfield, G.L. (1983). The relationship of seed filling to yield among dry beans with differing architectural forms. *J. Amer. Soc. Hort. Sci.* **108**, 106–11.
Jackson, D.M., Grant, W.R. & Shafer, C.E. (1980). *US Sorghum Industry.* USDA Agric. Econ. Rep. No. 457.
Jacob, F. (1982). *The Possible and the Actual.* University of Washington Press, Seattle.
Jacobsen, T. & Adams, R.M. (1958). Salt and silt in ancient Mesopotamian agriculture. *Science* **128**, 1251–8.
Jain, H.K. & Kulshrestha, V.P. (1976). Dwarfing genes and breeding for yield in bread wheat. *Zeitschr. f. Pflanzenzücht.* **76**, 102–12.
James, W.C. & Teng, P.S. (1979). The quantification of production constraints associated with plant diseases. *Applied Biol.* **4**, 201–67.
Janardhan, K.V. & Murty, K.S. (1978). Association of photosynthetic efficiency with various growth parameters and yield in rice. *Proc. Indian Natl. Sci. Acad. B*, **44**, 49–56.
Jansen, D.M. (1990). Potential rice yields in future weather conditions in different parts of Asia. *Neth. J. Agric. Sci.* **38**, 661–80.
Jarrige, J.F. & Meadow, R.H. (1980). The antecedents of civilization in the Indus Valley. *Sci. Amer.* **243**(2), 102–10.
Jefferies, R.A. & MacKerron, D.K.L. (1989). Radiation interception and growth of irrigated and droughted potato (*Solanum tuberosum*). *Field Crops Res.* **22**, 101–12.
Jeffers, D.L. & Shibles, R.M. (1969). Some effects of leaf area, solar radiation, air temperature, and variety on net photosynthesis in field grown soybeans. *Crop*

Sci. **9**, 762–4.

Jefferson, T. (1821). *The Jeffersonian Encyclopedia* (ed. J.P. Foley). Funk & Wagnalls, 1900.

Jenkinson, D.S. (1982). The nitrogen cycle in long-term field experiments. *Phil. Trans. R. Soc. Lond.* B**296**, 563–71.

Jenkinson, D.S. (1986). Nitrogen in UK arable agriculture. *J. Royal Agric. Soc. England* **147**, 178–89.

Jenkinson, D.S. (1991). The Rothamsted long-term experiments: are they still of use? *Agron. J.* **83**, 2–10.

Jenner, C.F. & Rathjen, A.J. (1975). Factors regulating the accumulation of starch in ripening wheat grain. *Aust. J. Plant Physiol.* **2**, 311–22.

Jennings, P.R. & Cock, J.H (1977). Centres of origin of crops and their productivity. *Econ. Bot.* **31**, 51–4.

Jensen, N.F. (1967). Agrobiology: specialization or systems analysis? *Science* **157**, 1405–9.

Jensen, N.F. (1978). Limits to growth in world food production. *Science* **201**, 317–20.

Johannessen, C.L. (1982). Domestication process of maize continues in Guatemala. *Econ. Bot.* **36**, 84–99.

Johannessen, C.L., Wilson, M.R. & Davenport, W.A. (1970). The domestication of maize: process or event? *Geograph. Rev.* **60**, 393–413.

Johnson, D.A., Richards, R.A. & Turner, N.C. (1983). Yield, water relations, gas exchange and surface reflectances of near-isogenic wheat lines differing in glaucousness. *Crop Sci.* **23**, 318–25.

Johnson, D.R. & Major, D.J. (1979). Harvest index of soybeans as affected by planting date and maturity rating. *Agron. J.* **71**, 538–41.

Johnson, D.R. & Tanner, J.W. (1972). Comparisons of corn (*Zea mays* L.) inbreds and hybrids grown at equal leaf area index, light penetration and population. *Crop Sci.* **12**, 482–5.

Johnson, E.C., Fischer, K.S., Edmeades, G.O. & Palmer, A.F.E. (1986). Recurrent selection for reduced plant height in lowland tropical maize. *Crop Sci.* **26**, 253–60.

Johnson, H.W., Borthwick, H.A. & Leffel, R.C. (1960). Effects of photoperiod and time of planting on rates of development of the soybean in various stages of the life cycle. *Bot. Gaz.* **122**, 77–95.

Johnson, R.C. & Bassett, L.M. (1991). Carbon isotope discrimination and water use efficiency in four cool-season grasses. *Crop Sci.* **31**, 157–62.

Johnson, R.C., Kebede, H., Mornhinweg, D.W., Carver, B.F., Rayburn, A.L. & Nguyen, H.T. (1987a). Photosynthetic differences among *Triticum* accessions at tillering. *Crop Sci.* **27**, 1046–50.

Johnson, R.C., Mornhinweg, D.W., Ferris, D.M. & Heitholt, J.J. (1987b). Leaf photosynthesis and conductance of selected *Triticum* species at different water potentials. *Plant Physiol.* **83**, 1014–17.

Johnson, S.K., Helsel, D.B. & Frey, K.J. (1983). Direct and indirect selection for grain yield in oats (*Avena sativa* L.). *Euphytica* **32**, 407–13.

Johnson, S.S. & Geadelmann, J.L. (1989). Influence of water stress on grain yield response to recurrent selection in maize. *Crop Sci.* **29**, 558–64.

Johnson, W.A., Stoltzfus, V. & Craumer, P. (1977). Energy conservation in Amish agriculture. *Science* **198**, 373–8.

Jones, F.W.G., Dunning, R.A. & Humphries, K.P. (1955). The effects of defoliation and loss of stand upon yield of sugar beet. *Ann. Appl. Biol.* **43**, 63–70.

Jones, H., Martin, R.V. & Porter, H.K. (1959). Translocation of ^{14}Carbon in tobacco following assimilation of ^{14}carbon dioxide by a single leaf. *Ann. Bot. N.S.* **23**, 493–508.

Jones, H.G. (1977). Transpiration in barley lines with differing stomatal frequencies. *J. Exp. Bot.* **28**, 162–8.

Jones, I.T. (1977). The effect on grain yield of adult plant resistance to mildew in oats. *Ann. Appl. Biol.* **86**, 267–77.

Jones, P., Allen, L.H., Jones, J.W. & Valle, R. (1985). Photosynthesis and transpiration responses of soybean canopies to short- and long-term CO_2 treatments. *Agron. J.* **77**, 119–26.

Jones, R.J., Nelson, C.J. & Sleper, D.A. (1979). Seedling selection for morphological characters associated with yield of tall fescue. *Crop Sci.* **19**, 631–4.

Jong, S.K., Brewbaker, J.L. & Lee, C.H. (1982). Effects of solar radiation on the performance of maize in 41 successive monthly plantings in Hawaii. *Crop Sci.* **22**, 13–18.

Jordan, T.N. & Bridge, R.R. (1979). Tolerance of cotton to the herbicide glyphosate. *Agron. J.* **71**, 927–8.

Jordan, W.R. & Miller, F.R. (1980). Genetic variability in sorghum root systems: Implications for drought tolerance. In *Adaptation of Plants to Water and High Temperature Stress* (ed. N.C. Turner & P.J. Kramer), pp. 383–99. Wiley, New York.

Joseph, M.C., Randall, D.D. & Nelson, C.J. (1981). Photosynthesis in polyploid tall fescue. II. Photosynthesis and ribulose-1,5-bisphosphate carboxylase of polyploid tall fescue. *Plant Physiol.* **68**, 894–8.

Jutsum, A.R. (1988). Commercial application of biological control: status and prospects. *Phil. Trans. R. Soc. Lond.* **B318**, 357–73.

Kabaki, N., Akita, S., Tanaka, I. & Amamiya, A. (1976). (Photosynthesis and photorespiration of cultivated rice and F_1 hybrid.) *Japanese J. Crop Sci.* **45** (extra issue 2), 177–8. (In Japanese.)

Kadkol, G.P., Beilharz, V.C., Halloran, G.M. & Macmillan, R.H. (1986). Anatomical basis of shatter-resistance in the oilseed Brassicas. *Aust. J. Bot.* **34**, 595–601.

Kadkol, G.P., Halloran, G.M. & MacMillan, R.H. (1989). Shatter resistance in crop plants. *CRC Crit. Rev. Plant Sci.* **8**, 169–88.

Kallarackal, J. & Milburn, J.A. (1984). Specific mass transfer and sink-controlled phloem translocation in castor bean. *Aust. J. Plant Physiol.* **11**, 483–90.

Kaneda, C. & Beachell, H.M. (1974). Response of *indica-japonica* rice hybrids to low temperature. *SABRAO J.* **7**(1), 17–32.

Kanemasu, E.T. & Hiebsch, C.K. (1975). Net carbon dioxide exchange of wheat, sorghum and soybean. *Can. J. Bot.* **53**, 382–9.

Kaplan, S.L. (1965). Archaeology and domestication in American *Phaseolus* (beans). *Econ. Bot.* **19**, 358–68.

Kaplan, S.L. & Koller, H.R. (1977). Leaf area and CO_2 exchange rate as determinants of the rate of vegetative growth in soybean plants. *Crop Sci.* **17**, 35–8.

Kapoor, R.L., Yadav, H.P., Singh, P., Khairwal, I.S. & Dahiya, B.N. (1982). Genetics of harvest index, grain yield and biological yield of pearl millet. *Indian J. Agric. Res.* **52**, 630–3.

Kariya, K. & Tsunoda, S. (1972). Relationship of chlorophyll content, chloroplast area index and leaf photosynthesis rate in *Brassica*. *Tohoku J. Agric. Res.* **23**, 1–14.

Kassam, A.H. & Andrews, D.J. (1975). Effects of sowing date on growth, development and yield of photosensitive sorghum at Samaru, Northern Nigeria.

Exptl. Agric. **11**, 227–40.

Katayama, T.C. (1977). Studies on the photoperiodism in the genus *Oryza*. *Japan Agric. Res. Quart.* **11**, 12–17.

Kato, K. & Yamashita, S. (1991). Varietal variation in photoperiodic response, chilling requirement and narrow-sense earliness and their relation to heading time in wheat (*Triticum aestivum* L.). *Jap. J. Breeding* **41**, 457–84.

Kaushik, K.P., Sharma, K.D. (1986). Extent of heterosis in rice (*Oryza sativa* L.) under cold stress conditions – yield and its components. *Theor. Appl. Genet.* **73**, 136–40.

Kawanabe, S. (1968). Temperature responses and the systematics of the Gramineae. *Proc. Jap. Soc. Plant Taxon.* **2**, 17–20.

Kawano, K. (1978). Genetic improvement of cassava (*Manihot esculenta* Crantz.) for productivity. *Trop. Agric. Res. Ser.* **11**, 9–21.

Kawano, K. (1990). Harvest index and evaluation of major food crop cultivars in the tropics. *Euphytica* **46**, 195–202.

Kawano, K. & Jennings, P.R. (1983). Tropical crop breeding-achievements and challenges. In *Potential Productivity of Field Crops under Different Environments*, pp. 81–99. IRRI, Los Baños.

Kawano, K., Kurosawa, K. & Takahashi, M. (1969). Heterosis in vegetative growth of the rice plant. Genetical studies on rice plant. *Jap. J. Breed.* **19**, 335–42.

Kawase, M. & Sakamoto, S. (1987). Geographical distribution of land race groups classified by hybrid pollen sterility in foxtail millet, *Setaria italica* (L.) P. Beauv. *Jap. J. Breed.* **37**, 1–9.

Kellogg, C.E. & Orvedal, A.C. (1969). Potentially arable soils of the world and critical measures for their use. *Adv. Agron.* **21**, 109–70.

Kenworthy, W.J. & Brim, C.A. (1979). Recurrent selection in soybeans. I. Seed yield. *Crop Sci.* **19**, 315–18.

Khan, A. & Sagar, G.R. (1969). Alteration of the pattern of distribution of photosynthetic products in the tomato by manipulation of the plant. *Ann. Bot.* NS **33**, 753–62.

Khan, M.A. & Tsunoda, S. (1970*a*). Evolutionary trends in leaf photosynthesis and related leaf characters among cultivated wheat species and its wild relatives. *Jap. J. Breed.* **20**, 133–40.

Khan, M.A. & Tsunoda, S. (1970*b*). Growth analysis of cultivated wheat species and their wild relatives with special reference to dry matter distribution among different plant organs and to leaf area expansion. *Tohoku J. Agric. Res.* **21**, 47–59.

Khan, M.A. & Tsunoda, S. (1970*c*). Leaf photosynthesis and transpiration under different levels of air flow rate and light intensity in cultivated wheat species and its wild relatives. *Jap. J. Breed.* **20**, 305–14.

Khanna-Chopra, R. (1982). Photosynthesis, photosynthetic enzymes and leaf area development in relation to hybrid vigour in *Sorghum vulgare* L. *Photosynth. Res.* **3**, 113–22.

Khush, G.S. (1987). Rice breeding: past, present and future. *J. Genet.* **66**, 195–216.

Kidambi, S.P., Krieg, D.R. & Rosenow, D.T. (1990). Genetic variation for gas exchange rates in grain sorghum. *Plant Physiol.* **92**, 1211–14.

Kimball, B.A. (1983). Carbon dioxide and agricultural yield: An assemblage and analysis of 430 prior observations. *Agron. J.* **75**, 779–88.

King, J.W., Hurst, E., Slater, A.J., Smith, P.A. & Tamkin, B. (1974). Agriculture and sunspots. *Nature (Lond.)* **252**, 2–3.

King, K.M. & Greer, D.H. (1986). Effects of carbon dioxide enrichment and soil water on maize. *Agron. J.* 78, 515–21.

King, R.W. (1976). Abscisic acid in developing wheat grains and its relationship to grain growth and maturation. *Planta* 132, 43–51.

King, R.W. & Chadim, H. (1983). Ear wetting and pre-harvest sprouting of wheat. In *Third International Symposium on Pre-Harvest Sprouting in Cereals* (ed. J.E. Kruger & D.E. La Berge), pp. 36–42. Westview, Boulder, Colorado.

King, R.W. & Evans, L.T. (1967). Photosynthesis in artificial communities of wheat, lucerne, and subterranean clover plants. *Aust. J. Biol. Sci.* 20, 623–35.

King, R.W., Wardlaw, I.F. & Evans, L.T. (1967). Effect of assimilate utilization on photosynthetic rate in wheat. *Planta* 77, 261–76.

King, R.W., Gale, M.D. & Quarrie, S.A. (1983). Effects of Norin 10 and Tom Thumb dwarfing genes on morphology, physiology and abscisic acid production in wheat. *Ann. Bot.* 51, 201–8.

Kinghorn, A.D. & Soejarto, D.D. (1986). Sweetening agents of plant origin. *CRC Crit. Rev. Plant Sci.* 4, 79–120.

Kiniry, J.R. (1988). Kernel weight increase in response to decreased kernel number in sorghum. *Agron. J.* 80, 221–6.

Kiniry, J.R., Jones, C.A., O'Toole, J.C., Blanchet, R., Cabelguenne, M. & Spanel, D.A. (1989). Radiation-use efficiency in biomass accumulation prior to grain-filling for five grain-crop species. *Field Crops Res.* 20, 51–64.

Kirby, E.J.M. (1969). The effect of sowing date and plant density on barley. *Ann. Appl. Biol.* 63, 513–21.

Kirch, P.V. (1989). Second millennium B.C. aboriculture in Melanesia: archaeological evidence from the Mussau Islands. *Econ. Bot.* 43, 225–40.

Kirkby, A.V.T. (1973). *The use of land and water resources in the past and present Valley of Oaxaca, Mexico.* Memoir No. 5, Museum of Anthropology, University of Michigan, Ann Arbor.

Kishitani, S. & Tsunoda, S. (1981). Physiological aspects of domestication in diploid wheat. *Euphytica* 30, 247–52.

Kislev, M.E. (1985). Early Neolithic horsebean from Yiftah'el, Israel. *Science* 228, 319–20.

Kislev, M.E., Bar-Josef, O. & Gopher, A. (1986). Early Neolithic domesticated and wild barley from the Netiv Hagdud regions in the Jordan Valley. *Israel J. Bot.* 35, 197–201.

Klepper, B., Taylor, H.M., Huck, M.G. & Fiscus, E.L. (1973). Water relations and growth of cotton in drying soil. *Agron. J.* 65, 307–10.

Kloppenburg, J. & Kleinman, D.L. (1987). The plant germplasm controversy. *BioScience* 37, 190–8.

Knight, R. (1970). The measurement and interpretation of genotype-environment interactions. *Euphytica* 19, 225–35.

Knott, D.R. (1986). Effect of genes for photoperiodism, semidwarfism, and awns on agronomic characters in a wheat cross. *Crop Sci.* 26, 1158–62.

Kobza, J. & Edwards, G.E. (1987). Control of photosynthesis in wheat by CO_2, O_2, and light intensity. *Plant Cell Physiol.* 28, 1141–52.

Kobza, J. & Seemann, J.R. (1988). Mechanisms for light-dependent regulation of ribulose-1,5-bisphosphate carboxylase activity and photosynthesis in intact leaves. *Proc. Natl. Acad. Sci. USA* 85, 3815–19.

Kobza, J. & Seemann, J.R. (1989). Regulation of ribulose-1,5-bisphosphate carboxylase activity in response to diurnal changes in irradiance. *Plant Physiol.* 89,

918–24.

Kodama, S., Chuman, K. & Tanoue, M. (1970). On the growth differentials of sweet potato to different soil fertility. *Bull. Kyushu Agric. Expt. Sta., Japan* **15**, 493–514.

Kokobun, M. & Wardlaw, I.F. (1988). Temperature adaptation of *Glycine* species as expressed by germination, photosynthesis, photosynthate accumulation and growth. *Jap. J. Crop Sci.* **57**, 211–19.

Kolderup, F. (1979). Application of different temperatures in three growth phases of wheat. II. Effects on ear size and seed setting. *Acta Agric. Scand.* **29**, 11–16.

Kølster, P., Munk, L., Stølen, O. & Løhde, J. (1986). Near-isogenic barley lines with genes for resistance to powdery mildew. *Crop Sci.* **26**, 903–7.

Kondo, K., Yoshida, Z. & Teranaka, K. (1975). (Recent wet rice yields: trends in the NE areas centering on Iwate prefecture.) *Nogyo fukyu* **27** (11), 17–19.

Koul, A.K. (1974). Job's tears. In *Evolutionary Studies in World Crops: Diversity and Change in the Indian Subcontinent* (ed. J. Hutchinson), pp. 63–6. Cambridge University Press.

Kovda, V.A. (1974). *Biosphere, Soils and their Utilization.* Moscow.

Krenzer, E.G. & Moss, D.N. (1975). Carbon dioxide enrichment effects upon yield and yield components in wheat. *Crop Sci.* **15**, 71–4.

Krieg, D.R. & Dalton, G. (1990). Analyses of genetic and cultural improvements in crop production: Sorghum. In *Proc. Intl. Congr. Plant Physiol., New Delhi, India* (ed. S.K. Sinha, P.V. Sane, S.C. Bhargava & P.K. Agrawal), vol. 1, pp. 134–41. Indian Agricultural Research Institute, New Delhi.

Krueger, R.W. & Miles, D. (1981). Photosynthesis in fescue. III. Rates of electron transport in a polyploid series of tall fescue plants. *Plant Physiol.* **68**, 1110–14.

Kudo, K. (1975). Economic yield and climate. *Jap. Intl. Biol. Progr. Synthesis (Tokyo).* **11**, 199–220.

Kueneman, E.A., Wallace, D.H. & Ludford, P. (1979). Photosynthetic measurements of field grown dry beans and their relation to selection for yield. *J. Amer. Soc. Hort. Sci.* **104**, 480–2.

Kühbauch, W. & Thome, U. (1989). Non-structural carbohydrates of wheat stems as influenced by sink-source manipulations. *J. Plant Physiol.* **134**, 243–50.

Kuhr, S.L., Johnson, V.A., Peterson, C.J. & Mattern, P.J. (1985). Trends in winter wheat performance as measured in international trials. *Crop Sci.* **25**, 1045–9.

Kulkarni, B.S. & Pandit, S.N.N. (1988). A discrete step in the technology trend for sorghum yields in Parbhani, India. *Agric. For. Meteorol.* **42**, 157–65.

Kulshrestha, V.P. & Jain, H.K. (1982). Eighty years of wheat breeding in India: past selection pressures and future prospects. *Zeitschr. f. Pflanzenzücht.* **89**, 19–30.

Kumazawa, K. (1984). Beneficial effects of organic matter on rice growth and yield in Japan. In *Organic Matter and Rice*, pp. 431–44. IRRI, Los Baños.

Kuo, C., Lindberg, C. & Thomson, D.J. (1990). Coherence established between atmospheric carbon dioxide and global temperature. *Nature (Lond.).* **343**, 709–14.

Kuroda, E. & Kumura, A. (1990*a*). (Difference in single leaf photosynthesis between old and new rice varieties. I. Single-leaf photosynthesis and its dependence on stomatal conductance.) *Proc. Crop Sci. Soc. Japan* **59**, 283–92. (In Japanese.)

Kuroda, E. & Kumura, A. (1990*b*). Ibid. II. (A physiological basis for the difference in stomatal conductance between varieties.) *Proc. Crop Sci. Soc. Japan* **59**, 293–7. (In Japanese.)

Kuroda, E. & Kumura, A. (1990*c*). Ibid. III. (Physiological basis of varietal difference in single-leaf photosynthesis between varieties viewed from nitrogen content and the nitrogen-photosynthesis relationship.) *Proc. Crop Sci. Soc. Japan* **59**, 298–302. (In Japanese.)

Kwapata, M.B. & Hall, A.E. (1990). Determinants of cowpea (*Vigna unguiculata*) seed yield at extremely high plant density. *Field Crops Res.* **24**, 23–32.

Ladizinsky, G. (1975). Collection of wild cereals in the Upper Jordan Valley. *Econ. Bot.* **29**, 264–7.

Ladizinsky, G. (1985). Founder effect in crop plant evolution. *Econ. Bot.* **39**, 191–9.

Ladizinsky, G. (1987). Pulse domestication before cultivation. *Econ. Bot.* **41**, 60–5.

Ladizinsky, G. (1989). Pulse domestication: fact and fiction. *Econ. Bot.* **43**, 131–2.

Ladizinsky, G. & Adler, A. (1976). The origin of chickpea, *Cicer arietinum* L. *Euphytica* **25**, 211–17.

Lafitte, H.R. & Loomis, R.S. (1988). Calculation of growth yield, growth respiration and heat content of grain sorghum from elemental and proximal analyses. *Ann. Bot.* **62**, 353–61.

Lafitte, H.R. & Travis, R.L. (1984). Photosynthesis and assimilate partitioning in closely related lines of rice exhibiting differences in sink:source relationships. *Crop Sci.* **24**, 447–52.

Laing, D.R. & Fischer, R.A. (1977). Adaptation of semi-dwarf wheat cultivars to rainfed conditions. *Euphytica* **26**, 129–39.

Laing, D.R., Kretchmer, P.J., Zuluaga, S. & Jones, P.G. (1983). Field bean. In *Potential Productivity of Field Crops under Different Environments*, pp. 227–48. IRRI, Los Banos.

Laing, D.R., Jones, P.G. & Davis, J.H.C. (1984). Common bean (*Phaseolus vulgaris* L.) In *The Physiology of Tropical Field Crops* (ed. P.R. Goldsworthy & N.M. Fisher), pp. 305–51. John Wiley, New York.

Laird, C.M., Laird, R.D., Zeller, E.J. & Dreschhoff, G.A.M. (1990). World grain yields, snow cover, solar activity and quasi-biennial oscillation relationships. *Agric. For. Meteorol.* **52**, 263–74.

Lambers, H. (1982). Cyanide-resistant respiration: a non-phosphorylating electron transport pathway acting as an energy overflow. *Physiol. Plantar.* **55**, 478–85.

Lambers, H., Szaniawski, R.K. & de Visser, R. (1983). Respiration for growth, maintenance and ion uptake. An evaluation of concepts, methods, values and their significance. *Physiol. Plantar.* **58**, 556–63.

Lamkey, K.R. & Smith, O.S. (1987). Performance and inbreeding depression of populations representing seven eras of maize breeding. *Crop Sci.* **27**, 695–9.

Lamkey, K.R., Hallauer, A.R. & Kahler, A.L. (1987). Allelic differences in enzyme loci and hybrid performance in maize. *J. Hered.* **78**, 231–4.

Landivar, J.A., Baker, D.N., Jenkins, J.N. (1983). Application of GOSSYM to genetic feasibility studies. I. Analyses of fruit abscission and yield in okra-leaf cottons. *Crop Sci.* **23**, 497–504.

Larson, E.M., Hesketh, J.D., Woolley, J.T. & Peters, D.B. (1981). Seasonal variations in apparent photosynthesis among plant stands of different soybean cultivars. *Photosynth. Res.* **2**, 3–20.

Lathrap, D.W. (1968). The hunting economies of the tropical forest zone of South America: an attempt at historical perspective. In *Man the Hunter* (ed. R.B. Lee & I. De Vore), pp. 23–9. Aldine, Chicago.

Lathrap, D.W. (1977). Our father the cayman, our mother the gourd: Spinden revisited, or a unitary model for the emergence of agriculture in the New World. In *Origins of Agriculture* (ed. C.A. Reed), pp. 713–51. Mouton, The Hague.

Lauer, M.J. & Shibles, R. (1987). Soybean leaf photosynthetic response to changing sink demand. *Crop Sci.* **27**, 1197–201.

Lavergne, D., Bismuth, E. & Champigny, M.L. (1979). Physiological studies on two cultivars of *Pennisetum: P. americanum* 23 DB, a cultivated species and *P. mollissimum*, a wild species. I. Photosynthetic carbon metabolism. *Z. Pflanzenphysiol.* **91**, 291–303.

Lawes, D.A. (1977). Yield improvement in spring oats. *J. Agric. Sci. (Camb.)* **89**, 751–7.

Lawes, D.A. & Treharne, K.J. (1971). Variation in photosynthetic activity in cereals and its implications in a plant breeding programme. I. Variation in seedling leaves and flag leaves. *Euphytica.* **20**, 86–92.

Lawlor, D.W. (1979). Effects of water and heat stress on carbon metabolism of plants with C_3 and C_4 photosynthesis. In *Stress Physiology in Crop Plants* (ed. H. Mussell & R.C. Staples), pp. 304–26. Wiley, New York.

Lawlor, D.W., Kontturi, M. & Young, A.T. (1989). Photosynthesis by flag leaves of wheat in relation to protein, ribulose bisphosphate carboxylase activity and nitrogen supply. *J. Exp. Bot.* **40**, 43–52.

Lawn, R.J. & Byth, D.E. (1973). Response of soybeans to planting date in Southeastern Queensland. I. Influence of photoperiod and temperature on phasic development patterns. *Aust. J. Agric. Res.* **24**, 67–80.

Lawn, R.J. & Troedson, R.J. (1990). Pigeon pea: physiology of yield formation. In *The Pigeon Pea.* (ed. Y.L. Nene, S.D. Hall & V.K. Sheila), pp. 179–208. CABI, Wallingford.

Lawton, J.H. (1972). The energy cost of 'food-gathering'. In *Resources and Population. 9th Annual Symposium*, pp. 59–76. Eugenics Society, London.

Leach, G. (1975). *Energy and Food Production.* Intl. Inst. for Environment and Development, London.

Leavitt, J.R.C., Dobrenz, A.K. & Stone, J.E. (1979). Physiological and morphological characteristics of large and small leaflet alfalfa genotypes. *Agron. J.* **71**, 529–32.

LeBaron, H.M. & Gressel, J. (eds). (1982). *Herbicide Resistance in Plants.* Wiley, New York.

Lebsock, K.L., Joppa, L.R. & Walsh, D.E. (1973). Effect of daylength response on agronomic and quality performance of durum wheat. *Crop Sci.* **13**, 670–4.

LeCain, D.R., Morgan, J.A. & Zerbi, G. (1989). Leaf anatomy and gas exchange in nearly isogenic semi-dwarf and tall winter wheat. *Crop Sci.* **29**, 1246–51.

Ledent, J.R. & Stoy, V. (1988). Yield of winter wheat: A comparison of genotypes from 1910 to 1976. *Cereal Res. Commun.* **16**, 151–6.

Lee, B.T., Martin, P. & Bangerth, F. (1988). Phytohormone levels in the florets of a single wheat spikelet during pre-anthesis development and relationships to grain set. *J. Expt. Bot.* **39**, 927–33.

Lee, M., Godshalk, E.B., Lamkey, K.R. & Woodman, W.W. (1989). Association of restriction fragment length polymorphisms among maize inbreds with agronomic performance of their crosses. *Crop Sci.* **29**, 1067–71.

Lee, R.B. (1968). What hunters do for a living, or how to make out on scarce resources. In *Man the Hunter* (ed. R.B., Lee & I. De Vore), pp. 30–48. Aldine, Chicago.

Le Gros Clark, F. (1951). The Malthusian heritage. In *Four Thousand Million Mouths* (ed. F. Le Gros Clark & N.W. Pirie), pp. 6–29. Oxford University Press.

Le Gros Clark, F. & Pirie, N.W. (eds). (1951). *Four Thousand Million Mouths. Scientific Humanism and the Shadow of World Hunger.* Oxford University Press.

Lemcoff, J.H. & Loomis, R.S. (1986). Nitrogen influences on yield determination in maize. *Crop Sci.* **26**, 1017–22.

Levi-Strauss, C. (1968). The concept of primitiveness. In *Man the Hunter*. (ed. R.B. Lee & I. De Vore), pp. 349–52, Aldine, Chicago.

Levy, M. & Robinson, J.L. (eds). (1984). *Energy and Agriculture: their Interacting Futures. Policy Implications of Global Models*. Harwood, United Nations University.

Lewis, D.A. & Tatchell, J.A. (1979). Energy in UK Agriculture. *J. Food Sci. Agric.* 30, 449–57.

Lewis, N. (1983). *Life in Egypt under Roman Rule*. Clarendon Press, Oxford.

Li, H.-L. (1970). The origin of cultivated plants in Southeast Asia. *Econ. Bot.* 24, 3–19.

Li, H.-L. (1983). The domestication of plants in China: ecogeographical considerations. In *The Origins of Chinese Civilization* (ed. D.N. Keightley), pp. 21–63. University of California Press, Berkeley.

Lichtner, F.T. & Spanswick, R.M. (1981). Sucrose uptake by developing soybean cotyledons. *Plant Physiol.* 68, 693–8.

Lin, C.S., Binns, M.R. & Lefkovitch, L.P. (1986). Stability analysis: where do we stand? *Crop Sci.* 26, 894–900.

Linnaeus, C. (1737). *Critica Botanica* (cited by Wichler, 1961).

Linnemann, A.R. (1991). Preliminary observations on photoperiod regulation of phenological development in bambara groundnut (*Vigna subterranea*). *Field Crops Res.* 26, 295–304.

Lipton, M. (1979). The technology, the system and the poor: the case of the new cereal varieties. In *Development of Societies: The Next Twenty Five Years*, pp. 1–16. Inst. Social Studies, Martinus Nijhoff, The Hague.

Lipton, M. (1983). *Poverty, Undernutrition and Hunger*. World Bank Staff Working Paper 5697. The World Bank, Washington, D.C.

Lipton, M. (1988). Attacking under-nutrition and poverty: some issues of adaptation and sustainability. IFPRI Reprint No. 153, pp. 1–35.

Liu, P., Wallace, D.H. & Ozbun, J.L. (1973). Influence of translocation on photosynthetic efficiency of *Phaseolus vulgaris* L. *Plant Physiol.* 52, 412–15.

Lloyd, N.D.H. & Canvin, D.T. (1977). Photosynthesis and photorespiration in sunflower selections. *Can. J. Bot.* 55, 3006–12.

Lockeretz, W. (1989). Problems in evaluating the economics of ecological agriculture. *Agriculture, Ecosys. Env.* 27, 67–75.

Lockeretz, W., Shearer, G. & Kohl, D.H. (1981). Organic farming in the Corn Belt. *Science* 211, 540–7.

Loffler, C.M., Rauch, T.L. & Busch R.H. (1985). Grain and plant protein relationships in hard red spring wheat. *Crop Sci.* 25, 521–4.

Long, S.P. (1983). C_4 photosynthesis at low temperatures. *Plant, Cell Environment* 6, 345–63.

Longden, P.C. (1980). Weed beet. *Agric. Progress* 55, 17–25.

Loomis, R.S. (1984). Traditional agriculture in America. *Ann. Rev. Ecol. System.* 15, 449–78.

Loomis, R.S. & Adams, S.S. (1983). Integrative analysis of host-pathogen relations. *Ann. Rev. Phytopathol.* 21, 341–62.

Loomis, R.S. & Gerakis, P.A. (1975). Productivity of agricultural ecosystems. In *Photosynthesis and Productivity in Different Environments* (ed. J.P. Cooper), pp. 145–72. Cambridge University Press.

Loomis, R.S. & Lafitte, H.R. (1987). The carbon economy of a maize crop exposed to elevated CO_2 concentrations and water stress, as determined from elemental analyses. *Field Crops Res.* 17, 63–74.

Loomis, R.S. & Worker, G.F. (1963). Responses of the sugar beet to low soil moisture at two levels of nitrogen nutrition. *Agron. J.* **55**, 509–15.

Lorimer, G.M. & Andrews, T.J. (1973). Plant photorespiration – an inevitable consequence of the existence of atmospheric oxygen. *Nature (Lond.)* **243**, 359–60.

Loss, S.P., Kirby, E.J.M., Siddique, K.H.M. & Perry, M.W. (1989). Grain growth and development of old and modern Australian wheats. *Field Crops Res.* **21**, 131–46.

Lu, Y.-C., Cline, P. & Quance, L. (1979). Prospects for productivity growth in US Agriculture. USDA Economics, Statistics and Cooperatives Service, Agric. Econ. Rept. No. 87.

Ludlow, M.M. & Muchow, R.C. (1990). A critical evaluation of traits for improving crop yields in water-limited environments. *Adv. Agron.* **43**, 107–53.

Ludlow, M.M., Santamaria, J.M. & Fukai, S. (1990). Contribution of osmotic adjustment to grain yield in *Sorghum bicolor* (L.) Moench under water-limited conditions. II. Water stress after anthesis. *Aust. J. Agric. Res.* **41**, 67–78.

Ludwig, L.J., Saeki, T. & Evans, L.T. (1965). Photosynthesis in artificial communities of cotton plants in relation to leaf area. I. Experiments with progressive defoliation of mature plants. *Aust. J. Biol. Sci.* **18**, 1103–18.

Luedders, V.D. (1977). Genetic improvement in yield of soybeans. *Crop Sci.* **17**, 971–2.

Lupton, F.G.H. (1976). Physiological basis of heterosis in wheat. In *Heterosis in Plant Breeding* (ed. A. Janossy & F.G.H. Lupton), pp. 71–80. Elsevier, Amsterdam.

Lupton, F.G.H., Oliver, R.H., Ellis, F.B., Barnes, B.T., Howse, K.R., Welbank, P.J. & Taylor, P.J. (1974). Root and shoot growth of semi-dwarf and taller winter wheats. *Ann. Appl. Biol.* **77**, 129–44.

Lush, W.M. (1979). Floral morphology of wild and cultivated cowpeas. *Econ. Bot.* **33**, 442–7.

Lush, W.M. & Evans, L.T. (1974). Translocation of photosynthetic assimilate from grass leaves, as influenced by environment and species. *Aust. J. Plant Physiol.* **1**, 417–31.

Lush, W.M. & Evans, L.T. (1980*a*). The seedcoats of cowpeas and other grain legumes: structure in relation to function. *Field Crops Res.* **3**, 267–86.

Lush, W.M. & Evans, L.T. (1980*b*). Photoperiodic regulation of flowering in cowpeas (*Vigna unguiculata* (L.) Walp.). *Ann. Bot.* **46**, 719–25.

Lush, W.M. & Evans, L.T. (1981). The domestication and improvement of cowpeas (*Vigna unguiculata* (L.) Walp.). *Euphytica* **30**, 579–87.

Lush, W.M. & Rawson, H.M. (1979). Effects of domestication and region of origin on leaf gas exchange in cowpea (*Vigna unguiculata* (L.) Walp.). *Photosynthetica* **13**, 419–27.

Lush, W.M. & Wien, H.C. (1980). The importance of seed size in early growth of wild and domesticated cowpeas. *J. Agric. Sci. (Camb.)* **94**, 177–82.

Lush, W.M., Evans, L.T. & Wien, H.C. (1980). Environmental adaptation of wild and domesticated cowpeas (*Vigna unguiculata* (L.) Walp.). *Field Crops Res.* **3**, 173–87.

Luttrell, C.B. & Gilbert, R.A. (1976). Crop yield: random, cyclical or bunchy? *Amer. J. Agric. Econ.* **58**, 521–31.

Ma, Y.Z., MacKown, C.T. & Van Sanford, D.A. (1990). Sink manipulation in wheat: compensatory changes in kernel size. *Crop Sci.* **30**, 1099–105.

McAlister, D.F. & Krober, D.A. (1958). Response of soybeans to leaf and pod removal. *Agron. J.* **50**, 674–7.

McCarthy, F. & McArthur, M. (1960). The food quest and the time factor in Aboriginal economic life. In *Records of the America-Australian Scientific*

Expedition to Arnhem Land, vol. 2. (*Anthropology and Nutrition*) (ed. C.P. Mountford). pp. 145–94 Melbourne.

McCashin, M.G., Cossins, E.A. & Canvin, D.T. (1988). Dark respiration during photosynthesis in wheat leaf slices. *Plant Physiol.* 87, 155–61.

McCree, K.J. (1970). An equation for the rate of respiration of white clover plants grown under controlled conditions. In *Prediction and Measurement of Photosynthetic Productivity*, pp. 221–9. Pudoc, Wageningen.

McCree, K.J. (1972). The action spectrum, absorptance and quantum yield of photosynthesis in crop plants. *Agric. Meteorol.* 9, 191–216.

McCree, K.J. (1988). Sensitivity of sorghum grain yield to ontogenetic changes in respiration coefficients. *Crop Sci.* 28, 114–20.

McCree, K.J. & Keener, M.E. (1974). Simulations of the photosynthetic rates of three selections of grain sorghum with extreme leaf angles. *Crop Sci.* 14, 584–7.

McCree, K.J. & Troughton, J.H. (1966a). Prediction of growth rate at different light levels from measured photosynthesis and respiration rates. *Plant Physiol.* 41, 559–66.

McCree, K.J. & Troughton, J.H. (1966b). Non-existence of an optimum leaf area index for the production rate of white clover grown under constant conditions. *Plant Physiol.* 41, 1615–22.

McCree, K.J., Kallsen, C.E. & Richardson, S.G. (1984). Carbon balance of sorghum plants during osmotic adjustment to water stress. *Plant Physiol.* 76, 898–902.

McDaniel, R.G. (1986). Biochemical and physiological basis of heterosis. *CRC Crit. Rev. Plant Sci.* 4, 227–46.

McDermitt, D.K. & Loomis, R.S. (1981). Elemental composition of biomass and its relation to energy content, growth efficiency, and growth yield. *Ann. Bot.* 48, 275–90.

McDonald, D.J. & Woodward, R.G. (1977). Control of plant size for maximum efficiency of rice production in temperate areas. *SABRAO, 3rd Intl. Congress. Plant Breeding Papers* 3(a), 17–22.

McDonald, G.K. (1990). The growth and yield of uniculm and tillered barley over a range of sowing rates. *Aust. J. Agric. Res.* 41, 449–61.

MacKenzie, D. (1990). Famines before the floods? *New Scientist* 125 (1710), 10–11.

MacKenzie, D.R., Ho, L., Liu, T.D., Wu, H.B.F. & Oyer, E.B. (1975). Photoperiodism of mung bean and four related species. *Hort. Science* 10, 486–7.

MacKey, J. (1979a). Genetic potentials for improved yield. In *Proc. Workshop on Agricultural Potentiality Directed by Nutritional Needs* (ed. S. Rajki), pp. 121–43. Akad. Kiado, Budapest.

MacKey, J. (1979b). Wheat domestication as a shoot:root interrelation process. *Proc. 5th Wheat Genetics Symposium*, vol. 2, pp. 875–90 Indian Soc. Genet. Plant Breed., New Delhi.

McLaren, D. (1974). The great protein fiasco. *Lancet* ii, 93–6.

MacNeish, R.S. (1958). Preliminary archaeological investigations in the Sierra de Tamaulipas, Mexico. *Trans. Amer. Philosoph. Soc.* 48(6), 1–210.

MacNeish, R.S. (1964). The origins of new world civilization. *Sci. Amer.* 211 (5), 29–37.

MacNeish, R.S. (1985). The archaeological record on the problem of the domestication of corn. *Maydica* 30, 171–8.

McRae, D.H. (1985). Advances in chemical hybridization. *Plant Breeding Rev.* 3, 169–91.

McWhirter, M. & McWhirter, R. (1987). *Guinness Book of Records*. Guinness

Superlatives, London.

McWilliam, J.R. & Griffing, B. (1964). Temperature-dependent heterosis in maize. *Aust. J. Biol. Sci.* **18**, 569–83.

McWilliam, J.R., Latter, B.D.H. & Mathison, M.J. (1969). Enhanced heterosis and stability in the growth of an interspecific *Phalaris* hybrid at high temperature. *Aust. J. Biol. Sci.* **22**, 493–504.

Mahalakshmi, V. & Bidinger, F.R. (1985). Water stress and time of floral initiation in pearl millet. *J. Agric. Sci. (Camb.)* **105**, 437–45.

Mahalakshmi, V., Bidinger, F.R. & Rao, G.D.P. (1988). Timing and intensity of water deficits during flowering and grain-filling in pearl millet. *Agron. J.* **80**, 130–5.

Mahalakshmi, V., Bidinger, F.R. & Rao, G.D.P. (1990). Line-source vs. irrigated/non-irrigated treatments for evaluation of genotype drought response. *Agron. J.* **82**, 841–4.

Mahon, J.D. (1982). Field evaluation of growth and nitrogen fixation in peas selected for high and low photosynthetic CO_2 exchange. *Can. J. Plant Sci.* **62**, 5–17.

Mahon, J.D. (1990). Photosynthetic carbon dioxide exchange, leaf area, and growth of field-grown pea genotypes. *Crop Sci.* **30**, 1093–8.

Mahon, J.D. & Hobbs, S.L.A. (1981). Selection of peas for photosynthetic CO_2 exchange rate under field conditions. *Crop Sci.* **21**, 616–21.

Mahon, J.D., Lowe, S.B. & Hunt, L.A. (1977). Variation in the rate of photosynthetic CO_2 uptake in cassava cultivars and related species of *Manihot*. *Photosynthetica* **11**, 131–8.

Mahon, J.D., Hobbs, S.L.A. & Salminen, S.O. (1983). Characteristics of pea leaves and their relationships to photosynthetic CO_2 exchange in the field. *Can. J. Bot.* **61**, 3283–92.

Maiden, J.H. (1889). *The Useful Native Plants of Australia*. Treubner, London.

Makhijani, A. (1975). *Energy and Agriculture in the Third World*. Ballinger, Cambridge, Massachusetts.

Makino, A., Mae, T. & Ohira, K. (1983). Photosynthesis and ribulose-1,5-bisphosphate carboxylase in rice leaves. *Plant Physiol.* **73**, 1002–7.

Makino, A., Mae, T. & Ohira, K. (1987). Variations in the contents and kinetic properties of ribulose-1,5-bisphosphate carboxylases among rice species. *Plant Cell Physiol.* **28**, 799–804.

Makino, A., Mae, T. & Ohira, K. (1988). Differences between wheat and rice in the enzymic properties of ribulose-1,5-bisphosphate carboxylase/oxygenase and the relationship to photosynthetic gas exchange. *Planta* **174**, 30–8.

Malthus, T.R. (1798/1830). I. *An Essay on the Principle of Population* (1798). II. *A Summary View of the Principle of Population* (1830). (ed. A. Flew). Penguin, London (1970).

Mangelsdorf, P.C. (1952). Review of *Agricultural Origins and Dispersals* by Carl O. Sauer *American Antiquity* **19**, 87–90.

Mangelsdorf, P.C. (1966). Genetic potentials for increasing yields of food crops and animals. *Proc. Natl. Acad. Sci. USA.* **56**, 370–5.

Mangelsdorf, P.C. (1986). The origin of corn. *Sci. Amer.* **255**(2), 72–8.

Mangelsdorf, P.C. & Reeves, R.G. (1938). The origin of maize. *Proc. Natl. Acad. Sci. USA.* **24**, 303–12.

Mangelsdorf, P.C., MacNeish, R.S. & Galinat, W.C. (1967). Prehistoric wild and cultivated maize. In *The Prehistory of the Tehuacan Valley*, vol. 1 (*Environment and Subsistence*) (ed. D.S. Byers), pp. 178–200. University of Texas, Austin.

Marc, J. & Gifford, R.M. (1984). Floral initiation in wheat, sunflower and sorghum under carbon dioxide enrichment. *Can. J. Bot* **62**, 9–14.

Marshall, D.R. (1991). Alternative approaches and perspectives in breeding for higher yields. *Field Crops Res.* **26**, 171–90.

Marshall, L., Busch, R., Cholick, F., Edwards, I. & Frohberg, R. (1989). Agronomic performance of spring wheat isolines differing for daylength response. *Crop Sci.* **29**, 752–7.

Marstrand, P.K. & Pavitt, K.L.R. (1973). The agricultural subsystem. In *Thinking About the Future: A Critique of The Limits to Growth* (ed. H.S.D. Cole), pp. 56–65. Chatto & Windus, London.

Marstrand, P.K. & Rush, H. (1978). Food and agriculture: when enough is not enough – the world food paradox. In *World Futures: The Great Debate* (ed. C. Freeman & M. Jahoda), pp. 79–112. Universe Books, New York.

Martin, B. & Thorstenson, Y.R. (1988). Stable carbon isotope composition (δ ^{13}C), water use efficiency, and biomass productivity of *Lycopersicon esculentum*, *Lycopersicon pennellii*, and the F_1 hybrid. *Plant Physiol.* **88**, 213–17.

Martiniello, P., Delogu, G., Odoardi, M., Boggini, G. & Stanca, A.M. (1987). Breeding progress in grain yield and selected agronomic characters of winter barley (*Hordeum vulgare* L.) over the last quarter of a century. *Plant Breeding* **99**, 289–94.

Marzola, D.L. & Bartholomew, D.P. (1979). Photosynthetic pathway and biomass energy production. *Science* **205**, 555–9.

Masle, J., Doussinault, G. & Sun, B. (1989). Response of wheat genotypes to temperature and photoperiod in natural conditions. *Crop Sci.* **29**, 712–21.

Matlick, D. (1974). *Michigan Farmer*, 2 March.

Matsuda, T. (1978). Studies on the breeding of high yield variety in air-cured tobacco. 4. Inheritance of apparent photosynthetic rate, rate of photorespiration, and rate of respiration. *Bull. Utsunomiya Tobacco Expt. Station* **16**, 9–18.

Matsuo, T. (1959). *Rice Culture in Japan*. Min. Agric. For. Japan, Tokyo.

Matthews, G.A. (1981). Improved systems of pesticide application. *Phil. Trans. R. Soc. Lond.* B**295**, 163–73.

Mattingley, G.E.G., Slope, D. & Gutteridge, T. (1980). Effects of phosphate and potassium manuring on take-all and yield of winter wheat. *Ann. Rep. Rothamsted for 1979.* **1**, 227–9.

Mauney, J.R., Guinn, G., Fry, K.E. & Hesketh, J.D. (1979). Correlation of photosynthetic carbon dioxide uptake and carbohydrate accumulation in cotton, soybean, sunflower and sorghum. *Photosynthetica* **13**, 260–6.

Mayer, A. & Mayer, J. (1974). Agriculture, the island empire. *Daedalus* **103**(3), 83–95.

Mayoral, M.L., Plaut, Z. & Reinhold, L. (1985). Effect of sink-source manipulations on the photosynthetic rate and carbohydrate content of cucumber cotyledons. *J. Exp. Bot.* **36**, 1551–8.

Mazur, B.J. & Falco, S.C. (1989). The development of herbicide-resistant crops. *Ann. Rev. Plant Physiol.* **40**, 441–70.

Meadows, D.H., Meadows, D.L., Randers, J. & Behrens, W.W. (1972). *The Limits to Growth*. Universe Books, New York.

Mearns, L.O., Katz, R.W. & Schneider, S.H. (1984). Extreme high temperature events: changes in their probabilities with changes in mean temperature. *J. Climate Appl. Meteorol.* **23**, 1601–13.

Medawar, P.B. (1967). *The Art of the Soluble*. Methuen, London.

Mederski, H.J. & Jeffers, D.L. (1973). Yield response of soybean varieties grown at two soil moisture stress levels. *Agron. J.* **65**, 410–12.

Meghji, M.R., Dudley, J.W., Lambert, R.J. & Sprague, G.F. (1984). Inbreeding depression, inbred and hybrid grain yields, and other traits of maize genotypes representing three eras. *Crop Sci.* **24**, 545–9.

Mehra, S. (1981). Instability in Indian agriculture in the context of the new technology. *IFPRI Res. Rep.* No. 25, pp. 1–55.

Meindl, J.D. (1987). Chips for advanced computing. *Sci. Amer.* **257**(4), 54–62.

Mendoza, H.A. & Estrada, R.N. (1979). Breeding potatoes for tolerance to stress: heat and frost. In *Stress Physiology in Crop Plants* (ed. H. Mussell & R.C. Staples), pp. 227–62. Wiley, New York.

Meredith, W.R. & Bridge, R.R. (1984). Genetic contributions to yield changes in upland cotton. In *Genetic Contributions to Yield Gains of Five Major Crop Plants* (ed. W.R. Fehr), pp. 75–87. Crop Sci. Soc. Amer., Madison, Wisconsin.

Meredith, W.R. & Wells, R. (1987). Sub-okra leaf influence on cotton yield. *Crop Sci.* **27**, 47–8.

Meredith, W.R. & Wells, R. (1989). Potential for increasing cotton yields through enhanced partitioning to reproductive structures. *Crop Sci.* **29**, 636–9.

Merwe, N.J. van der (1982). Carbon isotopes, photosynthesis, and archaeology. *American Sci.* **70**, 596–606.

Mesarovic, M. & Pestel, E. (1974). *Mankind at the Turning Point*. Hutchinson, London.

Metcalf, R.L. (1984). Trends in the use of chemical insecticides. In *Judicious and Efficient Use of Insecticides on Rice*, pp. 69–91. IRRI, Los Baños.

Metzger, D.D., Czaplewski, S.J. & Rasmusson, D.C. (1984). Grain filling duration and yield in spring barley. *Crop Sci.* **24**, 1101–5.

Meyers, K.B., Simmons, S.R. & Stuthman, D.D. (1985). Agronomic comparison of dwarf and conventional height oat genotypes. *Crop Sci.* **25**, 964–6.

Meyers, S.P., Nichols, S.L., Baer, G.R., Molin, W.T. & Schrader, L.E. (1982). Ploidy effects in isogenic populations of alfalfa. I. Ribulose-1-5-bisphosphate carboxylase, soluble protein, chlorophyll, and DNA in leaves. *Plant Physiol.* **70**, 1704–9.

Michaels, P.J. (1981). The climatic sensitivity of 'Green Revolution' wheat culture in Sonora, Mexico. *Environmental Conservation* **8**, 307–12.

Midmore, D.J. (1980). Effects of photoperiod on flowering and fertility of sugar cane (*Saccharum* spp.). *Field Crops Res.* **3**, 65–81.

Midmore, D.J., Cartwright, P.M. & Fischer, R.A. (1982). Wheat in tropical environments. I. Phasic development and spike size. *Field Crops Res.* **5**, 185–200.

Miksicek, C.H., Bird, R.M., Pickersgill, B., Donaghey, S., Cartwright, J. & Hammond, N. (1981). Preclassic lowland maize from Cuello, Belize. *Nature (Lond.)* **289**, 56–9.

Milford, G.F.J. (1976). Sugar concentration in sugar beet: varietal differences and the effects of soil type and planting density on the size of the root cells. *Ann. Appl. Biol.* **83**, 251–7.

Milford, G.F.J. & Pearman, I. (1975). The relationship between photosynthesis and the concentration of carbohydrates in the leaves of sugar beet. *Photosynthetica* **9**, 78–83.

Milford, G.F.J., Biscoe, P.V., Jaggard, K.W., Scott, R.K. & Draycott, A.P. (1980). Physiological potential for increasing yields of sugar beet. In *Opportunities for Increasing Crop Yields* (ed. R.G. Hurd, P.V. Biscoe & C. Dennis), pp. 71–83. Pitman, Boston.

Miller, D.S., Baker, J., Bowden, M., Evans, E., Holt, J., McKeag, R.J., Meinertzhagen, I., Mumford, P.M., Oddy, D.J., Rivers, J.P.W.R., Sevenhuyzen, G., Stock, M.J.,

Watts, M., Wolde-Gabriel, Y. & Wolde-Gabriel, Z. (1976). The Ethiopia applied nutrition project. *Proc. R. Soc. Lond.* **B194**, 23–48.

Miller, F.R. & Kebede, Y. (1984). Genetic contributions to yield gains in sorghum, 1959–1980. In *Genetic Contributions to Yield Gains in Five Major Crops* (ed. W.R. Fehr), pp. 1–14. Amer. Soc. Agron., Madison, Wisconsin.

Milthorpe, F.L. & Moorby, J. (1969). Vascular transport and its significance in plant growth. *Ann. Rev. Plant Physiol.* **20**, 117–38.

Mitchell, J.F.B., Senior, C.A. & Ingram, W.J. (1989). CO_2 and climate: a missing feedback? *Nature (Lond.).* **341**, 132–4.

Mitchell, T. (1848). *Journal of an Expedition into the Interior of Tropical Australia.* Longmans, London.

Mochida, O. (1974). Effect of insecticides and fungicides on yield of paddy rice in Kyushu, Japan. *FAO Plant Protection Bulletin* **22**(4), 87–91.

Mock, J.J. & Pearce, R.B. (1975). An ideotype of maize. *Euphytica* **24**, 613–23.

Moen, H.J. & Beek, K.J. (1974). *Literature Study on the Potential Irrigated Acreage in the World.* International Institute for Land Reclamation and Improvement, Wageningen.

Mohamed, H.A., Clark, J.A. & Ong, C.K. (1988). Genotypic differences in the temperature responses of tropical crops. II. Seedling emergence and leaf growth of groundnut (*Arachis hypogaea* L.) and pearl millet (*Pennisetum typhoides* S & H). *J. Exp. Bot.* **39**, 1129–35.

Moldau, H. & Karolin, A. (1977). Effect of the reserve pool on the relationship between respiration and photosynthesis. *Photosynthetica* **11**, 38–47.

Moll, A. & Henniger, W. (1978). Genotypische Photosyntheserate von Kartoffeln und ihre mögliche Rolle für die Ertragsbildung. *Photosynthetica* **12**, 51–61.

Mondal, M.H., Brun, W.A. & Brenner, M.L. (1978). Effects of sink removal on photosynthesis and senescence in leaves of soybean (*Glycine max* L.) plants. *Plant Physiol.* **61**, 394–7.

Monma, E. & Tsunoda, S. (1979). Photosynthetic heterosis in maize. *Jap. J. Breed.* **29**, 159–65.

Monsi, M. & Murata, Y. (1970). Development of photosynthetic systems as influenced by distribution of matter. In *Prediction and Measurement of Photosynthetic Productivity*, pp. 115–29. Pudoc, Wageningen.

Monteith, J.L. (1977). Climate and the efficiency of crop production in Britain. *Phil. Trans. R. Soc. Lond.* **B281**, 277–94.

Monteith, J.L. (1978). Reassessment of maximum growth rates for C_3 and C_4 crops. *Exptl. Agric.* **14**, 1–5.

Moody, K. (1981). Weed-fertilizer interactions in rice. *IRRI Res. Pap. Ser. No. 68*, pp. 1–35.

Moore, N.W. (1977). Arable land. In *Conservation and Agriculture* (ed. J. Davidson, R. Lloyd), pp. 23–43. Wiley, Chichester.

Moore, P.H. (1987). Physiology and control of flowering. *Copersucar Intl. Sugarcane Breeding Workshop*, pp. 103–127. Copersucar, Piraciciba, Sao Paulo, Brasil.

Morandi, E.W., Casano, L.M. & Reggiardo, L.M. (1988). Post-flowering photoperiodic effect on reproductive efficiency and seed growth in soybean. *Field Crops Res.* **18**, 227–41.

Morgan, J.A., Le Cain, D.R. & Wells, R. (1990). Semidwarfing genes concentrate photosynthetic machinery and affect leaf gas exchange of wheat. *Crop Sci.* **30**, 602–8.

Morgan, J.M. (1977). Changes in diffusive conductance and water potential of wheat plants before and after anthesis. *Aust. J. Plant Physiol.* **4**, 75–86.

Morgan, J.M. (1983). Osmoregulation as a selection criterion for drought tolerance in wheat. *Aust. J. Agric. Res.* **34**, 607–14.

Morgan, J.M. (1988). The use of coleoptile responses to water stress to differentiate wheat genotypes for osmoregulation, growth and yield. *Ann. Bot.* **62**, 193–8.

Morgan, J.M. (1991). A gene controlling differences in osmoregulation in wheat. *Aust. J. Plant Physiol.* **18**, 249–57.

Morgan, J.M. & Condon, A.G. (1986). Water use, grain yield, and osmoregulation in wheat. *Aust. J. Plant Physiol.* **13**, 523–32.

Morgan, J.M., Hare, R.A. & Fletcher, R.J. (1986). Genetic variation in osmoregulation in Bread and Durum wheats and its relationship to grain yield in a range of field environments. *Aust. J. Agric. Res.* **37**, 449–57.

Morison, J.I.L. (1985). Sensitivity of stomata and water use efficiency to high CO_2. *Plant, Cell Environ.* **8**, 467–74.

Morison, J.I.L. & Gifford, R.M. (1983). Stomatal sensitivity to carbon dioxide and humidity. *Plant Physiol.* **71**, 789–96.

Morris, R.A., Nataatmadja, H., Bagyo, A.S. & Hurun, A.M. (1977). Subang 1975–1976, Indonesia. In *Constraints to High Yields on Asian Rice Farms: An Interim Report*, pp. 45–119. IRRI, Los Baños.

Morrod, R.S. (1981). Lead generation: designing the right approach. *Phil. Trans. R. Soc. Lond.* B**295**, 35–44.

Morrow, D.T. (1980). The economics of the international stockholding of wheat. *IFPRI Research Rep.* No. 18, pp. 1–45.

Moss, D.N. (1962). Photosynthesis and barrenness. *Crop Sci.* **2**, 366–7.

Moss, D.N. & Stinson, H.T. (1961). Differential response of corn hybrids to shade. *Crop Sci.* **1**, 416–18.

Mozingo, R.W., Coffelt, T.A. & Wynne, J.C. (1987). Genetic improvement in large-seeded Virginia-type peanut cultivars since 1944. *Crop Sci.* **27**, 228–31.

Muchow, R.C. (1988*a*). Effect of nitrogen supply on the comparative productivity of maize and sorghum in a semi-arid tropical environment. I. Leaf growth and leaf nitrogen. *Field Crops Res.* **18**, 1–16.

Muchow, R.C. (1988*b*). Ibid. III. Grain yield and nitrogen accumulation. *Field Crops Res.* **18**, 31–43.

Muchow, R.C. (1989*a*). Comparative productivity of maize, sorghum and pearl millet in a semi-arid tropical environment. I. Yield potential. *Field Crops Res.* **20**, 191–205.

Muchow, R.C. (1989*b*). Ibid. II. Effect of water deficits. *Field Crops Res.* **20**, 207–19.

Muchow, R.C. (1990*a*). Effect of high temperature on grain growth in field-grown maize. *Field Crops Res.* **23**, 145–58.

Muchow, R.C. (1990*b*). Effect of high temperature on the rate and duration of grain growth in field-grown *Sorghum bicolor* (L.) Moench. *Aust. J. Agric. Res.* **41**, 329–37.

Muchow, R.C. & Carberry, P.S. (1990). Phenology and leaf area development in a tropical grain sorghum. *Field Crops Res.* **23**, 221–37.

Muchow, R.C. & Davis, R. (1988). Effect of nitrogen supply on the comparative productivity of maize and sorghum in a semi-arid tropical environment. II. Radiation interception and biomass accumulation. *Field Crops Res.* **18**, 17–30.

Mudahar, M.S. & Hignett, T.P. (1985). Energy efficiency in nitrogen fertilizer production. *Energy in Agric.* **4**, 159–77.

Mullen, J.A. & Koller, H.R. (1988). Daytime and night time carbon balance and assimilate export in soybean leaves at different photon flux densities. *Plant Physiol.* **86**, 880–4.

Mumford, J.D. & Norton, G.A. (1984). Economics of decision making in pest management. *Ann. Rev. Entomol.* **29**, 157–74.

Munakata, K. (1976). Effects of temperature and light on the reproductive growth and ripening of rice. In *Climate and Rice*, pp. 187–207. IRRI, Los Baños.

Munson, P.J. (1976). Archaeological data on the origins of cultivation in the southwestern Sahara and their implications for West Africa. In *Origins of African Plant Domestication* (ed. J.R. Harlan, J.M.J. de Wet & A.B.L. Stemler), pp. 187–209. Mouton, The Hague.

Muramoto, H., Hesketh, J. & El-Sharkawy, M. (1965). Relationships among rate of leaf area development, photosynthetic rate, and rate of dry matter production among American cultivated cottons and other species. *Crop Sci.* **5**, 163–6.

Murata, Y. (1961). Studies on the photosynthesis of rice plants and its culture significance. *Bull. Nat. Inst. Agric. Sci. D.* No. 9, pp. 1–169.

Murata, Y. (1975). Estimation and simulation of rice yield from climatic factors. *Agric. Meteorol.* **15**, 117–31.

Murata, Y. (1981). Dependence of potential productivity and efficiency for solar energy utilization on leaf photosynthetic capacity in crop species. *Jap. J. Crop Sci.* **50**, 223–32.

Murayama, S., Norihama, Y., Miyazato, K. & Nose, A. (1982). (Studies on productivity of F_1 hybrid rice.) *Jap. J. Crop Sci.* **51**, (suppl. 2), 85–6. (In Japanese.)

Murty, K.S. & Sahu, G. (1987). Impact of low-light stress on growth and yield of rice. In *Weather and Rice*, pp. 93–101. IRRI, Los Baños.

Muruli, B.I. & Paulsen, G.M. (1981). Improvement of nitrogen use efficiency and its relationship to other traits in maize. *Maydica* **26**, 63–73.

Musgrave, M.E., Murfet, I.C. & Siedow, J.N. (1986*a*). Inheritance of cyanide-resistant respiration in two cultivars of pea (*Pisum sativum* L.). *Plant Cell Environ.* **9**, 153–6.

Musgrave, M.E., Strain, B.R. & Siedow, J.N. (1986*b*). Response of two pea hybrids to CO_2 enrichment: a test of the energy overflow hypothesis for alternative respiration. *Proc. Natl. Acad. Sci. USA* **83**, 8157–61.

Nabhan, G. & de Wet, J.M.J. (1984). *Panicum sonorum* in Sonoran desert agriculture. *Econ. Bot.* **38**, 65–82.

Nafziger, E.D. & Koller, H.R. (1976). Influence of leaf starch concentration on CO_2 assimilation in soybean. *Plant Physiol.* **57**, 560–3.

Nagarajah, S. (1975). Effect of debudding on photosynthesis in leaves of cotton. *Physiol. Plantar.* **33**, 28–31.

Nakamoto, K. & Yamazaki, K. (1988). Quantitative relationships among vegetative organs and their conductive tissues in several millets. II. Quantitative observations on conductive tissues in leaves. *Jap. J. Crop Sci.* **57**, 482–9.

Narain, A. (1974). Castor. In *Evolutionary Studies in World Crops. Diversity and Change in the Indian Subcontinent* (ed. J.B. Hutchinson), pp. 71–80. Cambridge University Press.

Narain, D. (1977). Growth of productivity in Indian agriculture. *Indian J. Agric. Econ.* **32**, 1–44.

Nass, H.G. & Reiser, B. (1975). Grain filling period and grain yield relationships in spring wheat. *Can. J. Plant Sci.* **55**, 673–8.

Nath, V., Bhardwaj, S.N. (1987). Influence of sink on photosynthesis in field pea (*Pisum sativum* Linn. var. *arvensis*). *Indian J. Plant Physiol.* **30**, 398–99.

National Academy of Sciences, Washington, D.C. (1972). *Genetic Vulnerability of Major Crops*. Washington, DC.

National Research Council, Board on Agriculture (1989). *Alternative Agriculture*. US National Acad. Sci., Washington DC.

Natr, L. (1966). (Varietal differences in the intensity of photosynthesis.) *Rostlinna Vyroba* **12**, 163–78. (In Czech.)

Natr, L. (1972). Influence of mineral nutrients on photosynthesis of higher plants. *Photosynthetica* **6**, 80–99.

Natr, L. (1975). Influence of mineral nutrition on photosynthesis and the use of assimilates. In *Photosynthesis and Productivity in Different Environments* (ed. J.P. Cooper), pp. 537–55. Cambridge University Press.

Natr, L. & Ludlow, M.M. (1970). Influence of glucose absorption and photosynthate accumulation on gas exchange of barley leaf segments. *Photosynthetica* **4**, 288–94.

Neales, T.F. & Davies, J.A. (1966). The effect of photoperiod duration upon the respiratory activity of the roots of wheat seedlings. *Aust. J. Biol. Sci.* **19**, 471–80.

Neales, T.F. & Incoll, L.D. (1968). The control of leaf photosynthesis rate by the level of assimilate concentration in the leaf: a review of the hypothesis. *Bot. Rev.* **34**, 107–25.

Nelson, C.J., Asay, K.H. & Horst, G.L. (1975). Relationship of leaf photosynthesis to forage yield of tall fescue. *Crop Sci.* **15**, 476–8.

Nelson, C.J., Asay, K.H. & Sleper, D.A. (1977). Mechanism of canopy development in tall fescue genotypes. *Crop Sci.* **17**, 449–52.

Nelson, W.L. (1967). Nitrogen, phosphorus, and potassium – needs and balance for high yields. In *Maximum Yields – The Challenge*, pp. 57–67. Amer. Soc. Agron. Spec. Publicn., Madison, Wisconsin.

Nevins, D.J. & Loomis, R.S. (1970). Nitrogen nutrition and photosynthesis in sugar beet (*Beta vulgaris* L.) *Crop Sci.* **10**, 21–5.

Newman, J.E. (1978). Drought impacts on American agricultural productivity. In *North American Droughts* (ed. N.J. Rosenberg), pp. 43–62. Westview Press, Boulder, Colorado.

Ng, T.K., Busche, R.M., McDonald, C.C. & Hardy, R.W.F. (1983). Production of feedstock chemicals. *Science* **219**, 733–40.

Nguyen, H.P. & Anderson, J.R. (1991). Local adaption, varietal diversity and the skewness of regional crop yields. *Agric. Systems* **36**, 159–71.

Nichiporovich, A.A. (1956). (Photosynthesis and the theory of obtaining high crop yields.) USSR Academy of Sciences, Moscow. (In Russian.)

Nickell, L.G. (1982). *Plant Growth Regulators*. Springer, Berlin.

Nishiyama, I. (1985). Relation between rice yield and photosynthetically active solar radiation during seed ripening stage in selected Prefectures in Japan. *Jap. J. Crop Sci.* **54**, 8–14.

Nissly, C.R., Bernard, R.L. & Hittle, C.N. (1981). Variation in photoperiod sensitivity for time of flowering and maturity among soybean strains of Maturity Group III. *Crop Sci.* **21**, 833–6.

Nix, H.A. & Fitzpatrick, E.A. (1969). An index of crop water stress related to wheat and grain sorghum yields. *Agric. Meteorol.* **6**, 321–37.

Norby, R.J. & O'Neill, E.G. (1991). Leaf area compensation and nutrient interactions in CO_2-enriched seedlings of yellow-poplar (*Liriodendron tulipifera* L.). *New Phytol.* **17**, 515–28.

Norse, D. (1976). Development strategies and the world food problem. *J. Agric. Econ.* **27**, 137–58.

Nosberger, J. & Humphries, E.C. (1965). The influence of removing tubers on dry-matter production and net assimilation rate of potato plants. *Ann. Bot. N.S.* 29, 579–88.

Nour, A-E.M. & Weibel, D.E. (1978). Evaluation of root characteristics in grain sorghum *Agron. J.* 70, 217–18.

Nye, P.H. & Greenland, D.J. (1960). *The Soil under Shifting Cultivation.* Commonwealth Agric. Bur. (Tech. Comm. No. 51), Farnham Royal, UK.

Obenland, D., Hiser, C., McIntosh, L., Shibles, R. & Stewart, C.R. (1988). Occurrence of alternative respiratory capacity in soybean and pea. *Plant Physiol.* 88, 528–31.

O'Brien, L. (1982). Victorian wheat yield trends, 1898–1977. *J. Aust. Inst. Agric. Sci.* 48, 163–8.

O'Brien, L. & Ronalds, J.A. (1984). Yield and quality interrelationships amongst random F_3 lines and their implications for wheat breeding. *Aust. J. Agric. Res.* 35, 443–5.

O'Connell, J.F., Latz, P.K. & Barnett, P. (1983). Traditional and modern plant use among the Alyawara of Central Australia. *Econ. Bot.* 37, 80–109.

Odhiambo, M.O. & Compton, W.A. (1987). Twenty cycles of divergent mass selection for seed size in corn. *Crop Sci.* 27, 1113–16.

Odum, H.T. (1967). Energetics of world food production. In *The World Food Problem.* The White House, Washington, DC.

Odum, H.T. (1971). *Environment, Power, and Society.* Wiley, New York.

Oelhaf, R.C. (1978). *Organic Agriculture. Economic and Ecological Comparisons with Conventional Methods.* Halsted Press (Wiley), New York.

Ogata, C., Hatada, M., Tomlinson, G., Shin, W.-C. & Kim, S.-H. (1987). Crystal structure of the intensely sweet protein monellin. *Nature (Lond.).* 328, 739–42.

Ohkawa, K. & Rosovsky, H. (1960). The role of agriculture in modern Japanese economic development. In *Economic Development and Cultural Change* 9 (2), 43–67.

Ohno, Y. (1976). Varietal differences of photosynthetic efficiency and dry matter production in *Indica* rice. *Tech. Bull. Tropic. Agric. Res. Centre* no. 9, pp. 1–72.

Ojehomon, O.O., Rathjen, A.S. & Morgan, D.G. (1968). Effects of daylength on the morphology and flowering of five determinate varieties of *Phaseolus vulgaris* L. *J. Agric. Sci. (Camb.)* 71, 209–14.

Ojehomon, O.O., Zehni, M.S. & Morgan, D.G. (1973). The effects of photoperiod on flower bud development in *Phaseolus vulgaris. Ann. Bot.* 37, 871–84.

Ojima, M. (1972). (Improvement of leaf photosynthesis in soybean varieties). *Bull. Natl. Inst. Agric. Sci. Ser. D.* 23, 97–154. (In Japanese with English abstract.)

Ojima, M. (1974). Improvement of photosynthetic capacity in soybean variety. *Japan Agric. Res. Quart.* 8, 6–12.

Ojima, M. & Kawashima, R. (1968). Studies on the seed production of soybean. V. Varietal differences in photosynthetic rates of soybeans. *Proc. Crop Sci. Soc. Japan* 37, 667–75.

Ojima, M. & Kawashima, R. (1970). Ibid. VIII. The ability of photosynthesis of F_3 lines having different photosynthesis in their F_2 generations. *Proc. Crop Sci. Soc. Japan* 39, 440–5.

Oka, H. (1955). Phylogenetic differentiation of cultivated rice. *Jap. J. Breed.* 4, 213–21.

Oka, H.I. (1988). *Origin of Cultivated Rice.* Japan Scientific Soc./Elsevier, Amsterdam.

Oldeman, L.R., Seshu, D.V. & Cady, F.B. (1987). Response of rice to weather variables. In *Weather and Rice*, pp. 5–39. IRRI, Los Baños.

Olson, R.A. (1987). The use of fertilizers and soil amendments. In *Land Transformation in Agriculture* (ed. M.G. Wolman & F.G.A. Fournier), pp. 203–26. John Wiley, Chichester.

Olugbemi, L.B. & Bush, M.G. (1987). The influence of temperature on the contribution of awns to yield in wheat. *Aust. J. Plant Physiol.* **14**, 299–310.

Ong, C.K. & Everard, A. (1979). Short day induction of flowering in pearl millet (*Pennisetum typhoides*) and its effect on plant morphology. *Exptl. Agric.* **15**, 401–10.

Opeña, R.T. (1985). Development of tomato and Chinese cabbage cultivars adapted to the hot, humid tropics. *Acta Horticulturae* **153**, 421–36.

Oritani, T., Enbutsu, T. & Yoshida, R. (1979). Studies on nitrogen metabolism in crop plants. XVI. Changes in photosynthesis and nitrogen metabolism in relation to leaf area growth of several rice varieties. *Jap. J. Crop Sci.* **48**, 10–16.

Osada, A. (1963). Studies on the photosynthesis of indica rice. *Proc. Crop Sci. Soc. Japan* **33**, 69–76.

Osborne, B.A. & Garrett, M.K. (1983). Quantum yields for CO_2 uptake in some diploid and tetraploid plant species. *Plant, Cell Environ.* **6**, 135–44.

Osman, A.M., Goodman, P.J. & Cooper, J.P. (1977). The effects of nitrogen, phosphorus and potassium on rates of growth and photosynthesis of wheat. *Photosynthetica*, **11**, 66–75.

Osmond, C.B. & Bjorkman, O. (1972). Effects of O_2 on photosynthesis. *Carnegie Inst. Washington Yearbook* **71**, 141–8.

O'Toole, J.C. (1982). Adaptation of rice to drought-prone environments. In *Drought Resistance in Crops, with Emphasis on Rice*, pp. 195–213. IRRI, Los Baños.

O'Toole, J.C. & Bland, W.L. (1987). Genotypic variation in crop plant root systems. *Adv. Agron.* **41**, 91–145.

O'Toole, J.C. & Chang, T.T. (1979). Drought resistance in cereals: Rice, a case study. In *Stress Physiology in Crop Plants* (ed. H. Mussell & R.C. Staples), pp. 373–405. Wiley Interscience, New York.

O'Toole, J.C., Cruz, R.T. & Singh, T.N. (1979). Leaf rolling and transpiration. *Plant Sci. Lett.* **16**, 111–14.

Ottaviano, E. & Camussi, A. (1981). Phenotypic and genetic relationships between yield components in maize. *Euphytica* **30**, 601–9.

O'Type, G., O'Type, P. & O'Type, K. (1984). A new food, fiber and fuel crop. *Bioscience* **34**, 213.

Overman, A.J. (1976). Efficacy of soil fumigants applied via a drip irrigation system. *Proc. Florida State Hort. Soc.* **89**, 143–5.

Ovington, J.D. (1957). Dry matter production by *Pinus sylvestris* L. *Ann. Bot. N.S.* **21**, 287–314.

Owen, P.C. (1957). The effect of infection with tobacco etch virus on the rates of respiration and photosynthesis of tobacco leaves. *Ann. Appl. Biol.* **45**, 327–31.

Oxholm, M. & Chase, S.S. (1974). Description of maize. How it is planted and cultivated in North America, and the various uses of this grain, by Peter Kalm. *Econ. Bot.* **28**, 105–17.

Paccaud, F.X., Fossati, A. & Hong, S.C. (1985). Breeding for yield and quality in winter wheat: consequences for nitrogen uptake and partitioning efficiency. *Zeitschr. f. Pflanzenzücht.* **94**, 89–100.

Paddock, W. and Paddock, P. (1968). *Famine – 1975!* Weidenfeld & Nicolson, London.

Palfi, G. & Dezsi, L. (1960). The translocation of nutrients between fertile and sterile shoots of wheat. *Acta Bot. Acad. Sci. Hung.* **6**, 65–74.

Palit, P., Kundu, A., Mandal, R.K. & Sircar, S. (1979*a*). Productivity of rice plant in relation to photosynthesis, photorespiration and translocation. *Indian. J. Plant Physiol.* **22**, 66–74.

Palit, P., Kundu, A., Mandal, R.K. & Sircar, S.M. (1979*b*). Source:sink control of dry matter production and photosynthesis in the rice plant after flowering. *Indian J. Plant Physiol.* **22**, 87–91.

Palmer, E. (1883). On plants used by the natives of North Queensland, Flinders and Mitchell Rivers, for food, medicine etc. *J. Proc. Roy. Soc. NSW* **17**, 93–113.

Palta, J.A. (1982). Gas exchange of four cassava cultivars in relation to light intensity. *Expt. Agric.* **18**, 375–82.

Papendick, R.I., Young, D.L., McCool, D.K. & Krauss, H.A. (1985). Regional effects of soil erosion on crop productivity – the Palouse area of the Pacific Northwest. In *Soil Erosion and Crop Productivity* (ed. R.F. Follett & B.A. Stewart), pp. 305–20. Amer. Soc. Agron., Madison, Wisconsin.

Parker, C. & Fryer, J.D. (1975). Weed control problems causing major reductions in world food supplies. *FAO Plant Protection Bull.* **23**, 83–95.

Parker, M.L. & Ford, M.A. (1982). The structure of the mesophyll of flag leaves in three *Triticum* species. *Ann. Bot.* **49**, 165–76.

Parry, M. (1990). *Climatic Change and World Agriculture.* Earthscan/UNEP, London.

Parry, M.L., Porter, J.H. & Carter, T.R. (1990). Climatic change and its implications for agriculture. *Outlook on Agriculture* **19**, 9–15.

Parthasarathy, B., Munot, A.A. & Kothawale, D.R. (1988). Regression model for estimation of Indian foodgrain production from summer monsoon rainfall. *Agric. Forest Meteorol.* **42**, 167–82.

Passioura, J.B. (1972). The effect of root geometry on the yield of wheat growing on stored water. *Aust. J. Agric. Res.* **23**, 745–52.

Passioura, J.B. (1977). Grain yield, harvest index and water use by wheat. *J. Aust. Inst. Agric. Sci.* **43**, 117–20.

Passioura, J.B. & Ashford, A.E. (1974). Rapid translocation in the phloem of wheat roots. *Aust. J. Plant Physiol.* **1**, 521–7.

Patefield, W.M. & Austin, R.B. (1971). A model for the simulation of the growth of *Beta vulgaris* L. *Ann. Bot.* **35**, 1227–50.

Patrick, J.W. (1972). Distribution of assimilate during stem elongation in wheat. *Aust. J. Biol. Sci.* **25**, 455–67.

Patrick, J.W. & Wardlaw, I.F. (1984). Vascular control of photosynthate transfer from the flag leaf to the ear of wheat. *Aust. J. Plant Physiol.* **11**, 235–41.

Patrick, J.W. & Wareing, P.F. (1980). Hormonal control of assimilate movement and distribution. In *Aspects and Prospects of Plant Growth Regulators*, pp. 65–84. Monograph No. 6, British Plant Growth Regulator Group, Long Ashton, UK.

Patterson, B.D. (1988). Genes for cold resistance from wild tomatoes. *Hort. Sci.* **23**, 794, 947.

Patterson, F.L. & Ohm, H.W. (1975). Compensating ability of awns in soft red winter wheat. *Crop Sci.* **15**, 403–7.

Patterson, F.L., Compton, L.E., Caldwell, R.M. & Schafer, J.F. (1962). Effect of awns on yield, test weight, and kernel weight of soft red winter wheats. *Crop Sci.* **2**, 199–200.

Paulino, L.A. (1986). Food in the Third World: past trends and projections to 2000. *IFPRI Res. Rep.* No. 52, pp. 1–78.

Paulino, L.A. & Tseng, S.S. (1980). A comparative study of FAO and USDA data on production, area, and trade of major food staples. *IFPRI Res. Rep.* No. 19, pp. 1–76.

Payne, P.R. (1978). Human protein requirements. In *Plant Proteins* (ed. G. Norton), pp. 247–63. Butterworths, London.

Payne, T.S., Stuthman, D.D., McGraw, R.L. & Bregitzer, P.P. (1986). Physiological changes associated with three cycles of recurrent selection for grain yield improvement in oats. *Crop Sci.* **26**, 734–6.

Payton, F.V., Rhue, R.D. & Hensel, D.R. (1989). Mitserlich-Bray equation used to correlate soil phosphorus and potato yields. *Agron. J.* **81**, 571–6.

Pearce, R.B., Carlson, G.E., Barnes, B.K., Hart, R.H. & Hanson, C.H. (1969). Specific leaf weight and photosynthesis in alfalfa. *Crop Sci.* **9**, 423–6.

Pearman, I., Thomas, S.M. & Thorne, G.N. (1978). Effect of nitrogen fertilizer on growth and yield of semi-dwarf and tall varieties of winter wheat. *J. Agric. Sci. (Camb.).* **91**, 31–46.

Pearman, I., Thomas, S.M. & Thorne, G.N. (1979). Effect of nitrogen fertilizer on photosynthesis of several varieties of winter wheat. *Ann. Bot.* **43**, 613–21.

Pearse, A. (1980). *Seeds of Plenty, Seeds of Want. Social and Economic Implications of the Green Revolution.* Clarendon Press, Oxford.

Pearson, C.J., Larson, E.M., Hesketh, J.D. & Peters, D.B. (1984). Development and source-sink effects on single leaf and canopy carbon dioxide exchange in maize. *Field Crops Res.* **9**, 391–402.

Pederson, D.G. & Rathjen, A.J. (1981). Choosing trial sites to maximize selection response for grain yield in spring wheat. *Aust. J. Agric. Res.* **32**, 411–24.

Pedigo, L.P., Hutchins, S.H. & Higley, L.G. (1986). Economic injury levels in theory and practice. *Ann. Rev. Entomol.* **31**, 341–68.

Peel, A.J. & Ho, L.C. (1970). Colony size of *Tuberolachnus salignus* (Gmelin) in relation to mass transport of ^{14}C-labelled assimilates from the leaves in willow. *Physiol. Plantar.* **23**, 1033–8.

Peet, M.M., Bravo, A., Wallace, D.H. & Ozbun, J.L. (1977). Photosynthesis, stomatal resistance, and enzyme activities in relation to yield of field grown dry bean varieties. *Crop Sci.* **17**, 287–93.

Peeters, J.P. (1988). The emergence of new centres of diversity: evidence from barley. *Theoret. Appl. Genet.* **76**, 17–24.

Peltonen-Sainio, P. (1990). Genetic improvements in the structure of oat stands in northern growing conditions during this century. *Plant Breeding* **104**, 340–5.

Peng, S., Krieg, D.R. & Girma, F.S. (1991). Leaf photosynthetic rate is correlated with biomass and grain production in grain sorghum lines. *Photosynth. Res.* **28**, 1–7.

Penning de Vries, F.W.T. (1974). Substrate utilization and respiration in relation to growth and maintenance in higher plants. *Neth. J. Agric. Sci.* **22**, 40–4.

Penning de Vries, F.W.T. (1975a). Use of assimilates in higher plants. In *Photosynthesis and Productivity in Different Environments* (ed. J.P. Cooper), pp. 459–80. Cambridge University Press.

Penning de Vries, F.W.T. (1975b). The cost of maintenance processes in plant cells. *Ann. Bot.* **39**, 77–92.

Penuelas, J. & Matamala, R. (1990). Changes in N and S leaf content, stomatal density and specific leaf area of 14 plant species during the last three centuries of CO_2 increase. *J. Expt. Bot.* **41**, 1119–24.

Pepe, J.F. & Welsh, J.R. (1979). Soil water depletion patterns under dryland field conditions of closely related height lines of winter wheat. *Crop Sci.* **19**, 677–80.

Perfect, T.J. (1986). Irrigation as a factor influencing the management of agricultural pests. *Phil. Trans. R. Soc. Lond.* A316, 347–54.

Perry, M.W. & D'Antuono, M.F. (1989). Yield improvement and associated characteristics of some Australian spring wheat cultivars introduced between 1860 and 1982. *Aust. J. Agric. Res.* 40, 457–72.

Peters, D.B., Pendleton, J.W., Hageman, R.H. & Brown, C.M. (1971). Effect of night air temperature on grain yield of corn, wheat and soybeans. *Agron. J.* 63, 809.

Peterson, C.J., Johnson, V.A., Schmidt, J.W. & Mumm, R.F. (1989). Genetic improvement and the variability in wheat yields in the Great Plains. In *Variability in Grain Yields* (ed. J.R. Anderson & P.B.R. Hazell), pp. 175–84. Johns Hopkins University, Baltimore.

Peterson, D.M., Housley, T.L. & Luk, T.M. (1982). Oat stem vascular size in relation to kernel number and weight. II. Field environment. *Crop Sci.* 22, 274–78.

Peterson, R.B. & Zelitch, I. (1982). Relationship between net carbon dioxide assimilation and dry weight accumulation in field grown tobacco. *Plant Physiol.* 70, 677–85.

Pheloung, P.C. & Siddique, K.H.M. (1991). Contribution of stem dry matter to grain yield in wheat cultivars. *Aust. J. Plant Physiol.* 18, 53–64.

Phillips, R.E. & Phillips, S.H (eds). (1984). *No-Tillage Agriculture. Principles and Practices.* Van Nostrand, New York.

Phillips, R.E., Blevins, R.L., Thomas, G.W., Frye, W.W. & Phillips, S.H. (1980). No-Tillage agriculture. *Science* 208, 1108–13.

Pickersgill, B. (1989). Cytological and genetical evidence on the domestication and diffusion of crops within the Americas. In *Foraging and Farming: The Evolution of Plant Exploitation* (ed. D.R. Harris & G.C. Hillman), pp. 426–39. Unwin Hyman, London.

Pickersgill, B. & Heiser, C.B. (1977). Origins and distributions of plants domesticated in the New World tropics. In *Origins of Agriculture* (ed. C.A. Reed), pp. 803–35. Mouton, The Hague.

Pickett, J.A. (1988). Integrating use of beneficial organisms with chemical crop protection. *Phil. Trans. R. Soc. Lond.* B318, 203–11.

Pierce, J.T. (1990). *The Food Resource.* Longman, London.

Pilbeam, C.J. & Robson, M.J. (1992). Response of populations of *Lolium perenne* cv. S23 with contrasting rates of dark respiration to nitrogen supply and defoliation regime. 2. Grown as mixtures. *Ann. Bot.* 69, 79–86.

Pimentel, D. (1976). World food crisis: energy and pests. *Bull. Entomol. Soc. Amer.* 22, 20–6.

Pimentel, D. (1979). *Energy Resources and the World Food Problem.* Marine Sciences Distinguished Lecture Series. Louisiana State University, Baton Rouge.

Pimentel, D. (ed). (1980). *Handbook of Energy Utilization in Agriculture.* CRC Press, Boca Raton, Florida.

Pimentel, D. & Pimentel, M. (1979). *Food, Energy and Society.* Edward Arnold, London.

Pimentel, D. & Terhune, E.C. (1977a). Energy and food. *Ann. Rev. Energy* 2, 171–95.

Pimentel, D. & Terhune, E.C. (1977b). Energy use in food production. In *Dimensions of World Food Problems* (ed. E.R. Duncan), pp. 67–89. Iowa State University, Ames.

Pimentel, D., Hurd, L.E., Bellotti, A.C., Forster, M.J., Oka, I.N, Sholes, O.D. & Whitman, R.J. (1973). Food production and the energy crisis. *Science* 182, 443–9.

Pimentel, D., Berardi, G. & Fast, S. (1983). Energy efficiency of farming systems:

organic and conventional agriculture. *Agriculture, Ecosys. Env.* **9**, 359–72.

Pimentel, D., Berardi, G. & Fast, S. (1984*a*). Energy efficiencies of farming wheat, corn and potatoes organically. In *Organic Farming: Current Technology and Its Role in a Sustainable Agriculture* (ed. D.F. Bezdicek, J.F. Power, D.R. Keeney & M.J. Wright), pp. 151–61. Amer. Soc. Agron., Madison, Wisconsin.

Pimentel, D., Fried, C., Olson, L., Schmidt, S., Wagner-Johnson, K., Westman, A., Whelan, A., Foglia, K., Poole, P., Klein, T., Sobin, R. & Bochner, A. (1984*b*). Environmental and social costs of biomass energy. *Bioscience* **34**, 89–94.

Pimentel, D., Culliney, T.W., Buttler, I.W., Reinemann, D.J. & Beckman, K.B. (1989). Low-input sustainable agriculture using ecological management practices. *Agriculture, Ecosys. & Environ.* **27**, 3–24.

Pimentel, D., Dazhong, W. & Giampietro, M. (1990). Technological changes in energy use in US agricultural production. In *Agroecology. Researching the Ecological Basis for Sustainable Agriculture* (ed. S.R. Gliessman), pp. 305–21. Springer, New York.

Pinstrup-Andersen, P. & Hazell, P.B.R. (1985). The impact of the Green Revolution and prospects for the future. *Food Reviews Intl.* **1**, 1–25.

Planchon, C. (1969). Photosynthetic activity and yield in soft wheat (*Triticum aestivum*). *Genet. Agrar.* **23**, 485–90.

Planchon, C. (1979). Photosynthesis, transpiration, resistance to CO_2 transfer, and water efficiency of flag leaf of bread wheat, durum wheat and Triticale. *Euphytica* **28**, 403–8.

Planchon, C. & Fesquet, J. (1982). Effect of the D genome and of selection on photosynthesis in wheat. *Theoret. Appl. Genet.* **61**, 359–65.

Plucknett, D.L. & Smith, N.J.H. (1986). Sustaining agricultural yields. *Bioscience* **36** (1), 40–5.

Plucknett, D.L., Smith, N.J.H., Williams, J.T. & Anishetty, N.M. (1987). *Gene Banks and the World's Food*. Princeton University Press, Princeton, New Jersey.

Poleman, T.T. & Freebairn, D.K. (eds). (1973). *Food, Population and Employment: The Impact of the Green Revolution*. Praeger, New York.

Pollard, B. (1977). 'New' farmer sets world corn yield record. *Michigan Farmer*, 5 Nov., pp. 4–6.

Polson, D.E. (1972). Day neutrality in soybeans. *Crop Sci.* **12**, 773–6.

Pomeranz, Y. (1987). *Modern Cereal Science and Technology*. VCH Publ., New York.

Ponnuthurai, S., Virmani, S.S. & Vergara, B.S. (1984). Comparative studies on the growth and grain yield of some F_1 rice (*Oryza sativa* L.) hybrids. *Philipp. J. Crop Sci.* **9**, 183–93.

Poonyarit, M., Mackill, D.J. & Vergara, B.S. (1989). Genetics of photoperiod sensitivity and critical daylength in rice. *Crop Sci.* **29**, 647–52.

Poorter, H. (1989). Growth analysis: towards a synthesis of the classical and the functional approach. *Physiol. Plantar.* **75**, 237–44.

Poorter, H., Remkes, C. & Lambers, H. (1990). Carbon and nitrogen economy of 24 wild species differing in relative growth rate. *Plant Physiol.* **94**, 621–7.

Portères, R. (1962). Berceaux agricoles primaires sur le continent africain. *J. African History* **3**, 195–210.

Potter, J.R. & Breen, P.J. (1980). Maintenance of high photosynthetic rates during the accumulation of high leaf starch levels in sunflower and soybean. *Plant Physiol.* **66**, 528–31.

Powlson, D.S., Pruden, G., Johnston, A.E. & Jenkinson, D.S. (1986). The nitrogen cycle in the Broadbalk Wheat Experiment: recovery and losses of ^{15}N-labelled fertilizer applied in spring and inputs of nitrogen from the atmosphere. *J. Agric. Sci. (Camb.)* **107**, 591–609.

President's Science Advisory Committee (1967). *Report: The World Food Problem.* White House, Washington, DC.

Press, M.C., Nour, J.J., Bebawo, F.F. & Stewart, G.R. (1989). Anti-transpirant-induced heat stress in the parasitic plant *Striga hermonthica* – a novel method of control. *J. Expt. Bot.* **40**, 585–91.

Price, D.J. de Solla (1963). *Big Science, Little Science.* Columbia, New York.

Price, S.C., Kahler, A.L., Hallauer, A.R., Charmley, P. & Giegel, D.A. (1986). Relationship between performance and multilocus heterozygosity at enzyme loci in single cross hybrids of maize. *J. Hered.* **77**, 341–4.

Prokofyev, A.A., Zhdanova, L.P. & Sobolyev, A.M. (1957). (Some regularities pertaining to the flow of substances from leaves to reproductive organs.) *Fiziol. Rasten.* **4**, 425–31. (In Russian with English summary.)

Puckridge, D.W. (1971). Photosynthesis of wheat under field conditions. III. Seasonal trends in carbon dioxide uptake of crop communities. *Aust. J. Agric. Res.* **22**, 1–9.

Puri, Y.P., Qualset, C.O., Miller, M.F., Baghott, K.G., Jan, C.C. & de Pace, C. (1985). Barley, wheat, and Triticale grain yield in relation to solar radiation and heat units. *Crop Sci.* **25**, 893–90.

Purohit, A.N. (1970). The qualititative and quantitative photoperiodic response of Indian potato varieties. *New Phytol.* **69**, 521–7.

Purseglove, J.W. (1957). History and functions of Botanic Gardens with special reference to Singapore. *Trop. Agric. (Trinidad)* **34**, 165–89.

Purseglove, J.W. (1963). Some problems of the origin and distribution of tropical crops. *Genet. Agrar.* **17**, 105–22.

Purseglove, J.W. (1968). *Tropical Crops. Dicotyledons.* Longman, London.

Pyke, K.A., Jellings, A.J. & Leech, R.M. (1990). Variation in mesophyll cell number and size in wheat leaves. *Ann. Bot.* **65**, 679–83.

Qualset, C.O., Schaller, C.W. & Williams, J.C. (1965). Performance of isogenic lines of barley as influenced by awn length, linkage blocks, and environment. *Crop Sci.* **5**, 489–94.

Quebedeaux, B. & Hardy, R.W.F. (1973). Oxygen as a new factor controlling reproductive growth. *Nature (Lond.)* **243**, 477–9.

Quebedeaux, B. & Hardy, R.W.F. (1975). Reproductive growth and dry matter production of *Glycine max* (L.) Merr. in response to oxygen concentration. *Plant Physiol.* **55**, 102–7.

Quinby, J.R. (1974). *Sorghum Improvement and the Genetics of Growth.* Texas A and M University Press, College Station.

Rabbinge, R., Goudriaan, J., van Keulen, H., Penning de Vries, F.W.T. & van Laar, H.H. (eds). (1990). *Theoretical Production Ecology: Reflections and Prospects.* Pudoc, Wageningen.

Radin, J.W., Kimball, B.A., Hendrix, D.L. & Mauney, J.R. (1987). Photosynthesis of cotton plants exposed to elevated levels of carbon dioxide in the field. *Photosynthesis Res.* **12**, 191–203.

Rai, K.N. & Rao, A.S. (1991). Effect of d_2 dwarfing gene on grain yield and yield components in pearl millet near-isogenic lines. *Euphytica* **52**, 25–31.

Ramanujam, T. (1985). Leaf density profile and efficiency in partitioning dry matter among high and low yielding cultivars of cassava (*Manihot esculenta* Crantz). *Field Crops Res.* **10**, 291–303.

Ramanujam, T. (1987). Source-sink relationship in cassava (*Manihot esculenta*

Crantz). *Indian J. Plant Physiol.* **30**, 297–9.

Ramanujam, T. & Ghosh, S.P. (1990). Investigations of source-sink relations in cassava using reciprocal grafting. *Exptl. Agric.* **26**, 189–95.

Ramey, H.H. (1972). Yield response of six cultivars of upland cotton, *Gossypium hirsutum* L., in two cultural regimes. *Crop Sci.* **12**, 353–4.

Ranade, C.G. (1986). Growth of productivity in Indian agriculture. *Econ. Political Weekly* **21** (Nos. 25/26), pp. A75–A80.

Randall, D.D., Nelson, C.J. & Asay, K.H. (1977). Ribulose bisphosphate carboxylase: altered genetic expression in tall fescue. *Plant Physiol.* **59**, 38–41.

Randall, P.J., Spencer, K., Freney, J.R. (1981). Sulfur and nitrogen fertilizer effects on wheat. *Aust. J. Agric. Res.* **32**, 203–12.

Rangeley, W.R. (1986). Scientific advances most needed for progress in irrigation. *Phil. Trans. R. Soc. Lond.* A**316**, 355–68.

Rao, A.N., Rama Das, V.S. (1981). Leaf photosynthetic characters and crop growth rates in six cultivars of groundnut (*Arachis hypogea* L.). *Photosynthetica* **15**, 97–103.

Rao, A.R. & Whitcombe, J.R. (1977). Genetic adaptation for vernalization requirement in Nepalese wheat and barley. *Ann. Appl. Biol.* **85**, 121–30.

Rao, M. & Terry, N. (1989). Leaf phosphate status, photosynthesis, and carbon partitioning in sugar beet. *Plant Physiol.* **90**, 814–19.

Rao, R.C.N., Williams, J.H. & Singh, M. (1989). Genotypic sensitivity to drought and yield potential of peanut. *Agron. J.* **81**, 887–93.

Rao, P.S. (1977). Effects of flowering on yield and quality of sugar cane. *Expt. Agric.* **13**, 381–7.

Rapoport, H.F. & Loomis, R.S. (1985). Interaction of storage root and shoot in grafted sugarbeet and chard. *Crop Sci.* **25**, 1079–84.

Rapoport, H.F. & Loomis, R.S. (1986). Structural aspects of root thickening in *Beta vulgaris* L.: comparative thickening in sugar beet and chard. *Bot. Gaz.* **147**, 270–7.

Rappaport, B.D. & Axzley, J.H. (1984). Potassium chloride for improved urea fertilizer efficiency. *J. Soil Sci. Soc. Amer.* **48**, 399–401.

Rappaport, R.A. (1968). *Pigs for the Ancestors.* Yale University Press, New Haven, Connecticut.

Rappaport, R.A. (1971). The flow of energy in an agricultural society. *Sci. Amer.* **225**(3), 117–32.

Rasmussen, V.P. & Hanks, R.J. (1978). Model for predicting spring wheat yields with limited climatological and soil data. *Agron. J.* **70**, 940–4.

Rasmusson, D.C. (1987). An evaluation of ideotype breeding. *Crop Sci.* **27**, 1140–6.

Rasmusson, D.C. & Cannell, R.Q. (1970). Selection for grain yield and components of yield in barley. *Crop Sci.* **10**, 51–4.

Rasmusson, D.C. & Crookston, R.K. (1977). Role of multiple awns in determining barley yields. *Crop Sci.* **17**, 135–40.

Rasmusson, D.C., McLean, I. & Tew, T.L. (1979). Vegetative and grain filling periods in barley. *Crop Sci.* **19**, 5–9.

Raven, J.A. (1977). Ribulose biphosphate carboxylase activity in terrestrial plants; significance of O_2 and CO_2 diffusion. *Curr. Adv. Plant Sci.* **9**, 579–90.

Rawson, H.M. (1986). High-temperature-tolerant wheat: a description of variation and a search for some limitations to productivity. *Field Crops Res.* **14**, 197–212.

Rawson, H.M. (1988). Effects of high temperatures on the development and yield of wheat and practices to reduce deleterious effects. In *Wheat Production Constraints in Tropical Environments* (ed. A.R. Klatt), pp. 44–62. CIMMYT, Mexico.

Rawson, H.M. & Bagga, A.K. (1979). Influence of temperature between floral initiation and flag leaf emergence on grain number in wheat. *Aust. J. Plant Physiol.* **6**, 391–400.

Rawson, H.M. & Constable, G.A. (1980). Carbon production of sunflower cultivars in field and controlled environments. I. Photosynthesis and transpiration of leaves, stems and heads. *Aust. J. Plant Physiol.* **7**, 555–73.

Rawson, H.M. & Constable, G.A. (1981). Gas exchange of pigeon pea: a comparison with other crops and a model of carbon production and its distribution within the plant. In *Proc. Intl. Workshop on Pigeon Peas*, vol. 1, (ed. Y.L. Nene), pp. 175–89. ICRISAT, Patancheru.

Rawson, H.M. & Evans, L.T. (1970). The pattern of grain growth within the ear of wheat. *Aust. J. Biol. Sci.* **23**, 753–64.

Rawson, H.M. & Evans, L.T. (1971). The contribution of stem reserves to grain development in a range of wheat cultivars of different height. *Aust. J. Agric. Res.* **22**, 851–63.

Rawson, H.M. & Hackett, C. (1974). An exploration of the carbon economy of the tobacco plant. III. Gas exchange of leaves in relation to position on the stem, ontogeny and nitrogen content. *Aust. J. Plant Physiol.* **1**, 551–60.

Rawson, H.M. & Hofstra, G. (1969). Translocation and remobilization of ^{14}C assimilated at different stages by each leaf of the wheat plant. *Aust. J. Biol. Sci.* **22**, 321–31.

Rawson, H.M., Gifford, R.M. & Bremner, P.M. (1976). Carbon dioxide exchange in relation to sink demand in wheat. *Planta* **132**, 19–23.

Rawson, H.M., Hindmarsh, J.H., Fischer, R.A. & Stockman, Y.M. (1983). Changes in leaf photosynthesis with plant ontogeny and relationships with yield per ear in wheat cultivars and 120 progeny. *Aust. J. Plant Physiol.* **10**, 503–14.

Rawson, H.M., Gardner, P.A. & Long, M.J. (1987). Sources of variation in specific leaf area in wheat grown at high temperature. *Aust. J. Plant Physiol.* **14**, 287–98.

Reboud, N. & Till-Bottraud, I. (1991). The cost of herbicide resistance measured by a competition experiment. *Theor. Appl. Genet.* **82**, 690–6.

Reddy, V.R. & Baker, D.N. (1990). Application of GOSSYM to analysis of the effects of weather on cotton yields. *Agric. Systems* **32**, 83–95.

Reddy, V.R., Baker, D.N., Whisler, F.D. & McKinion, J.M. (1990). Analysis of the effects of herbicides on cotton yield trends. *Agric. Systems* **33**, 347–59.

Redman, C.L. (1977). Man, domestication and culture in Southwestern Asia. In *Origins of Agriculture* (ed. C.H. Reed), pp. 523–41. Mouton, The Hague.

Reed, A.J., Singletary, G.W., Schussler, J.R., Williamson, D.R. & Christy, A.L. (1988). Shading effects on dry matter and nitrogen partitioning, kernel number, and yield of maize. *Crop Sci.* **28**, 819–25.

Reeves, T. (1987). Technical progress on and off the farm: some principles but mainly practice. *J. Aust. Inst. Agric. Sci.* **53**, 154–7.

Rehm, G.W. & Wiese, R.A. (1975). Effect of method of nitrogen application on corn (*Zea mays* L.) grown on irrigated sandy soils. *Soil Sci. Soc. Amer. Proc.* **39**, 1217–20.

Reifsnyder, W.E. (1989). A tale of ten fallacies: the skeptical enquirer's view of the carbon dioxide/climate controversy. *Agric. Forest Meteorol.* **47**, 349–71.

Reitz, L.B. (1967). World distribution and importance of wheat. In *Wheat and Wheat Improvement* (ed. K.S. Quisenberry & L.P. Reitz), pp. 1–18. Amer. Soc. Agron., Madison, Wisconsin.

Renfrew, C. (1988). *Archaeology and Language: The Puzzle of IndoEuropean Origins.* Jonathan Cape/Cambridge University Press, London.

Renfrew, C. (1989). The origins of Indo-European languages. *Sci. Amer.* **261**(4), 82–90.

Retta, A. & Hanks, R.J. (1980). Corn and alfalfa production as influenced by limited irrigation. *Irrig. Sci.* **1**, 135–47.

Reuveni, J. & Gale, J. (1985). The effect of high levels of carbon dioxide on dark respiration and growth of plants. *Plant, Cell Environ.* **8**, 623–8.

Revelle, R. (1976). Energy use in rural India. *Science* **192**, 969–75.

Reynolds, P. (1981). Deadstock and livestock. In *Farming Practice in British Prehistory* (ed. R. Mercer), pp. 97–122. Edinburgh University Press.

Rhoads, F.M. & Stanley, R.L. (1984). Yield and nutrient utilization efficiency of irrigated corn. *Agron. J.* **76**, 219–23.

Rhodes, I. (1972). Yield, leaf-area index and photosynthetic rate in some perennial ryegrass (*Lolium perenne* L.) selections. *J. Agric. Sci. (Camb.)* **78**, 509–11.

Richards, R.A. (1991). Crop improvement for temperate Australia: future opportunities. *Field Crops Res.* **26**, 141–69.

Richards, R.A. & Passioura, J.B. (1981). Seminal root morphology and water use of wheat. II. Genetic variation. *Crop Sci.* **21**, 253–5.

Richards, R.A. & Passioura, J.B. (1989). A breeding program to reduce the diameter of the major xylem vessel in the seminal roots of wheat and its effect on grain yield in rain-fed environments. *Aust. J. Agric. Res.* **40**, 943–50.

Richards, R.A., Rawson, H.M. & Johnson, D.A. (1986). Glaucousness in wheat: its development and effect on water-use efficiency, gas exchange and photosynthetic tissue temperatures. *Aust. J. Plant Physiol.* **13**, 465–73.

Riches, N. (1937). *The Agricultural Revolution in Norfolk*. University North Carolina, Chapel Hill.

Richey, C.B., Griffith, D.R. & Parsons, S.D. (1977). Yields and cultural energy requirements for corn and soybeans with various tillage-planting systems. *Adv. Agron.* **29**, 141–82.

Richey, J.E. (1983). The phosphorus cycle. In *The Major Biogeochemical Cycles and their Interactions* (ed. B. Bolin & R.B. Cook), pp. 51–2. Wiley, New York.

Rick, C.M. (1988). Evolution of mating systems in cultivated plants. In *Plant Evolutionary Biology* (ed. L.D. Gottlieb & S.K. Jain), pp. 133–47. Chapman & Hall, London.

Rick, C.M. & Dempsey, W.H. (1969). Position of the stigma in relation to fruit setting of the tomato. *Bot. Gaz.* **130**, 180–6.

Riggs, T.J. & Hayter, A.M. (1975). A study of the inheritance and interrelationships of some agronomically important characters in spring barley. *Theoret. Appl. Genet.* **46**, 257–64.

Riggs, T.J., Hanson, P.R., Start, N.D., Miles, D.M., Morgan, C.L. & Ford, M.A. (1981). Comparison of spring barley varieties grown in England and Wales between 1880 and 1980. *J. Agric. Sci. (Camb.)* **97**, 599–610.

Rindos, D. (1984). *The Origins of Agriculture: an Evolutionary Perspective*. Academic Press, New York.

Ritchie, G.A. (ed.) (1979). *New Agricultural Crops*. Westview Press, Boulder.

Robbins, N.S. & Pharr, D.M. (1988). Effect of restricted root growth on carbohydrate metabolism and whole plant growth of *Cucumis sativus* L. *Plant Physiol.* **87**, 409–13.

Roberts, E. (1847). On the management of wheat. *J. Roy. Agric. Soc.* pp. 60–77.

Roberts, H.A. (1976). Weed competition in vegetable crops. *Ann. Appl. Biol.* **83**, 321–4.

Robertson, V.C. (1986). World setting: economic and social constraints. *Phil. Trans. Roy. Soc. Lond.* A316, 197–209.

Robson, M.J. (1980). A physiologist's approach to raising the potential yield of the grass crop through breeding. In *Opportunities for Increasing Crop Yields* (ed. R.G., Hurd, P.V., Biscoe & C. Dennis), pp. 33–49. Pitman, Boston.

Robson, M.J. (1982*a*). The growth and carbon economy of selection lines of *Lolium perenne* cv S23 with differing rates of dark respiration. 1. Grown as simulated swards during a regrowth-period. *Ann. Bot.* **49**, 321–9.

Robson, M.J. (1982*b*). Ibid. 2. Grown as young plants from seed. *Ann. Bot.* **49**, 331–9.

Robson, M.J., Stern, W.R. & Davidson, I.A. (1983). Yielding ability in pure swards and mixtures of perennial ryegrass with contrasting rates of 'mature tissue' respiration. In *Efficient Grassland Farming* (ed. A.J. Corrall), pp. 291–2. Occasional Symposium 14, British Grassland Soc., Hurley.

Rocher, J.P. (1988). Comparison of carbohydrate compartmentation in relation to photosynthesis, assimilate export and growth in a range of maize genotypes. *Aust. J. Plant Physiol.* **15**, 677–86.

Rocher, J.P. & Prioul, J.L. (1987). Compartmental analysis of assimilate export in a mature maize leaf. *Plant Physiol. Biochem.* **25**, 531–40.

Rocher, J.P., Prioul, J.L., Lechamy, A. & Reyss, A., Joussaume, M. (1989). Genetic variability in carbon fixation, sucrose-P-synthase and ADP glucose pyrophosphorylase in maize plants of differing growth rate. *Plant Physiol.* **89**, 416–20.

Rodgers, D.M., Murphy, J.P. & Frey, K.J. (1983). Impact of plant breeding on the grain yield and genetic diversity of spring oats. *Crop Sci.* **23**, 737–40.

Rodriguez-Maribona, B., Tenorio, J.L., Conde, J.R. & Ayerbe, L. (1992). Correlation between yield and osmotic adjustment of peas (*Pisum sativum* L.) under drought stress. *Field Crops Res.* **29**, 15–22.

Rood, S.B. & Major, D.J. (1980). Responses of early corn inbreds to photoperiod. *Crop Sci.* **20**, 679–82.

Rood, S.B., Buzzell, R.I. & MacDonald, M.D. (1988). Influence of temperature on heterosis for maize seedling growth. *Crop Sci.* **28**, 283–6.

Rosario, E.L. & Musgrave, R.B. (1974). The relationship of sugar yield and its components to some physiological and morphological characters. *Proc. Intl. Sugar Cane Technol.* **15**, 1011–20.

Rosenthal, W.D. & Gerik, T.J. (1991). Radiation use efficiency among cotton cultivars. *Agron. J.*, **83**, 655–8.

Rosielle, A.A. & Frey, K.J. (1975*a*). Estimates of selection parameters associated with harvest index in oat lines derived from a bulk population. *Euphytica* **24**, 121–31.

Rosielle, A.A. & Frey, K.J. (1975*b*). Application of restricted selection indices for grain yield improvement in oats. *Crop Sci.* **15**, 544–7.

Rosswall, T. (1983). The nitrogen cycle. In *The Major Biogeochemical Cycles and their Interactions* (SCOPE 21) (ed. B. Bolin & R.B. Cook), pp. 46–50. Wiley, New York.

Rostow, W.W. (1956). The take-off into self-sustained growth. *Econ. J.* **66**, 25–48.

Rowland-Bamford, A.J., Allen, L.H., Baker, J.T. & Boote, K.J. (1990). Carbon dioxide effects on carbohydrate status and partitioning in rice. *J. Exp. Bot.* **41**, 1601–8.

Rowley-Conwy, P. (1981). Slash and burn in the temperate European Neolithic. In *Farming Practice in British Prehistory* (ed. R. Mercer), pp. 85–96. Edinburgh University Press.

Roy, N.N. & Murty, B.R. (1970). A selection procedure in wheat for stress environment. *Euphytica* **19**, 509–21.

Ruckenbauer, P. (1971). Keimfähiger Winter-weizen aus dem Jahre 1877 Beobachtungen und Versuche. *K.v.K. Hochschule fur Bodenkultur, Wien* pp. 372–86.

Rufty, T.W., Raper, C.D. & Huber, S.C. (1984). Alterations in internal partitioning of carbon in soybean plants in response to nitrogen stress. *Canad. J. Bot.* **62**, 501–8.

Runge, E.C.A. (1968). Effects of rainfall and temperature interactions during the growing season on corn yield. *Agron. J.* **60**, 503–7.

Russell, J.S. (1968). Nitrogen fertilizer and wheat in a semi-arid environment. II. Climatic factors affecting response. *Aust. J. Expt. Agric. An. Hus.* **8**, 223–31.

Russell, J.S. (1973). Yield trends of different crops in different areas and reflections on the sources of crop yield improvement in the Australian environment. *J. Aust. Inst. Agric. Sci.* **39**, 156–66.

Russell, W.A. (1972). Effect of leaf angle on hybrid performance in maize (*Zea mays* L.). *Crop Sci.* **12**, 90–2.

Russell, W.A. (1974). Comparative performance for maize hybrids representing different eras of maize breeding. *Proc. 29th Ann. Corn & Sorghum Res. Conf*, pp. 81–101.

Russell, W.A. (1984). Agronomic performance of maize cultivars representing different eras of breeding. *Maydica* **29**, 375–90.

Russell, W.A. (1985). Evaluations for plant, ear and grain traits of maize cultivars representing different eras of breeding. *Maydica* **30**, 85–96.

Russell, W.A. (1991). Genetic improvement of maize yields. *Adv. Agron.* **46**, 245–98.

Russell, W.K. & Stuber, C.W. (1983) Inheritance of photosensitivity in maize. *Crop Sci.* **23**, 935–9.

Ruttan, V. (1982). *Agricultural Research Policy*. University of Minnesota, Minneapolis.

Ryan, C.A. (1988). Proteinase inhibitor gene families: tissue specificity and regulation. In *Temporal and Spatial Regulation of Plant Genes* (ed. D.P.S. Verma & R.B. Goldberg), pp. 223–33. Springer, Vienna.

Ryan, J.G. & Asokan, M. (1977). Effect of Green Revolution in wheat on production of pulses and nutrients in India. *Indian J. Agric. Econ.* **32**, 8–15.

Ryle, G.J.A. & Powell, C.E. (1975). Defoliation and regrowth in the Graminaceous plant: the role of current assimilate. *Ann. Bot.* **39**, 297–310.

Sage, R.F. (1990). A model describing the regulation of ribulose-1,5-bisphosphate carboxylase, electron transport, and triose phosphate use in response to light intensity and CO_2 in C_3 plants. *Plant Physiol.* **94**, 1728–34.

Sage, R.F. & Sharkey, T.D. (1987). The effect of temperature on the occurrence of O_2 and CO_2 insensitive photosynthesis in field grown plants. *Plant Physiol.* **84**, 658–64.

Sahlins, M. (1968). Notes on the original affluent society. In *Man the Hunter* (ed. R.B. Lee & I. De Vore), pp. 85–9. Aldine, Chicago.

Saitoh, H., Shimoda, H., Ishihara, K. (1990). Characteristics of dry matter production process in high yielding rice varieties. I. Canopy structure and light intercepting characteristics. *Jap. J. Crop Sci.* **59**, 130–9.

Sakamoto, S. (1987). Origin and dispersal of common millet and foxtail millet. *Jap. Agric. Res. Quart.* **21**, 84–9

Salado-Navarro, L.R., Hinson, K. & Sinclair, T.R. (1985). Nitrogen partitioning and dry matter allocation in soybeans with different seed protein concentration. *Crop Sci.* **25**, 451–5.

Salaman, R.N. (1949). *The History and Social Influence of the Potato*. Cambridge University Press.

Sale, P.J.M. (1973). Productivity of vegetable crops in a region of high solar input. II. Yields and efficiencies of water use and energy conversion by the potato (*Solanum tuberosum* L.). *Aust. J. Agric. Res.* **24**, 751–62.

Salisbury, F.B. & Bugbee, B.G. (1988). Space farming in the 21st Century. *21st Century Science & Technology* **1**(1), 32–41.

Salman, A.A. & Brinkman, M.A. (1992). Association of pre- and post-heading growth traits with grain yield in oats. *Field Crops Res.* **28**, 211–21.

Salvucci, M.E. (1989). Regulation of rubisco activity *in vivo*. *Physiol. Plantar.* **77**, 164–71.

Samoto, S. (1971). (Alteration of the important characteristics in the breeding programmes of high yield rice varieties). *Hokkaido Nat. Agric. Expt. Sta. Rep.* **78**, 23–73. (In Japanese.)

Sanchez, P.A. (1983). Productivity of soils in rainfed farming systems: examples of long term experiments. In *Potential Productivity of Field Crops Under Different Environments*, pp. 441–65. IRRI, Los Baños.

Sanchez, P.A. & Benites, J.R. (1987). Low-input cropping for acid soils of the humid tropics. *Science* **238**, 1521–7.

Sandfaer, J. & Haahr, V. (1975). Barley stripe mosaic virus and the yield of old and new barley varieties. *Zeitschr. f. Pflanzenzücht.* **74**, 211–22.

Sanford, D.A. van (1985). Variation in kernel growth characters among soft red winter wheats. *Crop Sci.* **25**, 626–30.

Sanford, J.C. & Johnston, S.A. (1985). The concept of parasite-derived resistance – deriving resistance genes from the parasite's own genome. *J. Theoret. Biol.* **113**, 395–405.

Sano, Y. & Morishima, H. (1982). Variation in resource allocation and adaptive strategy of a wild rice, *Oryza perennis* Moench. *Bot. Gaz.* **143**, 518–23.

Santamaria, J.M., Ludlow, M.M. & Fukai, S. (1990). Contribution of osmotic adjustment to grain yield in *Sorghum bicolor* (L.) Moench under water-limited conditions. I. Water stress before anthesis. *Aust. J. Agric. Res.* **41**, 51–65.

Sarma, J.S. (1986). Cereal feed use in the Third World: Past trends and projections to 2000. *IFPRI Research Rep.* No. 57, pp. 1–68.

Sarma, J.S. (1987). Improvements in basic agricultural statistics in support of African food strategies and policies and the role of donor agencies. In *Proc. Workshop on Statistics in Support of African Food Policies and Strategies*, pp. 274–281. (*Eurostat. News Spec. Ed.*)

Sasahara, T. (1971). Genetic variations in cell and tissue forms in relation to plant growth. II. Total cell surface area in the palisade parenchyma and total cell surface area: total nitrogen content ratio in relation to photosynthetic activity in *Brassica*. *Jap. J. Breed.* **21**, 61–8.

Sasahara, T. (1984). Varietal variations in leaf anatomy as related to photosynthesis in soybean (*Glycine max* (L.) Merr.). *Jap. J. Breed.* **34**, 295–303.

Sasahara, T. & Itoh, Y. (1989). Comparison of the effect of fertilizer application at and after the stage of panicle-base initiation on yield and yield components of semi-dwarf and standard rice cultivars. *Field Crops Res.* **20**, 157–64.

Satake, T. (1976). Sterile-type cool injury in paddy rice plants. In *Climate and Rice*, pp. 281–98. IRRI, Los Baños.

Satoh, M. & Hazama, K. (1971). (Studies on photosynthesis and translocation of photosynthate in Mulberry tree. I. Photosynthetic rate of remained leaves after shoot pruning.) *Proc. Crop Sci. Soc. Japan* **40**, 7–11. (In Japanese.)

Sauer, C.O. (1952). *Agricultural Origins and Dispersals*. Amer. Geograph. Soc., New York.

Sauer, J.D. (1950). The Grain Amaranths: A survey of their history and classification. *Ann. Missouri Bot. Gardens*, **37**, 561–632.

Sauer, J. (1967). The grain amaranths and their relatives: a revised taxonomic and geographic survey. *Ann. Missouri Bot. Gardens*. **54**, 103–37.

Sauer, J.D. (1969). Identity of archaeological grain amaranths from the valley of the Tehuacan, Puebla, Mexico. *American Antiquity* **34**, 80–1.

Sawada, S., Hayakawa, T., Fukushi, K. & Kasai, M. (1986). Influence of carbohydrates on photosynthesis in single, rooted soybean leaves used as a source-sink model. *Plant Cell Physiol.* **27**, 591–600.

Sawada, S., Usuda, H., Hasegawa, Y. & Tsukui, T. (1990). Regulation of ribulose-1,5-bisphosphate carboxylase activity in response to changes in the source/sink balance in single-rooted soybean leaves: the role of inorganic orthophosphate in activation of the enzyme. *Plant Cell Physiol.* **31**, 697–704.

Saxena, N.P. (1987). Screening for adaptation to drought: case studies with chickpea and pigeonpea. In *Adaptation of Chickpea and Pigeonpea to Abiotic Stresses* (ed. N.P. Saxena & C. Johansen), pp. 63–76. ICRISAT, Patancheru, India.

Saxena, N.P., Natarajan, M. & Reddy, R.S. (1983). Chickpea, pigeonpea and groundnut. In *Potential Productivity of Field Crops under Different Environments*, pp. 281–305. IRRI, Los Banos.

Schaller, C.W., Qualset, C.O. & Rutger, J.N. (1972). Isogenic analysis of the effects of the awn on productivity of barley. *Crop Sci.* **12**, 531–5.

Schapaugh, W.T. & Wilcox, J.R. (1980). Relationships between harvest indices and other plant characteristics in soybeans. *Crop Sci.* **20**, 529–33.

Scharen, A.L., Krupinsky, J.M. & Reid, D.A. (1983). Photosynthesis and yield of awned versus awnless isogenic lines of winter barley. *Can. J. Plant Sci.* **63**, 349–55.

Schiemann, E. (1939). Gedanken zur Genzentrentheorie Vavilovs. *Naturwiss.* **27**, 377–83.

Schmidt, J.W. (1984). Genetic contributions to yield gains in wheat. In *Genetic Contributions to Yield Gains of Five Major Crop Plants* (ed. W.R. Fehr), pp. 89–101. Crop Sci. Soc. Amer., Madison, Wisconsin.

Schneider, S.H. & Mesirow, L.E. (1976). *The Genesis Strategy: Climate and Global Survival*. Plenum, New York.

Schnell, F.W. & Becker, H.C. (1986). Yield and yield stability in a balanced system of widely differing population structures in *Zea mays* L. *Plant Breeding* **97**, 30–8.

Schön, H.G., Mengel, K. & De Datta, S.K. (1985). The importance of initial exchangeable ammonium in the nitrogen nutrition of lowland rice soils. *Plant Soil* **86**, 403–13.

Schoper, J.B., Johnson, R.R. & Lambert, R.J. (1982). Maize yield response to increased assimilate supply. *Crop Sci.* **22**, 1184–9.

Schultz, T.W. (1964). *Transforming Traditional Agriculture*. Yale University Press, New Haven.

Schultz, T.W. (1980). The economics of being poor. In *The Nobel Prizes 1979*, pp. 242–51. Nobel Foundation, Stockholm.

Schwanitz, F. (1966). *The Origin of Cultivated Plants*. Harvard University Press, Cambridge, Massachusetts.

Schwanitz, F. (1967). *Die Evolution der Kulturpflanzen*. Bayerischer Landwirtschafts-verlag, Munchen.

Schwarzbach, E. (1976). The pleiotropic effects of the ml-o gene and their implications in breeding. *Barley Genetics* **III**, 440–5.

Schweitzer, L.E. & Harper, J.E. (1985). Effect of hastened flowering on seed yield and dry matter partitioning in diverse soybean genotypes. *Crop Sci.* **25**, 995–8.

Scott, R.K. & Wilcockson, S.J. (1974). The effect of sowing date on the critical period for weed control in sugar beet. *Proc. 12th Brit. Weed Control Conf*, pp. 461–8.

Scott, R.K. & Wilcockson, S.J. (1976). Weed biology and the growth of sugar beet. *Ann. Appl. Biol.* **83**, 331–5.

Scott, R.K., Harper, F., Wood, D.W. & Jaggard, K.W. (1974) Effects of seed size on growth, development and yield of monogerm sugar beet. *J. Agric. Sci. (Camb.)* **82**, 517–30.

Scriber, J.M. (1984). Nitrogen nutrition of plants and insect invasion. In *Nitrogen in Crop Production* (ed. R.D. Hauck), pp. 441–60. Amer. Soc. Agron., Madison, Wisconsin.

Secor, J., McCarty, D.R., Shibles, R. & Green, D.E. (1982). Variability and selection for leaf photosynthesis in advanced generations of soybeans. *Crop Sci.* **22**, 255–9

Seddigh, M. & Jolliff, G.D. (1984). Physiological responses of field-grown soybean leaves to increased reproductive load induced by elevated night temperatures. *Crop Sci.* **24**, 952–7.

Seemann, J.R. & Berry, J.A. (1982). Interspecific differences in the kinetic properties of RuBPC'ase protein. *Carnegie Inst. Wash. Yearbook* **81**, 78–83.

Seemann, J.R., Badger, M.R. & Berry, J.A. (1984). Variations in the specific activity of ribulose-1,5-bisphosphate carboxylase between species utilizing different photosynthetic pathways. *Plant Physiol.* **74**, 791–4.

Seetharama, N., Subba Reddy, B.V., Peacock, J.M. & Bidinger, F.R. (1982). Sorghum improvement for drought resistance. In *Drought Resistance in Crops with Emphasis on Rice*, pp. 317–38. IRRI, Los Baños.

Seetharama, N., Mahalakshmi, V., Bidinger, F.R. & Singh, S. (1984). Response of sorghum and pearl millet to drought stress in semi-arid India. In *Agrometeorology of Sorghum and Millet in the Semi-Arid Tropics* (ed. V. Kumble), pp. 159–73. ICRISAT, Patancheru.

Seigler, D.S. (*ed.*) (1977). *Crop Resources.* Academic Press, New York.

Sen, A. (1981). *Poverty and Famines. An essay on entitlement and deprivation.* Clarendon Press, Oxford.

Senadhira, D. & Fu, L.G. (1989). Variability in rice grain filling duration. *Intl. Rice Res. Newsletter* **14**(1), 8–9.

Sengbusch, R. von & Zimmermann, K. (1937). Die Auffindung der ersten gelben und blauen Lupinen (*Lupinus luteus* und *Lupinus angustifolius*) mit nichtplatzenden Hulsen und die damit zusammenhangenden Probleme, insbesondere die der Susslupinenzüchtung. *Züchter* **9**, 57–65.

Servaites, J.C. & Geiger, D.R. (1974). Effects of light intensity and oxygen on photosynthesis and translocation in sugar beet. *Plant Physiol.* **54**, 575–8.

Seshu, D.V. & Cady, F.B. (1984). Response of rice to solar radiation and temperature estimated from international yield trials. *Crop Sci.* **24**, 649–54.

Setter, T.L., Brun, W.A. & Bremner, M.L. (1980). Stomatal closure and photosynthetic inhibition in soybean leaves induced by petiole girdling and pod removal. *Plant Physiol.* **65**, 884–7.

Shah, D.M., Horsch, R.B., Klee, H.J., Kishore, G.M., Winter, J.A., Tumer, N.E., Hironaka, C.M., Sanders, P.R., Gasser, C.S., Aykent, S., Siegel, N.R., Rogers, S.G. & Fraley, R.T.. (1986). Engineering herbicide tolerance in transgenic plants. *Science* **233**, 478–81.

Shakespeare, W. (1612). *The Tempest*, Act V, Scene 1.

Shanan, L. (1987). The impact of irrigation. In *Land Transformation in Agriculture* (ed. M.G. Wolman & F.G.A. Fournier), pp. 115–31. John Wiley, Chichester.

Shanmugasundaram, S. (1979). Variation in the photoperiodic response on several characters in soybean, *Glycine max* (L.) Merrill. *Euphytica* **28**, 495–507.

Shanmugasundaram, S. (1981). Varietal differences and genetic behaviour for the photoperiodic responses in soybeans. *Bull. Inst. Trop. Agric. Kyushu Univ.* **4**, 1–61.

Sharkey, T.D. (1985). Photosynthesis in intact leaves of C_3 plants: physics, physiology and rate limitations. *Bot. Rev.* **51**, 53–105.

Sharkey, T.D. (1989). Evaluating the role of Rubisco regulation in photosynthesis of C_3 plants. *Phil. Trans. R. Soc. Lond.* B**323**, 435–48.

Sharkey, T.D., Berry, J.A. & Sage, R.F. (1988). Regulation of photosynthetic electron-transport in *Phaseolus vulgaris* L., as determined by room-temperature chlorophyll a fluorescence. *Planta* **176**, 415–24.

Sharma, H.C., Waines, J.G. & Foster, K.W. (1981). Variability in primitive and wild wheats for useful genetic characters. *Crop Sci.* **21**, 555–9.

Sharma, R.C. & Smith, E.L. (1986). Selection for high and low harvest index in three winter wheat populations. *Crop Sci.* **26**, 1147–50.

Shaw, R.H. (1977). Climatic requirement. In *Corn and Corn Improvement* (ed. G.F. Sprague), pp. 591–623. Amer. Soc. Agron., Madison, Wisconsin.

Shaw, T. (1976). Early crops in Africa: a review of the evidence. In *Origins of African Plant Domestication* (ed. J.R. Harlan, J.M.J. de Wet & A.B.L. Stemler), pp. 107–53. Mouton, The Hague.

Shi, G. & Akita, S. (1988). Biomass production and grain yield of IR cultivars in high nitrogen water culture. *Jap. J. Crop Sci.* **57**, (Extra Issue 1), 23–4.

Shibles, R.M. & Weber, C.R. (1965). Leaf area, solar radiation interception and dry matter production by soybeans. *Crop Sci.* **5**, 575–7.

Shibles, R., Secor, J. & Ford, D.M. (1987). Carbon assimilation and metabolism. In *Soybeans: Improvement, Production and Uses* (ed. J.R. Wilcox), pp. 538–88. Amer. Soc. Agron., Madison, Wisconsin.

Shibles, R., Ford, D.M. & Secor, J. (1989). Regulation of soybean leaf photosynthesis. In *Proc. World Soybean Conference IV, Buenos Aires* (ed. A.J. Pascale), pp. 40–6.

Shimshi, D. & Ephrat, J. (1975). Stomatal behaviour of wheat cultivars in relation to their transpiration, photosynthesis and yield. *Agron. J.* **67**, 326–31.

Shiroya, M., Lister, G.R., Nelson, C.D. & Krotkov, G. (1961). Translocation of ^{14}C in tobacco at different stages of development following assimilation of $^{14}CO_2$ by a single leaf. *Can. J. Bot.* **39**, 855–64.

Siddique, K.H.M., Belford, R.K., Perry, M.W. & Tennant, D. (1989a). Growth, development and light interception of old and modern wheat cultivars in a Mediterranean-type environment. *Aust. J. Agric. Res.* **40**, 473–87.

Siddique, K.H.M., Kirby, E.J.M. & Perry, M.W. (1989b). Ear:stem ratio in old and modern wheat varieties; relationship with improvement in number of grains per ear and yield. *Field Crops Res.* **21**, 59–78.

Siddique, K.H.M., Belford, R.K. & Tennant, D. (1990a). Root:shoot ratios of old and modern, tall and semi-dwarf wheats in a Mediterranean environment. *Plant Soil* **121**, 89–98.

Siddique, K.H.M., Tennant, D., Perry, M.W. & Belford, R.K. (1990b). Water use and water use efficiency of old and modern wheat cultivars in a Mediterranean type environment. *Aust. J. Agric. Res.* **41**, 431–7.

Siegenthaler, V.L., Stepanich, J.E. & Briggle, L.W. (1986). Distribution of the varieties and classes of wheat in the United States, 1984. *USDA Stat. Bull.* No. 789, pp. 1–106.

Silvey, V. (1978). The contribution of new varieties to increasing cereal yield in England and Wales. *J. Natl. Inst. Agric. Bot.* **14**, 367–84.

Silvey, V. (1981). The contribution of new wheat, barley and oat varieties to increasing yield in England and Wales 1947–78. *J. Natl. Inst. Agric. Bot.* **15**, 399–412.

Silvey, V. (1986). The contribution of new varieties to cereal yields in England and Wales between 1947 and 1983. *J. Natl. Inst. Agric. Bot.* **17**, 155–68.

Simmonds, N.W. (1962). *The Evolution of the Bananas*. Longman, London.

Simmonds, N.W. (1981). Genotype (G), environment (E) and GE components of crop yields. *Exptl. Agric.* **17**, 355–62.

Simmonds, N.W. (1991). Genetics of horizontal resistance to diseases of crops. *Biol. Rev.* **66**, 189–241.

Simmons, A.H., Kohler-Rollefson, I., Rollefson, G.O., Mandel, R. & Kafafi, Z. (1988). 'Ain Ghazal: a major Neolithic settlement in Central Jordan. *Science*, **240**, 35–9.

Simmons, I.G. (1988). The environmental impact of pre-industrial agriculture. *Outlook Agric.* **17**(3), 90–6.

Simmons, S.R. & Jones, R.J. (1985). Contributions of pre-silking assimilate to grain yield in maize. *Crop Sci.* **25**, 1004–6.

Simon, J.L. & Kahn, H. (*eds*). (1984). *The Resourceful Earth. A Response to Global 2000*. Blackwell, Oxford.

Simpson, R.J., Lambers, H. & Dalling, M.J. (1983). Nitrogen redistribution during grain growth in wheat (*Triticum aestivum* L.). *Plant Physiol.* **71**, 7–14.

Sims, H.J. (1963). Changes in the hay production and the harvest index of Australian oat varieties. *Aust. J. Exp. Agric. An. Hus.* **3**, 198–201.

Sinclair, T.R. & Horie, T. (1989). Leaf nitrogen, photosynthesis, and crop radiation use efficiency: a review. *Crop Sci.* **29**, 90–8.

Sinclair, T.R. & de Wit, C.T. (1975). Photosynthate and nitrogen requirements for seed production by various crops. *Science* **189**, 565–7.

Sinclair, T.R. & de Wit, C.T. (1976). Analysis of the carbon and nitrogen limitations to soybean yield. *Agron. J.* **68**, 319–24.

Sinclair, T.R., Tanner, C.B. & Bennett, J.M. (1984). Water-use efficiency in crop production. *Bioscience* **34**(1), 36–40.

Singh, A.J. & Byerlee, D. (1990). Relative variability in wheat yields across countries and over time. *J. Agric. Econ.* **41**, 21–32.

Singh, H. & Gill, H.S. (1987). Interacting effects of magnesium with nitrogen on the growth and yield of wheat. *J. Agric. Sci. (Camb.)* **109**, 399–403.

Singh, M.K. & Tsunoda, S. (1978). Photosynthetic and transpirational response of a cultivated and a wild species of *Triticum* to soil moisture and air humidity. *Photosynthetica* **12**, 280–3.

Single, W.V. (1961). Studies on frost injury to wheat. I. Laboratory freezing tests in relation to the behaviour of varieties in the field. *Aust. J. Agric. Res.* **12**, 767–82.

Sinha, S.K. & Khanna, R. (1975). Physiological, biochemical and genetic basis of heterosis. *Adv. Agron.* **27**, 123–74.

Sinha, S.K., Aggarwal, P.K., Chaturvedi, G.S., Koundal, K.R. & Khanna-Chopra, R. (1981). A comparison of physiological and yield characters in old and new wheat varieties. *J. Agric. Sci. (Camb.)* **97**, 233–6.

Sionit, N. (1983). Response of soybean to two levels of mineral nutrition in CO_2-enriched atmosphere. *Crop Sci.* **23**, 329–33.

Sionit, N., Hellmers, H. & Strain, B.R. (1980). Growth and yield of wheat under CO_2 enrichment and water stress. *Crop Sci.* **20**, 687–90.

Sionit, N., Mortensen, D.A., Strain, B.R. & Hellmers, H. (1981). Growth response of wheat to CO_2 enrichment and different levels of mineral nutrition. *Agron. J.* **73**, 1023–7.

Slafer, G.A. & Andrade, F.H. (1989). Genetic improvement in bread wheat (*Triticum aestivum*) yield in Argentina. *Field Crops Res.* **21**, 289–96.

Slafer, G.A., Andrade, F.H. & Satorre, E.H. (1990). Genetic-improvement effects on pre-anthesis physiological attributes related to wheat grain-yield. *Field Crops Res.* **23**, 255–63.

Slesser, M. (1984). Energy use in the food producing sector of the European Economic Community. In *Energy and Agriculture* (ed. G. Stanhill), pp. 132–53. Springer, Berlin.

Slesser, M. & Lewis, C. (1979). *Biological Energy Resources.* E. & F.N. Spon (Wiley), New York.

Slicher van Bath, B.H. (1963). *The Agrarian History of Western Europe. AD 500–1850.* (Transl. by O. Ordish.) Edward Arnold, London.

Slicher van Bath, B.H. (1977). Agriculture in the vital revolution. In *Cambridge Economic History of Europe*, vol. 5. (*The economic organization of early modern Europe*) (ed. E.E. Rich & C.H. Wilson), pp. 42–133. Cambridge University Press.

Smartt, J. (1976). *Tropical Pulses.* Longman, London..

Smartt, J. (1978). Makulu Red – a Green Revolution groundnut variety. *Euphytica* **27**, 605–8.

Smartt, J. (1980). Evolution and evolutionary problems in food legumes. *Econ. Bot.* **34**, 219–35.

Smil, V., Nachman, P. & Long, T.V. (1983). *Energy Analysis and Agriculture: An Application to US Corn Production.* Westview Press, Boulder, Colorado.

Smillie, R.M., Hetherington, S.E., Ochoa, C. & Malagamba, P. (1983). Tolerances of wild potato species from different altitudes to cold and heat. *Planta* **159**, 112–18.

Smith, B.D. (1989). Origin of agriculture in Eastern North America. *Science* **246**, 1566–71.

Smith, C.E. (1966). Archeological evidence for selection in avocado. *Econ. Bot.* **20**, 169–75.

Smith, P.R. & Neales, T.F. (1977). Analysis of the effects of virus infection on the photosynthetic properties of peach leaves. *Aust. J. Plant Physiol.* **4**, 723–32.

Smith, R.J. (1988). Weed thresholds in southern US rice, *Oryza sativa. Weed Technol.* **2**, 232–41.

Snaydon, R.W. (1970). Rapid population differentiation in a mosaic environment. I. The response of *Anthoxanthum odoratum* populations to soils. *Evolution* **24**, 257–69.

Snogerup, S. (1980). The wild forms of the *Brassica oleracea* group (2n = 18) and their possible relations to the cultivated ones. In *Brassica Crops and Wild Allies* (ed. S. Tsunoda, K. Hinata & C. Gomez-Campo), pp. 121–32. Japan Sci. Soc. Press, Tokyo.

Snyder, F.W. & Carlson, G.E. (1978). Photosynthate partitioning in sugar beet. *Crop Sci.* **18**, 657–61.

Snyder, F.W. & Carlson, G.E. (1984). Selecting for partitioning of photosynthetic products in crops. *Adv. Agron.* **37**, 47–72.

Sobrado, M.A. (1983). *Influence of water deficits on the water relations and growth of wild and cultivated sunflowers.* Ph.D. thesis, Australian National University, Canberra.

Sobrado, M.A. & Turner, N.C. (1986). Photosynthesis, dry matter accumulation and distribution in the wild sunflower *Helianthus petiolaris* and the cultivated sunflower *Helianthus annuus* as influenced by water deficits. *Oecologia* 69, 181–7.

Socrates, cited by Kahn, E.J., *New Yorker*, 17 Dec 1984, p. 106.

Soejitno, J. (1977). Relation between damage by rice stem borer *Tryporyza incertulas* and yield of the rice variety Pelita 1-1. *Intl. Rice Res. Newsl.* 2(4), 6.

Sofield,, I., Evans, L.T., Cook, M.G. & Wardlaw, I.F. (1977). Factors influencing the rate and duration of grain filling in wheat. *Aust. J. Plant Physiol.* 4, 785–97.

Soja, G. & Haunold, E. (1991). Leaf gas exchange and tuber yield in Jerusalem artichoke (*Helianthus tuberosus*) cultivars. *Field Crops Res.* 26, 241–52.

Sokal, R.R., Oden, N.L. & Wilson, C. (1991). Genetic evidence for the spread of agriculture in Europe by demic diffusion. *Nature (London)* 351, 143–5.

Soliman, K.M. & Allard, R.W. (1991). Grain yield of composite cross populations of barley: effects of natural selection. *Crop Sci.* 31, 705–8.

Spagnoletti-Zeuli, P.L. & Qualset, C.O. (1987). Geographical diversity for quantitative spike characters in a world collection of durum wheat. *Crop Sci.* 27, 235–41.

Specht, J.E. & Williams, J.H. (1984). Contribution of genetic technology to soybean productivity – retrospect and prospect. In *Genetic Contributions to Yield Gains of Five Major Crop Plants* (ed. W.R. Fehr), pp. 49–74. Crop Sci. Soc. Amer., Madison, Wisconsin.

Spence, J.A. & Humphries, E.C. (1972). Effect of moisture supply, root temperature, and growth regulators on photosynthesis of isolated rooted leaves of sweet potato (*Ipomoea batatas*). *Ann. Bot.* 36, 115–21.

Spencer, J.E. (1966). *Shifting Cultivation in South Eastern Asia*. University of California Press, Berkeley.

Sperber, P. (1978). *Roman Palestine 200–400. The Land*. Bar-Ilan University Studies in Near Eastern Language and Culture. pp. 1–249.

Spiertz, J.H.J. (1978). Grain production and assimilate utilization of wheat in relation to cultivar characteristics, climatic factors and nitrogen supply. *Versl. Landbouwkund. Onderz.* 881, 1–35.

Spoor, G., Carillon, R., Bournas, L. & Brown, E.H. (1987). The impact of mechanization. In *Land Transformation in Agriculture* (ed. M.G. Wolman & F.G.A. Fournier), pp. 133–52. Wiley, Chichester.

Sprague, M.A. & Triplett, G.B (eds.) (1986). *No-tillage and Sub-surface Tillage Agriculture. The Tillage Revolution*. Wiley, New York.

Srinivasan, P.S., Chandrababu, R., Natarajaratnam, N. & Rangaswamy, S.R.S. (1985). Leaf photosynthesis and yield potential in green gram (*Vigna radiata* (L.) Wilczek) cultivars. *Trop. Agric.* 62, 222–4.

Srivastava, H.K. (1983). Heterosis and intergenomic complementation: Mitochondria, chloroplast and nucleus. In *Heterosis: Reappraisal of Theory and Practice* (ed. R. Frankel), pp. 260–86. Springer, Berlin.

Stahl, R.S. & McCree, K.J. (1988). Ontogenetic changes in the respiration coefficients of grain sorghum. *Crop Sci.* 28, 111–13.

Stalker, H.T. (1980). Utilization of wild species for crop improvement. *Adv. Agron.* 33, 112–47.

Stanford, G. & Legg, J.O. (1984). Nitrogen and yield potential. In *Nitrogen in Crop Production* (ed. R.D. Hauck), pp. 263–72. Amer. Soc. Agron., Madison, Wisconsin.

Stanhill, G. (1976). Trends and deviations in the yield of the English wheat crop during the last 750 years. *Agroecosystems* 3, 1–10.

Stanhill, G. (1980). The energy cost of protected cropping: a comparison of six systems of tomato production. *J. Agric. Eng. Res.* **25**, 145–54.

Stanhill, G. (1981*a*). The Egyptian agro-ecosystem at the end of the eighteenth century – an analysis based on the 'Description de l'Egypte'. *Agroecosystems* **6**, 305–14.

Stanhill, G. (1981*b*). Efficiency of water, solar energy and fossil fuel use in crop production. In *Physiological Processes Limiting Plant Productivity* (ed. C.B. Johnson), pp. 39–51. Butterworths, London.

Stanhill, G. (1984). Agricultural labour: from energy source to sink. In *Energy and Agriculture* (ed. G. Stanhill), pp. 113–30. Springer, Berlin.

Stanhill, G. (1985). The water resource for agriculture. *Phil. Trans. R. Soc. Lond.* **B310**, 161–73.

Stanhill, G. (1986). Irrigation in arid lands. *Phil. Trans. R. Soc. Lond.* **A316**, 261–273.

Stanhill, G. (1990). The comparative productivity of organic agriculture. *Agriculture, Ecosys. & Env.* **30**, 1–26.

Stapper, M. & Fischer, R.A. (1990). Genotype, sowing date and plant spacing influence on high-yielding irrigated wheat in Southern New South Wales. III. Potential yields and optimum flowering dates. *Aust. J. Agric. Res.* **41**, 1043–56.

Starr, C. & Rudman, R. (1973). Parameters of technological growth. *Science* **182**, 358–64.

Staswick, P.E. (1989). Developmental regulation and the influence of plant sinks on vegetative storage protein gene expression in soybean leaves. *Plant Physiol.* **89**, 309–15.

Stebbins, G.L. (1950). *Variation and Evolution in Plants*. Columbia University Press, New York.

Steinberg, R.A. (1953). Low temperature induction of flowering in a *Nicotiana rustica* tabacum hybrid. *Plant Physiol.* **28**, 131–4.

Steinhart, J.S. & Steinhart, C.E. (1974). Energy use in the US food system. *Science* **184**, 307–16.

Stephens, S.G. (1976). Some observations on photoperiodism and the development of annual forms of domesticated cottons. *Econ. Bot.* **30**, 409–18.

Stephenson, R.A., Brown, R.A. & Ashley, D.A. (1976). Translocation of ^{14}C labelled assimilate and photosynthesis in C_3 and C_4 species. *Crop Sci.* **16**, 285–8.

Stern, V.M. (1973). Economic thresholds. *Ann. Rev. Entomol.* **18**, 259–80.

Stern, V.M., Smith, R.F., van den Bosch, R. & Hagen, K.S. (1959). The integrated control concept. *Hilgardia* **29**, 81–101.

Stern, W.R. & Kirby, E.J.M. (1979). Primordium initiation at the shoot apex in four contrasting varieties of spring wheat in response to sowing date. *J. Agric. Sci. (Camb.)* **93**, 203–15.

Stevenson, J.C. & Goodman, M.M. (1972). Ecology of exotic races of maize. I. Leaf number and tillering of 16 races under four temperatures and two photoperiods. *Crop Sci.* **12**, 864–8.

Stewart, B.A. & Musick, J.T. (1982). Conjunctive use of rainfall and irrigation in semi-arid regions. *Adv. Irrig.* **1**, 1–24.

Stewart, G.A. & Lucas, S.M. (eds). (1986). *Potential Production of Natural Rubber from Guayule in Australia*. CSIRO Division of Chemical and Wood Technology, Canberra.

Stewart, G.A., Rawlins, W.H.M., Quick, G.R., Begg, J.E. & Peacock, W.J. (1981). Oilseeds as a renewable source of diesel fuel. *Search* **12**, 107–15.

Stewart, G.A., Hawker, J.S., Nix, H.A., Rawlins, W.H.M. & Williams, L.R. (1982). *The potential for production of hydrocarbon fuels from crops in Australia*. CSIRO, Melbourne.

Stitt, M. (1986). Limitation of photosynthesis by carbon metabolism. I. Evidence for excess electron transport capacity in leaves carrying out photosynthesis in saturating light and CO_2. *Plant Physiol.* **81**, 1115–22.

Stoy, V. (1965). Photosynthesis, respiration, and carbohydrate accumulation in spring wheat in relation to yield. *Physiol. Plantar.* Suppl. IV, pp. 1–125.

Stutler, R.K., James, D.W., Fullerton, T.M., Wells, R.F. & Shipe, E.R. (1981). Corn yield functions of irrigation and nitrogen in Central America. *Irrigation Sci.* **2**, 79–88.

Suetsugu, I. (1975). Records on high rice yields in India. *Nogyo Gyitsu* **30**, 212–15.

Sugiyama, T., Mizuno, M. & Hayashi, M. (1984). Partitioning of nitrogen among ribulose-1,5-bisphosphate carboxylase/oxygenase, phosphoenol pyruvate carboxylase, and pyruvate orthophosphate dikinase as related to biomass productivity in maize seedlings. *Plant Physiol.* **75**, 665–9.

Sullivan, C.Y. (1972). Mechanisms of heat and drought resistance in grain sorghum and methods of measurement. In *Sorghum in the Seventies* (ed. N.G.P. Rao & L.R. House), pp. 247–64. Oxford and IBH Publishing, New Delhi.

Summerfield, R.J. & Roberts, E.H. (1985). Grain legume species of significant importance in world agriculture. In *CRC Handbook of Flowering* (ed. A.H. Halevy), vol. 1, pp. 61–73. CRC Press, Boca Raton, Florida.

Suneson, C.A. (1956). An evolutionary plant breeding method. *Agron. J.* **48**, 188–91.

Suzuki, M. (1983). Growth characteristics and dry matter production of rice plants in the warm region of Japan. *Jap. Agric. Res. Quart.* **17**, 98–105.

Swank, J.C., Below, F.E., Lambert, R.J. & Hageman, R.H. (1982). Interaction of carbon and nitrogen metabolism in the productivity of maize. *Plant Physiol.* **70**, 1185–90.

Swiecicki, W.K., Blazczak, P., Hauke, J. & Mejza, S. (1981). Inheritance of protein content in pea. III. Correlation between protein content and seed yield. *Pisum Newsletter* **13**, 52–3.

Syme, J.R. (1968). Ear emergence of Australian, Mexican and European wheats in relation to time of sowing and their response to vernalization and daylength. *Aust. J. Exp. Agric. An. Hus.* **8**, 578–81.

Syme, J.R. (1972). Single plant characters as a measure of field plot performance of wheat cultivars. *Aust. J. Agric. Res.* **23**, 753–60.

Syme, J.R. (1973). Quantitative control of flowering time in wheat cultivars by vernalization and photoperiod sensitivities. *Aust. J. Agric. Res.* **24**, 657–65.

Syme, J.R. & Thompson, J.P. (1981). Phenotypic relationships among Australian and Mexican wheat cultivars. *Euphytica* **30**, 467–81.

Tainter, J.A. (1988). *The Collapse of Complex Societies*. Cambridge University Press.

Takano, Y. & Tsunoda, S. (1971). Curvilinear regression of the leaf photosynthetic rate on leaf nitrogen content among strains of *Oryza* species. *Jap. J. Breed.* **21**, 69–76.

Takeda, G. (1978). Photosynthesis and dry matter reproduction system in winter cereals. I. Photosynthetic function. *Bull. Natl. Inst. Agric. Sci. Ser. D.* **29**, 1–66.

Takeda, K. & Frey, K.J. (1976). Contributions of vegetative growth rate and harvest index to grain yield of progenies from *Avena sativa* × *A. sterilis* crosses. *Crop Sci.* **16**, 817–21.

Takeda, K. & Frey, K.J. (1985). Increasing grain yield of oats by independent culling for harvest index and vegetative growth index or unit straw weight. *Euphytica* **34**, 33–41.

Takeda, K. & Frey, K.J. (1987). Improving grain yield in backcross populations from *Avena sativa* × *A. sterilis* matings by using independent culling for harvest index and vegetative growth index or unit straw weight. *Theoret. Appl. Genet.* **74**, 659–65.

Takeda, K., Saito, K., Yamazaki, K. & Mikami, T. (1987). Environmental response of yielding capacity in isogenic lines for grain size of rice. *Jap. J. Breed.* **37**, 309–17.

Takeda, T., Oka, M. & Agata, W. (1980). Characteristics of dry matter and grain production of rice cultivars in the warmer part of Japan. I. Comparison of dry matter production between old and new types of rice cultivars. *Jap. J. Crop Sci.* **52**, 299–306.

Takeda, T., Oka, M. & Agata, W. (1984). Ibid. II. Comparison of grain production between old and new rice cultivars. *Jap. J. Crop Sci.* **53**, 12–21.

Takei, E. & Sakamoto, S. (1989). Further analysis of geographical variation of heading response to daylength in foxtail millet (*Setaria italica* P. Beauv.) *Jap. J. Breed.* **39**, 285–98.

Tammes, P.M.L. (1961). Studies of yield losses. II. Injury as a limiting factor of yield. *Neth. J. Plant Pathol.* **67**, 257–63.

Tanaka, A. (1983). Physiological aspects of productivity in field crops. In *Potential Productivity of Field Crops under Different Environments*, pp. 61–80. IRRI, Los Baños.

Tanaka, A., Kawano, K. & Yamaguchi, J. (1966). Photosynthesis, respiration, and plant type of the tropical rice plant. *IRRI Tech. Bull No.* 7, pp. 1–46.

Tanaka, A., Yamaguchi, J., Shimazaki, Y. & Shibata, K. (1968). Historical changes in plant type of rice varieties in Hokkaido. *J. Sci. Soil Manure Japan* **39**, 526–34.

Tanaka, I. (1976). Climatic influence on photosynthesis and respiration of rice. In *Climate and Rice*, pp. 223–47. IRRI, Los Baños.

Tanaka, T., El Sahrigi, A.F., Kamel, O., Sugawara, S., El Nemr, F., Namba, T., El Tanga, A.E.K., Kimura, Y., Abbas, M. & Miura, K. (1987). Establishment of a mechanised rice cultivation system in Egypt. In *Intl. Sympos. on Technology for Double Cropping of Rice in the Tropics*, pp. 72–81. Trop. Agric. Res. Centre, Yatabe, Japan.

Tandon, J.P. & Saini, J.P. (1979). Rooting patterns in wheat genotypes. In *Proc. 5th Intl. Wheat Genetics Sympos.* (ed. S. Ramanujam), pp. 845–851. Indian Soc. Genet. Plant Breeding, New Delhi.

Tanner, C.B. & Sinclair, T.R. (1983). Efficient water use in crop production: research or re-search? In *Limitations to Efficient Water Use in Crop Production* (ed. H.M. Taylor, W.R. Jordan & T.R. Sinclair), pp. 1–27. Amer. Soc. Agron., Madison, Wisconsin.

Terry, N. & Farquhar, G.D. (1984). Photochemical capacity and photosynthesis. In *Control of Crop Productivity* (ed. C.J. Pearson), pp. 43–57. Academic Press, Sydney.

Terry, N. & Ulrich, A. (1973). Effects of potassium deficiency on the photosynthesis and respiration of leaves of sugar beet. *Plant Physiol.* **51**, 783–6.

The, T.T., Latter, B.D.H. McIntosh, R.A., Ellison, F.W., Brennan, P.S., Fisher, J., Hollamby, G.J., Rathjen, A.J. & Wilson, R.E. (1988). Grain yields of near-isogenic lines with added genes for stem rust resistance. In *Proc. 7th Intl. Wheat Genetics Sympos.* (ed. T.E. Miller & R.M.D. Koebner), pp. 901–6. Inst. Plant Sci. Res., Cambridge.

Thomas, J.F. & Raper, C.D. (1976). Photoperiodic control of seed filling for soybeans. *Crop Sci.* **16**, 667–72.

Thompson, F.M.L. (1968). The second agricultural revolution, 1815–1880. *Econ. Hist. Rev.* **21** (1), 62–77.

Thompson, L.M. (1966). *Weather variability and the need for a food reserve*. Center for Agric. Econ. Development Rept. No. 26. Iowa State University, Ames.

Thompson, L.M. (1969). Weather and technology in the production of corn in the US corn belt. *Agron. J.* **61**, 453–456.

Thompson, L.M. (1970). Weather and technology in the production of soybeans in the central United States. *Agron. J.* **62**, 232–6.

Thompson, L.M. (1975). Weather variability, climatic change, and grain production. *Science* **188**, 535–41.

Thompson, L.M. (1986). Climatic change, weather variability, and corn production. *Agron. J.* **78**, 649–53.

Thomson, N.J. (1986). Relationships between climate and relative performance of cotton in New South Wales. *Aust. J. Agric. Res.* **37**, 23–30.

Thorne, G.N. & Evans, A.F. (1964). Influence of tops and roots on net assimilation rate of sugar beet and spinach beet and grafts between them. *Ann. Bot. N.S.* **28**, 499–508.

Thorne, J.H. (1982). Characterization of the active sucrose transport system of immature soybean embryos. *Plant Physiol.* **70**, 953–8.

Thorne, J.H. & Koller, H.R. (1974). Influence of assimilate demand on photosynthesis diffusive resistances, translocation, and carbohydrate levels of soybean leaves. *Plant Physiol.* **54**, 201–7.

Thornley, J.H.M. (1970). Respiration, growth and maintenance in plants. *Nature (Lond.)* **227**, 304–5.

Thornley, J.H.M. (1972). A model to describe the partitioning of photosynthate during vegetative plant growth. *Ann. Bot.* **36**, 419–30.

Tiedje, J.M., Colwell, R.K., Grossman, Y.L., Hodson, R.E., Lenski, R.E., Mack, R.N. & Regal, P.J. (1989). The planned introduction of genetically engineered organisms: ecological considerations and recommendations. *Ecology* **70**, 298–315.

Timothy, D.H., Harvey, P.H. & Dowswell, C.R. (1988). Development and spread of improved maize varieties and hybrids in developing countries. Bureau for Sci. and Technol. Agency for Intl. Development, Washington DC, pp. 1–71.

Tinker, P.B. (1983). Nutrient and micronutrient requirement of cereals. In *The Yield of Cereals: An international seminar on the technology of cereals production, fields and productivity* (ed. D.W. Wright), pp. 59–67. Monograph Series No. 1. Roy. Agric. Soc. of England, London.

Tinker, P.B. (1985). Crop nutrients: control and efficiency of use. *Phil. Trans. R. Soc. Lond.* **B310**, 175–91.

Tinker, P.B. & Widdowson, F.V. (1982). Maximizing wheat yields and some causes of yield variation. *Proc. Fert. Soc.* **211**, 149–84.

Titow, J.Z. (1972). *Winchester Yields. A study in medieval agricultural productivity.* Cambridge University Press.

Tollenaar, M. (1989). Genetic improvement in grain yield of commercial maize hybrids grown in Ontario from 1959 to 1988. *Crop Sci.*, **29**, 1365–71.

Tollenaar, M. (1991). Physiological basis of genetic improvement of maize hybrids in Ontario from 1959 to 1988. *Crop Sci.* **31**, 119–24.

Tollenaar, M. & Daynard, T.B. (1982). Effect of source-sink ratio on dry matter accumulation and leaf senescence of maize. *Can. J. Plant Sci.* **62**, 855–60.

Tolley-Henry, L. & Raper, C.S. (1986). Nitrogen and dry matter partitioning in soybean plants during onset of and recovery from nitrogen stress. *Bot. Gaz.* **147**, 392–9.

Topark-Ngarm, A., Carlson, I.T. & Pearce, R.B. (1977). Direct and correlated responses to selection for specific leaf weight in reed canarygrass. *Crop Sci.* **17**, 765–9.

Toriyama, K. (1962). Improvements of rice varieties adapted to wetland direct seeding. *Agric. Technol.* **17**, 305–9.

Trenbath, B.R. (1985). Weeds and agriculture: A question of balance. In *Studies on Plant Demography* (ed. J. White), pp. 171–83. Academic Press, London.

Trigger, B.G. (1989). *A History of Archaeological Thought.* Cambridge University Press.

Tripp, R. (1990). Does nutrition have a place in agricultural research? *Food Policy* **15**, 467–74.

Troughton, J.H., Chang, F.H. & Currie, B.G. (1974). Estimates of a mean speed of translocation in leaves of *Oryza sativa* L. *Plant Sci. Letters* **3**, 49–54.

Troughton, J.H., Currie, B.G. & Chang, F.H. (1977). Relations between light level, sucrose concentration, and translocation of carbon-11 in *Zea mays* leaves. *Plant Physiol.* **59**, 808–20.

Trow-Smith, R. (1982). Hants wheat yield beats world record. *Farmers Weekly*, 20 August 1982.

Troyer, A.F. & Larkins, J.R. (1985). Selection for early flowering in corn: 10 late synthetics. *Crop Sci.* **25**, 695–7.

Tsuchiya, T. (1987). Physiological and genetic analysis of pod shattering in soybeans. *Japan. Agric. Res. Quart.* **21**, 166–75.

Tsuno, J. & Fujise, K. (1965). Studies on the dry matter production of sweet potato. *Bull. Natl. Int. Agric. Sci., Nishigahara Ser. D.* (No. 13), pp. 1–131.

Tsunoda, S. (1980). Eco-physiology of wild and cultivated forms in Brassica and allied genera. In *Brassica Crops and Wild Allies* (ed. S. Tsunoda, K. Hinata & C. Gomez-Campo), pp. 109–20. Japan Sci. Soc. Press, Tokyo.

Tungland, L., Chapko, L.B., Wiersma, J.V. & Rasmusson, D.C. (1987). Effect of erect leaf angle on grain yield in barley. *Crop Sci.* **27**, 37–40.

Turgeon, R. (1984). Termination of nutrient import and development of vein loading capacity in albino tobacco leaves. *Plant Physiol.* **76**, 45–8.

Turner, N.C. (1979). Drought resistance and adaptation to water deficits in crop plants. In *Stress Physiology in Crop Plants* (ed. H. Mussell & R.C. Staples), pp. 343–72. Wiley, New York.

Turnipseed, S.G. (1972). Response of soybeans to foliage losses in South Carolina. *J. Econ. Entomol.* **60**, 224–9.

Uchijima, Z. (1981). Yield variability of crops in Japan. *GeoJournal* **5**, 151–64.

Uchimiya, H., Handa, T. & Brar, D.S. (1989). Transgenic plants. *J. Biotechnol.* **12**, 1–20.

Udagawa, I. (1976). (Estimation of energy inputs in paddy cultivation.) *Kankyo Johokagaku* **5** (2), 73–9. (In Japanese.)

Ugent, D., Pozorski, S. & Pozorski, T. (1982). Archaeological potato tuber remains from the Casma valley of Peru. *Econ. Bot.* **36**, 182–92.

Ugent, D., Dillehay, T. & Ramirez, C. (1987). Potato remains from a late Pleistocene settlement in South Central Chile. *Econ. Bot.* **41**(1), 17–27.

Ulrich, A. (1955). Influence of night temperature and nitrogen nutrition on the growth, sucrose accumulation and leaf minerals of sugar beet plants. *Plant Physiol.* **30**, 250–7.

Ulrich, A. (1961). Variety climate interactions of sugar beet varieties in simulated climates. *J. Amer. Soc. Sugar Beet Technol.* **11**, 376–87.

Unger, P.W. & McCalla, T.M. (1980). Conservation tillage systems. *Adv. Agron.* **33**, 1–58.

Unger-Hamilton, R. (1989). The Epi-Palaeolithic Southern Levant and the origins of cultivation. *Curr. Anthropol.* **30**, 88–103.

Upmeyer, D.J. & Koller, H.R. (1973). Diurnal trends in net photosynthetic rate and carbohydrate levels of soybean leaves. *Plant Physiol.* **51**, 871–4.

Urban, F. & Vollrath, T. (1984). *Patterns and Trends in World Agricultural Land Use.* USDA Foreign Agric. Econ. Rep. No. 198. US Govt. Printing Office, Washington, DC.

Vaeck, M., Reynaerts, A., Hofle, H., Jansens, S., De Beuckeleer, M., Dean, C., Zabeau, M., Van Montagu, M. & Leemans, J. (1987). Transgenic plants protected from insect attack. *Nature (Lond.)* **328**, 33–7.

Valladas, H., Reyss, J.L., Joron, J.L., Valladas,, G., Bar-Yosef, O. & Vandermeersch, B. (1988). Thermoluminescence dating of Mousterian 'Proto-Cro-Magnon' remains from Israel and the origin of modern man. *Nature (Lond.)* **331**, 614–16.

Vavilov, N.I. (1926). Studies on the origin of cultivated plants. *Bull. Appl. Bot. Plant Breeding* **16**, 139–245.

Vavilov, N.I. (1951). The Origin, Variation, Immunity and Breeding of Cultivated Plants. (Transl. K.S. Chester.) *Chron. Botan.* **13**, 1–366.

Veltkamp, H.J. (1985). Physiological causes of yield variation in cassava (*Manihot esculenta* Crantz). *Agric. University Wageningen Papers* **85–6**, 1–103.

Venkateswarlu, B., Rao, J.S. & Rao, A.V. (1969). Relationship between growth duration and yield parameters in irrigated rice (*Oryza sativa* L.) *Indian J. Plant Physiol.* **20**, 69–76.

Vergara, B.S. (1985). Growth and development of the deep water rice plant. *IRRI Res. Paper Ser.* No. 103, pp. 1–38.

Vergara, B.S. & Chang, T.T. (1985). *The flowering response of the rice plant to photoperiod. A review of the literature.* IRRI, Los Baños. (61 pp.)

Vergara, B.S., Lilis, R. & Tanaka, A. (1964). Relationship between length of growing period and yield of rice plants under limited nitrogen supply. *Soil Sci. Plant Nutr.* **10** (2), 15–21.

Vergara, B.S., Tanaka, A., Lilis, R. & Puranabhavung, S. (1966). Relationship between growth duration and grain yield of rice plants. *Soil Sci. Plant Nutr.* **12**, 31–9.

Vermeij, G.J. (1987). *Evolution and Escalation. An Ecological History of Life.* Princeton University Press, Princeton, N.J.

Vietmeyer, N.D. (*ed.*) (1989). *Lost Crops of the Incas.* Natl. Acad. Press, Washington DC.

Vietor, D.M. & Musgrave, R.B. (1979). Photosynthetic selection of *Zea mays* L. II. The relationship between CO_2 exchange and dry matter accumulation of canopies of two hybrids. *Crop Sci.* **19**, 70–5.

Vietor, D.M., Ariyanayagam, R.P. & Musgrave, R.B. (1977). Photosynthetic selection of *Zea mays* L. I. Plant age and leaf position effects and a relationship between leaf and canopy rates. *Crop Sci.* **17**, 567–73.

Villareal, R.L., Lai, S.H. & Wong, S.H. (1978). Screening for heat tolerance in the genus *Lycopersicon*. *Hort. Sci.* **13**, 479–81.

Villareal, R.L., Rajaram, S. & Nelson, W. (1985). Breeding wheat for more tropical environments at CIMMYT. In *Wheats for More Tropical Environments*, pp. 89–99. CIMMYT, Mexico.

Virmani, S.M. (1971). Rooting patterns of dwarf wheats. *Indian J. Agron.* **16**, 33–6.

Virmani, S.S. & Edwards, I.B. (1983). Current status and future prospects for breeding hybrid rice and wheat. *Adv. Agron.* **36**, 145–214.

Vishnu-Mittre (1974). Palaeobotanical evidence in India. In *Evolutionary Studies in World Crops* (ed. J. Hutchinson), pp. 3–30. Cambridge University Press.

Volenec, J.J., Nelson, C.J. & Sleper, D.A. (1984*a*). Influence of temperature on leaf dark respiration of diverse tall fescue genotypes. *Crop Sci.* **24**, 907–12.

Volenec, J.J., Nguyen, H.T., Nelson, C.J. & Sleper D.A. (1984*b*). Potential for genetically modifying dark respiration of tall fescue leaves. *Crop Sci.* **24**, 938–43.

Voss, J. (1979). Effects of temperature and nitrogen on carbon exchange rates and on growth of wheat during kernel filling. In *Crop Physiology and Cereal Breeding* (ed. J.H.J. Spiertz & T. Kramer), pp. 80–9. Pudoc, Wageningen.

Wada, G. & Cruz, P.C.S. (1989). Varietal difference in nitrogen response of rice plants with special reference to growth duration. *Jap. J. Crop. Sci.* **58**, 732–9.

Waddington, S.R., Ransom, J.K., Osmanzai, M. & Saunders, D.A. (1986). Improvement in the yield potential of bread wheat adapted to Northwest Mexico. *Crop Sci.* **26**, 698–703.

Waddington, S.R., Osmanzai, M., Yoshida, M. & Ransom, J.K. (1987). The yield of durum wheats released in Mexico between 1960 and 1984. *J. Agric. Sci. (Camb.)* **108**, 469–77.

Waggoner, P.E. (1979). Variability of annual wheat yields since 1909 and among nations. *Agric. Meteorol.* **20**, 41–5.

Waggoner, P.E. & Berger, R.D. (1987). Defoliation, disease and growth. *Phytopathol.* **77**, 393–8.

Walden, H.T. (1966). *Native Inheritance. The Story of Corn in America.* Harper and Row, New York.

Waldron, J.C., Glasziou, K.T. & Bull, T.A. (1967). The physiology of sugar-cane. IX. Factors affecting photosynthesis and sugar storage. *Aust. J. Biol. Sci.* **20**, 1043–52.

Walker-Simmons, M., Kudrna, D.A. & Warner, R.L. (1989). Reduced accumulation of ABA during water stress in a molybdenum cofactor mutant of barley. *Plant Physiol.* **90**, 728–33.

Wall, P.C. & Cartwright, P.M. (1974). Effects of photoperiod, temperature and vernalization on the phenology and spikelet numbers of spring wheats. *Ann. Appl. Biol.* **76**, 299–309.

Walton, M. & Helentjaris, T. (1987). Applications of restriction fragment length polymorphism (RFLP) technology to maize breeding. *Proc. 42nd Ann. Corn & Sorghum Res. Conf.*, pp. 48–75.

Wang, J., McBlain, B.A., Hesketh, J.D., Woolley, J.T. & Bernard, R.L. (1987). A data base for predicting soybean phenology. *Biotronics* **16**, 25–38.

Wanjura, D.F. & Barker, G.L. (1988). Simulation analysis of declining cotton yields. *Agric. Systems* **27**, 81–98.

Wardlaw, I.F. (1967). The effect of water stress on translocation in relation to photosynthesis and growth. I. Effect during grain development in wheat. *Aust. J. Biol. Sci.* **20**, 25–39.

Wardlaw, I.F. (1968). The control and pattern of movement of carbohydrates in plants. *Bot. Rev.* **34**, 79–105.

Wardlaw, I.F. (1976). Assimilate movement in *Lolium* and *Sorghum* leaves. I. Irradiance effects on photosynthesis, export and the distribution of assimilates. *Aust. J. Plant Physiol.* **3**, 377–87.

Wardlaw, I.F. (1990). The control of carbon partitioning in plants. *New Phytol.* **116**, 341–81.

Wardlaw, I.F. & Moncur, L. (1976). Source, sink and hormonal control of translocation in wheat. *Planta* **128**, 93–100.

Wardlaw, I.F. & Porter, H.K. (1967). The redistribution of stem sugars in wheat during grain development. *Aust. J. Biol. Sci.* **20**, 309–18.

Wardlaw, I.F., Dawson, I.A., Munibi, P. & Fewster, R. (1989*a*). The tolerance of wheat to high temperatures during reproductive growth. I. Survey procedures and general response patterns. *Aust. J. Agric. Res.* **40**, 1–13.

Wardlaw, I.F., Dawson, I.A. & Munibi, P. (1989*b*). The tolerance of wheat to high temperatures during reproductive growth. II. Grain development. *Aust. J. Agric. Res.* **40**, 15–24.

Warner, J.N. (1962). Sugar cane: an indigenous Papuan cultigen. *Ethnology* **1**, 405–11.

Warren-Wilson, J. (1972). Control of crop processes. In *Crop Processes in Controlled Environments* (ed. A.R. Rees, K.E. Cockshull, D.W. Haul & R.G. Hurd), pp. 7–30. Academic Press, London.

Warrington, I.J., Dunstone, R.L. & Green, L.M. (1977). Temperature effects at three development stages on the yield of the wheat ear. *Aust. J. Agric. Res.* **28**, 11–27.

Watanabe, I. (1973). Mechanism of varietal differences in photosynthetic rate of soybean leaves. I. Correlations between photosynthetic rates and some chloroplast characters. *Proc. Crop Sci. Soc. Japan* **42**, 377–86.

Waterbolk, H.T. (1968). Food production in prehistoric Europe. *Science* **162**, 1093–102.

Waterlow, J.C. & Payne, R. (1975). The protein gap. *Nature (Lond.)* **258**, 113–17.

Watson, D.J. (1958). The dependence of net assimilation rate on leaf area index. *Ann. Bot. N.S.* **22**, 37–54.

Watson, D.J. (1963). Climate, weather and plant yield. In *Environmental Control of Plant Growth* (ed. L.T. Evans), pp. 337–50. Academic Press, New York.

Watson, P.J. (1989). Early plant cultivation in the eastern woodlands of North America. In *Foraging and Farming. The Evolution of Plant Exploitation* (ed. D.R. Harris & G.C. Hillman), pp. 555–71. Unwin Hyman, London.

Webb, L.J. (1973). "Eat, die and learn" – the botany of the Australian aborigines. *Austral. Natural Hist.* **17**, 290–5.

Weber, A. (1979). *Langfristige Energiebilanz in der Landwirtschaft*. Schriftenreihe des Bundesministers für Ernährung Landwirtschaft und Forsten, Reihe A: Landwirtschaft-Angewandte Wissenschaft, Heft 221, Münster Hiltrup. (Cited by Slesser, 1984.)

Welbank, P.J., French, S.A.W. & Witts, K.J. (1966). Dependence of yields of wheat varieties on their leaf area durations. *Ann. Bot. N.S.* **30**, 291–9.

Welbank, P.J., Witts, K.J. & Thorne, G.N. (1968). Effect of radiation and temperature on efficiency of cereal leaves during grain growth. *Ann. Bot.* **32**, 79–95.

Welch, L.F. (1979). Nitrogen use and behaviour in crop production. *Agric. Expt. Sta., Urbana Champaign. Illinois, Bull.* **761**, 1–56.

Wells, R. (1988). Response of leaf ontogeny and photosynthetic activity to reproductive growth in cotton. *Plant Physiol.* **87**, 274–9.

Wells, R. & Meredith, W.R. (1984*a*). Comparative growth of obsolete and modern cotton cultivars. II. Reproductive dry matter partitioning. *Crop Sci.* **24**, 863–8.

Wells, R. & Meredith, W.R. (1984*b*). Ibid. III. Relationship of yield to observed growth characteristics. *Crop Sci.* **24**, 868–72.

Wells, R., Schulze, L.L., Ashley, D.A., Boerma, H.R. & Brown, R.H. (1982). Cultivar differences in canopy apparent photosynthesis and their relationship to seed yield in soybeans. *Crop Sci.* **22**, 886–90.

Wells, R., Meredith, W.R. & Williford, J.R. (1986). Canopy photosynthesis and its relationship to plant productivity in near-isogenic cotton lines differing in leaf morphology. *Plant Physiol.* **82**, 635–40.

Wells, R., Meredith, W.R. & Williford, J.R. (1988). Heterosis in upland cotton. II. Relationship of leaf area to plant photosynthesis. *Crop Sci.* **28**, 522–5.

Wells, R., Bi, T., Anderson, W.F. & Wynne, J.C. (1991). Peanut yield as a result of fifty years of breeding. *Agron. J.* **83**, 957–61.

Wendel, J.F. (1989). New World tetraploid cottons contain Old World cytoplasm. *Proc. Natl. Acad. Sci. USA* **86**, 4132–6.

Wendorf, F. & Schild, R. (1976). The use of ground grain during the late Palaeolithic of the lower Nile Valley, Egypt. In *Origins of African Plant Domestication* (ed. J.R. Harlan, J.M. de Wet & A.B.L. Stemler), pp. 269–88. Mouton, The Hague.

Wendorf, F., Schild, R., Close, A.E., Donahue, D.J., Jull, A.J.T., Zabel, T.H., Wieckowska, H., Kobusiewicz, M., Issawi, B. & el Hadidi, N. (1984). New radiocarbon dates on the cereals from Wadi Kubbaniya. *Science* **225**, 645–6.

Weng, J.H., Takeda, T., Agata, W. & Hakoyama, S. (1982). Studies on dry matter and grain production of rice plants. I. Influence of the reserved carbohydrate until heading stage and the assimilation products during the ripening period on grain production. *Jap. J. Crop. Sci.* **51**, 500–9.

Went, F.W. (1939). Some experiments on bud growth. *Amer. J. Bot.* **26**, 109–17.

Went, F.W. (1959). Effects of environment of parent and grandparent generations on tuber production by potatoes. *Amer. J. Bot.* **46**, 277–8.

Wet, J.M.J. de & Harlan, J.R. (1975). Weeds and domesticates: evolution in the manmade habitat. *Econ. Bot.* **29**, 99–107.

Wet, J.M.J. de, Harlan, J.R. & Price, E.G. (1976). Variability in *Sorghum bicolor*. In *Origins of African Plant Domestication* (ed. J.R. Harlan, J.M.J.de Wet & A.B.L. Stemler), pp. 453–63. Mouton, The Hague.

Wharton, C.R. (1969). Green Revolution. Cornucopia or Pandora's Box. *Foreign Affairs* **47**(3), 464–76.

Wheatley, G.A. (1987). Changing scenes of crop protection. *Ann. Appl. Biol.* **111**, 1–20.

Wheeler, A.W. (1972). Changes in growth substance contents during growth of wheat grains. *Ann. Appl. Biol.* **72**, 327–34.

Whigham, D.K. (1983). Soybean. In *Potential Productivity of Field Crops under Different Environments*, pp. 205–25. IRRI, Los Baños.

Whigham, D.K. & Minor, H.C. (1978). Agronomic characteristics and environmental stress. In *Soybean Physiology, Agronomy and Utilization* (ed. A.G. Norman), pp. 77–118. Academic Press, New York.

White, G.A., Willingham, B.C., Skrdla, W.H., Massey, J.H., Higgins, J.J., Calhoun, W., Davis, A.M., Dolan, D.D. & Earle, F.R. (1971). Agronomic evaluation of prospective new crop species. *Econ. Bot.* **25**, 22–43.

White, G.A., Willingham, B.C., Calhoun, W. & Miller, R.W. (1976). Agronomic evaluation of prospective new crop species. VI. *Briza humilis* – source of galactolipids. *Econ. Bot.* **30**, 193–7.

White, J.W. & Gonzalez, A. (1990). Characterization of the negative association between seed yield and seed size among genotypes of common bean. *Field Crops Res.* **23**, 159–75.

White, J.W. & Laing, D.R. (1989). Photoperiod response of flowering in diverse genotypes of common bean (*Phaseolus vulgaris*). *Field Crops Res.* **22**, 113–28.

White, J.W., Singh, S.P., Pino, C., Rios, B., M.J. & Buddenhagen, I. (1992). Effects of seed size and photoperiod response on crop growth and yield of common bean. *Field Crops Res.* **28**, 295–307.

White, K.D. (1963). Wheat-farming in Roman times. *Antiquity* **37** (147), 207–12.

Whiteman, P.C. (1968). The effects of temperature on the vegetative growth of six tropical legume pastures. *Aust. J. Exptl. Agric. An. Hus.* **8**, 528–32.

Whitten, R.C., Borucki, W.J., Capone, L.A., Riegel, C.A. & Turco, R.P. (1980). Nitrogen fertilizer and stratospheric ozone: latitudinal effects. *Nature (Lond.)* **283**, 191–2.

Whyte, R.O. (1977). The Botanical Neolithic Revolution. *Human Ecology* **5**, 209–22.

Wichler, G. (1961). *Charles Darwin, the Founder of the Theory of Evolution and Natural Selection*. Pergamon, Oxford.

Widdowson, F.W. & Penny, A. (1965). Effects of formalin on the yields of spring wheat. *Rep. Rothamsted Expt. Station* 1964, pp. 65–6.

Widdowson, F.V., Darby, R.J., Dewar, A.M., Jenkyn, J.F., Kerry, B.R., Lawlor, D.W., Plumb, R.T., Ross, G.J.S., Scott, G.C., Todd, A.D. & Wood, D.W. (1986). The effects of sowing date and other factors on growth, yield and nitrogen uptake, and on the incidence of pests and diseases, of winter barley at Rothamsted from 1981 to 1983. *J. Agric. Sci. (Camb.)* **106**, 551–74.

Widdowson, F.V., Penny, A., Darby, R.J., Bird, E. & Hewitt, M.V. (1987). Amounts of NO_3-N and NH_4-N in soil, from autumn to spring, under winter wheat and their relationship to soil type, sowing date, previous crop and N uptake at Rothamsted, Woburn and Saxmundham, 1979–1985. *J. Agric. Sci. (Camb.)* **108**, 73–95.

Wiebold, W.J., Shibles, R. & Green, D.E. (1981). Selection for apparent photosynthesis and related leaf traits in early generations of soybeans. *Crop Sci.* **21**, 969–73.

Wiegand, C.L. & Cuellar, J.A. (1981). Duration of grain filling and kernel weight of wheat as affected by temperature. *Crop Sci.* **21**, 95–101.

Wiegand, C.L., Gerbermann, A.H. & Cuellar, J.A. (1981). Development and yield of hard red winter wheats under semitropical conditions. *Agron. J.* **73**, 29–37.

Wilcox, J.R., Schapaugh, W.T., Bernard, R.L., Cooper, R.L., Fehr, W.R. & Niehaus, M.H. (1979). Genetic improvement of soybeans in the midwest. *Crop Sci.* **19**, 803–5.

Wildman, S.G. (1983). Tobacco revisited: a food from Fraction I protein plus a safer smoking material. In *Structure and Function of Plant Genomes* (ed. O. Ciferri & L. Dure III), pp. 481–92. Plenum Press, New York.

Wilhelm, W.W. & Nelson, C.J. (1978*a*). Irradiance response of tall fescue genotypes with contrasting levels of photosynthesis and yield. *Crop Sci.* **18**, 405–8.

Wilhelm, W.W. & Nelson, C.J. (1978*b*). Leaf growth, leaf aging, and photosynthetic rate of tall fescue genotypes. *Crop Sci.* **18**, 769–72.

Wilhelm, W.W. & Nelson, C.J. (1979). Growth analysis of tall fescue genotypes differing in yield and leaf photosynthesis. *Crop Sci.* **18**, 951–4.

Wilkerson, G.G., Jones, J.W., Boote, K.J. & Buol, G.S. (1989). Photoperiodically sensitive interval in time to flower of soybean. *Crop Sci.* **29**, 721–6.

Wilkes, H.G. (1972). Maize and its wild relatives. *Science* **177**, 1071–77.

Willey, R.W. & Holliday, R. (1971*a*). Plant population and shading studies in barley. *J. Agric. Sci. (Camb.)* **77**, 445–52.

Willey, R.W. & Holliday, R. (1971*b*). Plant population, shading and thinning studies in wheat. *J. Agric. Sci. (Camb.)* **77**, 453–61.

Williams, C.N. (1972). Growth and productivity of tapioca (*Manihot utilissima*). III. Crop ratio, spacing and yield of tubers. *Exptl. Agric.* **8**, 15–23.

Williams, J.H., Wilson, J.H.H. & Bate, G.C. (1975). The growth and development of four groundnut (*Arachis hypogaea* L.) cultivars in Rhodesia. *Rhod. J. Agric. Res.* **13**, 131–44.

Williams, W.A. Loomis, R.S., Duncan, W.G., Dovrat, A. & Nunez, A.F. (1968). Canopy architecture at various population densities and the growth and grain yield of corn. *Crop Sci.* **8**, 303–8.

Willman, M.R., Below, F.E., Lambert, R.J., Howey, A.E. & Mies, D.W. (1987). Plant traits related to productivity of maize. I. Genetic variability, environmental variation, and correlation with grain yield and stalk lodging. *Crop Sci.* **27**, 1116–21.

Wilmsen, E.N. (1989). *Land Filled with Flies. A political economy of the Kalahari.* University of Chicago, Chicago.

Wilson, D. (1975). Variation in leaf respiration in relation to growth and photosynthesis of *Lolium*. *Ann. Appl. Biol.* **80**, 323–38.

Wilson, D. (1982). Response to selection for dark respiration rate of mature leaves in *Lolium perenne* and its effects on growth of young plants and simulated swards. *Ann. Bot.* **49**, 303–12.

Wilson, D. (1984). Identifying and exploiting genetic variation in the physiological components of production. *Ann. Appl. Biol.* **104**, 527–36.

Wilson, D. & Cooper, J.P. (1969*a*). Effect of light intensity and CO_2 on apparent photosynthesis and its relationship with leaf anatomy in genotypes of *Lolium perenne* L. *New Phytol.* **68**, 627–44.

Wilson, D. & Cooper, J.P. (1969*b*). Diallel analysis of photosynthetic rate and related leaf characters among contrasting genotypes of *Lolium perenne*. *Heredity* **24**, 633–49.

Wilson, D. & Cooper, J.P. (1970). Effect of selection for mesophyll cell size on growth and assimilation in *Lolium perenne* L. *New Phytol.* **69**, 233–45.

Wilson, D. & Jones, J.G. (1982). Effect of selection for dark respiration rate of mature leaves on crop yields of *Lolium perenne* cv. S23. *Ann. Bot.* **49**, 313–20.

Wit, C.T. de (1975). Substitution of labour and energy in agriculture and options for growth. *Neth. J. Agric. Sci.* **23**, 145–62.

Wit, C.T. de & van Heemst, H.D.J. (1976). Aspects of agricultural resources. In *Chemical Engineering in a Changing World* (ed. W.T. Koetsier), pp. 125–45. Elsevier, Amsterdam.

Wit, C.T., de & Penning de Vries, F.W.T. (1983). Crop growth models without hormones. *Neth. J. Agric. Sci.* **31**, 313–23.

Wittenbach, V.A. (1983). Effect of pod removal on leaf photosynthesis and soluble protein composition of field grown soybeans. *Plant Physiol.* **73**, 121–4.

Wittwer, S.H. (1975). Food production: technology and the resource base. *Science* **188**, 579–84.

Wolfe, M.S. (1985). The current status and prospects of multiline cultivars and variety mixtures for disease resistance. *Ann. Rev. Phytopathol.* **23**, 251–73.

Wolfe, M.S. & Barrett, J.A. (1980). Can we lead the pathogen astray? *Plant Dis.* **64**, 148–55.

Wolman, M.G. & Fournier, F.G.A. (*eds*). (1987). *Land Transformation in Agriculture*. Wiley, Chichester.

Wong, J.H.H. & Randall, D.D. (1985). Translocation of photoassimilate from leaves of two polyploid genotypes of tall fescue differing in photosynthetic rates. *Physiol. Plantar.* **63**, 445–50.

Wong, S.C. (1979). Elevated atmospheric partial pressure of CO_2 and plant growth. I.

Interactions of nitrogen nutrition and photosynthetic capacity in C_3 and C_4 plants. *Oecologia* **44**, 68–74.

Woodburn, J. (1968). An introduction to Hadza ecology. In *Man the Hunter* (ed. R.B. Lee & I. De Vore), pp. 49–55. Aldine, Chicago.

Woodruff, D.R. & Tonks, J. (1983). Relationship between time of anthesis and grain yield of wheat genotypes with differing developmental patterns. *Aust. J. Agric. Res.* **34**, 1–11.

Woodward, F.I. (1987). Stomatal numbers are sensitive to increases in CO_2 from pre-industrial levels. *Nature (Lond.)* **327**, 617–18.

Woodward, R.G. (1976). Photosynthesis and expansion of leaves of soybean grown in two environments. *Photosynthetica* **10**, 274–9.

World Bank. (1989). *World Development Report*. Oxford University Press.

Wright, G.C., Hubick, K.T. & Farquhar, G.D. (1988). Discrimination in carbon isotopes of leaves correlates with water-use efficiency of field-grown peanut cultivars. *Aust. J. Plant Physiol.* **15**, 815–25.

Wright, H.E. (1976). The environmental setting for plant domestication in the Near East. *Science* **194**, 385–9.

Wright, H.E. (1977). Environmental change and the origin of agriculture in the Old and New Worlds. In *Origins of Agriculture* (ed. C.A. Reed), pp. 281–318. Mouton, The Hague.

Wych, R.D. & Rasmusson, D.C. (1983). Genetic improvement in malting barley cultivars since 1920. *Crop Sci.* **23**, 1037–40.

Wych, R.D. & Stuthman, D.D. (1983). Genetic improvement in Minnesota-adapted oat cultivars released since 1923. *Crop Sci.* **3**, 879–81.

Wych, R.D., McGraw, R.L. & Stuthman, D.D. (1982). Genotype × year interaction for length and rate of grain filling in oats. *Crop Sci.* **22**, 1025–8.

Wyse, R.E. (1979). Parameters controlling sucrose content and yield of sugar beet roots (genetic aspects). *J. Amer. Soc. Sugar Beet Technol.* **20**, 368–85.

Wyse, R.E., Theurer, J.C. & Doney, D.L. (1978). Genetic variability in post-harvest respiration rates of sugarbeet roots. *Crop Sci.* **18**, 264–6.

Xu, H., Pan, Q. & Qi, Y. (1984). The Ganhua No. 2 population with a yield potential beyond 1,800 jin per mu and its control technique. *Scientia Agric. Sinica* **1984**(5), 12–17.

Yamada, M., Ishige, T. & Ohkawa, Y. (1985). Reappraisal of Ashby's hypothesis on heterosis of physiological traits in maize, *Zea mays* L. *Euphytica* **34**, 593–8.

Yamaguchi, J. (1978). Respiration and the growth efficiency in relation to crop productivity. *J. Fac. Agric. Hokkaido University* **59**, 59–129.

Yamashita, A. (1984). Inheritance of photosynthetic rate and its selection for crop improvement. In *Selection in Mutation Breeding*, pp. 67–83. Intl. Atomic Energy Agency, Vienna.

Yamauchi, M. & Yoshida, S. (1985). Heterosis in net photosynthetic rate, leaf area, tillering and some physiological characters of 35 F_1 rice hybrids. *J. Expt. Bot.* **36**, 274–80.

Yan, W. (1991). China's earliest rice agriculture remains. *Bull. Indo-Pacific Prehistory Assoc.* **10**, 118–26.

Yates, F. & Cochran, W.G. (1938). The analysis of groups of experiments. *J. Agric. Sci. (Camb.)* **28**, 556–80.

Yates, P. Lamartine (1968). The future of farming and food. *Misc. Papers Landbouwhogesch., Wageningen* **3**, 92–111.

Yelle, S., Beeson, R.C., Trudel, M.J. & Gosselin, A. (1989a). Acclimation of two tomato species to high atmospheric CO_2. I. Sugar and starch concentrations. *Plant Physiol.* **90**, 1465–72.

Yelle, S., Beeson, R.C., Trudel, M.J. & Gosselin, A. (1989*b*). Acclimation of two tomato species to high atmospheric CO_2. II. Ribulose-1,5-bisphosphate carboxylase/oxygenase and phosphoenol-pyruvate carboxylase. *Plant Physiol.* **90**, 1473–7.

Yen, D.E. (1977). Hoabinhian horticulture? The evidence and the questions from northwest Thailand. In *Sunda and Sahul* (ed. J. Allen, J. Golson & R. Jones), pp. 567–99. Academic Press, London.

Yen, D.E., (1982). The history of cultivated plants. In *Melanesia: Beyond Diversity* (ed. R.J. May & H. Nelson), vol. 1, pp. 281–95. Australian National University, Canberra.

Yeoh, H.H., Badger, M.R. & Watson, L. (1980). Variations in K_m (CO_2) of ribulose-1,5-bisphosphate carboxylase among grasses. *Plant Physiol.* **66**, 1110–12.

Yoshida, S. (1972). Physiological aspects of grain yield. *Ann. Rev. Plant Physiol.* **23**, 437–64.

Yoshida, S. (1981). *Fundamentals of Rice Crop Science.* IRRI, Los Baños.

Yoshida, S. & Hara, T. (1977). Effects of air temperature and light on grain filling of an *indica* and a *japonica* rice (*Oryza sativa* L.) under controlled environmental conditions. *Soil Sci. Plant Nutr. (Tokyo)* **23**, 93–107.

Yoshida, S. & Hasegawa, S. (1982). The rice root system: its development and function. In *Drought Resistance in Crops with Emphasis on Rice*, pp. 97–114. IRRI, Los Baños.

Yoshida, S. & Parao, F.T. (1976). Climatic influence on yield and yield components of lowland rice in the tropics. In *Climate and Rice*, pp. 471–94. IRRI, Los Baños.

Yoshida, S. & Reyes, E. de los (1976). Leaf cuticular resistance of rice varieties. *Soil Sci. Plant Nutr.* **22**, 95–8.

Yoshida, S., Cock, J.H. & Parao, F.T. (1972*a*). Physiological aspects of high yields. In *Rice Breeding*, pp. 455–69. IRRI, Los Baños.

Yoshida, S., Parao, F.T. & Beachell, H.M. (1972*b*). A maximum annual rice production trial in the tropics. *Intl. Rice Comm. Newslett.* **21**(3), 27–30.

Yoshida, S., Bhattacharjee, D.P. & Cabuslay, G.S. (1982). Relationship between plant type and root growth in rice. *Soil Sci. Plant Nutr.* **28**, 473–82.

Young, Arthur (1771). *A Farmer's Tour.*

Young, R.A., Ozbun, J.L., Bauer, A. & Vasey, E.H. (1967). Yield response of spring wheat and barley to nitrogen fertilizer in relation to soil and climatic factors. *Soil. Sci. Soc. Amer. Proc.* **31**, 407–10.

Zadoks, J.C. (1985). On the conceptual basis of crop loss assessment: the threshold theory. *Ann. Rev. Phytopathol.* **23**, 455–73.

Zeist, W. van & Bottema, S. (1982). Vegetational history of the eastern Mediterranean and the Near East during the last 20,000 years. In *Palaeoenvironments and Human Communities in the Eastern Mediterranean Regions in Later Prehistory* (ed. J.L. Bintliff & W. van Zeist), pp. 271–321. British Archaeol. Rep., Intl. Ser. 133.

Zelitch, I. (1975). Improving the efficiency of photosynthesis. *Science* **188**, 626–33.

Zelitch, I. (1982). The close relationship between net photosynthesis and crop yield. *Bioscience* **32**, 796–802.

Zelitch, I. & Day, P.R. (1973). The effect on net photosynthesis of pedigree selection for low and high rates of photorespiration in tobacco. *Plant Physiol.* **52**, 33–7.

Zeven, A.C. (1973). Dr Th. H. Engelbrecht's views on the origin of cultivated plants. *Euphytica* **22**, 279–86.

Zeven, A.C. & de Wet, J.M.J. (1982). *Dictionary of Cultivated Plants and their Regions of Diversity*. (3rd edn.) Pudoc, Wageningen.

Zeven, A.C., Knott, D.R. & Johnson, R. (1983). Investigation of linkage drag in near-isogenic lines of wheat by testing for seedling reaction to races of stem rust, leaf rust and yellow rust. *Euphytica* **32**, 319–27.

Zhukovsky, P.M. (1968). (New centres of the origin and new gene centres of cultivated plants including specifically endemic microcentres of species closely allied to cultivated species.) *Bot. Zhur.* **53**, 430–60. (In Russian.)

Ziman, J. (1976). *The Force of Knowledge. The Scientific Dimension of Society*. Cambridge University Press.

Zohary, D. (1969). The progenitors of wheat and barley in relation to domestication and agricultural dispersal in the Old World. In *The Domestication and Exploitation of Plants and Animals* (ed. P.J. Ucko & G.W. Dimbleby), pp. 47–65. Duckworth, London.

Zohary, D. (1972). The wild progenitor and the place of origin of the cultivated lentil: *Lens culinaris*. *Econ. Bot.* **26**, 326–32.

Zohary, D. (1989). Domestication of the southwest Asian Neolithic crop assemblage of cereals, pulses and flax: the evidence from the living plants. In *Foraging and Farming: The Evolution of Plant Exploitation* (ed. D.R. Harris & G.C. Hillman), pp. 358–73. Unwin Hyman, London.

Zohary, D. & Hopf, M. (1973). Domestication of pulses in the Old World. *Science* **182**, 887–94.

Zohary, D. & Hopf, M. (1988). *Domestication of Plants in the Old World*. Clarendon Press, Oxford.

Zohary, D. & Spiegel-Roy, P. (1975). Beginnings of fruit growing in the Old World. *Science* **187**, 319–27.

Zohary, D., Harlan, J.R. & Vardi, A. (1969). The wild diploid progenitors of wheat and their breeding value. *Euphytica* **18**, 58–65.

Zuño-Altoveros, C., Loresto, S.G., Obien, M. & Chang, T.T. (1990). Differences in root volume of selected upland and lowland rice varieties. *Intl. Rice Res. Newsl.* **15**(2), 8.

Appendix: List of acronyms and abbreviations

BP	before present (in years)
CER	CO_2 exchange rate per unit leaf area
CGR	crop growth rate (per unit ground area)
CIAT	Centro Internacional de Agricultura Tropical, Cali, Colombia
CIMMYT	Centro International de Mejoramiento de Maiz y Trigo, El Batan, Mexico
cv.	cultivar
C_3	plants with Calvin cycle photosynthesis
C_4	Plants with C_4 photosynthetic pathway
2,4-D	2,4-dichlorophenoxyacetic acid
DNA	deoxyribose nucleic acid
EEC	European Economic Community
ER	energy return ratio
FAO	United Nations Food and Agriculture Organization, Rome
GCM	general circulation model
GGR	grain growth rate
GNP	gross national product
ha	hectare
HI	harvest index
HSPA	Hawaian Sugar Planter's Association
IBPGR	International Board for Plant Genetic Resources, Rome
ICRISAT	International Crops Research Institute for the Semi-Arid Tropics, Hyderabad, India
IFIAS	International Federation of Institutes for Advanced Study
IFPRI	International Food Policy Research Institute, Washington DC
IRRI	International Rice Research Institute, Los Baños, Philippines
LAD	leaf area duration
LAI	leaf area index
NIAB	National Institute of Agricultural Botany, Cambridge, England
PAR	photosynthetically active radiation
RFLP	restriction fragment length polymorphism

484

RGR	relative growth rate
Rht	reduced height genes
r_s	stomatal resistance
rubisco	ribulose-1, 5-bisphosphate carboxylase-oxygenase
RuBP	ribulose-1, 5-bisphosphate
SLW	specific leaf weight
TMV	tobacco mosaic virus
UNU	United Nations University
USAID	United States Agency for International Development
USDA	United States Department of Agriculture
WHO	World Health Organization, Geneva
WUE	Water use efficiency

Index

486